DYNAMICS OF SOLAR AND STELLAR CONVECTION ZONES AND ATMOSPHERES

IAU SYMPOSIUM 365

COVER ILLUSTRATION: INTERNAL STRUCTURE OF THE SUN

This illustration is adapted from Figure 1 (A) of "A simple but powerful model of the solar cycle" by Dário Passos, CSEI2012 – Conferência Nacional sobre Computação Simbólica no Ensino e na Investigação, Lisbon, 2012.

The white central sphere is the core of the Sun, the yellow spherical shell surrounding the core is the radiative zone and the outermost shell is the convection zone. In the latter, helioseismologically mapped constant-rotation-rate contours are shown, with the fastest rotation in the equatorial areas (warm colours) and slower rotation at higher latitudes (cold dark colours). Closed curves in the other meridional section of the convection zone (light brown) conventionally represent convective circulation of the solar plasma. The radiative zone rotates as a solid body.

IAU SYMPOSIUM PROCEEDINGS SERIES

INTERNATIONAL ASTRONOMICAL UNION

UNION ASTRONOMIQUE INTERNATIONALE

International Astronomical Union

IAU

DYNAMICS OF SOLAR AND STELLAR CONVECTION ZONES AND ATMOSPHERES

PROCEEDINGS OF THE 365th SYMPOSIUM OF THE INTERNATIONAL ASTRONOMICAL UNION YEREVAN, REPUBLIC OF ARMENIA 21–25 AUGUST 2023

Edited by

ALEXANDER V. GETLING
Skobeltsyn Institute of Nuclear Physics,
Lomonosov Moscow State University, Moscow 119991, Russia

LEONID L. KITCHATINOV
Institute of Solar–Terrestrial Physics, Irkutsk 664033, Russia

CAMBRIDGE
UNIVERSITY PRESS

CAMBRIDGE UNIVERSITY PRESS
University Printing House, Cambridge CB2 8BS, United Kingdom
1 Liberty Plaza, Floor 20, New York, NY 10006, USA
10 Stamford Road, Oakleigh, Melbourne 3166, Australia

First published 2024

Printed in Great Britain by Henry Ling Limited, The Dorset Press, Dorchester, DT1 1HD

Typeset in System LaTeX 2ε

A catalogue record for this book is available from the British Library of Congress Cataloguing in Publication data

This journal issue has been printed on FSC™-certified paper and cover board. FSC is an independent, non-governmental, not-for-profit organization established to promote the responsible management of the world's forests. Please see www.fsc.org for information.

ISBN 9781108490771 hardback
ISSN 1743-9213

Table of Contents

Preface

Impressive events of magnetic-energy release, such as solar and stellar flares or coronal mass ejections, take place in the atmospheric layers above the solar/stellar photosphere. The magnetic-activity events are closely related to the dynamics of subsurface convection zones inside the stars. The last few decades have been marked by substantial progress in understanding the physics of internal and external processes in the Sun and solar-type stars, but the scientific communities making this progress have not worked closely together. The idea to bring together specialists in the physics of convection zones and atmospheres materialised in a proposal to organise a symposium on the dynamics of solar and stellar convection zones and atmospheres, which was approved by the IAU Executive Committee in May 2019.

IAU Symposium 365 was originally planned to be held at the Lomonosov Moscow State University in 2020. However, the COVID pandemic and unfortunate political developments led to a long postponement and finally to a search for an alternative venue. Eventually, the Byurakan Astrophysical Observatory, National Academy of Sciences of the Republic of Armenia, kindly agreed to host the Symposium in the city of Yerevan in August 2023.

The Sun as a complex hydrodynamic and magnetohydrodynamic object has attracted much researchers' attention over the past few decades. Modern observational instrumentation aboard orbiting observatories, such as the Solar Dynamics Observatory (SDO), have provided an invaluable wealth of diverse data. Substantial progress has been achieved in theoretical studies of solar convection, differential rotation, meridional circulation, the global solar dynamo, and local processes of interaction between plasma flows and magnetic fields in the upper convection zone and photosphere, which are responsible for the local phenomena of solar activity. On the other hand, the progress in the development of observational techniques has stimulated investigations aimed at understanding the dynamics of stellar plasmas. Hydrodynamic and magnetohydrodynamic studies of the Sun and stars show a remarkable convergence, approaching a unified description of all these objects. In the numerical simulations of convection and magnetoconvection, which have now greatly increased their coverage in the parameter space, the solar convection regime plays the role of a 'reference point' for the set of models. The theories of differential rotation and the global dynamo have now reached a level that allows these two components of the global dynamics to be considered in a single model. Helioseismology offers a previously unattainable insight into the dynamics of layers hidden from our eyes. The ideas of helioseismology are now being extended to stellar physics, leading to the successful development of asteroseismology.

A particular point of importance is related to magnetic-flux-emergence processes producing active regions and to the link between global and local magnetohydrodynamic phenomena. The extensive data from, e.g., SDO can be used for detailed analyses and investigations of the underlying physics on the basis of an adequate magnetohydrodynamic description, which is of paramount importance for the elaboration of techniques of active-phenomena predictions.

An important application of the study of global solar processes is the prediction of solar activity cycles. Such predictions will be all the more reliable the more complete our understanding of the global solar dynamics is, and the study of stellar dynamics can contribute significantly to the completeness of the general view of the physical processes.

Further progress in the study of this multifaceted complex of phenomena requires co-ordination of the efforts of researchers working on different problems, and extensive discussion and exchange of views. Particular attention must be paid to processes in the dense plasmas of the solar and stellar convection zones and photospheres, where the complex of active phenomena originates. At the same time, since the atmosphere is strongly dynamically coupled to the underlying photosphere and convection zone, a strict separation between the photosphere and the layers above it would not be justified. For this reason, although the scope of the Symposium was mainly focused on the dynamics of the convection zones and photospheres, it was also intended to encompass those processes in the lower atmospheres that can be regarded as a direct continuation of the processes in the underlying layers.

The Symposium was devoted to observational and theoretical aspects of solar and stellar hydrodynamics and magnetohydrodynamics, both global and local, including numerical studies as a special branch of theoretical research. With this scope, the meeting stimulated the emergence of new ideas and the development of new techniques in this field of active research.

It is our great pleasure to acknowledge the enormous role of the LOC chair Areg Mickaelian and the members of the LOC team in the realisation of the Symposium.

Alexander V. Getling and Leonid L. Kitchatinov, co-chairs SOC,
Yerevan, 25 August 2023

The Organising Committee

Scientific

Valentina Abramenko (Russia)
Axel Brandenburg (Sweden)
Paul Charbonneau (Canada)
Arnab Rai Choudhuri (India)
Yuhong Fan (United States)
Alexander Getling (Co-Chair, Russia)
Hideyuki Hotta (Japan)
Jie Jiang (China)
Maarit Korpi-Lagg (Finland)
Leonid Kitchatinov (Co-Chair, Russia)
Irina Kitiashvili (United States)
Areg Mickaelian (Armenia)
Åke Nordlund (Denmark)
Dmitry Sokoloff (Russia)

Local

Ruslan Akmetdinov
Meline Asryan
Gayane Baleyan
Lilit Darbinyan
Sona Farmanyan (Secretary)
Liana Hambardzumyan
Areg Mickaelian (Chair)
Gor Mikayelyan (Coordinator)
Varduhi Mkrtchyan
Sama Shamyar

Acknowledgements

The Local Organising Committee operated under the auspices of the Byurakan Astrophysical Observatory, National Academy of Sciences of the Republic of Armenia

Funding by the International Astronomical Union and sponsorship by Kaspersky Lab are gratefully acknowledged.

Conference Photographs

Figure 1. A group of participants after the closing ceremony.

Figure 2. A group of participants visiting the Byurakan Astrophysical Observatory, in the tower of the 2.6-m telescope.

List of Participants

Valentina **Abramenko**, Crimean astrophysical observatory, Nauchny, Russia vabramenko@gmail.com

Vardan **Adibekyan**, Institute of Astrophysics and Space Sciences (IA), Porto, Portugal vadibekyan@astro.up.pt

Olga **Andreeva**, Crimean astrophysical observatory, Nauchny, Russia olga@craocrimea.ru

Dipankar **Banerjee**, Aryabhatta Research Institute of Observational sciences, Nainital, India dipu@aries.res.in

Tanayveer Singh **Bhatia**, Max Planck Institute for Solar System Research, Göttingen, Germany bhatia@mps.mpg.de

Aaron **Birch**, Max-Planck-Institut für Sonnensystemforschung, Göttingen, Germany birch@mps.mpg.de

Akash **Biswas**, Indian Institute of Technology (BHU), Varanasi, India akashbiswas.rs.phy20@iitbhu.ac.in

Teresa **Braun**, Max Planck Institute for Astrophysics, Germany braun@mpa-garching.mpg.de

Ana **Brito**, Instituto Superior de Gestão/IST - CENTRA, Portugal anabrito@isg.pt

Maria **Camisassa**, Universitat Politècnica de Catalunya, Barcelona, Spain maria.camisassa@upc.edu

Gabriela **Carvalho Silva**, Universidade de São Paulo, Osasco, Brazil gabriela.carvalho.silva@usp.br

Ruizhu **Chen**, Stanford University, Stanford, United States rzchen@stanford.edu

Arnab Rai **Choudhuri**, Indian Institute of Science, India arnab@physics.iisc.ernet.in

Cannelle **Clavier**, CNRS – UVSQ – SU, LATMOS, France cannelle.clavier@latmos.ipsl.fr

Prithwitosh **Dey**, Max Planck Institute for Solar System Research, Germany dey@mps.mpg.de

Marianne **Faurobert**, University Cote d'Azur, Nice, France marianne.faurobert@oca.eu

Yuriy **Fursyak**, Crimean Astrophysical Observatory of RAS, Nauchny, Russia yuriy_fursyak@mail.ru

Alexander **Getling**, Lomonosov Moscow State University, Moscow, Russia a.getling@mail.ru

Laurent **Gizon**, MPI for Solar System Research, Göttingen, Germany gizon@mps.mpg.de

Anna Mária **Görgei**, CSFK, Konkoly Observatory, Budapest, Hungary gorgei.anna@csfk.org

Gustavo **Guerrero**, Universidade Federal de Minas Gerais, Belo Horizonte, Brazil guerrero@fisica.ufmg.br

Christopher **Hanson**, New York University Abu Dhabi, Abu Dhabi, United Arab Emirates hanson@nyu.edu

Xavier **Haubois**, European Southern Observatory (ESO), Santiago, Chile xhaubois@eso.org

Anne **Hedlund**, New Mexico State University, United States, ahedlund@lanl.gov

Rachel **Howe**, University of Birmingham, Sheffield, United Kingdom r.howe@bham.ac.uk

Egor **Illarionov**, Moscow State University, Moscow, Russia egor.mypost@gmail.com

Kiran **Jain**, National Solar Observatory, Boulder, United States kjain@nso.edu

Bibhuti Kumar **Jha**, Southwest Research Institute, New Delhi, India maitraibibhu@gmail.com

Jie **Jiang**, Beihang University, Beijing, China jiejiang@buaa.edu.cn

Petri **Käpylä**, Leibniz-Institute for Solar Physics (KIS), Freiburg im Breisgau, Germany pkapyla@leibniz-kis.de

Bidya Binay **Karak**, Indian Institute of Technology (BHU) Varanasi, India karak.phy@iitbhu.ac.in

Spiridon **Kasapis**, NASA Ames Research Center, San Francisco, United States skasapis@umich.edu
Sara **Khodairy**, National Research Institute of Astronomy and Geophysics, Cairo, Egypt sara_math20032004@yahoo.com
Leonid **Kitchatinov**, Institute of Solar-Terrestrial Physics, Irkutsk, Russia kit@iszf.irk.ru
Irina **Kitiashvili**, NASA, United States irina.n.kitiashvili@nasa.gov
Nathan **Kleeorin**, Ben-Gurion University of the Negev, Beer-Sheva, Israel nat@bgu.ac.il
Maarit **Korpi-Lagg**, Aalto University, Helsinki, Finland maarit.korpi-lagg@aalto.fi
Alexander **Kosovichev**, New Jersey Institute of Technology, Newark, United States alexander.g.kosovichev@njit.edu
Zsolt **Kővári**, Konkoly Observatory, Budapest, Hungary kovari@konkoly.hu
Levente **Kriskovics**, Research Centre for Astronomy and Earth Sciences, Budapest, Hungary kriskovics.levente@csfk.org
Pawan **Kumar**, Indian Institute of Technology (BHU) Varanasi, Indiai pawan.kumar.rs.phy18@itbhu.ac.in
Alexander **Kutsenko**, Crimean Astrophysical Observatory, Nauchny, Russia alex.s.kutsenko@gmail.com
Kirill **Kuzanyan**, IZMIRAN and ICMM, Moscow, Russia kuzanyan@gmail.com
Biji **Lekshmi**, Max Planck Institute for Solar System Research, Göttingen, Germany lekshmib@mps.mpg.de
Arthur **Le Saux**, University of Exeter, United Kingdom al598@exeter.ac.uk
Guifang **Lin**, Yunnan Observatories, Kunming, China linguifang@ynao.ac.cn
Daria **Litvishko**, Crimean Astrophysical Observatory, Nauchny, Russia 24dasha01@gmail.com
Yu-Qing **Lou**, Tsinghua University, Beijing, China louyq@tsinghua.edu.cn
Yukun **Luo**, Beihang University, Beijing, China luoyukun@buaa.edu.cn
Sushant **Mahajan**, Stanford University, Stanford, United States mahajans@stanford.edu
Yash **Mandowara**, Max Planck Institute for Solar System Research, Germany mandowara@mps.mpg.de
Angel **Martinez**, Monash University, Bogota, Colombia angel.martinez@monash.edu
Youhei **Masada**, Fukuoka University, Fukuoka, Japan ymasada@fukuoka-u.ac.jp
Savita **Mathur**, Instituto de Astrofisica de Canarias, Spain smathur@iac.es
Mustapha **Meftah**, CNRS – UVSQ – SU, LATMOS, Paris, France mustapha.meftah@latmos.ipsl.fr
Fabian **Menezes**, Universidade Presbiteriana Mackenzie, Brazil, São Paulo menezes.astroph@gmail.com
Dibya Kirti **Mishra**, Aryabhatta Research Institute of Observational Sciences, Nainital, India dibyakirtimishra@aries.res.in
David **Mkrtichian**, National Astronomical Research Institute of Thailand, Chiang Mai, Thailand davidmkrt@gmail.com
Galina **Motorina**, Astronomical Institute of the CAS, Czech Republic motorina@asu.cas.cz
Dibyendu **Nandi**, Indian Institute of Science Education and Research Kolkata, Mohanpur, India, dnandi@iiserkol.ac.in
Quentin **Noraz**, University of Oslo, Norway quentin.noraz@astro.uio.no
Savannah **Perez-Piel**, UC Berkeley, Emeryville, United States savannah_sky@berkeley.edu
Alexei **Pevtsov**, National Solar Observatory, United States apevtsov@nso.edu
Jordan **Philidet**, Max-Planck Institute for Solar System Research, Göttingen, Germany philidet@mps.mpg.de
Alexander **Pietrow**, AIP, Potsdam, Germany apietrow@aip.de
Valery **Pipin**, Institute for solar-terrestrial physics, Irkutsk, Russia valery.pipin@gmail.com
Andrei **Plotnikov**, Crimean Astrophysical Observatory, Nauchny, Russia plotnikov.andrey.alex@yandex.ru
Mariano **Poisson**, Instituto de Astronomía y Física del Espacio, Buenos Aires, Argentina marianopoisson@gmail.com

Paul **Rajaguru**, Indian Institute of Astrophysics, Bangalore, India rajaguru@iiap.res.in
Rebecca **Robinson**, University of Oslo, Norway rebecrob@uio.no
Ilan **Roth**, University of California, Berkeley, United States ilanr@berkeley.edu
Rayhane **Sadeghi**, PNU Physics Department, Tehran, Iran rayhane.sadeghi@gmail.com
Aidar **Sadykov**, Lomonosov Moscow State University, Moscow, Russia sadykov.am19@physics.msu.ru
Viacheslav **Sadykov**, Georgia State University, Atlanta, United States vsadykov@gsu.edu
Chitradeep **Saha**, Indian Institute of Science Education and Research Kolkata, India cs20rs018@iiserkol.ac.in
Alain **Sarkissian**, LATMOS / UVSQ / PARIS-SACLAY University, Paris, France alain.sarkissian@latmos.ipsl.fr
Markus **Schmassmann**, Leibniz-Institute for Solar Physics (KIS), Freiburg im Breisgau, Germany schmassmann@leibniz-kis.de
Victoria **Smirnova**, Crimean Astrophysical Observatory, Nauchny, Russia vvsvid.sm@gmail.com
Ulysses **Sofia**, Scripps College, Claremont, United States usofia@kecksci.claremont.edu
Dmitry **Sokoloff**, Moscow State University and IZMIRAN, Moscow, Russia
Sami **Solanki**, Max-Planck-Institute for Solar System Research, Göttingen, Germany solanki-office@mps.mpg.de
Anu B **Sreedevi**, Indian Institute of Technology (BHU), Varanasi, India anubsreedevi.rs.phy20@itbhu.ac.in
Sergey **Starchenko**, IZMIRAN, Moscow, Russia sstarchenko@mail.ru
John **Stefan**, New Jersey Institute of Technology, Newark, United States john.stefan@njit.edu
Regina **Suleymanova**, Crimean astrophysical observatory, Nauchny, Russia bictr97@gmail.com
Ehsan **Tavabi**, Payam e Noor university, Tehran, Iran etavabi@gmail.com
Andrey **Tlatov**, Pulkovo Observatory, Kislovodsk mountain astronomical station, Kislovodsk, Russia atlatov@gmail.com
Ksenia **Tlatova**, Pulkovo Observatory, Kislovodsk mountain astronomical station, St. Petersburg, Russia k.tlatova@mail.ru
Sushanta **Tripathy**, National Solar Obsrrvatory, Boulder, United States stripathy@nso.edu
Yuriy **Tsap**, Crimean Astrophysical Observatory, Nauchny, Russia yur_crao@mail.ru
Lisa **Upton**, Southwest Research Institute, Westminster, United States upton.lisa.a@gmail.com
Adriana **Valio**, Mackenzie Presbyterian University, São Paulo, Brazil adrivalio@gmail.com
M. Kumar **Varnana**, Mar Ivanious College, Kerala, Trivandrum, India mvarnana@gmail.com
Vindya **Vashishth**, Indian Institute of Technology, Uttar Pradesh, India vindyavashishth.rs.phy19@itbhu.ac.in
Krisztian **Vida**, CSFK, Konkoly Observatory, Budapest, Hungary vidakris@konkoly.hu
Matthias **Waidele**, Stanford University, United States mwaidele@stanford.edu
Zi-Fan **Wang**, National Astronomical Observatories, Beijing, China zfwang@nao.cas.cn
Ruihui **Wang**, Beihang University, Beijing, China wangruihui@buaa.edu.cn
Svetlana **Yakovleva**, IZMIRAN, Moscow, Russia svyakov@inbox.ru
Mai **Yamashita**, University of Hyogo, Japan yamashita@nhao.jp
Hanlin **Yang**, National Astronomical Observatories, Beijing, China yanghanlin20@mails.ucas.ac.cn
Zebin **Zhang**, Beihang university, Beijing, China zebinzhang@buaa.edu.cn
Anastasiya **Zhukova**, Crimean Astrophysical Observatory, Nauchny, Russia anastasiya.v.zhukova@gmail.com
Nadezhda **Zolotova**, St Petersburg State University, Saint Petersburg, Russia ned@geo.phys.spbu.ru

Section 1. Solar and Stellar Convection

Invited talks

Dynamics of Solar and Stellar Convection Zones and Atmospheres
Proceedings IAU Symposium No. 365, 2024
A. V. Getling & L. L. Kitchatinov, eds.
doi:10.1017/S1743921323005239

Effects of rotation and surface forcing on deep stellar convection zones

Petri J. Käpylä

Leibniz-Insitute for Solar Physics (KIS), Schöneckstraße 6, 79104 Freiburg im Breisgau, Germany
email: pkapyla@leibniz-kis.de

Abstract. The canonical undestanding of stellar convection has recently been put under doubt due to helioseismic results and global 3D convection simulations. This "convective conundrum" is manifested by much higher velocity amplitudes in simulations at large scales in comparison to helioseismic results, and the difficulty in reproducing the solar differential rotation and dynamo with global 3D simulations. Here some aspects of this conundrum are discussed from the view-point of hydrodynamic Cartesian 3D simulations targeted at testing the rotational influence and surface forcing on deep convection. More specifically, the dominant scale of convection and the depths of the convection zone and the weakly subadiabatic – yet convecting – Deardorff zone are discussed in detail.

Keywords. Convection, turbulence, Sun: rotation, Sun: interior

---

1. Introduction

The solar convective envelope rotates differentially, such that the rotation rate at the equator is about 40 per cent faster than at near the poles. Furthermore, helioseismology has revealed that the angular velocity Ω increases (decreases) with radius near the equator (high latitudes), with narrow shear layers at the base and near the surface of the convection zone (e.g. Thompson et al. 2003). This large-scale phenomenon is one of the principal observations that global 3D simulations seek to reproduce. Early 3D simulations of the late 1970s and early 1980s were able to capture this (e.g. Gilman 1977), although dynamo cycles in those simulations did not match that of the Sun (e.g. Gilman 1983; Glatzmaier 1985). However, it took another two decades for such simulations to become more mainstream (e.g. Brun et al. 2004; Ghizaru et al. 2010; Brown et al. 2011; Käpylä et al. 2012); see also Käpylä et al. (2023) for a recent review. Soon thereafter it was realized that obtaining solar-like differential rotation (fast equator, slow poles) with simulations with the nominal solar rotation rate and luminosity is highly non-trivial (e.g. Gastine et al. 2014; Käpylä et al. 2014; Fan and Fang 2014; O'Mara et al. 2016). This is thought to be due to too weak rotational influence on the dominant convective scales, or equivalently, a too low Coriolis (inverse Rossby) number.

At the same time, efforts were made to study the velocity amplitudes in the Sun using helioseismology (Hanasoge et al. 2010, 2012). These studies led to the realization that convective amplitudes at horizontal scales of the order of hundreds of Mm in the Sun appear to be several orders of magnitude weaker than in the global simulations, and that the velocity power spectrum in the Sun peaks at supergranular scale of 20-30 Mm. While the difference between helioseismic and simulation results has reduced somewhat in the meantime, a large discrepancy remains (e.g. Proxauf 2021). Adding to the puzzle are the results of Greer et al. (2015) from a ring-diagram analysis that shows high velocity amplitudes in the near-surface shear layer of the Sun consistent with global 3D convection simulations.

Several physical processes have been suggested as possible solutions of the convective conundrum. Rotationally constrained convection in the deep parts of the convection zone is one such possibility (Featherstone and Hindman 2016; Vasil et al. 2021). Linear stability analysis and non-linear simulations of convection indicate that the convective scale decreases with rotation. Given that the velocity power spectrum peaks at the supergranular scale in the Sun, it is has been conjectured that this scale coincides with the largest convectively driven scale in the deep convection zone. The question if convection in the Sun is indeed sufficiently constrained by rotation was studied systematically in Käpylä (2023a). These results are reviewed in more detail below.

Another possibility is that the solar convection zone is in fact largely subadiabatic, that is, the thermal stratification is formally weakly Schwarzschild stable. This can be enabled by plumes originating near the surface that transport cool low entropy material deep into the interior far beyond the formally unstable layer. This is related to the idea that convection in the Sun is driven by the cooling at the surface rather than by a superadiabatic temperature gradient throughout the convection zone (Stein and Nordlund 1989; Spruit 1997). Such non-local driving of convection due to surface cooling has been dubbed "entropy rain" (e.g. Brandenburg 2016). The convective flux in the stably stratified, but convecting, layer is carried by a counter-gradient term proportional to the variance of entropy fluctuations (Deardorff 1961, 1966). Hence this layer is referred to as the Deardorff zone. Simulations of overshooting convection routinely capture such subadiabatic layers if the transition between the radiative and convective regions is smooth enough (e.g. Roxburgh and Simmons 1993; Tremblay et al. 2015; Käpylä et al. 2017; Hotta 2017). Most of the previous works considered non-rotating cases, whereas here recent results of Käpylä (2023a), where the effects of rotation were included, are discussed.

Finally, the strength of the surface forcing depends on the physics near the surface of the star. In real stellar convection zones the density drops vertigineously near the surface and this cannot be directly reproduced in numerical simulations (e.g. Kupka and Muthsam 2017; Käpylä et al. 2023). Here preliminary results from an effort to study the effects of surface forcing by varying the (imposed) surperadiabatic temperature gradient at the surface are discussed based on earlier models presented in Käpylä et al. (2017). The novelty of these simulations is that they are constructed in such a way that the depth and structure of the convection zone are self-consistent results of the models instead of being fixed from the outset.

2. The model

The set-up is the same as in Käpylä (2019), Käpylä (2021), and Käpylä (2023a), and the Pencil Code (Pencil Code Collaboration et al. 2021) was used to make the simulations. The simulation domain is a rectangular box with dimensions $(L_{\rm x}, L_{\rm y}, L_{\rm z}) = (4, 4, 1.5)d$, where d is the depth of the initially isentropic layer which is situated between $0 \leq z/d \leq 1$. Initially this layer is sandwiched between a radiative layer with polytropic index $n = 3.25$ $(-0.45 \leq z/d < 0)$ and an isothermal layer $(1 < z/d \leq 1.05)$. The equations for compressible hydrodynamics are solved:

$$\frac{D \ln \rho}{Dt} = -\boldsymbol{\nabla} \cdot \boldsymbol{u}, \tag{1}$$

$$\frac{D\boldsymbol{u}}{Dt} = \boldsymbol{g} - \frac{1}{\rho}(\boldsymbol{\nabla} p - \boldsymbol{\nabla} \cdot 2\nu\rho\mathbf{S}) - 2\boldsymbol{\Omega} \times \boldsymbol{u}, \tag{2}$$

$$T\frac{Ds}{Dt} = -\frac{1}{\rho}\left[\boldsymbol{\nabla}\cdot(\boldsymbol{F}_{\rm rad} + \boldsymbol{F}_{\rm SGS}) - \mathscr{C}\right] + 2\nu\mathbf{S}^2, \tag{3}$$

where $D/Dt = \partial/\partial t + \boldsymbol{u}\cdot\boldsymbol{\nabla}$ is the advective derivative, ρ is the density, $\boldsymbol{u}$ is the velocity, $\boldsymbol{g} = -g\hat{\boldsymbol{e}}_z$ with $g > 0$ is the acceleration due to gravity where $\hat{\boldsymbol{e}}_z$ is the unit vector along the vertical (z) direction, p is the gas pressure, ν is the viscosity, $\mathbf{S}$ is the traceless rate-of-strain tensor, $\boldsymbol{\Omega} = \Omega\hat{\boldsymbol{e}}_z$ is the rotation vector, T is the temperature, and s is the specific entropy. $\boldsymbol{F}_{\rm rad} = -K\boldsymbol{\nabla}T$ is the radiative flux where $K = K_0\rho^{-2}T^{6.5}$ is the heat conductivity following Kramers opacity law, and $\boldsymbol{F}_{\rm SGS} = -\chi_{\rm SGS}\rho T\boldsymbol{\nabla}s'$ is the subgrid-scale (SGS) entropy flux, where $\chi_{\rm SGS}$ is a constant SGS diffusivity and $s' = s - \overline{s}$ is the deviation of the entropy from its horizontally averaged profile which is denoted by the overbar. The gas obeys the ideal gas equation $p = \mathscr{R}\rho T$, where $\mathscr{R} = c_{\rm P} - c_{\rm V}$ is the gas constant and where $c_{\rm P}$ and $c_{\rm V}$ are the specific heats in constant pressure and volume, respectively. Finally, $\mathscr{C}$ describes cooling near the surface.

In Käpylä (2023a) it was shown that a Coriolis number based on a hypothetical velocity $u_\star = (F_{\rm tot}/\rho)^{1/3}$ is equivalent to

$$\mathrm{Co_F} = 2\Omega H_{\rm p}\left(\frac{\rho}{F_{\rm tot}}\right)^{1/3} = (\mathrm{Ra_F^\star})^{-1/3}, \tag{4}$$

where $F_{\rm tot}$ is the total energy flux, $H_{\rm p} = ({\rm d}\ln\overline{p}/{\rm d}z)^{-1}$ is the pressure scale height, and

$$\mathrm{Ra_F^\star} = \frac{gF_{\rm tot}}{8c_{\rm P}\rho T\Omega^3 H^2} = \frac{F_{\rm tot}}{8\rho\Omega^3 H_{\rm p}^3}, \tag{5}$$

is the flux-based diffusion-free modified Rayleigh number (e.g. Christensen 2002), where the length scale H was chosen such that $H = c_{\rm P}T/g = H_{\rm p}$, where $H_{\rm p}$ is taken at the base of the convection zone. The advantage of $\mathrm{Co_F}$ is that it does not depend on any dynamical velocity or length scale and it can be computed using observables (Ω, $F_{\rm tot}$) and quantities from stellar structure models (ρ, $H_{\rm p}$); see also the discussion in Käpylä (2023b). Further system parameters include the Taylor number $\mathrm{Ta} = 4\Omega^2 d^4/\nu^2$, and the Prandtl number related to the SGS diffusivity $\mathrm{Pr_{SGS}} = \nu/\chi_{\rm SGS}$. The energy flux is measured by the dimensionless flux $\mathscr{F}_{\rm n} = F_{\rm tot}/\rho c_{\rm s}^3$ at $z/d = -0.45$ in the initial non-convecting state. Diagnostic quantities include the Reynolds ($\mathrm{Re} = u_{\rm rms}/\nu k_1$) and Péclet number ($\mathrm{Pe} = u_{\rm rms}/\chi_{\rm SGS}k_1 = \mathrm{Pr_{SGS}Re}$), and the global Coriolis number $\mathrm{Co} = 2\Omega/(u_{\rm rms}k_1)$, where $u_{\rm rms}$ is the volume-averaged rms-velocity, and $k_1 = 2\pi/d$ is an estimate of the scale of the largest eddies. A more detailed description of the model is given in Käpylä (2023a).

Three main sets of runs (Sets A, B, and C) were made where Co was varied between 0 and about 17. The imposed flux $\mathscr{F}_{\rm n}$ was varied between the sets to study the scaling of dynamical quantities with respect to it. The diffusivities were varied proportional to $\mathscr{F}_{\rm n}^{1/3}$ to achieve the same Re, Pe, and Co in each set (cf. Käpylä et al. 2020, for more details). The primary difference between the sets is that the Mach number $\mathrm{Ma} = u_{\rm rms}/c_{\rm s}$, where $c_{\rm s}$ is the sound speed, and therefore relative stability of the radiative layer below the convection zone vary. In Sets A to C, $\mathrm{Re} = \mathrm{Pe} \approx 30\ldots40$ and $\mathrm{Pr_{SGS}} = 1$. A subset of Set A, denoted as Set Am, was repeated at a higher resolution (576^3 instead of 288^3 grid points), and correspondingly higher values of Reynolds and Péclet numbers ($\mathrm{Re} = \mathrm{Pe} \approx 65\ldots84$), while keeping $\mathrm{Ra_F^\star}$ fixed.

3. Rotational scaling of convection

In Käpylä (2023a) the scaling of various quantities in rotating convection were studied. The numerical results were compared with scalings derived for slow rotation where a balance between inertial and buoyancy forces is assumed and for rapid rotation where a balance between Coriolis, inertial, and Archimedean (buoyancy) forces, or the CIA

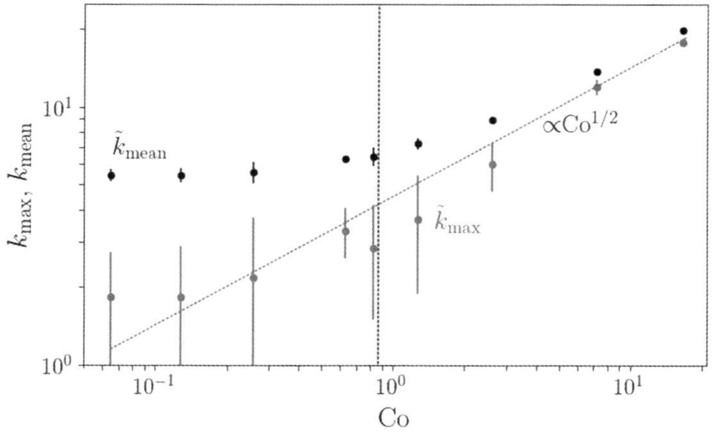

Figure 1. Mean ($k_{\rm mean}$) and peak ($k_{\rm max}$) wavenumbers from power spectra of velocity as a function of Co from simulations Set A from Käpylä (2023a). Data taken near the surface of the convectively unstable layer at $z/d = 0.85$. The dotted line shows the ${\rm Co}^{1/2}$ prediction from the CIA balance. The blue vertical dotted line indicates the solar value of ${\rm Co}_{\rm F}$. The tildes indicate normalization by $k_{\rm H}$.

balance (e.g. Stevenson 1979; Barker et al. 2014; Aurnou et al. 2020), is assumed. For the dominant convective scale this leads to:

$$\ell_{\rm conv} \sim H_{\rm p} \ \ \text{(slow rotation)}, \quad \text{and} \quad \ell_{\rm conv} \sim H_{\rm p}{\rm Co}^{-1/2} \ \ \text{(rapid rotation)}. \tag{6}$$

Similarly, the scalings for the convective velocity are:

$$u_{\rm conv} \sim u_\star \ \ \text{(slow rotation)}, \quad \text{and} \quad u_{\rm conv} \sim u_\star {\rm Co}^{-1/6} \ \ \text{(rapid rotation)}. \tag{7}$$

Finally, the local Coriolis number ${\rm Co}_\ell = 2\Omega\ell_{\rm conv}/u_{\rm conv}$, can be shown to depend on ${\rm Ra}_{\rm F}^\star$:

$${\rm Co}_\ell \sim ({\rm Ra}_{\rm F}^\star)^{-1/3} \ \ \text{(slow rotation)}, \quad \text{and} \quad {\rm Co}_\ell \sim ({\rm Ra}_{\rm F}^\star)^{-1/5} \ \ \text{(rapid rotation)}. \tag{8}$$

Here the convective length scale is estimated from the power spectrum of velocity, $E(k)$, for which $\boldsymbol{u}^2 = \int E(k)dk$, either by taking the wavenumber where the power has its maximum ($k_{\rm max}$) or the mean wavenumber $k_{\rm mean} = \int kE(k)dk / \int E(k)dk$. The length scales $\ell_{\rm max}$ and $\ell_{\rm mean}$ are given by $\ell_{\rm max,mean} = L_{\rm x}/\tilde{k}_{\rm max,mean}$, where $\tilde{k}_{\rm max,mean} = k_{\rm max,mean}/k_{\rm H}$, and where $k_{\rm H} = 2\pi/L_{\rm x} = \pi/2d$ is the wavenumber corresponding to the horizontal extent of the simulation domain. Figure 1 shows $\tilde{k}_{\rm max}$ and $\tilde{k}_{\rm mean}$ from Set A. For slow rotation (${\rm Co} \lesssim 1$) both $\tilde{k}_{\rm max}$ and $\tilde{k}_{\rm mean}$ are approximately constant, although the former is already consistent with the ${\rm Co}^{1/2}$ scaling due to the large error estimates which are taken to be the standard deviation of the mean values calculated from several snapshots. For rapid rotation the ${\rm Co}^{1/2}$ scaling from the CIA balance is recovered for both $\tilde{k}_{\rm max}$ and $\tilde{k}_{\rm mean}$.

The case that the deep parts of the solar convection zone is strongly rotationally constrained has been discussed recently by Featherstone and Hindman (2016) and Vasil et al. (2021). Both of these studies argue that the maximum horizontal scale of convection is reduced by rotation in the deep parts of the convection zone, such that the largest convectively driven scale coincides with supergranules (20–30 Mm) at a spherical harmonic degree $\ell \approx 100$. Using Eq. (4) it is possible to compute ${\rm Co}_{\rm F}$ at the base of the solar convection zone with $\Omega_\odot = 2.7 \cdot 10^{-6}$ s^{-1}, $H_{\rm p}^\odot \approx 5 \cdot 10^7$ m, $\rho^\odot \approx 200$ kg m^{-3}, and $F_{\rm tot}^\odot = L_\odot/(4\pi r_{\rm bot}^2)$, where $L_\odot = 3.83 \cdot 10^{26}$ W, and $r_{\rm bot} = 0.7R_\odot$, gives ${\rm Co}_{\rm F}^\odot \approx 3.1$. On the other hand,

$${\rm Co} = \frac{u_\star}{u_{\rm rms}} \frac{{\rm Co}_{\rm F}}{k_1 H_{\rm p}}. \tag{9}$$

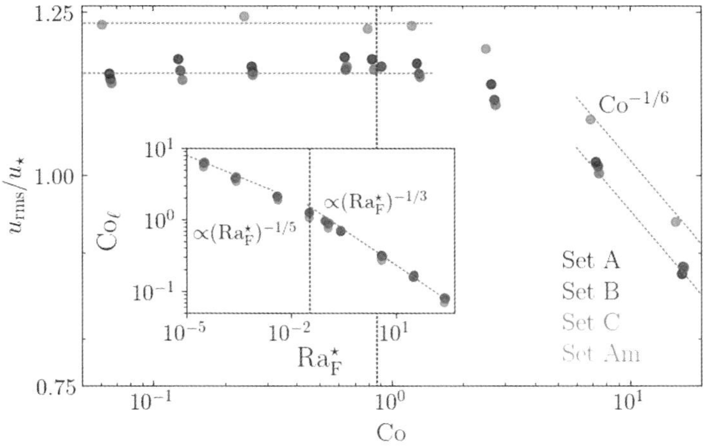

Figure 2. Volume-averaged rms-velocity as a function of Co from simulations in Sets A, B, C, and Am from Käpylä (2023a). The dotted lines are either constant (for $\mathrm{Co} \leq 1.5$) or proportional to $\mathrm{Co}^{-1/6}$ (for $\mathrm{Co} \geq 6$). The inset shows Co_ℓ as a function of $\mathrm{Ra}_\mathrm{F}^\star$ for the same runs with power laws proportional to $(\mathrm{Ra}_\mathrm{F}^\star)^{-1/5}$ for fast and $(\mathrm{Ra}_\mathrm{F}^\star)^{-1/3}$ for slow rotation corresponding to $\mathrm{Ra}_\mathrm{F}^\star \leq 3 \cdot 10^{-3}$ and $\mathrm{Ra}_\mathrm{F}^\star \geq 0.03$, respectively. The blue dotted vertical lines indicate the solar values of Co and $\mathrm{Ra}_\mathrm{F}^\star$, respectively.

For the current slowly rotating simulations in Sets A to C, $u_\star/u_{\rm rms} \approx 0.87$ (see, Figure 2), and $(k_1 H_\mathrm{p})^{-1} \approx 0.32$, such that the solar Co_F is achieved in a simulation with $\mathrm{Co} \approx 0.87$. Inspection of Figure 1 suggests that the Sun is somewhere in between the weakly rotationally influenced and the rotationally constrained regimes.

The dominant convective scale in a simulation with $\mathrm{Co} \approx 0.83$ and $\mathrm{Co}_\mathrm{F} = 3.1$ is again estimated from the power spectrum of the velocity. In this case the maximum power occurs at wavenumber $\tilde{k}_{\rm max} = 3$ and the mean wavenumber is $\tilde{k}_{\rm mean} = 7$, corresponding to length scales $\ell_{\rm max} = 1.33d$ and $\ell_{\rm mean} = 0.57d$. The pressure scale height at the base of of the convective layer in this run is $H_\mathrm{p} = 0.49d$. Assuming the simulations to represent the deep parts of the convection zone at the interface to the radiative layer, the pressure scale height corresponds to $H_\mathrm{p}^\odot \approx 5 \cdot 10^7$ m. This leads to $\ell_{\rm max} \approx 135$ Mm and $\ell_{\rm mean} \approx 58$ Mm, respectively. These results seem to refute the idea that rotationally constrained convection can explain the supergranular scale as the largest convectively driven scale. Furthermore, Käpylä (2023a) showed that in the simulations of Featherstone and Hindman (2016), where the supergranular scale is the dominant scale correspond to a value of Co_F that requires a rotation rate which is at least 15 times higher than in the Sun.

Figure 2 shows the time and volume-averaged rms-velocity from Sets A, B, C, and Am normalized by $u_\star$. The scalings for slow and rapid rotation from Eq. (7) are recovered for $\mathrm{Co} \lesssim 1.5$ and $\mathrm{Co} \gtrsim 6$, respectively. For Co exceeding the maximum values here ($\mathrm{Co} \approx 17$), the flow begins to develop a large-scale vortical component (see also Chan 2007; Käpylä et al. 2011) which is likely due to two-dimensionalization of turbulence, and extending the calculations to higher Co becomes challenging. Finally, the local Coriolis number Co_ℓ is shown in the inset of Figure 2. Co_ℓ adheres to the scalings given in Eq. (8) with respect to $\mathrm{Ra}_\mathrm{F}^\star$ for both slowly and rapidly rotating regimes.

4. Deardorff layer as a function of rotation

In the canonical models of stellar convection relying on mixing length theory (e.g. Böhm-Vitense 1958), the whole convection zone is unstably stratified and convection is

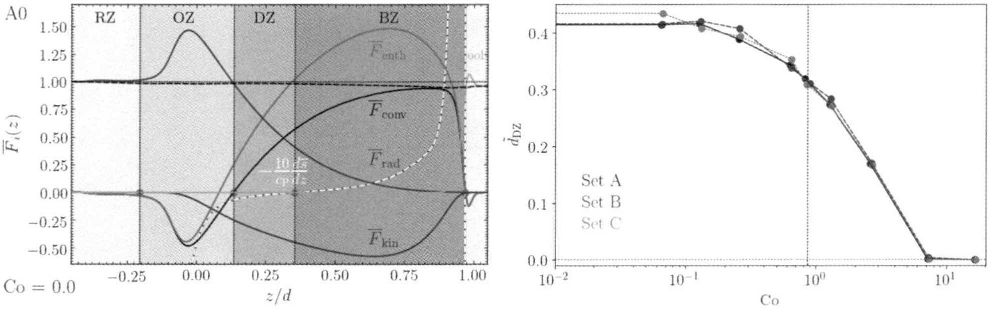

Figure 3. Left panel: Horizontally averaged energy fluxes from the non-rotating Run A0 from Käpylä (2023a). Grey (orange) areas indicate mixed (radiative) layers, and the red circles highlight the boundaries between the layers. The various zones are characterised as $\overline{F}_{\rm rad} \approx F_{\rm tot}$ (radiative; RZ), $\overline{F}_{\rm conv} < 0$, $\overline{F}_{\rm rad} > F_{\rm tot}$ (overshoot; OZ), $\overline{F}_{\rm conv} > 0$, $\Delta\nabla < 0$ (Deardorff; DZ), and $\overline{F}_{\rm conv} > 0$, $\Delta\nabla > 0$ (buoyancy; BZ). Right: Depth of the Deardorff layer ($d_{\rm DZ}$) as a function of Co from simulations in Sets A (black solid line), B (blue dashed), and C (red dotted). The blue dotted vertical line indicates the value of Co corresponding to the solar $\mathrm{Co_F}$. Adapted from Käpylä (2023a).

thought to be driven locally by a superadiabatic temperature gradient

$$\Delta\nabla = \nabla - \nabla_{\rm ad} = -\frac{1}{H_{\rm p}}\frac{{\rm d}s}{{\rm d}z} > 0, \tag{10}$$

where $\nabla = {\rm d}\ln\overline{T}/{\rm d}\ln p$ is the logarithmic temperature gradient, $\nabla_{\rm ad} = ({\rm d}\ln\overline{T}/{\rm d}\ln p)_{\rm ad} = 1 - 1/\gamma$ is the corresponding adiabatic gradient, and where $\gamma = c_{\rm P}/c_{\rm V}$. If this were the case in the Sun, giant cell convection on the scale of 200 Mm is expected to be prominent. This is not observed in the Sun and 3D hydrodynamic simulations suggest that the deep parts of convective layers are often weakly subadiabatic (e.g. Roxburgh and Simmons 1993; Tremblay et al. 2015; Käpylä et al. 2017; Hotta 2017). This is thought to be due to the inherently non-local nature of convection which is driven by cooling near the surface instead of heating from the base as has been shown, e.g., in Käpylä et al. (2017).

This subadiabatic but convecting layer is referred to as the Deardorff zone (DZ), and it is characterised by $\Delta\nabla < 0$ and $\overline{F}_{\rm conv} > 0$, where

$$\overline{F}_{\rm conv} = \overline{F}_{\rm enth} + \overline{F}_{\rm kin} = c_{\rm P}\overline{(\rho u_z)'T'} + \tfrac{1}{2}\overline{\rho\boldsymbol{u}^2 u_z}, \tag{11}$$

is the total convected flux (e.g. Cattaneo et al. 1991), which is the sum of the time and horizontal averages of the enthalpy and kinetic energy fluxes, and where the primes denote deviations from the horizontal average. The left panel of Figure 3 shows the flux balance from a non-rotating Run A0 from Käpylä (2023a), where in addition to $\overline{F}_{\rm conv}$, $\overline{F}_{\rm enth}$, and $\overline{F}_{\rm kin}$, also the radiative ($\overline{F}_{\rm rad} = -\overline{K}{\rm d}\overline{T}/{\rm d}z$) and cooling ($\overline{F}_{\rm cool} = -\int\overline{\mathscr{C}}{\rm d}z$) fluxes, as well as a quantity proportional to $\Delta\nabla$ are shown for reference. The DZ in this run is over 40 per cent of the pressure scale height at the base of the convection zone which is here defined as the lower boundary of the DZ. For the run closest to the solar value of $\mathrm{Co_F}$, $d_{\rm DZ} \approx 0.3H_{\rm p}$ which corresponds to 15 Mm in the Sun.

Figure 3 shows $d_{\rm DZ}$ as a fraction of the pressure scale height at the base of the convection zone as a function of rotation for Sets A, B, and C from Käpylä (2023a). The sets differ from each other in that the input energy flux is varied such that between the extreme cases (Sets A and C), $\mathscr{F}_{\rm n}$ decreases by factor of five. This has implications for the Mach number and also for the overshooting below the convection zone (e.g. Käpylä 2019). However, the depth of the DZ is virtually unaffected by the change of $\mathscr{F}_{\rm n}$. This is

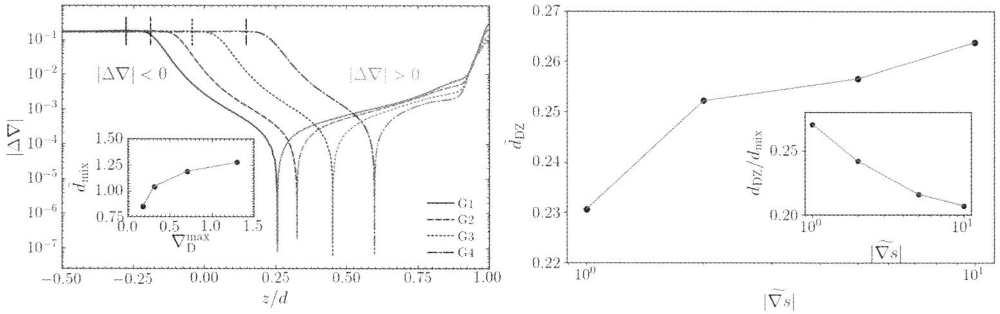

Figure 4. Left panel: Absolute value of the superadiabatic temperature gradient $\Delta\nabla$ for Runs G[1-4]. Red (blue) parts of the curves indicate regions where $\Delta\nabla > 0$ ($\Delta\nabla < 0$). The black vertical lines denote the bottom of the convectively mixed layer. The inset shows the depth of the mixed zone $\tilde{d}_{\rm mix} = d_{\rm mix}/d$ as a function $\nabla_{\rm D}^{\rm max}$. Right: depth of the Deardorff zone, $\tilde{d}_{\rm DZ} = d_{\rm DZ}/d$ as a function of $|\widetilde{\nabla s}|$. The inset shows the relative fraction of the Deardorff zone of the mixed zone as a function of $|\widetilde{\nabla s}|$.

because the cooling time at the surface is varied inversely proportional to $\mathscr{F}_{\rm n}$ such that the thermal forcing remains unaffected.

5. Effects of surface forcing

There are some a few into the effects of varying surface forcing using convection simulations. For example, Cossette and Rast (2016) varied the superadiabatic gradient at the surface and found that it had a substantial effect on the convective length scale and deep convection zone dynamics. On the other hand, Hotta et al. (2019) found only a weak influence of the surface in a simulation that encompassed nominally the entire convection zone of the Sun.

Here a similar approach as in Cossette and Rast (2016) is explored with a simulation set-up that was used in Käpylä et al. (2017). In distinction to the simulations discussed above, the upper cooling layer is replaced by an imposed entropy gradient at the upper boundary, and the z-coordinate runs between $-0.5 \le z/d \le 1$ such that the transition between the initially isentropic and radiative layers is at $z = 0$. Furthermore, the SGS diffusion term has an extra term proportional to the mean entropy gradient:

$$\boldsymbol{F}_{\rm SGS} = -(\chi_{\rm SGS}\rho T\boldsymbol{\nabla} s' + \chi_{\rm SGS}^{\rm m}\rho T\boldsymbol{\nabla}\overline{s}), \tag{12}$$

where $\chi_{\rm SGS}^{\rm m}$ is non-zero only above $z/d = 0.95$, such that the second term on the rhs of Eq. (12) transports the heat flux through the upper boundary. Here ${\rm Pr}_{\rm SGS} = 1$ and ${\rm Pr}_{\rm SGS}^{\rm m} = \nu/\chi_{\rm SGS}^{\rm m} = 0.5$. In Käpylä et al. (2017), the entropy gradient at the surface was fixed to $\widetilde{\nabla s} = (d/c_{\rm P})(\boldsymbol{e}_z \cdot \boldsymbol{\nabla}\overline{s}) = -10$. Here four values between -1 and -10 for $\widetilde{\nabla s}$ are explored in Set G. In distinction to Cossette and Rast (2016) where a spatially fixed Newtonian cooling term was used which does not allow the depth of the convection zone to change appreciably, the current simulations use Kramers opacity law which enables this.

In the absence of convection the hydrostatic solution with the Kramers opacity law is convectively unstable only in a shallow surface layer (e.g. Barekat and Brandenburg 2014; Käpylä 2023a). This solution is modified by the onset of convection and the final outcome is expected to be sensitive to the surface physics. Decreasing $|\widetilde{\nabla s}|$ leads to a shallower convection zone as can be seen from the left panel of Figure 4. Furthermore, $\Delta\nabla$ near the surface decreases and the surface temperature increases. In the updated mixing length model of Brandenburg (2016), the enthalpy flux was quantified in terms of gradient ($\overline{F}_{\rm G}$) and non-gradient ($\overline{F}_{\rm D}$) contributions

$$\overline{F}_{\rm enth} = \overline{F}_{\rm G} + \overline{F}_{\rm D} = -\tau\overline{\rho}\overline{T}(\tfrac{1}{3}u_{\rm rms}^2\nabla_z\overline{s} + \overline{s'^2}g/c_{\rm P}) = \tfrac{1}{3}\overline{\rho}c_{\rm P}\overline{T}(\tau u_{\rm rms}^2/H_{\rm p})(\Delta\nabla + \nabla_{\rm D}), \quad (13)$$

where τ is a relaxation time, and where the magnitude of the non-gradient is characterised by

$$\nabla_{\rm D} = (3/\gamma)(\overline{s'^2}/c_{\rm P}^2){\rm Ma}^{-2}. \quad (14)$$

Brandenburg (2016) argued that $\nabla_{\rm D}$ in deeper parts is proportional to its maximum value near the surface, $\nabla_{\rm D}^{\rm max}$. The depth of the mixed layer $d_{\rm mix}$, consisting of the buoyancy, Deardorff, and overshoot zones, is shown as a function of $\nabla_{\rm D}^{\rm max}$ in the inset of the left panel of Figure 4 for the runs in Set G. While $d_{\rm mix}$ increases monotonically with $\nabla_{\rm D}^{\rm max}$ (corresponding to increasing $|\widetilde{\nabla s}|$), there appears to be no straightforward relation between the two. The depth of the Deardorff zone is not very sensitive to $|\widetilde{\nabla s}|$, although its size relative to the mixed zone decreases somewhat as $|\widetilde{\nabla s}|$ increases; see the inset of the right panel of Figure 4.

6. Conclusions

Several ways to address the "convective conundrum," or the discrepancy between convective velocity amplitudes in simulations and solar observations, were reviewed based on results from recent hydrodynamic Cartesian convection simulations. First the effects of rotation from Käpylä (2023a) were considered. These results suggest that the convective scale in the deep convection zone of the Sun is not sufficiently affected by rotation to reduce the largest convectively driven scale to the supergranular scale of 20-30 Mm as has been conjectured earlier (Featherstone and Hindman 2016; Vasil et al. 2021). These simulations also suggest that the depth of the convective but formally stably stratified Deardorff zone is reduced as rotation increases, but that a substantial subadiabatic layer of about 15 Mm is still expected to be found at the base of the solar convection zone. Scaling laws of several dynamical quantities such as convective scale, velocity amplitude, and local Coriolis number were shown to follow scalings derived under the CIA balance (e.g. Stevenson 1979; Barker et al. 2014; Aurnou et al. 2020).

The effects of surface forcing were explored with a set of new simulations where the entropy gradient at the surface was imposed similarly as in Käpylä et al. (2017). Unlike in the previous studies in the literature that study the effects of the surface for the deep convection zone, the current simulations allow the depth of the convective layer to vary self-consistently. These preliminary results show that stronger surface forcing, in terms of a steeper entropy gradient, leads to a deeper convectively mixed layer. Although there is a monotonic dependence between the imposed entropy gradient and the depth of the convective layer, no clear relation between the two can be identified. However, the fraction of the Deardorff layer of the total depth of the convectively mixed layer decreases somewhat when the surface forcing is increased.

The results quoted above come with the caveat that the surface forcing of convection in the current simulations is assumed to be accurately modelled. This, however, cannot be guaranteed, and it is likely that much smaller scales need to be resolved to capture the effects of radiative cooling in the photosphere accurately (e.g. Kupka and Muthsam 2017). Furthermore, the effects of astrophysically relevant low Prandtl numbers (e.g. Spiegel 1962; Käpylä 2021) and vigorous magnetism (e.g. Hotta et al. 2022) are also likely to play important roles for solar and stellar convection.

Acknowledgements

The simulations were performed using the resources granted by the Gauss Center for Supercomputing for the Large-Scale computing project "Cracking the Convective Conundrum" in the Leibniz Supercomputing Centre's SuperMUC-NG supercomputer in Garching, Germany. This work was supported in part by the Deutsche Forschungsgemeinschaft (DFG, German Research Foundation) Heisenberg programme (grant No. KA 4825/4-1) and by the Munich Institute for Astro-, Particle and BioPhysics (MIAPbP) which is funded by the DFG under Germany's Excellence Strategy - EXC-2094 - 390783311.

References

Aurnou, J. M., Horn, S., & Julien, K. 2020, Connections between nonrotating, slowly rotating, and rapidly rotating turbulent convection transport scalings. *Physical Review Research*, 2(4), 043115.

Barekat, A. & Brandenburg, A. 2014, Near-polytropic stellar simulations with a radiative surface. *A&A*, 571, A68.

Barker, A. J., Dempsey, A. M., & Lithwick, Y. 2014, Theory and Simulations of Rotating Convection. *ApJ*, 791(1), 13.

Böhm-Vitense, E. 1958, Über die Wasserstoffkonvektionszone in Sternen verschiedener Effektivtemperaturen und Leuchtkräfte. Mit 5 Textabbildungen. *ZAp*, 46, 108.

Brandenburg, A. 2016, Stellar mixing length theory with entropy rain. *ApJ*, 832, 6.

Brown, B. P., Miesch, M. S., Browning, M. K., Brun, A. S., & Toomre, J. 2011, Magnetic Cycles in a Convective Dynamo Simulation of a Young Solar-type Star. *ApJ*, 731, 69.

Brun, A. S., Miesch, M. S., & Toomre, J. 2004, Global-Scale Turbulent Convection and Magnetic Dynamo Action in the Solar Envelope. *ApJ*, 614, 1073–1098.

Cattaneo, F., Brummell, N. H., Toomre, J., Malagoli, A., & Hurlburt, N. E. 1991, Turbulent compressible convection. *ApJ*, 370, 282–294.

Chan, K. L. 2007, Rotating convection in f-boxes: Faster rotation. *Astron. Nachr.*, 328, 1059.

Christensen, U. R. 2002, Zonal flow driven by strongly supercritical convection in rotating spherical shells. *Journal of Fluid Mechanics*, 470(1), 115–133.

Cossette, J.-F. & Rast, M. P. 2016, Supergranulation as the Largest Buoyantly Driven Convective Scale of the Sun. *ApJ*, 829, L17.

Deardorff, J. W. 1961, On the Direction and Divergence of the Small-Scale Turbulent Heat Flux. *J. Atmosph. Sci.*, 18, 540–548.

Deardorff, J. W. 1966, The Counter-Gradient Heat Flux in the Lower Atmosphere and in the Laboratory. *J. Atmosph. Sci.*, 23, 503–506.

Fan, Y. & Fang, F. 2014, A Simulation of Convective Dynamo in the Solar Convective Envelope: Maintenance of the Solar-like Differential Rotation and Emerging Flux. *ApJ*, 789, 35.

Featherstone, N. A. & Hindman, B. W. 2016, The Emergence of Solar Supergranulation as a Natural Consequence of Rotationally Constrained Interior Convection. *ApJ*, 830, L15.

Gastine, T., Yadav, R. K., Morin, J., Reiners, A., & Wicht, J. 2014, From solar-like to antisolar differential rotation in cool stars. *MNRAS*, 438, L76–L80.

Ghizaru, M., Charbonneau, P., & Smolarkiewicz, P. K. 2010, Magnetic Cycles in Global Large-eddy Simulations of Solar Convection. *ApJ*, 715, L133–L137.

Gilman, P. A. 1977, Nonlinear Dynamics of Boussinesq Convection in a Deep Rotating Spherical Shell. I. *Geophys. Astrophys. Fluid Dynam.*, 8, 93–135.

Gilman, P. A. 1983, Dynamically consistent nonlinear dynamos driven by convection in a rotating spherical shell. II - Dynamos with cycles and strong feedbacks. *ApJS*, 53, 243–268.

Glatzmaier, G. A. 1985, Numerical simulations of stellar convective dynamos. II - Field propagation in the convection zone. *ApJ*, 291, 300–307.

Greer, B. J., Hindman, B. W., Featherstone, N. A., & Toomre, J. 2015, Helioseismic Imaging of Fast Convective Flows throughout the Near-surface Shear Layer. *ApJ*, 803, L17.

Hanasoge, S. M., Duvall, Thomas L., J., & DeRosa, M. L. 2010, Seismic Constraints on Interior Solar Convection. *ApJ*, 712(1), L98–L102.

Hanasoge, S. M., Duvall, T. L., & Sreenivasan, K. R. 2012, Anomalously weak solar convection. *Proc. Natl. Acad. Sci.*, 109, 11928–11932.

Hotta, H. 2017, Solar Overshoot Region and Small-scale Dynamo with Realistic Energy Flux. *ApJ*, 843, 52.

Hotta, H., Iijima, H., & Kusano, K. 2019, Weak influence of near-surface layer on solar deep convection zone revealed by comprehensive simulation from base to surface. *Science Advances*, 5(1), 2307.

Hotta, H., Kusano, K., & Shimada, R. 2022, Generation of Solar-like Differential Rotation. *ApJ*, 933(2), 199.

Käpylä, P. J. 2019, Overshooting in simulations of compressible convection. *A&A*, 631, A122.

Käpylä, P. J. 2021, Prandtl number dependence of stellar convection: Flow statistics and convective energy transport. *A&A*, 655, A78.

Käpylä, P. J. 2023,a Convective scale and subadiabatic layers in simulations of rotating compressible convection. *arXiv e-prints,*, arXiv:2310.12855.

Käpylä, P. J. 2023,b Transition from anti-solar to solar-like differential rotation: Dependence on Prandtl number. *A&A*, 669b, A98.

Käpylä, P. J., Browning, M. K., Brun, A. S., Guerrero, G., & Warnecke, J. 2023, Simulations of Solar and Stellar Dynamos and Their Theoretical Interpretation. *Space Sci. Rev.*, 219(7), 58.

Käpylä, P. J., Gent, F. A., Olspert, N., Käpylä, M. J., & Brandenburg, A. 2020, Sensitivity to luminosity, centrifugal force, and boundary conditions in spherical shell convection. *Geophysical and Astrophysical Fluid Dynamics*, 114(1-2), 8–34.

Käpylä, P. J., Käpylä, M. J., & Brandenburg, A. 2014, Confirmation of bistable stellar differential rotation profiles. *A&A*, 570, A43.

Käpylä, P. J., Mantere, M. J., & Brandenburg, A. 2012, Cyclic Magnetic Activity due to Turbulent Convection in Spherical Wedge Geometry. *ApJ*, 755, L22.

Käpylä, P. J., Mantere, M. J., & Hackman, T. 2011, Starspots due to Large-scale Vortices in Rotating Turbulent Convection. *ApJ*, 742, 34.

Käpylä, P. J., Rheinhardt, M., Brandenburg, A., Arlt, R., Käpylä, M. J., Lagg, A., Olspert, N., & Warnecke, J. 2017, Extended Subadiabatic Layer in Simulations of Overshooting Convection. *ApJ*, 845, L23.

Kupka, F. & Muthsam, H. J. 2017, Modelling of stellar convection. *Liv. Rev. Comp. Astrophys.*, 3, 1.

O'Mara, B., Miesch, M. S., Featherstone, N. A., & Augustson, K. C. 2016, Velocity amplitudes in global convection simulations: The role of the Prandtl number and near-surface driving. *Adv. Space Res.*, 58, 1475–1489.

Pencil Code Collaboration, Brandenburg, A., Johansen, A., Bourdin, P., Dobler, W., Lyra, W., Rheinhardt, M., Bingert, S., Haugen, N., Mee, A., Gent, F., Babkovskaia, N., Yang, C.-C., Heinemann, T., Dintrans, B., Mitra, D., Candelaresi, S., Warnecke, J., Käpylä, P., Schreiber, A., Chatterjee, P., Käpylä, M., Li, X.-Y., Krüger, J., Aarnes, J., Sarson, G., Oishi, J., Schober, J., Plasson, R., Sandin, C., Karchniwy, E., Rodrigues, L., Hubbard, A., Guerrero, G., Snodin, A., Losada, I., Pekkilä, J., & Qian, C. 2021, The Pencil Code, a modular MPI code for partial differential equations and particles: multipurpose and multiuser-maintained. *The Journal of Open Source Software*, 6(58), 2807.

Proxauf, B. 2021,. *Observations of large-scale solar flows.* PhD thesis, Georg August University of Göttingen, Germany.

Roxburgh, L. W. & Simmons, J. 1993, Numerical studies of convective penetration in plane parallel layers and the integral constraint. *A&A*, 277, 93.

Spiegel, E. A. 1962, Thermal Turbulence at Very Small Prandtl Number. *J. Geophys. Res.*, 67(3), 3063.

Spruit, H. 1997, Convection in stellar envelopes: a changing paradigm. *Mem. Soc. Astron. Italiana*, 68, 397.

Stein, R. F. & Nordlund, A. 1989, Topology of convection beneath the solar surface. *ApJ*, 342, L95–L98.

Stevenson, D. J. 1979, Turbulent thermal convection in the presence of rotation and a magnetic field: A heuristic theory. *Geophysical and Astrophysical Fluid Dynamics*, 12(1), 139–169.

Thompson, M. J., Christensen–Dalsgaard, J., Miesch, M. S., & Toomre, J. 2003, The Internal Rotation of the Sun. *ARA&A*, 41, 599–643.

Tremblay, P.-E., Ludwig, H.-G., Freytag, B., Fontaine, G., Steffen, M., & Brassard, P. 2015, Calibration of the Mixing-length Theory for Convective White Dwarf Envelopes. *ApJ*, 799, 142.

Vasil, G. M., Julien, K., & Featherstone, N. A. 2021, Rotation suppresses giant-scale solar convection. *Proceedings of the National Academy of Science*, 118(31), e2022518118.

Contributed papers

Dynamics of Solar and Stellar Convection Zones and Atmospheres
Proceedings IAU Symposium No. 365, 2024
A. V. Getling & L. L. Kitchatinov, eds.
doi:10.1017/S1743921324000139

Analyzing Magnetic Network and its Variations over Solar Cycles 23, 24 and 25 Based on Magnetic Power Spectra

Yukun Luo[1,2], Jie Jiang[1,2] and Ruihui Wang[1,2]

[1]School of Space and Environment, Beihang University, Beijing, China
email: luoyukun@buaa.edu.cn

[2]Key Laboratory of Space Environment monitoring and Information Processing of MIIT, Beijing, China

Abstract. The magnetic network is a typical magnetic structure of the quiet Sun. Investigating its cycle dependence is crucial for understanding its evolution. We aim to identify and analyze the spatial scales of the magnetic network within magnetic power spectra derived from high-resolution Solar and Heliospheric Observatory (SOHO)/Michelson Doppler Imager (MDI) and Solar Dynamics Observatory (SDO)/ Helioseismic and Magnetic Imager (HMI) synoptic magnetograms. The data sets cover the entirety of solar cycles 23, 24, and part of cycle 25. We find that the identified magnetic network sizes identified range from 26 Mm to 41 Mm. There seems to be no obvious dependence on the solar cycle, and the sizes are distributed uniformly within the identification range.

Keywords. Magnetic Network Size, Solar Cycle Dependence

1. Introduction

The solar surface outside of active regions is mainly covered by the quiet sun. The magnetic network contributes most of the flux in the quiet sun (Gosic 2015). The network field is typically associated with supergranulation. This is due to the interaction between magnetic fields and convection flow fields (Rincon and Rieutord 2018). Additionally, the investigation of the magnetic network is important because it is also crucial to understand the origin of the magnetic field and its contribution to the solar atmosphere (Bellot Rubio and Orozco Suárez 2019). However, there is still a debate about the variation of the network.

Previous works investigating the chromospheric network and supergranulation produce divergent results. While Hagenaar et al. (1997) suggest no dependence on network sizes, most other studies propose that the size should be either anti-correlated or correlated to the magnetic activity (Rincon and Rieutord 2018). For instance, Singh and Bappu (1981) find that network sizes tend to be smaller during the solar maximum with respect to the chromosphere network. Muenzer et al. (1989) report that the network sizes should increase with the solar activity. Previous disparities may be due to various identification and analysis methods for different proxies Berrilli et al. (1999). Therefore, it is necessary to investigate the size variation of the network directly from magnetograms.

The power spectrum is a useful tool for investigating characteristic structures in the images. Therefore, in this study, we aim to derive magnetic power spectra from MDI and HMI synoptic maps. We then identify the network based on the power spectra and provide insights into the cycle dependence of network sizes.

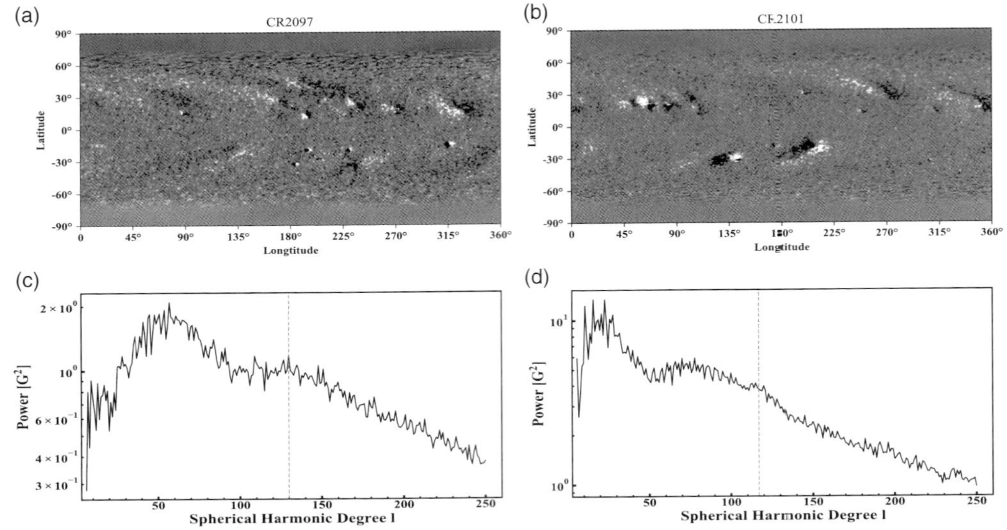

Figure 1. Example of network identification results. (a) & (b) are synoptic magnetograms for CR 2097 and CR 2101, respectively. (c) & (d) are corresponding magnetic power spectra for (a) and (b). The identified network is highlighted by the vertical dashed lines. Image adapted with permission from Luo et al. (2023), copyright by the American Astronomical Society.

2. Methods

For this study, we analyze continuous radial HMI and MDI synoptic magnetograms covering solar cycles 23, 24, and part of cycle 25. The data sets begin in Carrington Rotation (CR) 1911 (1996 July) and end in CR 2265 (2022 December). We perform the spherical harmonic decomposition to obtain the magnetic power spectra, after preprocessing the HMI and MDI synoptic magnetograms. Then we calibrate HMI power spectra based on the methods proposed by Luo et al. (2023) to ensure consistent and homogeneous data for the following analysis.

The magnetic power spectra show network features as peaks or knees, depending on the relative strength relationship between active region power and network power. We improve the morphology methods used in Luo et al. (2023) for the peak and knee, respectively. All of the data processing is done in Python. We successfully identified the network in the power spectra of 254 CRs, for a total of 355 CRs.

3. Results

Figure 1 displays two examples of network identification. The top panel shows two synoptic maps during the overlapping period for HMI and MDI. The bottom panel displays magnetic power spectra derived by spherical harmonic composition. In both power spectra, the small peaks in $l = 130$ for (c) and $l = 117$ for (d) are considered as the representation of the magnetic network. Their actual sizes are 33.7 Mm and 37.4 Mm, respectively.

The identified results for all CRs and their relationship with time are shown in Figure 2. The typical magnetic network sizes typically range from $l = 106$ ($\approx$26 Mm) to $l = 169$ ($\approx$41 Mm). During any phase of the solar cycle, a wide range of network sizes can be identified. The distribution of network sizes is also homogeneous over time, suggesting that the network can emerge randomly in the same size range during both solar maximum and minimum. Furthermore, this may imply that some intrinsic properties of the network

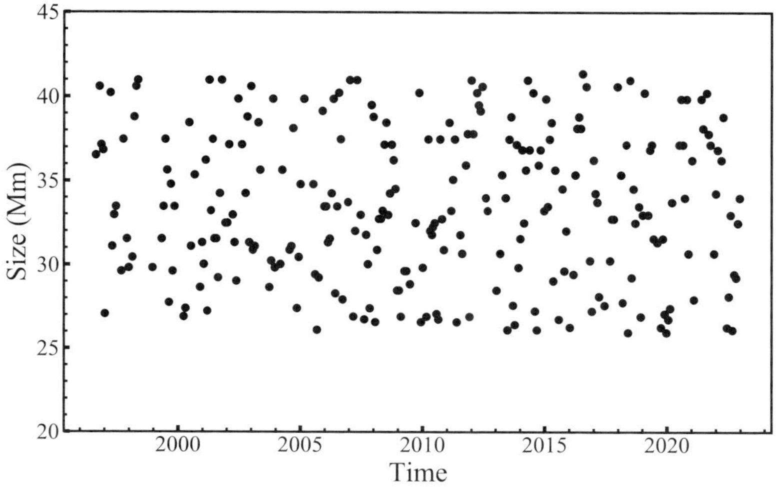

Figure 2. The network sizes identified by magnetic power spectra during 23rd, 24th and part of 25th solar cycles. Image reproduced with permission from Luo et al. (2024).

are unchanged for different solar cycles. Importantly, there is no significant temporal variation in the range and distribution of identified network sizes. This may suggest that magnetic network sizes are not dependent on cycles.

Our results are similar to those of Hagenaar et al. (1997), who focus on the chromosphere network. They propose no correlation between network sizes and local magnetic strength. Similarly, Hagenaar et al. (2003) study the temporal variation of network fields with different flux. They find that the network with fluxes $\leq 20 \times 10^{18}$ is independent of the solar cycle, while the network with more flux varies with solar activity. The study explains two different relationships based on the origin of the network. They suggest that the weaker network is contributed by ephemeral regions and active regions, showing cycle independence. As the network flux increases, the dominant role of decaying active regions is enhanced, causing the dependence on solar cycles. This may imply that the network we identified is related to both ephemeral regions and decaying active regions.

Figure 3 shows the statistical results of the identified network sizes. The histogram indicates that the counts of different network sizes do not change significantly. Additionally, the mean network size is 33.4 Mm ($l = 131$), and the standard deviation is 4.43 Mm. These values are similar to the characteristic parameters of uniform distribution in the same size range. The skewness and kurtosis are 0.058 and 1.87, respectively, which are close to the values for the uniform distribution (0 and 1.8). All of the above statistical parameters indicate that the identified network sizes are close to the uniform distribution in our identification range. This is different from the previous work about supergranulation sizes, which usually give an unimodal distribution like Simon and Leighton (1964). The difference in results may be due to various analysis methods and subjects, and further investigation is needed to resolve this.

Our identification results may contain possible misidentification due to similar scale structures, such as ephemeral regions. This misidentification contributes to the uncertain range of identified network sizes. It is possible that a weak cycle dependence of network sizes exists but is masked by the uncertain range. Therefore, our identification results may not show the possible weak cycle dependence.

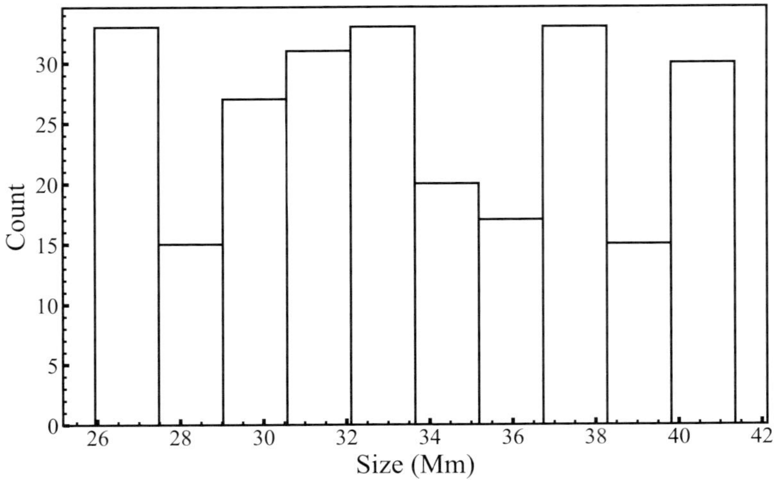

Figure 3. The histogram of identified network sizes. Image reproduced with permission from Luo et al. (2024).

4. Conclusions

In this study, we successfully identify magnetic network features in magnetic power spectra using morphology identification methods. The identified network sizes range from 26 Mm to 41 Mm. The statistical analysis of typical network sizes indicates that the size of the magnetic network is close to the uniform distribution and does not depend significantly on the solar cycle. Additionally, we can speculate that the supergranulation size may also have a weak cycle dependence according to the relationship between supergranulation and the magnetic network (Rincon and Rieutord 2018).

The magnetic network is an important source of the variation in the total solar irradiance (TSI). However, it is not well-studied in previous models (Solanki et al. 2013). Our results may provide a new constraint for the investigation of the TSI model.

Acknowledgements. The research is supported by the National Natural Science Foundation of China No. 12350004, No. 12173005 and National Key R&D Program of China No. 2022YFF0503800. The SDO/HMI data are courtesy of NASA and the SDO/HMI team. SOHO is a project of international cooperation between ESA and NASA. We would like to express our gratitude to the teams responsible for the development of Python toolkits such as 'pyshtools' and 'scipy'.

References

Bellot Rubio, L. & Orozco Suárez, D. 2019, Quiet Sun magnetic fields: an observational view. *Living Reviews in Solar Physics*, 16(1), 1.

Berrilli, F., Ermolli, I., Florio, A., & Pietropaolo, E. 1999, Average properties and temporal variations of the geometry of solar network cells. *Astron. Astrophys.*, 344, 965–972.

Gosic, M. 2015,. *The solar internetwork*. PhD thesis, University of Granada, Spain.

Hagenaar, H. J., Schrijver, C. J., & Title, A. M. 1997, The Distribution of Cell Sizes of the Solar Chromospheric Network. *Astrophys. J.*, 481(2), 988–995.

Hagenaar, H. J., Schrijver, C. J., & Title, A. M. 2003, The Properties of Small Magnetic Regions on the Solar Surface and the Implications for the Solar Dynamo(s). *Astrophys. J.*, 584(2), 1107–1119.

Luo, Y., Jiang, J., & Wang, R. 2023, The Sun's Magnetic Power Spectra over Two Solar Cycles. I. Calibration between SDO/HMI and SOHO/MDI Magnetograms. *Astrophys. J.*, 954(2), 199.

Luo, Y., Jiang, J., & Wang, R. 2024, The Sun's Magnetic Power Spectra over Two Solar Cycles. II. Cycle Dependence of Active Region and Magnetic Network. *in Prep.*,.

Muenzer, H., Schroeter, E. H., Woehl, H., & Hanslmeier, A. 1989, Pole-equator-difference of the size of the chromospheric CA II-K-network in quiet and active solar regions. *Astron. Astrophys.*, 213(1-2), 431–435.

Rincon, F. & Rieutord, M. 2018, The Sun's supergranulation. *Living Reviews in Solar Physics*, 15(1), 6.

Simon, G. W. & Leighton, R. B. 1964, Velocity Fields in the Solar Atmosphere. III. Large-Scale Motions, the Chromospheric Network, and Magnetic Fields. *Astrophys. J.*, 140, 1120.

Singh, J. & Bappu, M. K. V. 1981, A dependence on solar cycle of the size of the CA^+ network. *Solar Phys.*, 71(1), 161–168.

Solanki, S. K., Krivova, N. A., & Haigh, J. D. 2013, Solar Irradiance Variability and Climate. *ARA&A*, 51(1), 311–351.

Dynamics of Solar and Stellar Convection Zones and Atmospheres
Proceedings IAU Symposium No. 365, 2024
A. V. Getling & L. L. Kitchatinov, eds.
doi:10.1017/S174392132300529X

Imaging stellar surface features with optical interferometry

X. Haubois

European Organisation for Astronomical Research in the Southern Hemisphere (ESO), Casilla 19001, Santiago 19, Chile
email: xhaubois@eso.org

Abstract. In this contribution, I present a selected overview of optical interferometry imaging results that brought insights on stellar activity and mass loss in evolved stars. I briefly introduce the STELLIM project that aims to characterize stellar surfaces and circumstellar environments by producing fast and reliable interferometric images.

Keywords. Convection, techniques: interferometric, stars: fundamental parameters

1. Introduction on optical interferometers

Optical interferometers like VLTI (Cerro Paranal, Chile) or CHARA (Mount Wilson, US), measure the coherence of the light collected with different telescopes. As such, they do not directly produce images like a standard single telescope would do. However, they deliver information at the very high spatial resolution that is given by the on-sky projected distance between telescopes. For a 200m distance in the near-infrared domain, the achieved spatial resolution is about 1 milli arcsecond (1 mas), which represents a $\sim 1/20$ fraction of the radius of the angularly largest stellar surfaces like Betelgeuse's.

2. Signatures of surface features with optical interferometry

Interferometric observables can be of great help to detect and characterize surface features on a variety of objects. An early example of this type of works can be found in Wilson et al. (1997) with aperture masking observations of Betelgeuse performed on the Herschel Telescope. A modelling with less than 15 parameters allowed to fit a stellar disk and spots. Spots account for 20% of the total flux of the stellar surface and up to a third of the stellar surface. This is in line with theoretical prediction from Schwarzschild (1975) that "convective elements are so large that only a modest number of them exists at any one time on the entire surface of the star ". They were also able to detect variability in the spots flux and positions between the 2 epochs separated by 8 weeks which is compatible with convective activity.

Optical interferometers are sensitive to asymmetric brightness distribution via an observable called the closure phase that is not biased by atmospheric errors. Various estimators can be based on this quantity to probe for asymmetries. Cruzalèbes et al. (2015) studied the degree of asymmetries in 16 evolved stars as a function of spectral types using the VLTI instrument AMBER. Asymmetries were found to be located in the upper right corner of the Hertzsprung–Russell Diagram, and associated with an increase of atmospheric-pressure scale-height, which is compatible with 3D hydro-radiative simulations.

Combined with a relatively high (~10,000) resolution in spectral dispersion, optical interferometry is then a powerful tool to probe dynamics at high spatial resolution. On the AGB star R Dor, Ohnaka et al. (2019) produced images in spectral lines, exhibiting a bright central region. In the CO first overtone lines: outward motion of $7 - 15 km/s$ were detected at ~1.8 Rs The origin of this outflow is open: radiation pressure on dust grains, motion due to convection and/or pulsation?

3. Regularized imaging

Parametric imaging is limited when modelling complex stellar surfaces and environments. It is then possible to produce images that fit interferometric observables by using regularization algorithms This technique has the big advantage of rendering structures that are hard to intuit and model with a parametric approach. In general, the spatial frequencies that are not constrained by observations are filled with *a priori* information on the object. In the reconstruction process, a criteria is minimized to both fit the data and get close to the regularization. The image can be reconstructed with different types of norms, to favor sharp edges for example.

These images can be used as a sort of a guide to orient the parametric modelling. But they must be taken with caution as the reconstruction process can lead to artefacts. Fitting parameters directly on a reconstructed image can lead to biases. A common practice is to make images with various regularization algorithms and parameters and analyse the common features.

This technique has been frequently used to image stellar surfaces on evolved stars (e.g. Haubois et al. 2009; Climent et al. 2020; Montargès et al. 2018; Paladini et al. 2018) but also on solar-type stars (e.g. Roettenbacher et al. 2016). They can be repeated overtime to follow the morphology evolution and determine spots evolution timescales (Norris et al. 2021). Roettenbacher et al. (2017) presents an interesting comparison between photometric imaging, Doppler imaging and interferometric imaging in the detection of spots. They differ in particular in the estimation of the spot latitudes.

Blind imaging contests based on optical interferometry data allow to compare the images reconstructed with various regularized approaches. In Monnier et al. (2014), one can appreciate the variety of reconstructed morphologies that are based on the same dataset obtained from the AGB star R Car observed with the VLTI/PIONIER instrument. They show significant differences up to the highest levels of relative intensity, which points out the necessity to improve the reliability of this technique.

Operational constraints and low number of telescopes imposes to use super-synthesis (uv plane filling with earth rotation) and to move telescopes as quickly as possible. It is currently very difficult to reach a low-artefact image in a short amount of time to allow less than a ~2 week temporal resolution, which is probably key to determine the nature of surface features on many objects. Trying to go beyond the current status of interferometric imaging, we introduce the STELLIM project (Haubois et al. 2022) that aims to characterize stellar surfaces and circumstellar environments by producing fast and reliable interferometric images. Using a new array of 10+ small apertures optimized for imaging, we want to achieve low-artefact, model-independent, milli-arcsecond imaging at the timescale of the hour.

4. Summary

Optical interferometry is a powerful technique to observe convection and stellar surface dynamics at high spatial resolution. Combined with asteroseismology, it can be used to derive fundamental parameters like mass and effective temperature. Optical interferometric Imaging has boomed in the last decade. However some limits remain due to the small number of telescopes that are currently recombined. In particular, it stops us from reliably imaging phenomena with variability shorter than a few weeks. The STELLIM concept aims at delivering fast and reliable interferometric imaging in the visible, using VLTI as a testbed. It also represents a technological and operational precursor of a kilometric array with more and larger apertures. A 2-telescope demonstrator is being assembled and should soon demonstrate the critical functions of this new type of interferometers. This should push the limits of our current understanding of stellar surfaces and atmospheres, which is at the heart of many central topics in stellar physics.

References

Aurière, M., Donati, J.-F., Konstantinova–Antova, R., et al. 2010, *A&A*, 516, L2

Chiavassa, A., Haubois, X., Young, J. S., et al. 2010, *A&A*, 515, A12

Climent, J. B., Wittkowski, M., Chiavassa, A., et al. 2020, *A&A*, 635, A160

Cruzalèbes, P., Jorissen, A., Chiavassa, A., et al. 2015, *MNRAS*, 446, 3277.

Haubois, X., Perrin, G., Lacour, S., et al. 2009, *A&A*, 508, 923

Haubois, X., Schuhler, N., Bourget, P., et al. 2022, *Proceedings of the SPIE conference*, 12183, 121831T.

Höfner, S. 2008, *A&A*, 491, L1.

Josselin, E. & Plez, B. 2007, *A&A*, 469, 671.

Monnier, J. D., Berger, J.-P., Le Bouquin, J.-B., et al. 2014, *Proceedings of the SPIE*, 9146, 91461Q

Montargès, M., Norris, R., Chiavassa, A., et al. 2018, *A&A*, 614, A12

Norris, R. P., Baron, F. R., Monnier, J. D., et al. 2021,*ApJ*, 919, 124.

Ohnaka, K., Weigelt, G., & Hofmann, K.-H. 2019, *ApJ*, 883, 89

Paladini, C., Baron, F., Jorissen, A., et al. 2018, *Nature*, 553, 310

Roettenbacher, R. M., Monnier, J. D., Korhonen, H., et al. 2016, *Nature*, 533, 217

Roettenbacher, R. M., Monnier, J. D., Korhonen, H., et al. 2017, *ApJ*, 849, 120

Schwarzschild, M. 1975, *ApJ*, 195, 137

Wilson, R. W., Dhillon, V. S., & Haniff, C. A. 1997, *MNRAS*, 291, 819.

Woitke, P. 2006, *A&A*, 460, L9.

Section 2. Differential Rotation and Meridional Circulation

Contributed papers

Dynamics of Solar and Stellar Convection Zones and Atmospheres
Proceedings IAU Symposium No. 365, 2024
A. V. Getling & L. L. Kitchatinov, eds.
doi:10.1017/S1743921323005197

Differential Rotation of the Solar Chromosphere using multidecadal Ca II K Spectroheliograms

Dibya Kirti Mishra[1,2], Srinjana Routh[1,2], Bibhuti Kumar Jha[3], Subhamoy Chatterjee[3] and Dipankar Banerjee[1,4,5]

[1]Aryabhatta Research Institute of Observational Sciences, Nainital-263002, Uttarakhand, India
email: dibyakirtimishra@aries.res.in

[2]Mahatma Jyotiba Phule Rohilkhand University, Bareilly-243006, Uttar Pradesh, India

[3]Southwest Research Institute, Boulder, CO 80302, USA

[4]Indian Institute of Astrophysics, Koramangala, Bangalore 560034, India

[5]Center of Excellence in Space Sciences India, IISER Kolkata, Mohanpur 741246, West Bengal, India
email: dipu@aries.res.in

Abstract. The study of the differential rotation in the chromosphere of the Sun is of significant importance as it provides valuable insights into the rotational behaviour of the solar atmosphere at higher altitudes and the coupling mechanism between the various layers of the solar atmosphere. In this work, we employed the image correlation technique, explicitly focusing on plages, intending to estimate the chromospheric differential rotation. For this purpose, we have utilized Ca II K spectroheliograms (1907–2007) from the Kodaikanal Solar Observatory (KoSO), recently calibrated with a better technique to ensure accuracy. Our analysis indicates that plages in the chromosphere exhibit faster rotation and a smaller latitudinal gradient when compared to the rotation rate obtained through sunspot tracking. Furthermore, we investigate the temporal analysis of the chromospheric differential rotation parameters across various solar cycles.

Keywords. Sun: chromosphere, Sun: plages, Sun: solar cycle, Sun: differential rotation

1. Introduction

After the discovery of sunspots by Galileo and other astronomers, the concept of solar differential rotation first came into existence in 1630 by Christoph Scheiner. Since then, it has been the subject of interest in the scientific community and the extensive study to describe this with the help of sunspot observations led to the development of the well-established empirical relation

$$\Omega = A + B\sin^2\theta + C\sin^4\theta, \tag{1}$$

where Ω represents the angular velocity, θ denotes the heliographic latitude, while A represents the equatorial rotation rate, and B and C correspond to latitudinal gradients. Gradually, advancements in technology have enabled extensive investigations into the solar differential rotation, utilizing various methods such as sunspot tracking (Jha et al. 2021; Jha 2022), spectroscopy technique(Howard and Harvey 1970), and helioseismology (Komm et al. 2008). Despite the numerous studies conducted in both the photosphere and the interior of the Sun, our understanding of differential rotation behaviour in the higher atmospheric layers, including the chromosphere and corona, remains limited.

As previously mentioned, sunspots are employed as tracers to explore photospheric differential rotation owing to their prolonged persistence. Similarly, plages are also considered promising candidates for investigating differential rotation phenomena in the chromosphere due to their prolonged existence (Bertello et al. 2020). These extended chromospheric features are predominantly observed in Ca II K spectral line centered at 393.367 nm. However, it is important to note that plages differ from sunspots in that they undergo relatively rapid morphological changes, making it challenging to employ traditional tracking methods on them. Despite this limitation, there have been various attempts to analyze the rotation profile in the chromosphere by tracking plages over relatively short time spans (Antonucci et al. 1979). In addition to these, more recent efforts have employed advanced techniques such as Fast Fourier Transform (Singh and Prabhu 1985), image cross-correlation (Bertello et al. 2020), and autocorrelation method (Li et al. 2023) to gain insights into this phenomenon.

Despite the previous works, our comprehension of chromospheric differential rotation and its connection to the photosphere remains incomplete. The Kodaikanal Solar Observatory (KoSO), which possesses an extensive archive of Ca II K spectroheliograms (1904 – 2007), presents a distinctive opportunity for investigating chromospheric differential rotation at various phases of the solar cycle. In this article, we will discuss the image correlation method employed for finding the chromospheric rotation profile by utilizing newly calibrated Ca II K data of KoSO.

2. Data and Methodology

The archival Ca II K dataset over a century (1904–2007) collected by KoSO† provides extensive information about the chromospheric activity. These data were acquired on photographic plates/films using the diffraction grating system of the bandpass of 0.05 nm centered at 393.367 nm (Chatterjee et al. 2016; Jha 2022). To enhance their quality, the dataset has undergone digitization and advanced calibration procedures were applied to ensure accuracy, as described in Chatzistergos et al. (2020) with details. Furthermore, Jha (2022) devised a reliable approach to enhance the precise alignment of Ca II K images spanning from 1907 to 2007, a critical factor for the accurate determination of differential rotation. For our purpose of estimating differential rotation in the chromosphere, we employed the recently calibrated and rotation-corrected Ca II K data (1907 – 2007) of KoSO.

To account for the dynamic nature of plages, as elaborated in Section 1, we have implemented the image correlation technique to determine chromospheric differential rotation. In this process, we select a pair of observations while imposing a constraint on the time interval Δt having condition: $0.5\,\text{days} < \Delta t < 1.5\,\text{days}$. Subsequently, these images are transformed into the heliographic grid (Thompson 2006) with a resolution of 0.1° per pixel. Next, we partition the data into 5° latitudinal bands, as exemplified by A1 and A2 in Fig. 1(a), within the range of $\pm 55^\circ$ in both latitude and longitude. This approach mitigates the impact of projection effects near the solar limb, ensuring more accurate and robust measurements.

Following this, we employ a 2D image correlation technique on selected latitude bands, where we specify a range for the shift in both latitude $[\Delta\theta_0 \pm 1^\circ]$ and longitude $[\Delta\phi_0 \pm 2^\circ]$. In this context, $\Delta\theta_0$ is set to 0, and $\Delta\phi_0$ is determined as $\Omega(\theta)\Delta t$ based on the work by Jha et al. (2021). This process yields optimal values for the longitudinal shift, denoted as $\Delta\phi$, which corresponds to the highest Correlation Coefficient (CC_{max}) and is depicted

† The digitized data are available at `https://kso.iiap.res.in/data`.

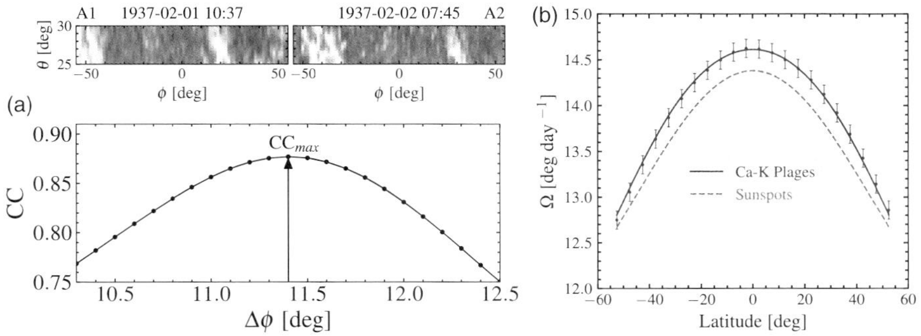

Figure 1. a) The latitude bands for two consecutive days observed on 1937-02-01 10:37 IST (A1) and 1937-02-02 07:45 IST (A2), within the latitude range of +25° to +30°. The brown curve in (a) illustrates the variation in CC with $\Delta\phi$ for a constant $\Delta\theta$, a product of a 2D cross-correlation technique applied on A1 and A2, showing the location of the maximum Cross-Correlation value (CC_{max}). b) The comparative representation of chromospheric rotation using plages (purple continuous curve) and sunspot rotation profile (olive dashed line) obtained from Jha et al. (2021).

in the brown curve of Fig. 1(a). These values are instrumental in computing $\Omega_{synodic}$, defined by the equation

$$\Omega_{synodic} = \frac{\Delta\phi}{\Delta t}. \tag{2}$$

In the process of computing $\Omega_{synodic}$, we apply a conditional criterion for which we consider values of $\Delta\phi$ where CC exceeds 0.2. In cases where CC falls below this threshold and there is no discernible variation in CC, such values are excluded from the analysis. After the computation of $\Omega_{synodic}$ as per Eq. 2, we perform a sidereal correction to account for the Earth's motion around the Sun. This correction is accomplished using the equation followed in Jha et al. (2021) for synodic to sidereal conversion. This procedure enables us to calculate Ω (Ωsidereal) successfully for each latitudinal band, encompassing the entire data period spanning from 1907 to 2007.

3. Result

To determine the chromospheric differential rotation, we calculate the angular velocity Ω for 22 latitude bands ranging from $-55°$ to $+55°$ by taking the mean value over the entire data period (purple dots; Fig. 1(b)). The associated errors in these Ω values encompass both statistical errors (σ_{SSE}) and least count errors (σ_{LCE}). Notably, σ_{LCE} due to the resolution of the heliographic grid ($0.1/\Delta t$) predominantly contributes to the total error. To quantify the differential rotation within the chromosphere, we employed a fitting procedure involving Eq. 1 applied to the average angular velocity (Ω). This analysis yielded the following parameter values: A = 14.61 deg day^{-1}, B = −2.18 deg day^{-1}, C = −1.10 deg day^{-1} with corresponding uncertainty of ±0.04, ±0.37, ±0.61 respectively (Mishra et al. 2023). We then compare the rotation rate of plages (purple continuous curve) with the sunspot rotation rate (olive dashed line; Jha et al. 2021) obtained using KoSO White-Light data spanning from 1923 to 2011, as can be seen in Fig. 1(b). The comparison reveals that the rotation rate of plages is greater than that of the sunspots, thus indicating faster rotation of the chromosphere using plage as a tracer compared to the photosphere based on sunspot tracking.

We also explore the temporal evolution of these differential rotation parameters over the course of solar cycles, as shown in Fig. 2. The analysis reveals that the chromospheric differential rotation remains insignificantly variant across solar cycles. Notably, when

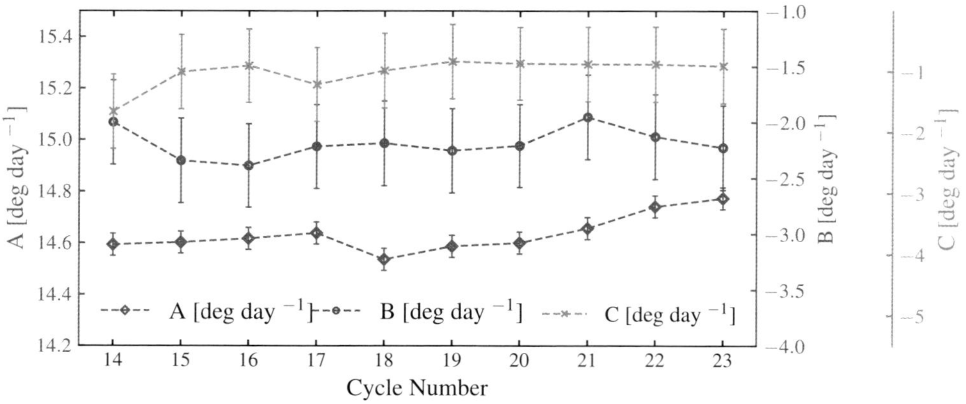

Figure 2. The variation in differential rotation parameters (A, B, and C) throughout solar cycles.

examining the variation of parameter A (red) concerning solar cycle number (Fig. 2), we observe an increasing trend in A values following solar cycle number 20. This could be attributed to a degradation in data quality during the latter half of the dataset, as discussed in previous studies (Mishra et al. 2023).

4. Conclusion

In summary, our analysis indicates that plages exhibit a higher rotation rate compared to sunspots, characterized by a less latitudinal gradient. Furthermore, it suggests that this differential rotation of plages does not change significantly throughout solar cycles. These findings align with recent studies by Li et al. (2023), which also support the observation of a faster rotation in the chromosphere than the photosphere. In addition to these, there are also a few studies done recently in the higher solar atmosphere (e.g. Sharma et al. 2020), which also indicate faster rotation in the higher solar atmosphere. However, the underlying reasons for this higher rotation rate in the chromosphere remain not fully understood. In the future, we intend to undertake a more comprehensive investigation into solar rotation across various heights within the solar atmosphere to deepen our understanding of this phenomenon.

5. Acknowledgment

We thank IAU for the travel grant awarded to us, making it possible to attend the symposium in person in Armenia.

References

Antonucci, E., Azzarelli, L., Casalini, P., Cerri, S., & Denoth, F. 1979, Chromospheric rotation. II. Dependence on the size of chromospheric features. *Solar Phys.*, 63(1), 17–30.

Bertello, L., Pevtsov, A. A., & Ulrich, R. K. 2020, 70 Years of Chromospheric Solar Activity and Dynamics. *Astrophys. J.*, 897(2), 181.

Chatterjee, S., Banerjee, D., & Ravindra, B. 2016, A Butterfly Diagram and Carrington Maps for Century-long CA II K Spectroheliograms from The Kodaikanal Observatory. *Astrophys. J.*, 827(1), 87.

Chatzistergos, T., Ermolli, I., Krivova, N. A., Solanki, S. K., Banerjee, D., Barata, T., Belik, M., Gafeira, R., Garcia, A., Hanaoka, Y., Hegde, M., Klimeš, J., Korokhin, V. V., Lourenço, A., Malherbe, J.-M., Marchenko, G. P., Peixinho, N., Sakurai, T., & Tlatov, A. G. 2020, Analysis of full-disc Ca II K spectroheliograms - III. Plage area composite series covering 1892–2019. *Astronomy & Astrophysics*, 639, A88.

Howard, R. & Harvey, J. 1970, Spectroscopic determinations of solar rotation. *Solar Physics 1970 12:1*, 12, 23–51.

Jha, B. K. 2022,. *Long-term Study of the Sun and Its Implications to Solar Dynamo Models*. PhD thesis, Department of Physics, Pondicherry University.

Jha, B. K., Priyadarshi, A., Mandal, S., Chatterjee, S., & Banerjee, D. 2021, Measurements of Solar Differential Rotation Using the Century Long Kodaikanal Sunspot Data. *Solar Phys.*, 296(1), 25.

Komm, R., Howe, R., Hill, F., & Hernández, I. G. 2008, Subsurface zonal flows. *Solar Physics 2008 254:1*, 254, 1–15.

Li, K. J., Wan, M., & Feng, W. 2023, The quiet chromosphere: differential rotation. *Mon. Not. Roy. Astron. Soc.*, 520(4), 5928–5937.

Mishra, D. K., Routh, S., Jha, B. K., Chatzistergos, T., Judhajeet, B., Chatterjee Subhamoy Banerjee, D., & Ermolli, I. 2024, Differential Rotation of the Solar Chromosphere: A Century-long Perspective from Kodaikanal Solar Observatory Ca II K Data. *Astrophys. J.*, 961, 40.

Sharma, J., Kumar, B., Malik, A. K., & Vats, H. O. 2020, On the variation of solar coronal rotation using SDO/AIA observations. *Monthly Notices of the Royal Astronomical Society*, 492(4), 5391–5398.

Singh, J. & Prabhu, T. P. 1985, Variations in the solar rotation rate derived from Ca^+ K plage areas. *Solar Phys.*, 97(1), 203–212.

Thompson, W. T. 2006, Coordinate systems for solar image data. *Astron. Astrophys.*, 449(2), 791–803.

Dynamics of Solar and Stellar Convection Zones and Atmospheres
Proceedings IAU Symposium No. 365, 2024
A. V. Getling & L. L. Kitchatinov, eds.
doi:10.1017/S1743921324000164

A Frequency-Dependent Method for Measuring the Sun's Interior Meridional Circulation

Ruizhu Chen and Junwei Zhao

W. W. Hansen Experimental Physics Laboratory, Stanford University, Stanford, CA, USA
email: rzchen@stanford.edu

Abstract. The Sun's meridional circulation is a crucial component for understanding the Sun's dynamo and its interior dynamics. However, the determination of meridional circulation is affected by a systematic center-to-limb (CtoL) effect, which introduces systematic errors 5–10 times stronger than the meridional-flow-induced travel-time shifts in deep-flow measurements. Recently, it was found that the CtoL effect has a significant acoustic-frequency dependence, while flow-induced travel-time shifts show little frequency dependence (Chen & Zhao 2018). This discovery forms the basis for designing a new method to remove the CtoL effect. We therefore propose a frequency-dependent approach to measure the CtoL effect and the flow-induced signals in the Fourier domain. In this work, we present this new method and compare time–distance measurements in different frequency bands with those obtained by previous time-domain methods. The results demonstrate consistency with conventional time-domain fitting methods in the dominant frequency range, promising the potential for conducting meridional flow inversion across a broader frequency spectrum.

Keywords. Sun: helioseismology, Sun: interior, Sun: oscillations

1. Introduction

The Sun's meridional circulation is a critical element in understanding the Sun's dynamo and internal dynamics. While direct observation of solar deep interior flows is not possible, time-distance helioseismology offers a way to infer these flows by measuring the acoustic travel times of p-mode waves traveling inside the Sun (Duvall *et al.* 1993). However, it was found that these time–distance measurements are complicated by a systematic center-to-limb variation (Zhao *et al.* 2012). This systematic effect, known as the Center-to-Limb (CtoL) effect, has been consistently observed across multiple instruments and by different authors (Zhao *et al.* 2013; Kholikov *et al.* 2014; Jackiewicz *et al.* 2015; Liang & Chou 2015b; Rajaguru & Antia 2015; Liang *et al.* 2017; Chen & Zhao 2017; Gizon *et al.* 2020), yet its underlying cause remains elusive. Of particular concern is the fact that the CtoL effect can be an order of magnitude larger than the travel-time shifts induced by flows for deep regions, significantly impacting the accurate inference of the Sun's meridional circulation. Addressing and removing this effect remains as a major challenge in the field of solar meridional flow measurements.

Recent findings by Chen & Zhao (2018) have shed light on the CtoL effect measured in time-distance helioseismology, revealing a strong acoustic-frequency dependence that

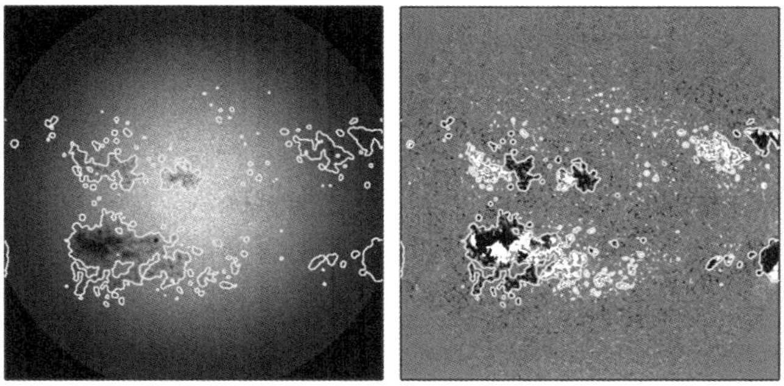

Figure 1. (Left) Daily acoustic power map of 2012 July 8, with green contours of 10 Gauss threshold on the daily-averaged magnetograms, inside which Doppler data are not used in our time–distance measurements. (Right) Line-of-sight magnetic field image captured on the same date, with identical contours overlaid.

reverses sign at a specific frequency. This unique frequency dependence, in contrast to the relatively frequency-independent nature of flow-caused travel-time shifts, not only provides potential insights into the origin of the systematic effect but also suggests a novel approach to disentangling the CtoL effect from meridional flow signals in the frequency domain. Traditional time-domain fitting of travel-time shifts primarily captures signals within dominant frequency ranges, overlooking the frequency-dependent differences between the CtoL effect and flow-induced travel-time shifts. Exploiting information across a wider frequency range holds the promise of extracting richer details from the observed signals.

In this study, we extend the investigation initiated by Chen & Zhao (2018) by conducting frequency-dependent measurements of both the CtoL effect and meridional-flow-induced travel-time shifts. We compare the meridional-flow travel-time shift signals measured across multiple frequency bands and in the time domain. The paper is organized as follows: Section 2 introduces the data preparation, Section 3 outlines the methodology and presents measurements, and Section 4 discusses our findings.

2. Data Preparation

We utilize HMI full-disk Doppler observations, covering the period from May 1, 2010, to April 30, 2021, a total of 11 years' worth of data. We first downsample full-disk images to 1024×1024 pixels, track them using the Snodgrass rotation rate at the disk center, and project them onto Postel's coordinates, with the disk center as the image center. Subsequently, we further downsample the tracked images to 425×425 pixels, resulting in a spatial resolution of $0.^{\circ}36$ pixel^{-1} (where 1° represents 1 heliographic degree) at the disk center. For the analysis, we employ running-difference data to eliminate solar convection, and no additional filtering is applied.

The influence of surface magnetic effects on travel time anomalies becomes apparent when measurements are taken in magtnetic regions (Gizon et al. 2009; Liang & Chou 2015a). To address this, we apply a threshold of 10 Gauss to mask magnetic regions on the daily-averaged and Gaussian-smoothed line-of-sight magnetic field maps. It's noteworthy that despite the seemingly small value of 10 Gauss, this threshold is considered appropriate as magnetic fields are substantially averaged. Figure 1 provides an illustrative example, depicting a daily acoustic oscillation power map along a magnetogram map, with the applied mask overlaid for reference.

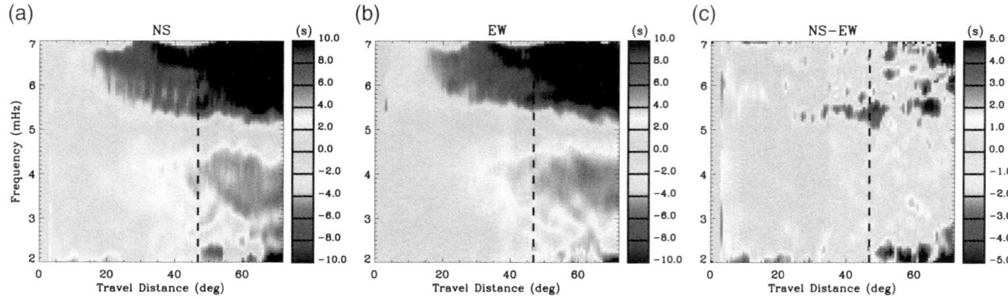

Figure 2. (a) Travel-time shifts $\delta\tau_{NS}$ measured along the NS direction at 32°.4 latitude on the central meridian, displayed as functions of travel distance and frequency. Black-dashed line indicates the approximate travel distance for waves passing the BCZ, beyond which the data are typically not robust enough for inversion and are generally not used in inversions. (b) Same but for $\delta\tau_{EW}$ measured along the EW direction at 32°.4 longitude on the Equator. (c) $\delta\tau_{NS} - \delta\tau_{EW}$, illustrating the meridional-flow-caused travel-time shifts.

3. Frequency-Dependent Measurement Approach

In previous time-distance helioseismic measurements, travel times were conventionally fitted from the cross-covariance functions in the time domain, often using techniques such as Gabor-wavelet fitting. The travel-time shifts used for flow inversion were then determined by taking the differences between the travel times of waves traveling in opposite directions. In our current study, we adopt the methodology developed by Chen & Zhao (2018) to derive frequency-dependent travel-time shifts in the Fourier domain. This involves calculating the relative phase differences between the positive- and negative-time-lag segments of the cross-covariance function, and then convert the phase differences to travel-time shifts. We have enhanced the method by imcoporating a refined whitening profile and windowing functions utilized in the calculation (refer to Section 2.3 of Chen & Zhao (2018)). These empirical profiles used are now tailored as functions of both disk location and travel distance, determined from 11 years of data. We have also modified the phase averaging strategy. Consequently, the current measurements are less noisy with little aliasing effect..

3.1. *Measurements of the CtoL Effect and Meridional Flow-Induced Travel-Time Shifts*

We proceed to conduct time-distance measurements in the equatorial region along the East-West (EW) direction and in the central meridian regions along the North-South (NS) direction. All measurements are performed between pairs of arcs positioned along the disk-radial direction, and the focus points of the arc pairs are situated within a narrow sector spanning $\pm 10°$ around the disk center (refer to the illustration of the EW scheme in Figure 1 of Chen & Zhao (2018)). The resulting series of cross-covariance functions, measured as a function of travel distance and disk location, are utilized to calculate the frequency-dependent travel-time shifts. The outcomes for NS and EW measurements, along with their difference, are illustrated in Figure 2.

The frequency-dependent travel-time shifts measured along both the NS and EW directions exhibit significant frequency dependence. The EW results presented in Figure 2(b) have already removed its zonal flow components by removing its symmetric component around disk center. Following this adjustment, the remaining EW signals are often considered as a proxy for the CtoL effect. By subtracting this CtoL proxy from the NS results, the residual signals in Figure 2(c) are considered as the meridional-flow-induced travel-time shifts.

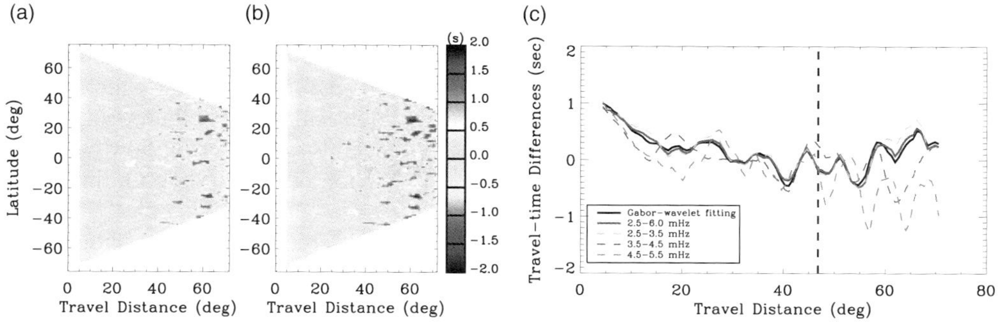

Figure 3. (a) Time-domain measurement of $\delta\tau_{NS} - \delta\tau_{EW}$ by Gabor-wavelet fitting. (b) Fourier-domain measurement of $\delta\tau_{NS} - \delta\tau_{EW}$ in 2.5-6mHz (power-weighted average). (c) Comparison of the time-domain and Fourier-domain measurements of $\delta\tau_{NS} - \delta\tau_{EW}$ for 10° - 30° latitudes. The black-dashed line indicates the approximate travel distance for waves passing the BCZ.

The flow-induced signal exhibits little frequency dependence for frequencies below 5 mHz and for travel distances smaller than the black-dashed line, representing the approximate travel distance for waves passing the base of the convection zone (BCZ). Typically, only measurements shallower than this are considered robust and will be utilized in inversions. Within this range, the results at different frequencies appear consistent, promising the potential of getting consistent results through inverting measurements of various frequency bands.

However, the measurements become considerably noisier beyond approximately 5 mHz or beyond about 40°travel distance. Notably, it still displays aliasing artifacts, appearing as alternating bands of positive and negative phases in the higher-frequency bands, despite our efforts to mitigate these artifacts through improved methods. Completely eliminating these artifacts proves challenging due to the weak oscillation signals in the high-frequency range. Nevertheless, it is essential to acknowledge that these artifacts are most severe for high frequencies and large distances, and the low-frequency and small-travel-distance results, on the other hand, remain useful despite these challenges.

3.2. *Comparative Analysis with Time-Domain Method*

In the range below about 5 mHz and for small travel distances, where the results exhibit robustness, we proceed to conduct a comparison of the travel-time shift curves in multiple frequency bands with those obtained through conventional time-domain fitting methods.

Figure 3 presents the time-domain result of $\delta\tau_{NS} - \delta\tau_{EW}$, along with the wide-band (2.5-6 mHz) power-weighted-average result, indicating a noticeable similarity between the two. Their values in the 10° - 30° range are further compared in Figure 3(c), along with those from 1-mHz narrow-band results around 3, 4, and 5 mHz. The curves for the wide-band and dominant frequency band (2.5-3.5 mHz) are almost identical due to the prominent dominance of oscillation power around the 3.3 mHz peak. Both of them are also nearly identical to the time-domain results, indicating consistency with the time-domain fitting method. The curve for 3.5-4.5 mHz remains consistent but exhibits some noise. Results for 4.5-5.5 mHz become noisier and show slight deviations, yet these deviations are not significant compared to their error level (~ 0.5 seconds).

4. Discussion

We have devised a new procedure for conducting frequency-dependent measurements of the meridional-flow-induced travel-time shifts and the associated CtoL effects. In the dominant frequency range, the method yields nearly identical results when compared to time-domain fitting. Yet its advantage lies in the potential for conducting flow inversions across a broader frequency spectrum, expanding the scope of our understanding across multiple frequency bands.

Ideally, the inversion of meridional flow using different frequency bands should yield consistent results since the interior flow remains the same regardless of acoustic wave frequencies. This consistency serves as a criterion to assess whether systematic effects are appropriately removed, whether inversions are conducted effectively, and whether the results are reliable. A recent study by Rajaguru & Antia (2020) also conducts frequency-dependent meridional flow measurements by applying frequency filters before fitting travel times. Their findings reveal a noticeable difference, showing a single-cell circulation profile in 3 mHz and a double-cell profile in 4 mHz, suggesting unresolved challenges in the measurement process.

In our results, we have also noted variations in travel-time shifts at higher frequencies. However, understanding how these differences translate into inverted flow profiles requires further investigation, as the sensitivity of travel-time shifts could also vary with frequency. In future studies, we plan to perform meridional flow inversions using frequency-dependent sensitivity kernels (Hartlep & Zhao 2021). Additionally, we acknowledge the presence of remaining artifacts in the high-frequency range above 5.0 mHz, which could potentially be further addressed in future works.

Acknowledgements. *SDO* is a NASA mission, and HMI project is supported by NASA contract NAS5-02139 to Stanford University. R.C. and J.Z. are partly supported by NASA DRIVE Center COFFIES grant 80NSSC22M0162.

References

Chen, R., & Zhao, J. 2017, *Astrophys. J.*, 849, 144
Chen, R., & Zhao, J. 2018, *Astrophys. J.*, 853, 161.
Duvall, T. L., Jr., Jefferies, S. M., Harvey, J. W., & Pomerantz, M. A. 1993, *Nature*, 362, 430
Gizon, L., Cameron, R. H., Pourabdian, M., *et al.* 2020, *Science*, 368, 1469.
Gizon, L., Schunker, H., Baldner, C. S., et al. 2009, *Space Sci. Rev.*, 144, 249.
Hartlep, T., & Zhao, J. 2021, *Astrophys. J.*, 909, 66.
Jackiewicz, J., Serebryanskiy, A., & Kholikov, S. 2015, *Astrophys. J.*, 805, 133
Kholikov, S., Serebryanskiy, A., & Jackiewicz, J. 2014, *Astrophys. J.*, 784, 145
Liang, Z.-C., Birch, A. C., Duvall, T. L., Jr., Gizon, L., & Schou, J. 2017, *A&A*, 601, A46
Liang, Z.-C., & Chou, D.-Y. 2015a, *Astrophys. J.*, 805, 165
Liang, Z.-C., & Chou, D.-Y. 2015b, *Astrophys. J.*, 809, 150
Rajaguru, S. P., & Antia, H. M. 2015, *Astrophys. J.*, 813, 114
Rajaguru, S. P. & Antia, H. M. 2020, *Dynamics of the Sun and Stars; Honoring the Life and Work of Michael J. Thompson*, 57, 107.
Zhao, J., Bogart, R. S., Kosovichev, A. G., Duvall, T. L., Jr., Hartlep, T. 2013, *Astrophys. J. Letters*, 774, L29
Zhao, J., Nagashima, K., Bogart, R. S., Kosovichev, A. G., Duvall, T. L., Jr. 2012, *Astrophys. J. Letters*, 749, L5

Dynamics of Solar and Stellar Convection Zones and Atmospheres
Proceedings IAU Symposium No. 365, 2024
A. V. Getling & L. L. Kitchatinov, eds.
doi:10.1017/S1743921323004234

Probing the Sun's Near Surface Shear Layer using HMI Spherical Harmonic Coefficients

S. C. Tripathy[1], K. Jain[1], S. Kholikov[1,2] and R. Komm[1]

[1]National Solar Observatory, 3665 Discovery Dr., Boulder, CO 80303, USA
email: stripathy@nso.edu

[2]Institute of Fundamental and Applied Research, National Research University, TIIAME, Uzbekistan

Abstract. We have measured zonal and meridional components of subsurface flows up to a depth of 30 Mm below the solar surface by applying the technique of ring diagram on Dopplergrams which are constructed from the spherical harmonic (SH) coefficients. The SH coefficients are obtained from the Helioseismic and Magnetic Imager (HMI) full-disk Dopplergrams. We find a good agreement and some differences between the flows obtained in this study with those from the traditional methods using direct Dopplergrams.

Keywords. Solar physics, Helioseismology, Solar interior, Solar oscillations, Solar activity, Solar convection zone, Solar rotation, Solar meridional circulation

1. Introduction

Helioseismic studies have illustrated that the most significant changes with the solar cycle occur in the near-surface shear layer (NSSL) occupying outer 5% or about 35 Mm of the Sun by radius. Within this relatively thin layer, the density changes by several orders of magnitude and the rotation rate shows a local maxima. It is also believed that a nonlinear $\alpha\Omega$ dynamo could be operating in this layer where the velocity shear converts a part of the poloidal magnetic field into the toroidal field. The existence of this secondary dynamo is in addition to the global dynamo operating in the tachocline region and may be responsible for the small-scale field of the quiet Sun (Brandenburg 2013). However, the dynamics of NSSL is poorly understood mainly because of the complex interactions between rotation, magnetism and convection. The advent of high-resolution Doppler observations from both space-borne and ground-based instruments, in conjunction with local helioseismic techniques such as ring-diagram (Hill 1988) and time-distance (Duvall et al. 1993), have made it possible to infer the subsurface flows in NSSL and their temporal evolution (Zhao & Kosovichev 2004; Basu & Antia 2010; Komm et al. 2018; Lin & Chou 2018). But local helioseismology techniques are limited in the depth range that can be probed. In particular most of the studies using ring-diagram analysis is confined to a depth of 15 Mm. Only a handful of studies have used large-aperture tiles to analyse flows up to a depth of 30 Mm (González Hernández *et al.* 2006; Komm 2021).

In this context, we measure zonal and meridional components of the horizontal flows up to a depth of 30 Mm below the surface during the first six and half years of the Helioseismic and Magnetic Imager (HMI) observations on board Solar Dynamics Observatory (SDO). We use a non-traditional method comprising of spherical harmonic (SH) coefficients derived from the full-disk Dopplergrams and the ring-diagram (RD) technique. This method has the advantage to infer flows from low-resolution as well as high-resolution Dopplergrams.

2. Data and Analysis Technique

We use HMI SH coefficient time series† for $\ell = 0$ to 400 obtained from the Joint Science Operation Center (JSOC) covering the period 2010 May 1 to 2016 December 31. Each time series is organized as series of individual (n, ℓ, m) coefficients and are 72 days long. We also download the window function of each series to track the duty cycle. Use of this SH time series has several advantages as these have been generated from the velocity images which are corrected for known systematics, e.g., image scale, cubic distortion, CCD misalignment, inclination error and CCD tilt *etc.* The Dopplergrams were further detrended and gap filled. A description of all the corrections applied to the time series along with other details are provided in Larson & Schou (2018).

Reconstruction of Dopplergrams using the SH coefficient time series (inverse SH decomposition) is carried out following the procedure described in Kholikov et al. (2014)

$$V(\theta, \phi) = \sum_{\ell=0}^{\ell=400} \sum_{m=0}^{\ell} C_l^m P_l^m(\theta) e^{im\phi+\delta\phi} \tag{1}$$

where C_ℓ^m are SH coefficients, P_ℓ^m is the associated Legendre polynomial of spherical degree ℓ and order m, θ and ϕ are latitude and longitude, respectively. We also use the differential rotation profile $(\delta\phi)$ of Libbrecht and Morrow (1991) to remove the surface differential rotation where tracking is carried out by introducing the surface differential rotation profile into the argument of the complex exponential in the right hand side of Eq. (1). The reconstructed images were then remapped to 187 patches of the size $30° \times 30°$ from disk center to $\pm 60°$ in latitude and $\pm 37.5°$ in longitude spaced by $7.5°$. All the patches were tracked for 1664 min (the time series are divided in to 1664 min time interval to produce the data cubes). A three-dimensional fast Fourier transform (FFT) is applied to each cube in both spatial and temporal directions to obtain a three-dimensional power spectrum which was subsequently processed through the RD pipeline (Corbard et al. 2003). The fitting algorithm in the RD pipeline uses the maximum-likelihood procedure described in Anderson, Duvall, and Jefferies (1990) and fits about 270 modes for n-values between $0 - 9$ and ℓ values between $110 - 400$ in the frequency range of $1400 - 5200$ μHz. Figure 1 displays a typical ℓ-ν diagram comparing modes fitted at the disk center patch for three different data sets. It is evident that we recover most of the input modes in the ℓ range of approximately 100 to 400 (the RD technique does not fit modes below approximately ℓ of 100). We also note that we fit many more modes in the low ℓ and higher frequency ranges compared to methods using direct Dopplergrams.

The subsurface zonal and meridional flows are finally derived using the Regularized Least Square (RLS) inversion method (Haber et al. 2002). The depth dependence in the inverted flows is known to have significant influence from the mode kernels used in the inversion. The kernels peak sharply near the surface owing to the availability of significant number of modes but are less localized with increasing depth. Thus the depth resolution turns out to be poor resulting in relatively larger errors in deeper layers (Jain et al. 2017). However, in this study we have a limited set of high-degree modes and thus we find that the errors associated with the flows measured closer to the surface are larger compared to deeper depths. We also correct the flows for the annual variation with the solar inclination angle towards Earth (B_0 angle) and other systematic effects following the procedure described in Komm et al. (2015).

† series name: hmi.ap_V_sht_gf_72d

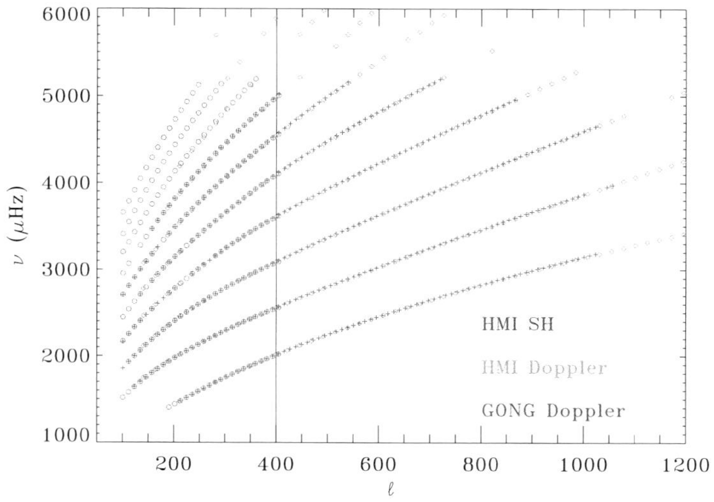

Figure 1. A typical ℓ-ν diagram showing the distribution of the fitted modes at the disk center using HMI spherical harmonic coefficients (red circles), HMI Dopplergrams (green diamonds) and the GONG Dopplergrams (blue plus). Vertical line at $\ell = 400$ defines the boundary of the fitted modes using spherical harmonic coefficients.

3. Results

Figure 2 shows the zonal flow at three depths (5.2 ± 2.2 Mm, 15 ± 7.2 Mm, and 25.6 ± 7.2 Mm) as a function of time and latitude where the mean is subtracted at each latitude and depth. The contours on top of the flows are plotted to depict the evolution of magnetic activity. The figure illustrates the bands of slower and faster rotation relative to the mean flow, known as torsional oscillations, of few m/s at all depths. These are comparable with those obtained in the traditional methods (e.g., Komm et al. 2018; Zhao et al. 2014, for ring-diagram and time-distance analyses, respectively). As is well known, we find that the bands of faster rotation at mid latitudes are well aligned with the migration of magnetic activity towards the equator but appears about a year earlier before the activity appears at the surface. Although the signature of the poleward branch is visible starting around 2012 during the ascending phase of the cycle, we do not clearly see the branch probably due to the weak cycle 24. Similar results are reported in torsional oscillations derived from the global modes (Howe et al. 2018). Also, the connection between the mid-latitude branch in this solar cycle with the polar branch of the previous cycle is not apparent in the figure due to the temporal coverage. Nevertheless, a close comparison of the zonal flows obtained in Komm et al. (2018) displays similarities as well as some differences in finer details. These differences probably arise due to the different patch sizes used in these studies covering different areas at each grid point and the different smoothing parameters. However the overall patterns are similar in both the studies. Further, we find a discontinuity in the mid-latitude branch in the Northern hemisphere around 2012 in the top and bottom panels of Figure 2 corresponding to 5 and 25 Mm (the band is barely visible in 5 Mm). While the traditional methods provide continuous branches in both hemispheres, our analysis shows the continuous branch to be present in the stronger hemisphere only. It is not clear if this effect is due to smaller temporal coverage which provides different mean values over the two hemispheres. It is also plausible that the corrections that is applied at deeper depths to remove the systematic variations are not optimized.

It is well recognized that the meridional flow is poleward on the surface and in shallow depths and increases gradually with depth. We also find similar results with a peak

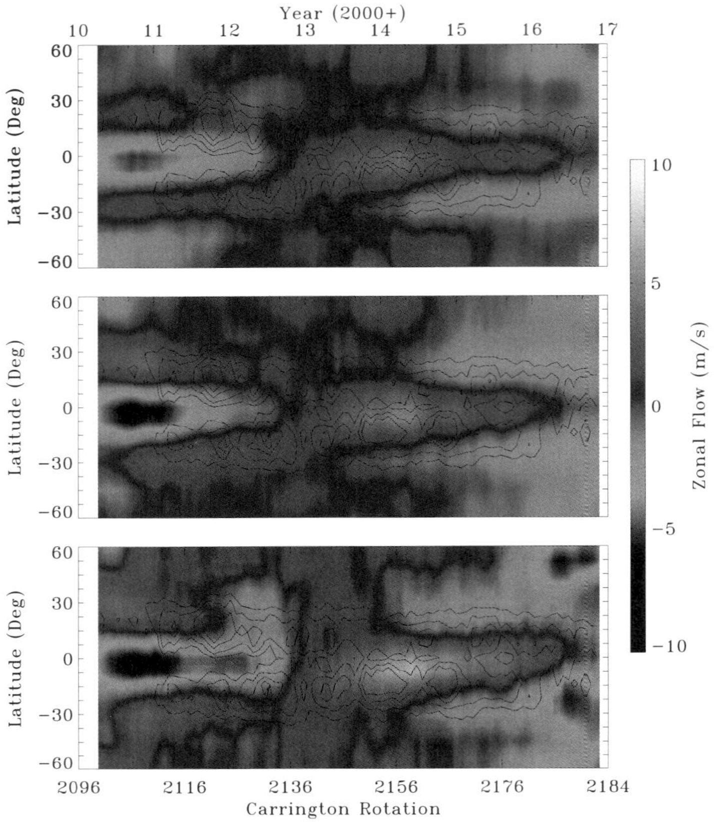

Figure 2. Temporal variation in zonal flow at three different depths: 5.2 ± 2.2 Mm (top), 15 ± 7.2 Mm (middle) and 25.6 ± 7.2 Mm (bottom). The temporal mean has been subtracted at every latitude and depth. Black contours in each panel indicate the magnetic activity.

amplitude of about 15 m/s near the surface and 30 m/s at a depth of 29 Mm. Figure 3 displays the temporal evolution of the meridional flow relative to the mean flow at each latitude and depth for the same depths as the zonal flows. We obtain bands of faster and slower meridional circulation residuals with time but with opposite signs in both hemispheres implying poleward flows as positive values indicate flows to the north and negative values for the flows to the south. At the shallow depths, the band of poleward meridional flow appears at about 60° latitude around the mid-2010 (beginning of our data) and slowly migrates towards the equator reaching there in about 2013. A similar result is reported in Komm et al. (2018). But at deeper depths, the flow pattern is reversed; the fast band has changed to the slow band and vice versa and differs from the result of Komm (2021) where results from HMI 30° tiles were used. However, we should note that the HMI RD result has a coarser resolution in latitude and longitude since the flow measurement are available only at a grid spacing of 15°.

The amplitude and direction of the meridional flow at the equator is an important topic in the context of cross-equatorial cancellation of magnetic flux and formation of polar flux. Early investigations that found flows crossing the equator believed it to be an error in the alignment of the telescope since its magnitude is one order smaller than the maximum amplitude of the meridional flow. But recent advances in the surface flux transport models indicate that it plays a crucial role in transporting magnetic flux across the equator from the dominant hemisphere to the other one and thus cancels the excess

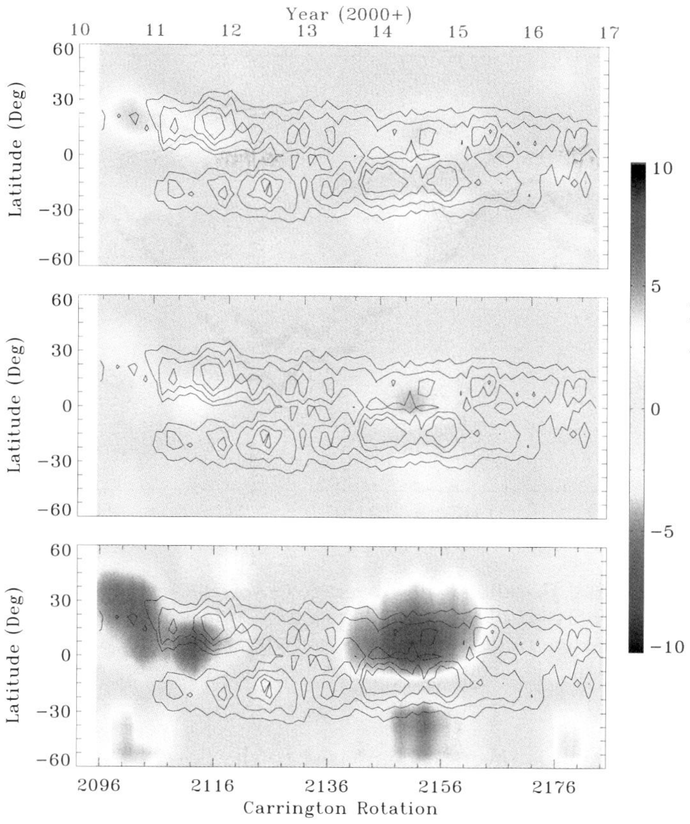

Figure 3. Same as Figure 2 but for meridional flows

flux. We further investigate, as shown in Figure 4, the meridional flows at the equator at a depth closer to the surface (averaged over 0.9 – 3.4 Mm). It is evident that the meridional flows at the equator are small but non-zero. It is mainly positive at the beginning of cycle and oscillates between positive and negative values as the solar cycle progresses which is believed to be a consequence of the imbalance of activity between the two hemispheres. To quantify this flux imbalance, we subtract the magnetic activity index (MAI) of 7.5° S tiles from 7.5° N tiles. The MAI values for each tile is calculated by tracking and re-mapping line-of-sight magnetograms in the same fashion as Dopplergrams for the same length of time. The absolute values of all pixels higher than a defined threshold are then averaged to compute the unsigned proxy for magnetic field (MAI). It is evident from Figure 4 that the cross equatorial flows follow the hemisphere with dominant activity. The Pearson's linear cross correlation is found to be 75% confirming the relation between the cross-equatorial flow and strength of the magnetic activity. These findings are in agreement with those reported in Komm (2022) where subsurface flow measurements were derived from 15° tiles.

4. Summary

Reconstructing Dopplergrams from the HMI SH coefficients and subsequently applying the ring-diagram technique, we have derived subsurface flows up to a depth of 30 Mm from mid-2010 to the end of 2016. Our preliminary results of zonal and meridional flow profiles mostly agree near the surface and intermediate depth ranges with other studies using observed Dopplergrams. However, the measured meridional flows at greater depths

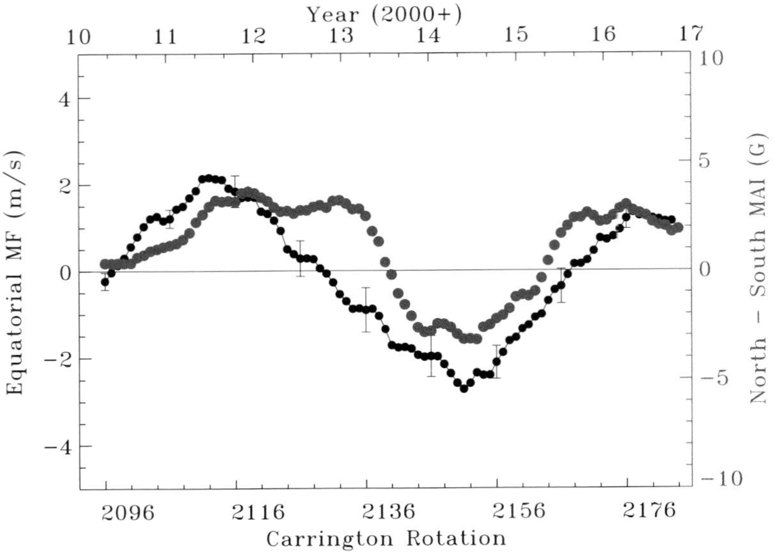

Figure 4. The equatorial meridional flow averaged over the depth range of 0.9 – 3.4 Mm. Positive values indicate northward flows, while negative values indicate southward flows. The blue line represents the strength of the magnetic activity where the MAI of 7.5° S tile is subtracted from the 7.5° N tile. For clarity, the error bars for flow measurements are shown only for few points.

show a flow pattern which is reversed compared to the flows observed near the surface. Although there are limited helioseismic studies which probes the flows at depths greater than 15 Mm using RD technique, our result appears to differ from these findings. It is plausible that the applied empirical correction for the systematic variations has not removed all artifacts. We plan to revisit the correction scheme when flow measurements for whole cycle 24 is computed.

The data used here are courtesy of NASA/SDO and the HMI Science Team. This work was supported by NASA grants 80NSSC20K0194, 80NSSC21K0735, and 80NSSC23K0404 to the National Solar Observatory. SCT, KJ and RK acknowledge the partial financial support from NASA Cooperative Agreement 80NSSC22M0162 to Stanford University for the COFFIES Drive Science Center.

References

Anderson, E.R., Duvall, T.L., and Jefferies, S.M.: 1990, *ApJ*, 364, 699

Basu, S., & Antia, H. M. 2010, *ApJ*, 717, 488

Brandenburg, A.: 2013, *Solar and Astrophysical Dynamos and Magnetic Activity* 294, 387

Corbard, T., Toner, C., Hill, F., et al. 2003, in ESA Special Publication, Vol. 517, GONG+ 2002. Local and Global Helioseismology: the Present and Future, ed. H. Sawaya–Lacoste, 255

Duvall, T. L., J., Jefferies, S. M., Harvey, J. W., & Pomerantz, M. A. 1993, *Nature*, 362, 430

González Hernández, I., Komm, R., Hill, F., Howe, R., Corbard, T., and Haber, D.A.: 2006, *ApJ* 638, 576

Haber, D. A., Hindman, B. W., Toomre, J., et al. 2002, *ApJ*, 570, 855

Hill, F. 1988, *ApJ*, 333, 996

Howe, R., Hill, F., Komm, R., Chaplin, W. J., Elsworth, Y., Davies, G. R., and, ...: 2018, *ApJ*, 862, L5

Jain, K., Tripathy, S. C., & Hill, F. 2017, *ApJ*, 849, 94

Kholikov, S., Serebryanskiy, A., & Jackiewicz, J. 2014, *ApJ*, 784, 145

Komm, R.: 2021, *Solar Physics* 296, 174

Komm, R.: 2022, *Solar Physics*, 297, 99
Komm, R., Howe, R., González Hernández, I., & Hill, F. 2015, *Solar Physics*, 290, 1081
Komm, R., Howe, R., & Hill, F. 2018, *Solar Physics*, 293, 145
Larson, T. P., & Schou, J. 2018, *Solar Physics*, 293, 29
Libbrecht, K.G. and Morrow, C.A.: 1991, *Solar Interior and Atmosphere*, 479
Lin, C.-H., & Chou, D.-Y. 2018, *ApJ*, 860, 48
Zhao, J., & Kosovichev, A. G. 2004, *ApJ*, 603, 776
Zhao, J., & Kosovichev, A. G. 2014, *ApJ*, 789, L7

Dynamics of Solar and Stellar Convection Zones and Atmospheres
Proceedings IAU Symposium No. 365, 2024
A. V. Getling & L. L. Kitchatinov, eds.
doi:10.1017/S174392132300546X

Rotational shear in the low photosphere of the Sun

T. Corbard, M. Faurobert, B. Gelly, R. Douet and D. Laforgue

University Cote d'Azur, Cote d'Azur Observatory, CNRS UMR7293 J.L. Lagrange laboratory
CNRS-IRL2009
email: thierry.corbard@oca.eu

Abstract. We present a new method to measure the rotational height gradient in the solar photosphere. The method is inspired from differential interferometric techniques, we applied it to spectroscopic observations in the FeI 630.15 nm obtained at the solar telescope THEMIS which is equipped with an efficient adaptive optics system. The spectroscopic data was used to obtain images of the granulation at different line cords formed at different heights in the photosphere. Cross-correlation allows us to measure small systematic shifts between similar images. When observations are performed out of the center of the solar disk, the perspective effect gives rise to a radial shift between images formed at different heights. The measurement of this shift provides us with their formation-height difference. At the center of the disk the perspective effect vanishes but we measured a systematic retrograde shift along the east/west direction of the images formed at higher heights. The measured shifts are proportional to the formation height of the images. We interpret these findings as the evidence of a decrease of the rotational velocity with height in the low photosphere of the Sun and we give an estimate of this gradient.

Keywords. Sun: photosphere, differential rotation, High angular resolution

1. Introduction

Rotational shear plays a major role in driving the stellar dynamo. In the Sun, helioseismic observations allow to probe the rotational profile in the convection zone. They have revealed two regions that show a strong height-gradient of the rotation rate, namely, the tachocline at the base of the the convective zone and the near surface shear layer (NSSL). By using observations of surface gravity modes (f-modes), observed by the Michelson Doppler Imager (MDI) onboard the Heliospheric Observatory (SoHO), Corbard and Thompson (2002) were able to obtained a clean measurement of the mean radial gradient of the angular velocity in the sub-surface layer, from about 4 Mm below the surface down to 9 Mm. Ring-diagram analysis by Zaatri and Corbard (2009) using Global Oscillation Network Group data over 7 years also showed that the amplitude of the radial gradient increases closer to the surface in layers above 5 Mm. More recent global and local helioseismology results (Reiter et al. 2020; Komm 2022) covering longer periods and a full solar cycle have confirmed that the gradient of rotation is not uniform in the NSSL and that the rotation rate exhibits a steep decrease close to the surface.

On the theoretical side, in a recent paper devoted to radiative hydrodynamic numerical simulations of the near-surface of the sun in the presence of rotation, Kitiashvili et al. (2023) derived a strong negative gradient in the angular velocity in a subsurface layer at depths smaller than 4 Mm.

In the present work, we are interested in the rotation profile in the photospheric layer. Depth variations in the rotational velocity have been investigated in Livingston and Milkey (1972) by measuring the Doppler effect in several photospheric lines. This did not show any evidence of a rotational depth gradient. However, using a new analysis of 5 mn solar p-mode limb oscillations from 3.5 years of the Helioseismic and Magnetic Imager (HMI) data, Cunnyngham et al. (2017) measured a steep radial decrease of the latitudinal average of the solar rotation rate in the upper photosphere.

In this work, we report the detection of a steep rotational shear at the equator in the low photosphere. The detection does not rely on the Doppler effect but on the measurement of systematic shifts between simultaneous images of the granulation taken at different altitudes. In the field of view (along the spectrograph slit and at many slit positions), we have observed a large number of granular structures. To extract statistical information on possible systematic effects we performed ensemble averages of the Fourier cross-spectra among simultaneous images at different levels in the FeI 630.15 nm line. If the images are similar but slightly shifted the phase of the cross-spectrum shows a linear behavior with respect to the spatial frequency, the slope of this linear trend is directly related to the image shift. Systematic shifts may arise from two different effects, namely, a perspective effect in the radial direction when the images are taken away from the center of the solar disk or horizontal shifts that are induced by a height-gradient of large scale horizontal velocities projected on the plane of the sky. Other kinds of turbulent motions would not lead to any measurable systematic shift. The horizontal shift of a structure is proportional to the duration of the dragging of the upper layer by the velocity gradient namely here, to the correlation time of the granulation pattern (see the sketch in Fig. 1). On average the shift Δx measured at the center of the disk along the equator is given by $\Delta x = \Delta v_{rot}(z)\tau$, where $\Delta v_{rot}(z)$ is the difference in the rotation velocity at the altitude z and τ is the coherence time of the granulation pattern. We used the perspective effect to measure the formation height of the images at different line-cords and an independent measurement of the correlation time by Title et al. (1988) to derive an estimate of the rotational-velocity height gradient. In the next Section we describe our observations and method, in the third Section we present the results, the last Section is devoted to some concluding remarks.

2. Observations and methods

We used spectroscopic observations obtained at the solar telescope THEMIS dating from July, 18 to July 22, 2022, in the FeI 630.15 nm line. We also used HINODE Solar Optical Telescope (SOT) spectropolarimetric (SP) data from the irradiance program, HOP79, performed on July 21, 2022. With THEMIS, we scanned 40"x 108" quiet solar regions at different positions along the west-east central axis and along the south-north polar axis. The HOP79 irradiance program performs montly systematic scans of 30"x130" regions at 20 positions along the polar axis. The pixel size along the slit is 0.235" at THEMIS and 0.32" on SOT-SP, the spectral pixel is 1.35 pm at Themis and 2.1 pm on SOT-SP. A clear advantage of HINODE is the absence of seeing-induced image degradation. With THEMIS, we were able to benefit from some good weather conditions and of the Adaptative Optic system (AO) (Thiébaut et al. 2022; Gelly et al. 2016). We further selected the best spectrograms with a quality criteria based on the root mean square (rms) contrast of the granulation in the continuum along the slit.

To measure a very small displacement between images in the continuum and in the line-wings, we used the phase of their cross-spectrum. This method requires us to consider the power spectra of intensity fluctuations in images taken rigorously at the same time in the continuum and in the line. Actually, 2D images are not needed as far as the spectrograph slit is oriented in the direction of the displacement. In practice, we used

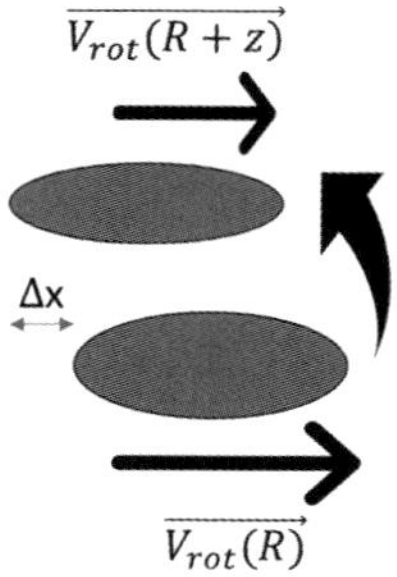

Figure 1. Sketch of the bending of granular structures in the equatorward direction due to a rotational shear.

1D power spectra of the intensity variations along the spectrograph slit and we summed over all the slit positions. Using $I_i(x)$ and $I_c(x)$ to denote the 1D cuts of the line-wing and continuum images, respectively, and assuming that the line-wing image is similar to the continuum one, but slightly displaced at a distance δ, we write $I_i(x) \sim I_c(x-\delta)$ and subsequently, $\widehat{I}_i(u) \sim \widehat{I}_c(u)\exp(2i\pi\delta u)$.

We used a series of spectrograms to estimate the cross-spectrum $\widehat{Q}_{ci}$ between $I_c(x)$ and $I_i(x)$ defined as

$$\widehat{Q}_{ci}(u) = < \widehat{I}_c(u)\widehat{I}_i^*(u) >, \tag{1}$$

where $\widehat{I}_i^*(u)$ represents the conjugate Fourier transform of $I_i(x)$ and the brackets refer to the ensemble average. For similar but shifted images we thus get

$$\widehat{Q}_{ci}(u) \sim < \mid \widehat{I}_c(u) \mid^2 > e^{-2i\pi\delta u}. \tag{2}$$

A linear fit of the phase variation with respect to the spatial frequency variable u allows us to measure δ. We note that a positive shift corresponds to a negative slope of the phase. We stress here that this method is not limited by the spatial resolution of the instrument. The main limitation is the signal-to-noise ratio (S/N) of the image spectrum and on the validity of the assumption of similarity between the line-wing and continuum images.

To apply this technics we first construct images formed on surfaces at roughly constant continuum optical depth. Our reconstruction method relies on the relation between the line and continuum absorption coefficients in the Lorentzian damping wings of a line. This is presented in detail in Faurobert et al. (2012) where the method is tested in the framework of the Milne–Eddington approach and applied to Hinode observations. Figure 2 shows a vertical cut of the granulation along the spectrograph slit obtained from 25 line cords in the FeI 630.15 nm line observed with SOT/SP onboard HINODE. The altitudes at disk center of the line-core formation depth and of the contrast inversion layer are derived in Faurobert et al. (2012) and in Faurobert et al. (2013), respectively.

3. Results

Figure 3 illustrates the measurement of the formation height of the line-cord images from the perspective effect. It shows the phase of the cross-spectrum of the images at line cords 16 and 25 (continuum) obtained along the north/south polar axis at an heliocentric angle of 42° in the southern and northern hemispheres both with THEMIS and HINODE data. We do observe a linear behavior of the phase at spatial frequencies smaller than 0.6 arcsecond^{-1}, at larger frequencies it becomes noisy due to a lower signal-to-noise ratio

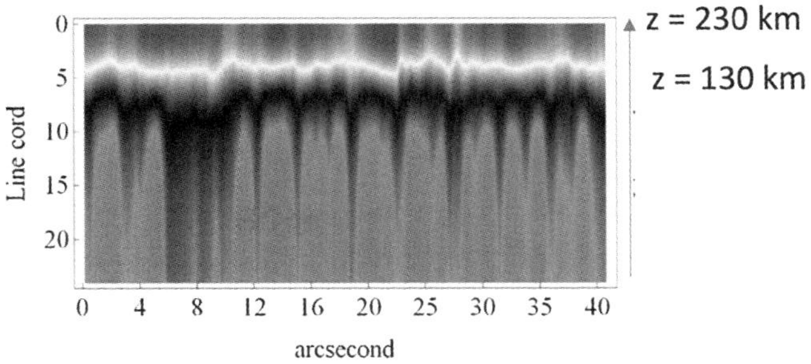

Figure 2. A vertical cut of the solar granulation along the spectrograph slit at the center of the solar disk obtained from SOT/SP data by using 25 line cords in the FeI 630.15 nm. The spectrograph slit is along the North-South axis. The contrast-inversion layer appears in black.

of the reconstructed images at small scales. We observe that the phase shows opposite values in the two hemispheres, in agreement with the perspective effect. A linear fit in the interval [-0.6, 0.6] yields $z = 21.6$ km $\pm$ 0.9 km for the image formation height above the continuum level.

Figure 4 shows, as an example, the cross-spectrum phase derived at the center of the disk between the images in the continuum and at line cord 15, with the spectrograph slit along the north/south and east/west directions. In the last case we show the results obtained for three different observing days. No shift of the images is observed along the polar axis (the phase is noisy but remains close to 0), whereas we do observe a linear behavior of the cross-spectrum phase when the slit is along the equator. This allows us to measure the shift Δx, that it is negative meaning that the rotational velocity decreases with height. For the line cord shown here we get $\Delta x = -22.3 \pm 1.3$ km. In Faurobert et al. (2023) it is shown, using several line cords that Δx is proportional to the height z (obtained from the perspective effect) in the low photosphere with $\Delta x \sim -0.79\, z$. The measurements are made on a very thin layer within 30 km above the continuum level. We interpret these observations as the signature of a negative height gradient of the angular velocity of the following form: $\Omega(R+z) = \Omega(R)(1 - \alpha z)$.

Estimating the value of α requires the knowledge of the coherence time of the granulation pattern. Its value was measured by Title et al. (1988) who found that $\tau = 440$ s. Then using the value of the rotational velocity at the continuum level $v_{rot}(R) = 2$ km/s, we obtain $\alpha \simeq 9.\,10^{-4}$ km^{-1}. This leads to a decrease of the angular velocity $\Delta\Omega$ of about -23μrd/s at $z = 100$ km.

4. Concluding remarks

The rotational shear we obtain here is steeper by a factor of 4 than the result found by Cunnyngham et al. (2017). We do expect to find a steeper gradient because they measured a latitudinal average of the angular velocity over some effective heights whereas we measure the rotational shear at the center of the disk close to the equator, and we have more precise values for the altitudes of the measurements. Nevertheless we want to stress that our result depends of the coherence time of the structures that give rise to the signal. We remark that the linear behavior of the cross-spectrum phase is observed at spatial scale smaller than 0.6 arcsecond^{-1}, i. e. at scales larger than 1.6 arcseconds. So it seems that the signal comes mainly from the largest granules that have longer lifetimes. Increasing the value of τ would decrease our estimate of the rotational shear. This issue has to be addressed in future works.

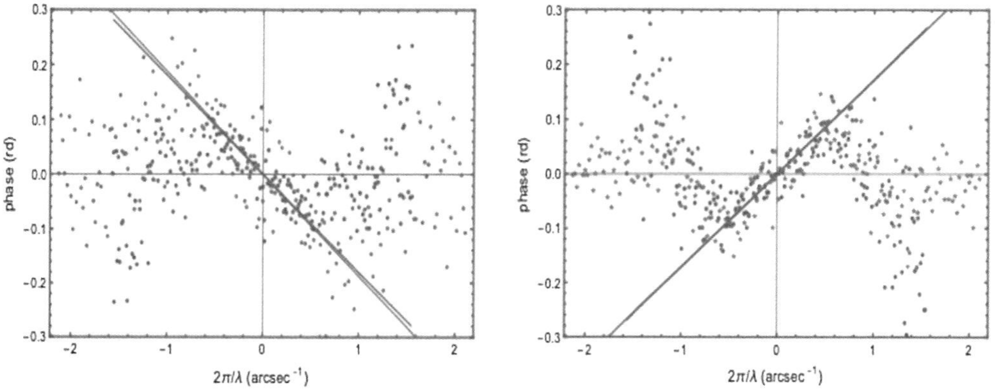

Figure 3. Phase of the cross-spectrum of images at cords 16 and 25 (continuum) with the slit along the north/south axis at two symmetrical position on the polar axis. Blue symbols: from Themis data, red symbols: from Hinode data. **Left:** In the southern hemisphere, **Right:** in the northern hemisphere. The straight lines show linear fits at spatial frequencies smaller than 0.6 arcsec^{-1}.

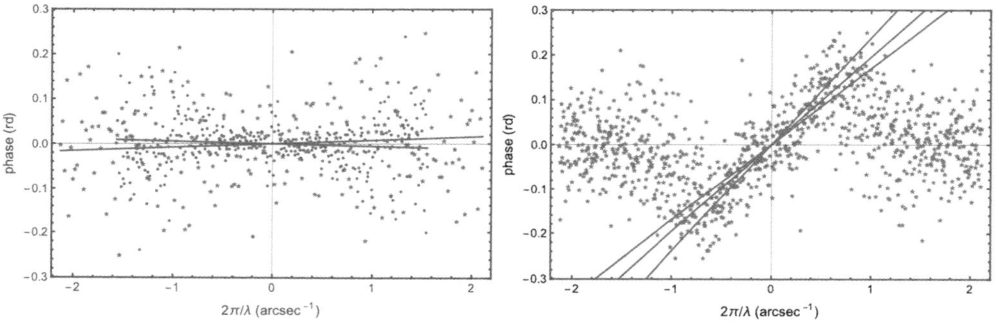

Figure 4. Left: Phase of the cross-spectrum of images at cords 15 and 25 (continuum) with the slit along the polar axis. Blue symbols: from Themis data, red symbols: from Hinode data. **Right:** Phase of the cross-spectrum with the slit along the equator for three different observing days. The straight lines show linear fits at spatial frequencies smaller than 0.6 arcsec^{-1}.

We also remark that a height decrease of the rotational velocity could be misinterpreted as a latitudinal decrease because measurements at high latitudes are performed at smaller limb-distances i.e. at higher altitudes than measurements at low latitudes.

References

Corbard, T. & Thompson, M. J. 2002, *Solar Phys.*, 205(2), 211–229.

Cunnyngham, I., Emilio, M., Kuhn, J., Scholl, I., & Bush, R. 2017, *Phys. Rev. Lett.*, 118, 051102.

Faurobert, M., Ricort, G., & Aime, C. 2012, *A& A*, 548, A80.

Faurobert, M., Ricort, G., & Aime, C. 2013, *A& A*, 554, A116.

Faurobert, M., Corbard, T., Gelly, B., Douet, R., & Laforgue, D. 2023, *A& A*, 676, L4.

Gelly, B., Langlois, M., Moretto, G., Douet, R., Lopez Ariste, A., Tallon, M., Thiébaut, E., Geyskens, N., Lorgeoux, G., Léger, J., & Le Men, C. In Hall, H. J., Gilmozzi, R., & Marshall, H. K., editors, *Ground-based and Airborne Telescopes VI* 2016,, volume 9906 of *Society of Photo-Optical Instrumentation Engineers (SPIE) Conference Series*, 99065A.

Kitiashvili, I. N., Kosovichev, A. G., Wray, A. A., Sadykov, V. M., & Guerrero, G. 2023, *MNRAS*, 518(1), 504–512.

Komm, R. 2022, *Frontiers in Astronomy and Space Sciences*, 9, 1017414.

Livingston, W., & Milkey, R.1972, *Solar. Phys.* 25, 267

Reiter, J., Rhodes, E. J., J., Kosovichev, A. G., Scherrer, P. H., Larson, T. P., & , S. F. Pinkerton, I. 2020, *ApJ*, 894(1), 80.

Thiébaut, É., Tallon, M., Tallon–Bosc, I., Gelly, B., Douet, R., Langlois, M., & Moretto, G. In Schreiber, L., Schmidt, D., & Vernet, E., editors, *Adaptive Optics Systems VIII* 2022,, volume 12185, 1218507. International Society for Optics and Photonics, SPIE.

Title, A., Tarbell, T., Topka, K., Acton, L., Duncan, D., Ferguson, S., Finch, M., Frank, Z., Kelly, G., Lindgren, R., Morrill, M., Pope, T., Reeves, R., Rehse, R., Shine, R., Simon, G., Harvey, J., Leibacher, J., Livingston, W., November, L., & Zirker, J. 1988, *Astrophysical Letters and Communications*, 27, 141.

Wöhl, H., Kučera, A., Rybák, J., & Hanslmeier, A. 2002, *A& A*, 394, 1077–1091.

Zaatri, A. & Corbard, T. In Dikpati, M., Arentoft, T., González Hernández, I., Lindsey, C., & Hill, F., editors, *Solar-Stellar Dynamos as Revealed by Helio- and Asteroseismology: GONG 2008/SOHO 21* 2009,, volume 416 of *Astronomical Society of the Pacific Conference Series*, 99.

Dynamics of Solar and Stellar Convection Zones and Atmospheres
Proceedings IAU Symposium No. 365, 2024
A. V. Getling & L. L. Kitchatinov, eds.
doi:10.1017/S174392132300491X

Near-surface shear layer of solar rotation: origin and significance

Leonid L. Kitchatinov[1,2]

[1]Institute of Solar-Terrestrial Physics, Lermontov Str. 126A, 664033, Irkutsk, Russia;
kit@iszf.irk.ru

[2]Pulkovo Astronomical Observatory, Pulkovskoe Sh. 65, 196140, St–Petersburg, Russia

Abstract. Helioseismology has discovered a thin layer beneath the solar surface where the rotation rate increases rapidly with depth. The normalized rotational shear in the upper 10 Mm of the layer is constant with latitude. Differential rotation theory explains such a rotational state by a radial-type anisotropy of the near-surface convection and a short correlation time of convective turbulence compared to the rotation period. The shear layer is the main driver of the global meridional circulation.

Keywords. Sun: interior, Sun: rotation, turbulence

1. Introduction

Helioseismology has detected a steep increase in the rotation rate with depth just below the surface (Thompson et al. 1996; Schou et al. 1998). The subsurface radial shear exceeds the latitudinal shear seen on the solar surface. The potential importance of the large differential rotation in the near-surface shear layer (NSSL) for the solar dynamo (Brandenburg 2005; Pipin and Kosovichev 2011) has triggered a discussion on the origin of the layer (see e.g. Kitchatinov 2013; Hotta et al. 2015; Gunderson and Bhattacharjee 2019; Jha and Choudhuri 2021). The discussion has not led to a consensus yet.

It is noteworthy that although the near-surface shear varies with latitude, it varies coherently with the rotation rate, so that the normalised shear

$$\frac{r}{\Omega}\frac{\partial\Omega}{\partial r} \simeq -1 \tag{1}$$

in the upper 10 Mm of the solar convection zone is essentially constant (Barekat et al. 2014). An adequate theory should reproduce this remarkable property.

The rotational shear proportional to the rotation rate is a robust prediction of differential rotation theory for the case of a short correlation time of the convective turbulence compared to the rotation period (Rüdiger 1989). The upper 10 Mm of the NSSL belong to this case. The constant normalised shear then follows from the standard stress-free condition for the upper boundary.

2. Differential rotation theory for NSSL

Thermal convection is driven by buoyancy forces pointing up or down the radius. Convective turbulence can therefore be anisotropic, with different intensities of radial and horizontal mixing. Lebedinsky (1941) was probably the first to point out that the influence of anisotropic turbulence on global rotation is not limited to the smoothing of rotational inhomogeneity by turbulent viscosity. Anisotropy allows the turbulence to generate differential rotation through non-diffusive fluxes of angular momentum (see

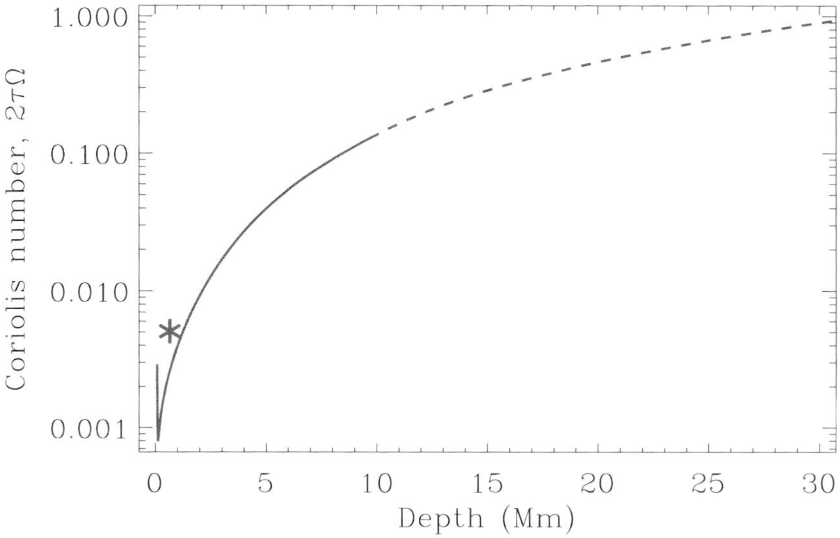

Figure 1. Depth profile of the Coriolis number (3) in the NSSL. The solid line shows the profile for the depth range where Barekat et al. (2014) found the constant shear of Eq. (1). The profile for greater depths is shown dashed. The star indicates an estimate for the solar granulation.

Kitchatinov 2005, for pictorial explanation). The Lebedinsky effect is now known as the Λ-effect (Rüdiger 1989, p.37).

The turbulent transport of momentum is accounted for in the mean field hydrodynamics by the Reynolds stress, $R_{ij} = -\rho\langle u_i u_j\rangle$, where the angular brackets mean averaging, ρ is the density and $\mathbf{u}$ is the turbulent velocity. The differential rotation theory distinguishes two parts in the stress tensor,

$$R_{ij} = R_{ij}^{\nu} + R_{ij}^{\Lambda}, \tag{2}$$

which are responsible for the turbulent viscosity and the Λ-effect.

The key parameter of the theory is the dimensionless factor

$$\Omega^* = 2\tau\Omega \tag{3}$$

of the Coriolis force in the normalised equation of motion; τ is the convective turnover time. The Coriolis number (3) measures the intensity of the interaction between convection and rotation. The main difficulty with the differential rotation theory was a large value of the Coriolis number in the depth of the solar/stellar convection zone. Therefore, the theory has to be non-linear in this number (Rüdiger et al. 2013). However, this difficulty does not apply to the depth range where Barekat et al. (2014) found the constant normalised shear of Eq. (1). Figure 1 shows small Coriolis number for these depths.

The total depth of the NSSL is conventionally assumed to be about 30 Mm. The radial shear changes sign around this depth (Schou et al. 1998). The shear value decreases and the relation (1) is violated as this depth is approached (Komm 2022). It is a matter of definition whether the depths near 30 Mm, where the radial shear is no longer large compared to the latitudinal shear, should be assigned to the NSSL. This paper focuses on the upper 10 Mm of the convection zone where the relation (1) constrains possible explanations of the NSSL.

A further complication to the theory is that not only the Reynolds stress but also the meridional flow can also cause differential rotation. However, the meridional flow does not enter the stress-free boundary condition,

$$R_{r\phi} = R^{\Lambda}_{r\phi} + R^{\nu}_{r\phi} = 0 \quad \text{at } r = R_\odot, \tag{4}$$

which controls the surface shear (Kitchatinov 2013).

The smallness of the Coriolis number and the condition (4) simplify matters to such an extent that the constancy of the normalised shear with latitude can be proved without addressing any specific approximation of mean-field theory. Since the dependence on angular velocity enters via the Coriolis number, the Reynolds stress can be linearised in angular velocity. The general structure for the real R^{Λ}_{ij} tensor linear in pseudo-vector $\mathbf{\Omega}$ reads

$$R^{\Lambda}_{ij} = -\rho\nu_\Lambda \left(\hat{r}_i\varepsilon_{jkl} + \hat{r}_j\varepsilon_{ikl}\right)\hat{r}_k\Omega_l, \tag{5}$$

where $\hat{\mathbf{r}} = \mathbf{r}/r$ is the radial unit vector, ν_Λ is a yet indefinite latitude-independent constant, ε_{jkl} is the fully antisymmetric unit tensor and repetition of subscripts means summation. The viscous part of the Reynolds stress is given by the viscosity tensor N_{ijkl}

$$R^{\nu}_{ij} = \rho N_{ijkl}(\nabla_l V_k), \tag{6}$$

where $\boldsymbol{V}$ is the large-scale velocity, which includes rotation and the meridional circulation. General expression for the viscosity tensor in the case of horizontally isotropic turbulence ($\langle u^2_\phi\rangle = \langle u^2_\theta\rangle \neq \langle u^2_r\rangle$) is

$$\begin{aligned} N_{ijkl} = \quad & \nu_1\left(\delta_{ik}\delta_{jl} + \delta_{jk}\delta_{il}\right) + \nu_2\delta_{ij}\delta_{kl} + \nu_3\left(\delta_{ik}\hat{r}_j\hat{r}_l + \delta_{jk}\hat{r}_i\hat{r}_l\right) \\ & + \nu_4\left(\delta_{il}\hat{r}_j\hat{r}_k + \delta_{jl}\hat{r}_i\hat{r}_k\right) + \nu_5\delta_{ij}\hat{r}_k\hat{r}_l + \nu_6\delta_{kl}\hat{r}_i\hat{r}_j + \nu_7\hat{r}_i\hat{r}_j\hat{r}_k\hat{r}_l. \end{aligned} \tag{7}$$

Substituting Eqs. (5–7) into Eq. (4) gives the expression for the normalised shear

$$\frac{r}{\Omega}\frac{\partial\Omega}{\partial r} = -\frac{\nu_\Lambda}{\nu_1 + \nu_3} \tag{8}$$

in terms of the eddy transport coefficients for the case of a horizontally isotropic background turbulence and small Coriolis number.

The normalised shear (8) is constant with latitude. This constancy is therefore a robust result of mean-field theory independent of any particular theoretical tool used to derive the shear. However, the value of the surface shear depends on the - necessarily approximate - theoretical tool used for the evaluation. Derivations with the quasi-linear approximation give (Kitchatinov 2023)

$$\frac{r}{\Omega}\frac{\partial\Omega}{\partial r} = \frac{\langle u^2_h\rangle}{\langle u^2_r\rangle} - 2, \tag{9}$$

where $u^2_h = u^2_\phi + u^2_\theta$ is the intensity of horizontal mixing. Shear is negative for anisotropy of radial type, $\langle u^2_r\rangle > \langle u^2_\phi\rangle = \langle u^2_\theta\rangle$, as it should be (Lebedinsky 1941).

Equation (9) reproduces the seismically detected shear (1) for $\langle u^2_h\rangle/\langle u^2_r\rangle = 1$. This intensity ratio is in agreement with the numerical experiment on the NSSL by Kitiashvili et al. (2023). The anisotropy parameter A_V of their fig. 4 corresponds to $\langle u^2_r\rangle \simeq \langle u^2_h\rangle$ *within* the convective zone.

3. NSSL and meridional flow

The global meridional flow is an important component of the flux transport dynamo models for solar activity. The importance of the NSSL for the solar dynamo can be mediated by the NSSL relation to the meridional flow. The flow is currently understood to result from a slight imbalance between strong centrifugal and baroclinic drivers of the flow (see review by Hazra et al. 2023, for further details).

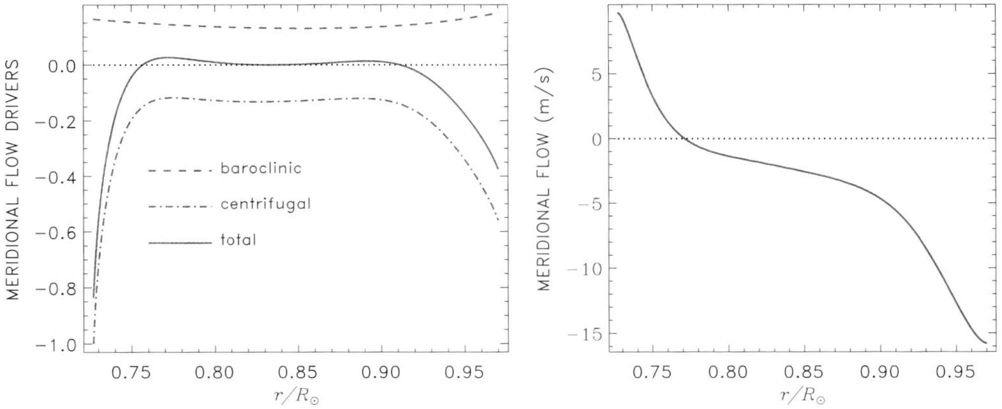

Figure 2. Centrifugal and baroclinic drivers of the meridional flow and their sum (left) and the resulting meridional flow (right) as functions of radius for the 45° latitude according to the mean-field model of Kitchatinov and Olemskoy (2011).

The profiles of the two drivers and the resulting meridional flow computed with a mean-field hydrodynamical model are shown in Fig. 2. The two drivers are almost balanced in the bulk of the convection zone. Close to the boundaries, a complete set of the boundary conditions does not allow the additional condition of the thermo-rotational balance to be satisfied. The increase of the centrifugal driving term near the boundaries violates the balance. Accordingly, the meridional flow of Fig. 2 reaches its maximum velocity at the boundaries and decreases inside the convection zone, in agreement with the seismological detections by Rajaguru and Antia (2015) and Gizon et al. (2020). The meridional flow is generated in the NSSL.

4. Conclusions

The NSSL is a consequence of the radial type of anisotropy of the near-surface convection, and the condition (4) that the external azimuthal force is zero. The constancy of the normalised surface shear (1) with latitude is a consequence of the short convective turnover time relative to the rotation period in the depth range where the constancy was found.

Global meridional flow is excited in the boundary layers near the top and bottom of the convection zone. The large rotational shear in the NSSL drives the meridional circulation by its non-conservative centrifugal force.

Acknowledgements. The author acknowledges financial support from the Ministry of Science and High Education of the Russian Federation.

References

Barekat, A., Schou, J., & Gizon, L. 2014, *A&A*, 570, L12.
Brandenburg, A. 2005, *ApJ*, 625, 539.
Gizon, L., Cameron, R. H., Pourabdian, M. *et al.* 2020, *Science*, 368, 1469.
Gunderson, L. M. & Bhattacharjee, A. 2019, *ApJ*, 870, 47.
Hazra, G., Nandy, D., Kitchatinov, L., & Choudhuri, A. R. 2023, *Space Sci. Review*, 219, 39.
Hotta, H., Rempel, M., & Yokoyama, T. 2015, *ApJ*, 798, 51.
Jha, B. K. & Choudhuri, A. R. 2021, *MNRAS*, 506, 2189.
Kitchatinov, L. L. 2005, *Physics Uspekhi*, 48, 449.
Kitchatinov, L. L. 2013, In Kosovichev, A. G. de Gouveia Dal Pino, E., & Yan, Y., editors. *Solar and Astrophysical Dynamos and Magnetic Activity*, IAU Symposium Vol.294, 399–410.

Kitchatinov, L. L. 2023, *Astronomy Letters*, 49(11), 754.
Kitchatinov, L. L. & Olemskoy, S. V. 2011, *MNRAS*, 411, 1059.
Kitiashvili, I. N., Kosovichev, A. G., Wray, A. A. *et al.* 2023, *MNRAS*, 518, 504.
Komm, R. 2022, *Frontiers in Astronomy and Space Sciences*, 9, 428.
Lebedinsky, A. I. 1941, *Astronomical Journal (USSR)*, 18, 10.
Pipin, V. V. & Kosovichev, A. G. 2011, *ApJ*, 727, L45.
Rajaguru, S. P. & Antia, H. M. 2015, *ApJ*, 813, 114.
Rüdiger, G. 1989, *Differential rotation and stellar convection. Sun and solar-type stars.* Akademie–Verlag, Berlin.
Rüdiger, G., Kitchatinov, L. L., & Hollerbach, R. 2013, *Magnetic processes in astrophysics: theory, simulations, experiments.* Wiley-VCH, Weinheim.
Schou, J., Antia, H. M., Basu, S. *et al.* 1998, *ApJ*, 505, 390.
Thompson, M. J., Toomre, J., Anderson, E. R. *et al.* 1996, *Science*, 272, 1300.

Dynamics of Solar and Stellar Convection Zones and Atmospheres
Proceedings IAU Symposium No. 365, 2024
A. V. Getling & L. L. Kitchatinov, eds.
doi:10.1017/S1743921323004969

The near-surface shear layer (NSSL) of the Sun: a theoretical model

Arnab Rai Choudhuri[1] **and Bibhuti Kumar Jha**[2]

[1]Department of Physics, Indian Institute of Science, Bangalore 560012, India
email: arnab@iisc.ac.in

[2]Southwest Research Institute, Boulder CO 80302, USA
email: maitraibibhu@gmail.com

Abstract. We present a theoretical model of the near-surface shear layer (NSSL) of the Sun. Convection cells deeper down are affected by the Sun's rotation, but this is not the case in a layer just below the solar surface due to the smallness of the convection cells there. Based on this idea, we show that the thermal wind balance equation (the basic equation in the theory of the meridional circulation which holds inside the convection zone) can be solved to obtain the structure of the NSSL, matching observational data remarkably well.

Keywords. differential rotation, meridional circulation, convection, near-surface shear layer

1. Introduction

There is a layer just below the solar surface of thickness $\approx 0.05 R_\odot$ within which the angular velocity of the Sun decreases with radius by a few percent. This near-surface shear layer (NSSL) is clearly visible in any map of solar internal differential rotation produced by helioseismology: see Figure 3 of Choudhuri (2021b). There is not yet a consensus on what gives rise to this layer, although several different ideas have been proposed (Guerrero et al. 2013; Hotta et al. 2015; Matilsky et al. 2019). Here we summarize our work reported by Choudhuri (2021a) and Jha and Choudhuri (2021).

The theoretical ideas of how the two large-scale flow patterns of the Sun—the differential rotation and the meridional circulation—are produced are inter-related. A basic equation in the theory of the meridional circulation is the thermal wind balance equation, which holds within the body of the solar convection zone. It is often supposed that this equation breaks down near the solar surface surface, giving rise to a boundary layer in the form of NSSL. We propose the opposite view that this equation holds even near the surface and the change in the nature of convection due to the small scale height at the top layer causes the NSSL, of which the structure can be obtained by applying the thermal wind balance equation.

In §2 we summarize the basic theoretical ideas about the thermal wind balance equation and its implications for the changed nature of convection in the top layer within the solar convection zone. After presenting our results in §3, we end with our concluding remarks in §4.

2. Basic theoretical ideas

We first write down the thermal wind balance equation before explaining its significance. One standard form of this is

$$r \sin\theta \frac{\partial}{\partial z}\Omega^2 = \frac{g}{\gamma T}\left(\frac{\partial}{\partial \theta}\Delta T\right), \tag{1}$$

where z is the distance from the equatorial plane and the other symbols have their usual meanings, ΔT being the temperature variation with latitude to be discussed in more detail later.

Let us give an idea how the two sides of (1) arise. Since convective motions are influenced by the Coriolis force, the nature of convection changes slightly with latitude. The effect of the Coriolis force on radially moving fluid blobs being the least near the poles, the poles of the Sun are expected to be slightly hotter than the equatorial region. This would tend to drive a flow from a pole to the equator near the surface, i.e a clockwise meridional circulation in the northern hemisphere—which is opposite of what is seen. This effect is captured in the thermal wind term in the RHS of (1). On the other hand, if the centrifugal force varies along a straight line parallel to the rotation axis, that also can drive a meridional circulation which is anti-clockwise in northern hemisphere for solar-like angular velocity distribution. The centrifugal term in the LHS of (1) corresponds to this effect. The full dynamical equation of the meridional circulation has other terms (Choudhuri 2021b). However, a straightforward order-of-magnitude estimate suggests that the centrifugal term is much larger than these terms and can only be balanced by the thermal wind term, leading to (1).

Convective motions are influenced by the Coriolis force only if the convective turnover time is comparable to or more than the rotation period ≈ 27 days. This is certainly not true for granules near the surface having turnover time of the order of a few minutes. We thus expect a layer at the top of the convection zone which is not affected by the Coriolis force. On the other hand, we believe that convective motions deeper down with much longer turnover times are affected by the Coriolis force. Although this is a gradual transition, we make the simplifying assumption that the Coriolis force is important when $r < r_c$ and negligible when $r > r_c$, where r_c is a critical radius.

Now we write

$$T(r,\theta) = T(r,0) + \Delta T(r,\theta), \tag{2}$$

where $T(r,0)$ is the radial distribution of temperature as given by the standard solar model and $\Delta T(r,\theta)$ is the departure from it. Since convective motions in the top layer $r > r_c$ are not affected by rotation, we expect the temperature gradient there to be independent of latitude, i.e.

$$\frac{d}{dr}T(r,\theta) = \frac{d}{dr}T(r,0). \tag{3}$$

On substituting (2) in (3), we arrive at

$$\frac{d}{dr}\Delta T(r,\theta) = 0. \tag{4}$$

This suggests that $\Delta T(r,\theta)$ does not vary with r in this top layer $r > r_c$ and we can write

$$\Delta T(r,\theta) = \Delta T(r_c,\theta). \tag{5}$$

In other words, the latitudinal variation of temperature at the critical radius r_c gets mapped to the visible surface.

We now look at the RHS of (1). In the top layer $r > r_c$, the temperature T drops sharply with increasing r as we approach the surface, but (4) ensures that other quantities

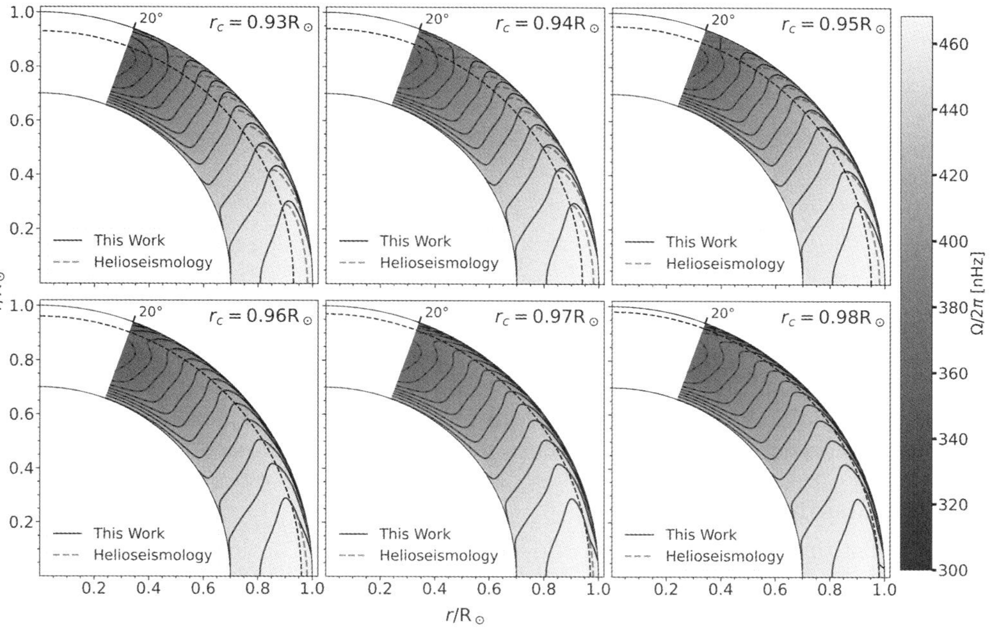

Figure 1. Contours of angular velocity Ω as given by helioseismolgy (dashed red curves) and by our model (solid black curves), for different assumed values of r_c. The different values of r_c given in different panels are indicated by dashed black circles. From Jha and Choudhuri (2021).

appearing in the RHS of (1) do not vary much with r within this top layer. This implies that the RHS of (1) has to become very large close to the solar surface due to the drop in T. If (1) has to be satisfied, then the RHS of (1) also has to become very large just below the solar surface. For this to happen, Ω^2 has to vary with z rapidly. We suggest that this is what gives rise to the NSSL. This basic idea was suggested by Choudhuri (2021a) based on order-of-magnitude estimates. Detailed calculations were presented by Jha and Choudhuri (2021).

3. Results

We calculate the structure of the NSSL in the top layer of the convection zone in the following manner. Assuming a particular value of r_c, we take Ω determined from helioseismology for $r < r_c$ and find ΔT in those deeper layers by using (1). This gives us $\Delta T(r_c, \theta)$ and we easily get $\Delta T(r, \theta)$ in the top layer $r > r_c$ from (5). Once we have $\Delta T(r, \theta)$ in the top layer $r > r_c$, we can use (1) to obtain Ω there, giving us the structure of NSSL. Figure 1 shows different contours of Ω we get by using different reasonable values of r_c. The solid black and dashed red lines show contours of Ω from our model and from helioseismology. They are the same below r_c because we started our calculations in our model by using Ω from helioseismology below r_c. In the layer above r_c, we find that the two different sets of curves are very close to each other and our theoretical model gives the NSSL very nicely. We get best results when we assume $r_c = 0.96R_\odot$. Figure 2 shows the percentile error in our theoretical model in matching helioseismology data in the top layer for $r_c = 0.96R_\odot$. Except at the very surface, we find the error to be much less than 5%. This fantastic fit between theory and observations gives us confidence that we are on the right track.

A side result of our theoretical model is that we get a temperature variation at the solar surface. For the case $r_c = 0.96R_\odot$, the pole is found to be about 3 K hotter than

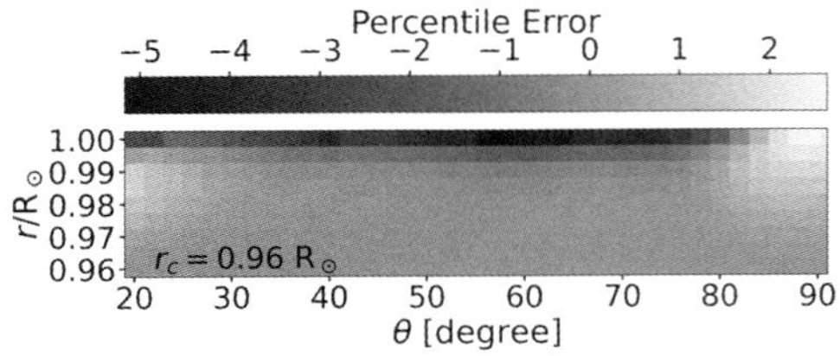

Figure 2. The percentile error of the theoretical model in matching the observational data of Ω for the case $r_c = 0.96R_\odot$. From Jha and Choudhuri (2021).

the equator. It may be noted that Kitchatinov and Ruediger (1995) found a temperature difference of 5 K in their theoretical model of large-scale flows in the solar convection zone. There are some observational studies (Kuhn et al. 1988; Rast et al. 2008) suggesting that poles of the Sun are indeed slightly hotter than the equator. We hope that a more accurate observational study with modern techniques will be carried out in near future.

4. Conclusion

For the fist time, we are able to provide a theoretical model of the NSSL which matches observational data in quantitative detail. This model is based on the idea that the thermal wind balance equation holds even in the top layers of the convection zone just below the solar surface. We get the best results on assuming that $r_c = 0.96R_\odot$ is the critical radius above which convective motions are not affected by the solar rotation, but below which they are affected. Perhaps it will be possible to check this through numerical simulations in future. One outcome of our model is that the poles of the Sun should be about 3 K hotter than the equator, another testable prediction.

References

Choudhuri, A. R. 2021a, A Theoretical Estimate of the Pole-Equator Temperature Difference and a Possible Origin of the Near-Surface Shear Layer. *Solar Phys.*, 296(2), 37.

Choudhuri, A. R. 2021b, The meridional circulation of the Sun: Observations, theory and connections with the solar dynamo. *Science China Physics, Mechanics, and Astronomy*, 64(3), 239601.

Guerrero, G., Smolarkiewicz, P. K., Kosovichev, A. G., & Mansour, N. N. 2013, Differential Rotation in Solar-like Stars from Global Simulations. *ApJ*, 779(2), 176.

Hotta, H., Rempel, M., & Yokoyama, T. 2015, High-resolution Calculation of the Solar Global Convection with the Reduced Speed of Sound Technique. II. Near Surface Shear Layer with the Rotation. *ApJ*, 798(1), 51.

Jha, B. K. & Choudhuri, A. R. 2021, A theoretical model of the near-surface shear layer of the Sun. *MNRAS*, 506(2), 2189–2198.

Kitchatinov, L. L. & Ruediger, G. 1995, Differential rotation in solar-type stars: revisiting the Taylor-number puzzle. *A&A*, 299, 446.

Kuhn, J. R., Libbrecht, K. G., & Dicke, R. H. 1988, The Surface Temperature of the Sun and Changes in the Solar Constant. *Science*, 242(4880), 908–911.

Matilsky, L. I., Hindman, B. W., & Toomre, J. 2019, The Role of Downflows in Establishing Solar Near-surface Shear. *ApJ*, 871(2), 217.

Rast, M. P., Ortiz, A., & Meisner, R. W. 2008, Latitudinal Variation of the Solar Photospheric Intensity. *ApJ*, 673(2), 1209–1217.

Dynamics of Solar and Stellar Convection Zones and Atmospheres
Proceedings IAU Symposium No. 365, 2024
A. V. Getling & L. L. Kitchatinov, eds.
doi:10.1017/S1743921324000206

Differential rotation of stars from spot transit mapping: dependence on rotation period and effective temperature

Alexandre Araújo and Adriana Valio

Centre for Radio Astronomy and Astrophysics, Mackenzie Presbyterian University, São Paulo, Brazil
email: adrivalio@gmail.com

Abstract. Just like the Sun, other stars also exhibit differential rotation. Currently, the rotation profile of a star that hosts a transiting planet can be estimated if during a transits, the planet occults a spot on the photosphere of the star, causing slight variations in its light curve. By detecting the same spot during a later transit, the stellar rotation period at that latitude is determined. Here, we present the results of differential rotation for 48 stars, 13 from the spot transit mapping method, while the remaining 35 stars from other techniques. The results show that the differential rotation is correlated with the stellar mean rotation period for fast rotating stars and strongly anti-correlated for slow rotators. The transition occuring at rotation period of 5 days. On the other hand, the differential shear increases with effective temperature for fast rotating stars, but the correlation is lost for the slow rotators.

Keywords. Starspot, Stellar rotation, Differential Rotation

1. Introduction

Stellar evolution studies rely on star rotation to understand age, angular momentum transfer, and magnetic fields. Differential rotation, similar to the Sun, is observed in other stars, accompanied by magnetic phenomena like spots, plages, loops, and flares (Reinhold et al. 2013). Detecting starspots at different latitudes helps track differential rotation. Understanding differential rotation and its correlations with other stellar parameters, such as rotation period and effective temperature, is important for the stellar dynamo (Küker and Rüdiger 2011), but conclusive results are lacking. Space missions provide data to study stellar rotation, flares, and starspots, enhancing our understanding of stellar activity.

2. Data and Starspot mapping

2.1. *Data*

This study explores correlations between the differential rotation shear with effective temperature and rotation period. Differential rotation using the transit mapping technique (Silva 2003) are presented for 13 stars. For comparison, we further analysed 35 stars which differential rotation measurements were available in the literature. The stars studied here and their differential rotation are listed in Table 1.

Table 1. Effective temperature, rotational period and differential rotation for all stars.

Star	T_{Teff} [K]	P_{rot} [d]	$\Delta\Omega$ [rad d^{-1}]	Ref	Star	T_{Teff} [K]	P_{rot} [d]	$\Delta\Omega$ [rad d^{-1}]	Ref
Kepler-17*	5781	12.4	$0.0409^{+0.0043}_{-0.0007}$	1	LQ Hya	5019	1.60	0.194 ± 0.002	17
Kepler-45*	3820	15.76	0.031 ± 0.014	2	LO Peg	4577	0.42	0.036 ± 0.007	18
Kepler-71*	5540	19.77	0.00098 ± 0.00010	3	HK Aqr	3697	0.43	0.00484 ± 0.0019	19
Kepler-210*	4559	12.35	$0.0436^{+0.011}_{-0.007}$	4	EY Dra	3489	0.45	0.00030 ± 0.003	20
Kepler-411*	4832	10.52	0.050 ± 0.0006	5	KIC 03541346	4194	0.908154	0.017	21
Kepler-1651*	3594	18.43	0.022 ± 0.004	6	KIC 04819564	4125	0.380794	0.0099	21
KOI-883*	4840	8.99	0.069 ± 0.011	7	KIC 04953358	3843	0.649015	0.0048	21
KapCeti	5677	8.77	$0.064 \pm {}^{0.006}_{-0.005}$	8	KIC 05791720	3533	0.765051	0.0098	21
EpsilonEri	5146	11.2	$0.062 \pm {}^{0.03}_{-0.004}$	9	KIC 06675318	4206	0.577727	0.0098	21
CoRoT-4*	6189	8.87	$0.03780.0261^{+0.0002}_{-0.0009}$	10	KIC 07592990	4004	0.442148	0.0071	21
CoRoT-5*	6100	26.63	$0.014^{+0.001}_{-0.011}$	10	KIC 08314902	4176	0.813534	0.0061	21
CoRoT-6*	6090	6.35	$0.105^{+0.011}_{-0.08}$	10	KIC 10515986	3668	0.746207	0.018	21
CoRoT-8*	5080	21.7	$0.014^{+0.003}_{-0.008}$	10	KIC 11087527	4303	0.41096	0.0076	21
CoRot-18*	5440	5.4	0.094 ± 0.003	10	KIC 10063343	3976	0.3326	0.0057	21
CoRoT-2*	5625	4.54	$0.042^{+0.005}_{-0.016}$	11	KIC 8429280	5055	1.20	0.26 ± 0.001	22
IL HYa	4500	12.73	0.035 ± 0.003	12	HD 35850	6138	1	0.256 ± 0.017	23
V889 Her	5750	1.34	0.042 ± 0.044	13	KIC 7985370	5815	2.856	0.194 ± 0.0002	24
R58	5880	0.56	0.08 ± 0.07	14	KIC 7765135	5835	2.485	0.18 ± 0.0002	24
HD 171488	5800	1.33	0.41 ± 0.16	14	HD 175726	6030	3.95	0.636	25
HD 106506	5900	1.39	0.24 ± 0.03	14	HD 181906	6360	2.71	0.25	25
HD 141943	5850	2.17	0.240 ± 0.03	14	HD 307938	5859	0.57	0.025 ± 0.015	26
LQ Lup	5750	0.31	0.12 ± 0.02	14	AF Lep	6100	0.97	0.259 ± 0.019	27
PZ Tel	5448	0.944	0.0113 ± 0.0018	15	HD 155555A	5400	1.67	0.078 ± 0.006	28
AB Dor	5386	0.51	0.046 ± 0.006	16	HD 155555B	5050	1.67	0.039 ± 0.006	28

(1) (Valio et al. 2017); (2) (Zaleski et al. 2022); (3) (Zaleski et al. 2019); (4) (Valio and Araújo 2022); (5) (Araújo and Valio 2021); (6) (Araújo et al. 2023); (7) (Zaleski et al. 2022); (8) (Walker et al. 2007); (9) (Croll et al. 2006); (10) (Valio 2013); (11) (Silva–Valio et al. 2010) (12) (Petit et al. 2013); (13) (Marsden et al. 2006); (14) (Marsden et al. 2011); (15) (Barnes et al. 2000); (16) (Cameron and Donati 2002); (17) (Donati et al. 2003); (18) (Karmakar et al. 2016); (19) (Barnes et al. 2004); (20) (Barnes et al. 2005); (21) (Vida et al. 2014); (22) (Frasca et al. 2011); (23) (Mengel 2005); (24) (Fröhlich et al. 2012); (25) (Mosser et al. 2009); (26) (Marsden et al. 2004); (27) (Järvinen et al. 2015); (28) (Dunstone et al. 2008).

2.2. *Starspot mapping*

The stellar rotation profile can be estimated if the star is eclipsed periodically by an orbiting planet. During one of these transits, the planet may occult a spot on the photosphere of the star. This occultation will cause a slight increase in the intensity of the transit light curve during a few minutes. The fit of the transit light curve by a model allows us to determine the physical characteristics of the spots. By monitoring the positions of these spots in subsequent transits, it is possible to estimate the rotation period of a star Silva–Valio (2008). The model used is described in Silva (2003), and its basic assumptions are:

- Star: synthesised 2D image of a star with limb darkening.
- Planet: completely dark disc of radius $\frac{R_{planet}}{R_{star}}$.
- Spot: round grey feature with a given size, intensity (or temperature), and location (longitude and latitude).

2.3. *Differential Rotation*

It is assumed that all stars exhibit differential rotation with latitude similar to that of the Sun:

$$\Omega = \Omega_{eq} - \Delta\Omega \sin^2(\alpha) \tag{1}$$

where the constants are: Ω_{eq} (rotation rate at the equator) and $\Delta\Omega$ (rotational shear). Two measurements of the rotation are considered: the mean angular rotation, $\bar{\Omega}$, and the angular rotation, Ω_1, at a certain latitude α_1, which are given by:

$$\bar{\Omega} = \Omega_{eq} - \frac{\Delta\Omega}{2} \quad \text{and} \quad \Omega_1 = \Omega_{eq} - \Delta\Omega \sin^2(\alpha_1) \tag{2}$$

Since, the rotation period is $P = 2\pi/\Omega$, $\bar{\Omega}$ is obtained from the mean rotation period, usually measured from a Lomb-Scargle periodogram of the entire light curve. From the spot transit mapping, one can obtain the rotation period at a certain latitude, and thus Ω_1 (Valio 2016).

3. Result and Discussion

3.1. *Differential rotation of slow and fast rotating stars*

Recently, Araújo and Valio (2023) have analysed the differential rotation of 13 stars obtained from the spot transit mapping. To compare our results on rotational shear with those of other stars, we expanded our sample to include stars with $\Delta\Omega$ values reported in the literature. The full sample, consisting of 48 stars, is listed in Table 1. Here, we investigate the dependence of stellar differential rotation for both slow and fast (≤ 5 days) rotating stars. We present the results of the differential rotation for all stars in Figure 1 as a function of the stellar rotational period and in Figure 2 as a function of the effective temperature of the stars.

Shown in Figure 1, the results reveal a strong anti-correlation (Pearson coefficient of approximately -0.80, middle panel) between $\Delta\Omega$ and the rotation period for stars with slow (black squares) rotation rates, similar to that found by Araújo and Valio (2023).

On the other hand, the fast rotating stars (blue squares) exhibit an opposite behavior, with the rotational shear increasing with the rotation period of the star (Pearson correlation coefficient of 0.70, shown in the right panel of Figure 1). As can be seen in the left panel of Figure 1, an abrupt change in behaviour occurs at stellar rotation periods around 5.0 days.

The relationship between $\Delta\Omega$ and effective temperature for all stars is shown in the left panel of Figure 2. Stars with fast rotation rates (right panel) tend to exhibit strong

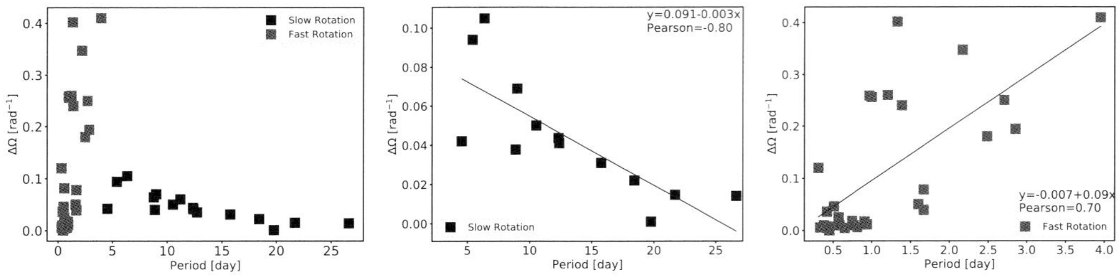

Figure 1. Differential rotation, $\Delta\Omega$, as a function of stellar average rotation period, for all stars (**Left**), for slowly rotating stars with average rotation periods $P_{\rm rot} \geq 4.5$ days, (**Middle**), and for rapidly rotating stars with average rotation periods of $P_{\rm rot} \leq 4.5$ (**Right**). Rapid rotators are shown as blue squares whereas the slow rotators are depicted as black squares.

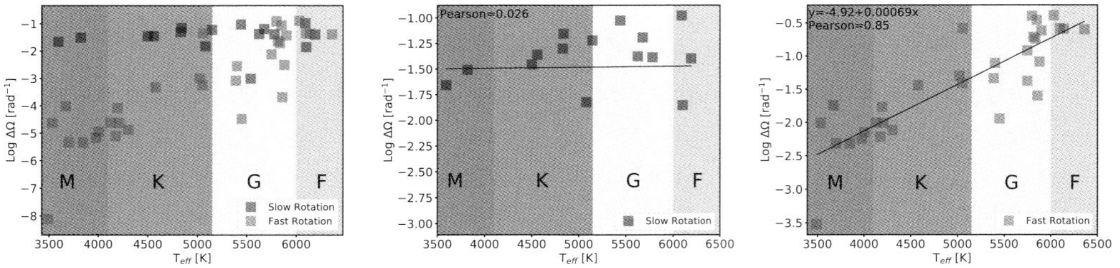

Figure 2. Relation between $\Delta\Omega$ and effective temperature, for all stars (**Left**), for slowly rotating stars with average rotation periods $P_{\rm rot} \geq 4.5$ days, (**Middle**), and for rapidly rotating stars with average rotation periods of $P_{\rm rot} \leq 4.5$ (**Right**). Rapid rotators are shown as blue squares whereas the slow rotators are depicted as black squares.

correlation (correlation index of 0.85) than their slow-rotating (middle panel) counterparts. In contrast, slowly rotating stars show significant dispersion, with no apparent correlation (Pearson correlation index of 0.026, middle panel of Figure 2). Thus, $\Delta\Omega$ tends to increase with temperature, but this trend is only evident in rapidly rotating stars. These results are consistent with the findings of Balona and Abedigamba (2016).

Studies have shown that there is a correlation between the magnetic activity of stars and their differential rotation, which is caused by the magnetic dynamo (Parker 1955). Young solar-type stars are known to have faster rotation periods and stronger magnetic activity compared to the Sun (Fröhlich et al. 2012). This relationship is evident in other phenomena associated with the stellar dynamo, including the presence of large stellar spots, such as those observed in Kepler-17 with mean radii of 49,000 $\pm$ 10,000 km (Valio et al. 2017), and the production of superflares and stellar spots with complex magnetic topology (Araújo and Valio 2023). Furthermore, our results suggest that there is an increasing difference in rotation rate, referred to as $\Delta\Omega$, between the equator and poles of stars, which increases with rotation periods up to $\sim$5 days and then decreases as the rotation rates of stars decrease, or conversely the rotation period increases.

4. Conclusion

In this study, we compared the differential rotation behaviour of fast and slow rotating stars, for a total of 48 stars, ranging from spectral types M to F. The data included differential rotation of 13 stars from measurements of spot transit mapping method. Our aim was to explore how differential rotation relates to the stars' mean rotation period and effective temperature.

We found that for young rotation stars ($P < 5$ days), the differential rotation increases with rotation period, and then decreases for stars with periods longer than 5 days, with a strong negative correlation between rotational shear and the average rotation period (coefficient = -0.80). As for the dependence of the differential shear with effective temperature (or stellar mass), our results showed that there is a strong correlation for the fast young rotating stars, but this correlation is lost altogether for the slowly rotators ($P > 5$ days).

A comprehensive understanding of the connection between rotational shear, effective temperature, and rotation rate is crucial to furthering our knowledge of stellar dynamos.

Acknowledgements: We thank the financial support from FAPESP grants #2013/10559-5 and #2018/04055-8, as well as CNPq grant #150817/2022-3.

References

Araújo, A., Valio, A., & Alexandre, T. 2023, Kepler-1651a: Superflares and starspot. *in prep.*,.

Araújo, A. & Valio, A. 2021, Kepler-411 differential rotation from three transiting planets. *The Astrophysical Journal Letters*, 907(1), L5.

Araújo, A. & Valio, A. 2023, The connection between starspots and superflares: a case study of two stars. *Monthly Notices of the Royal Astronomical Society: Letters*, 522(1), L16–L20.

Araújo, A. & Valio, A. 2023, Dependence of Stellar Differential Rotation on Effective Temperature and Rotation: An Analysis from Starspot Transit Mapping. *Astrophysical Journal*, 956(2), 141.

Balona, L. A. & Abedigamba, O. P. 2016, Differential rotation in k, g, f and a stars. *Monthly Notices of the Royal Astronomical Society*, 461(1), 497–506.

Barnes, J., Cameron, A. C., Donati, J.-F., James, D., Marsden, S., & Petit, P. 2005, The dependence of differential rotation on temperature and rotation. *Monthly Notices of the Royal Astronomical Society: Letters*, 357(1), L1–L5.

Barnes, J., Cameron, A. C., James, D., & Donati, J.-F. 2000, Doppler images from dual-site observations of southern rapidly rotating stars-i. differential rotation on pz tel. *Monthly Notices of the Royal Astronomical Society*, 314(1), 162–174.

Barnes, J., James, D., & Cameron, A. C. 2004, Differential rotation and star-spot evolution on hk aqr in 2001 and 2002. *Monthly Notices of the Royal Astronomical Society*, 352(2), 589–599.

Cameron, A. C. & Donati, J.-F. 2002, Doin'the twist: secular changes in the surface differential rotation on ab doradus. *Monthly Notices of the Royal Astronomical Society*, 329(1), L23–L27.

Croll, B., Walker, G. A., Kuschnig, R., Matthews, J. M., Rowe, J. F., Walker, A., Rucinski, S. M., Hatzes, A. P., Cochran, W. D., Robb, R. M., *et al.* 2006, Differential rotation of ε eridani detected by most. *The Astrophysical Journal*, 648(1), 607.

Donati, J.-F., Collier Cameron, A., & Petit, P. 2003, Temporal fluctuations in the differential rotation of cool active stars. *Monthly Notices of the Royal Astronomical Society*, 345(4), 1187–1199.

Dunstone, N., Hussain, G., Cameron, A. C., Marsden, S., Jardine, M., Barnes, J., Ramirez Velez, J., & Donati, J.-F. 2008, Differential rotation on both components of the pre-main-sequence binary system hd 155555. *Monthly Notices of the Royal Astronomical Society*, 387(4), 1525–1536.

Frasca, A., Fröhlich, H.-E., Bonanno, A., Catanzaro, G., Biazzo, K., & Molenda-Żakowicz, J. 2011, Magnetic activity and differential rotation in the very young star kic 8429280. *Astronomy & Astrophysics*, 532, A81.

Fröhlich, H.-E., Frasca, A., Catanzaro, G., Bonanno, A., Corsaro, E., Molenda-Żakowicz, J., Klutsch, A., & Montes, D. 2012, Magnetic activity and differential rotation in the young sun-like stars kic 7985370 and kic 7765135. *Astronomy & Astrophysics*, 543, A146.

Järvinen, S., Arlt, R., Hackman, T., Marsden, S., Küker, M., Ilyin, I., Berdyugina, S., Strassmeier, K., & Waite, I. 2015, Doppler images and the underlying dynamo-the case of af leporis. *Astronomy & Astrophysics*, 574, A25.

Karmakar, S., Pandey, J., Savanov, I., Taş, G., Pandey, S., Misra, K., Joshi, S., Dmitrienko, E., Sakamoto, T., Gehrels, N., *et al.* 2016, Lo peg: surface differential rotation, flares, and spot-topographic evolution. *Monthly Notices of the Royal Astronomical Society*, 459(3), 3112–3129.

Küker, M. & Rüdiger, G. 2011, Differential rotation and meridional flow on the lower zero-age main sequence: Reynolds stress versus barocinic flow. *Astronomische Nachrichten*, 332(9-10), 933–938.

Marsden, S., Donati, J., Semel, M., Petit, P., & Carter, B. 2006, Surface differential rotation and photospheric magnetic field of the young solar-type star hd 171488 (v889 her). *Monthly Notices of the Royal Astronomical Society*, 370(1), 468–476.

Marsden, S., Jardine, M., Ramírez Vélez, J., Alecian, E., Brown, C., Carter, B., Donati, J.-F., Dunstone, N., Hart, R., Semel, M., *et al.* 2011, Magnetic fields and differential rotation on the pre-main sequence–ii. the early-g star hd 141943–coronal magnetic field, hα emission and differential rotation. *Monthly Notices of the Royal Astronomical Society*, 413(3), 1939–1948.

Marsden, S., Waite, I., Carter, B., & Donati, J.-F. 2004, Doppler imaging of g-dwarfs in two young open clusters. *Astronomische Nachrichten*, 325(3), 246.

Mengel, M. W. 2005,. *The active young solar-type star HR 1817 (= HD 35850)*. PhD thesis, University of Southern Queensland.

Mosser, B., Baudin, F., Lanza, A. F., Hulot, J., Catala, C., Baglin, A., & Auvergne, M. 2009, Short-lived spots in solar-like stars as observed by corot. *Astronomy & Astrophysics*, 506(1), 245–254.

Parker, E. N. 1955, Hydromagnetic Dynamo Models. *Astrophysical Journal*, 122, 293.

Petit, P., Jardine, M., & Spruit, H. C. Magnetic fields throughout stellar evolution. In *Proc. IAU Symp* 2013,, volume 302, 142.

Reinhold, T., Reiners, A., & Basri, G. 2013, Rotation and differential rotation of active kepler stars. *Astronomy and Astrophysics*, 560, A4.

Silva, A. V. 2003, Method for spot detection on solar-like stars. *The Astrophysical Journal Letters*, 585(2), L147.

Silva–Valio, A. 2008, Estimating Stellar Rotation from Starspot Detection during Planetary Transits. *Astrophysical Journal, Letters*, 683(2), L179.

Silva–Valio, A., Lanza, A., Alonso, R., & Barge, P. 2010, Properties of starspots on corot-2. *Astronomy & Astrophysics*, 510, A25.

Valio, A. Stellar Activity from Analysis of Planetary Transits. In Chavez, M., Bertone, E., Vega, O., & De la Luz, V., editors, *New Quests in Stellar Astrophysics III: A Panchromatic View of Solar-Like Stars, With and Without Planets* 2013,, volume 472 of *Astronomical Society of the Pacific Conference Series*, 239.

Valio, A. 2016, Starspots properties and stellar activity from planetary transits. *Proceedings of the International Astronomical Union*, 12(S328), 69–76.

Valio, A. & Araújo, A. 2022, Stellar Obliquity from Spot Transit Mapping of Kepler-210. *Astrophysical Journal*, 940(2), 132.

Valio, A., Estrela, R., Netto, Y., Bravo, J., & de Medeiros, J. 2017, Activity and rotation of kepler-17. *Astrophysical Journal*, 835(2), 294.

Vida, K., Oláh, K., & Szabó, R. 2014, Looking for activity cycles in late-type kepler stars using time–frequency analysis. *Monthly Notices of the Royal Astronomical Society*, 441(3), 2744–2753.

Walker, G. A., Croll, B., Kuschnig, R., Walker, A., Rucinski, S. M., Matthews, J. M., Guenther, D. B., Moffat, A. F., Sasselov, D., & Weiss, W. W. 2007, The differential rotation of κ1 ceti as observed by most. *The Astrophysical Journal*, 659(2), 1611.

Zaleski, S., Valio, A., Carter, B., & Marsden, S. 2022, Dynamo activity of the k dwarf koi-883 from transit photometry mapping. *Monthly Notices of the Royal Astronomical Society*, 510(4), 5348–5361.

Zaleski, S., Valio, A., Marsden, S., & Carter, B. 2019, Differential rotation of kepler-71 via transit photometry mapping of faculae and starspots. *Monthly Notices of the Royal Astronomical Society*, 484(1), 618–630.

Dynamics of Solar and Stellar Convection Zones and Atmospheres
Proceedings IAU Symposium No. 365, 2024
A. V. Getling & L. L. Kitchatinov, eds.
doi:10.1017/S1743921324000152

Short and long term spot evolution on EI Eri

L. Kriskovics[1,2], Zs. Kővári[1,2], B. Seli[1,2], K. Oláh[1,2], K. Vida[1,2], G. W. Henry[3], T. Granzer[4] and A. Görgei[1,2,5]

[1]Konkoly Observatory, HUN-REN Research Centre for Astronomy and Earth Sciences, Budapest, Hungary
email: kriskovics@konkoly.hu

[2]CSFK, MTA Centre of Excellence, Budapest, Hungary

[3]Center of Excellence in Information Systems, Tennessee State University, Nashville

[4]Leibniz-Institut für Astrophysik Potsdam (AIP), An der Sternwarte 16, D14482 Potsdam, Germany

[5]Eötvös Loránd University, Hungary

Abstract. We investigate flare activity using photometric data obtained with the Transiting Exoplanet Survey Satellite (TESS). Long-term seasonal period analysis was applied on our APT (Automatic Photoelectric Telescopes af Fairborn observatory, Arizona) time series to study changes in the rotational period. We also looked for activity cycle-like changes with short-term Fourier-transform. We also studied the phase and frequency distribution of hand-selected flares on the available TESS data. The MUlti-SIte Continuous Spectroscopy (MUSICOS) campaign was designed in 1998 to achieve high-resolution, multi-wavelength spectroscopic observations from many sites around the globe, which meant that uninterrupted phase coverage of EI Eri became available. We use these data to reconstruct successive surface-temperature maps of the star in order to study the changes of starspots on a very short timescale. We applied our multi-line Doppler imaging code to reconstruct four consecutive Doppler images. These images were also used to measure surface differential rotation with our cross-correlation technique.

Keywords. close binaries, stellar magnetic activity, starspots, Doppler imaging, differential rotation

1. EI Eri: a single lined RS CVn system with an active sub-giant primary

EI Eridani = HD 26337 (G5 IV, $P_{\rm rot} = 1.945\,$d, $V = 7.1$) is a well-known, rapidly rotating (v sin i = 51 km/s), active, non-eclipsing, single-lined spectroscopic binary. Table 1 summarizes the astrophysical parameters derived with Spectroscopy Made Easy (Piskunov and Valenti 2017).

It was Fekel et al. (1982) who identified it as an RS CVn type variable. They also detected photometric variability with an amplitude of $V \approx 0.2$. Fekel et al. (1986) derived an orbital period of 1.9472 days, while Hall et al. (1987) detected a photometric period of 1.945 ± 0.005 days from UBV photometry. Oláh and Strassmeier (2002) reported photometric cycles of approximately 2.4 and 12.2 years, which were later confirmed and refined by Oláh et al. (2009) with values of $\approx 2.9 - 3.1$ and ≈ 14 years. They also found a cycle with a length of roughly 4.1–4.9 years. EI Eri is a well-known flaring variable. Pandey and Singh (2012) estimated the flare peak energies in the 0.3–10 keV energy band to be $\approx 10^{31} - 10^{32}$ erg using observations obtained by the XMM-Newton X-ray observatory.

Table 1. Astrophysical parameters for EI Eri derived in this work.

$T_{\rm eff}$	5580 K
$\log g$	3.75
[Fe/H]	-0.27
$v_{\rm mic}$	2 km/s
$v \sin i$	51 km/s

EI Eri has been a prime target for Doppler imaging since the first applications of the technique. Strassmeier (1990), Strassmeier et al. (1991) and Hatzes and Vogt (1992) reported a permanent polar spot with a radically changing appendage. The presence of the polar feature was confirmed by Washuettl et al. (2001) and later, Washuettl et al. (2009) and Kővári et al. (2009). Constantly changing low latitude features were reported as well by these authors. However, since the rotational period of EI Eri is very close to two days, covering a rotation with optimal phase coverage from a singular observing site takes around a month, which introduces serious blurring and renders the Doppler images unable to represent the rapid spot evolution with adequate time resolution. Consequently, EI Eri was selected as a target for the MUSICOS multi-site campaign. Our results are based on these data.

2. Activity cycles and flaring

To look for activity cycles, we used short-term Fourier transform (Kolláth and Oláh 2009) on our 41 years long dataset. The signal of the longest, high-amplitude feature on the upper time-frenquency plot is suppressed, but above about 13.6 years, which is one-third of the length of the data, all signals have the same amplification. Smoothly changing cycles can be seen between 4.5 and 5.5, and between 8.9 and 11.6 years, which are not harmonics, as the speed and the direction of their changes are not the same. A weak cycle-like feature of about 2.5 years is also present at the beginning of the covered time interval. For the STFT plot, see Fig 1.

We identified 41 flares in the two available TESS sectors. To calculate their energies in the TESS band, we followed similar procedures as in Oláh et al. (2022). The flare frequency distribution in the left panel of Fig 2 shows a broken power-law shape. Two-component fits of the FFD yield power-law indices of 1.41 ± 0.02 and 2.26 ± 0.08 (one minus the slope of the fitted line) for the first and second parts, respectively, while a fit for all datapoints gives 1.66 ± 0.04.

To test whether there is a significant increase in the occurrence of flares at given phases, we ran a two-sided Kuiper's test (Kuiper 1960). This test compares the measured data to a given distributiion, but is invariant under cyclic transformation, which makes it well-suited for comparing phase-distributions. The test rejects the null hypothesis of a uniform flare-phase distribution with a p-value of 0.006. The increase in flare rate appears on the side facing the secondary component.

3. Doppler imaging and differential rotation of EI Eri

The Doppler imaging code iMap (Carroll et al. 2012) used in this work carries out multi-line Doppler inversion on a list of photospheric lines. As the wavelength coverage of our dataset is limited, we used five (Fe I 6400, Fe I 6408, Fe I 6421, Fe I 6430 and Ca I 6439) lines with well-defined continuum, suitable line depth, and temperature sensitivity. The stellar surface was divided into 6x6 deg segments. Since our MUSICOS data comes from six different instruments, spectral resolutions differ, which may introduce artifacts since the profiles and distortions caused by the surface features are sampled differently. We counteract this by decreasing the spectral resolutions to the lowest value with a

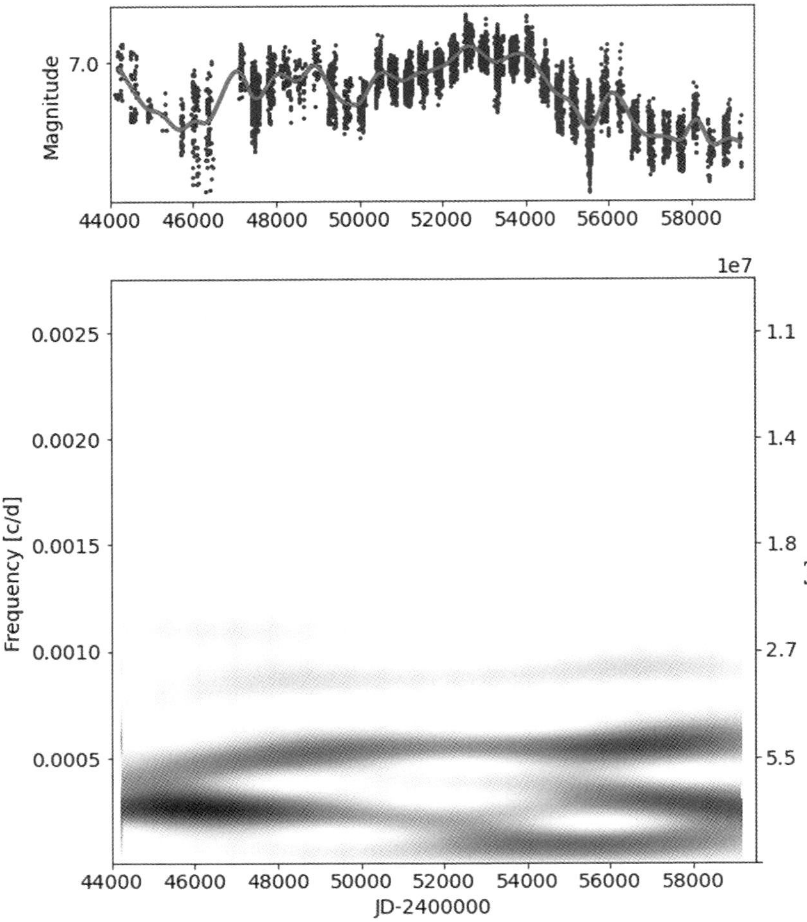

Figure 1. Short term Fourier transform of all the available V band data for EI Eri.

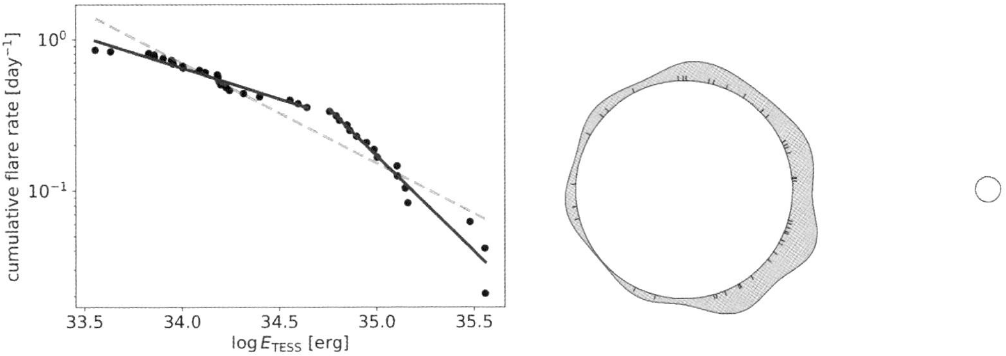

Figure 2. *Left*: Flare-frequency distribution of EI Eri from the two available TESS sectors. *Right*: Flare distribution in the reference frame of the secondary component.

Gaussian-kernel. The viability of this approach is thoroughly tested in Kriskovics et al. (2023). The resulting four consecutive Doppler images in equirectangular projection are shown in Fig 3 (clockwise, from the upper left). The four maps show a persistent polar feature with a temperature of roughly 1100 K below the temperature of the unspotted

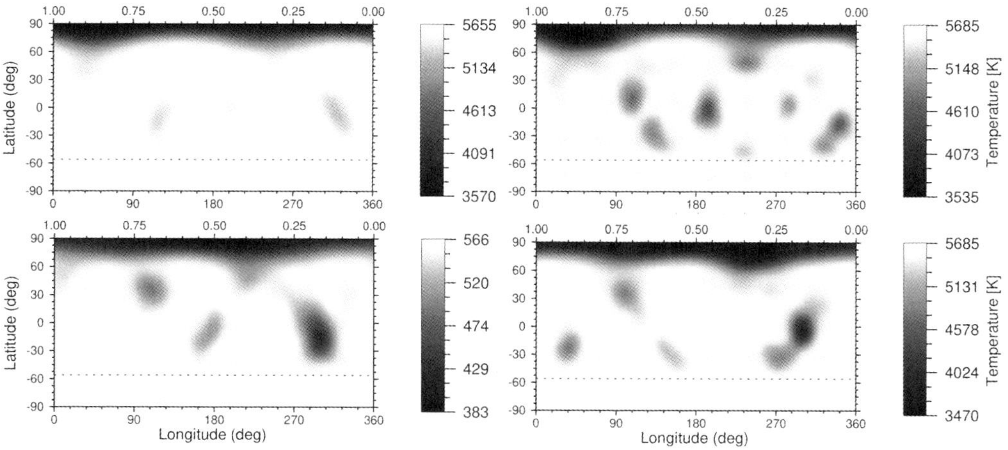

Figure 3. Doppler images of EI Eri in chronological order, clockwise from the upper left.

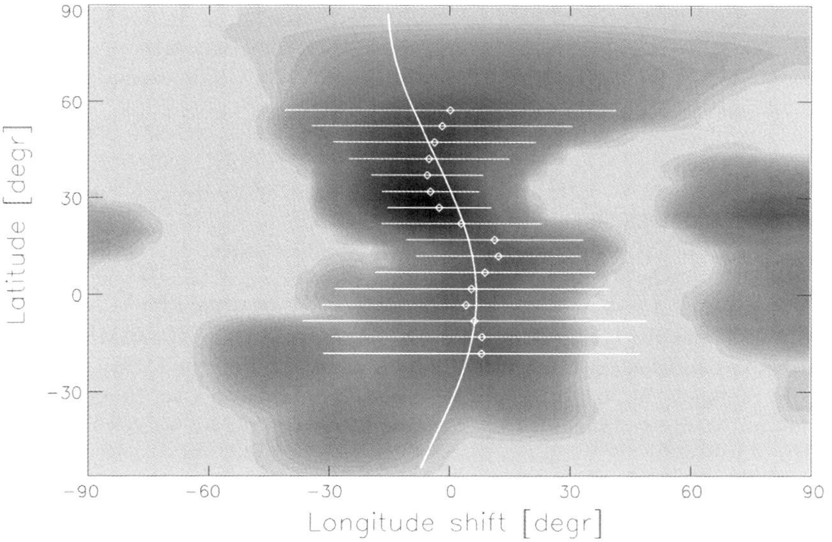

Figure 4. Average cross-correlation map for EI Eri.

surface. Associated with the dominant polar spot, a less prominent appendage appears on the second image around 120 deg longitude. This feature is further strengthened on the third image. On the fourth inversion, the appendage starts to fade, but is still present. The overall contrast of the polar feature seems to increase in time. In addi tion, several low-latitude features show rapid evolution as well, e.g. around 270 deg longitude, a spot starts to appear and strenghthens from the first image to the fourth. For more discussion, see Kriskovics et al. (2023).

We measured the surface differential rotation of EI Eri by the means of cross-correlation of consecutive Doppler images (Donati and Collier–Cameron 1997). In Fig 4, the average correlation pattern is fitted with a quadratic DR law, which yields $\alpha = 0.036 \pm 0.007$ ($\Omega_{\rm eq} = 186.922 \pm 0.384$ deg/day, $\Delta\Omega = 6.768 \pm 1.349$ deg/day). This fits well with previous measurements on rapidly rotating subgiants, as well as with the empirical relation of $|\alpha| \approx 0.013\, P_{\rm rot}$ for binaries originally suggested by Kővári et al. (2017). For more details on the surface differential rotation of different active stars, see Kővári et al. (2017), and their Fig 1.

Acknowledgements Authors from Konkoly Observatory acknowledge the Hungarian National Research, Development and Innovation Office (NKFIH) grants OTKA K-131508 and KKP-143986. LK acknowledges the NKFIH grant OTKA PD-134784. LK and KV are Bolyai János research fellows. KV is supported by the Bolyai+ grant ÚNKP-22-5-ELTE-1093, BS is supported by the ÚNKP-22-3 New National Excellence Program of the Ministry for Culture and Innovation from the source of the National Research, Development and Innovation Fund.

References

Carroll, T. A., Strassmeier, K. G., Rice, J. B., & Künstler, A. 2012, The magnetic field topology of the weak-lined T Tauri star V410 Tauri. New strategies for Zeeman–Doppler imaging. *A&A*, 548, A95.

Donati, J. F. & Collier–Cameron, A. 1997, Differential rotation and magnetic polarity patterns on AB Doradus. *MNRAS*, 291(1), 1–19.

Fekel, F. C., Hall, D. S., Henry, G. W., Landis, H. J., & Renner, T. R. 1982, HD 26337: a New RS CVn Variable Star. *Information Bulletin on Variable Stars*, 2110, 1.

Fekel, F. C., Moffett, T. J., & Henry, G. W. 1986, A Survey of Chromospherically Active Stars. *ApJS*, 60, 551.

Hall, D. S., Osborn, S. A. G., Seufert, E. R., Boyd, L. J., Genet, R. M., & Fried, R. E. 1987, Six Years of Photometry of the RS CVn Binary EI ERI = HD 26337. *AJ*, 94, 723.

Hatzes, A. P. & Vogt, S. S. 1992, Doppler images of the spotted RS CVn star HD 26337 - 1984-87. *MNRAS*, 258, 387.

Kővári, Zs., Oláh, K., Kriskovics, L., Vida, K., Forgács-Dajka, E., & Strassmeier, K. G. 2017, Rotation-differential rotation relationships for late-type single and binary stars from Doppler imaging. *Astronomische Nachrichten*, 338(8), 903–909.

Kővári, Zs., Washuettl, A., Foing, B. H., Vida, K., Bartus, J., Oláh, K., & MUSICOS 98 Team. Doppler maps and surface differential rotation of EI Eri from the MUSICOS 1998 observations. In Stempels, E., editor, *15th Cambridge Workshop on Cool Stars, Stellar Systems, and the Sun* 2009,, volume 1094 of *American Institute of Physics Conference Series*, pp. 676–679.

Kolláth, Z. & Oláh, K. 2009, Multiple and changing cycles of active stars. I. Methods of analysis and application to the solar cycles. *A&A*, 501(2), 695–702.

Kriskovics, L., Kővári, Z., Seli, B., Oláh, K., Vida, K., Henry, G. W., Granzer, T., & Görgei, A. 2023, EI Eridani: A star under the influence. The effect of magnetic activity in the short and long term. *A&A*, 674, A143.

Kuiper, N. H. Tests concerning random points on a circle. In *Nederl. Akad. Wetensch. Proc. Ser. A* 1960,, volume 63, pp. 38–47.

Oláh, K., Kolláth, Z., Granzer, T., Strassmeier, K. G., Lanza, A. F., Järvinen, S., Korhonen, H., Baliunas, S. L., Soon, W., Messina, S., & Cutispoto, G. 2009, Multiple and changing cycles of active stars. II. Results. *A&A*, 501(2), 703–713.

Oláh, K., Seli, B., Kővári, Z., Kriskovics, L., & Vida, K. 2022, Characteristics of flares on giant stars. *arXiv e-prints*,, arXiv:2210.09710.

Oláh, K. & Strassmeier, K. G. 2002, Starspot cycles from long-term photometry. *Astronomische Nachrichten*, 323, 361–366.

Pandey, J. C. & Singh, K. P. 2012, A study of X-ray flares - II. RS CVn-type binaries. *MNRAS*, 419(2), 1219–1237.

Piskunov, N. & Valenti, J. A. 2017, Spectroscopy Made Easy: Evolution. *A&A*, 597, A16.

Strassmeier, K. G. 1990, Photometric and Spectroscopic Modeling of Starspots on the RS Canum Venaticorum Binary HD 26337. *ApJ*, 348, 682.

Strassmeier, K. G., Rice, J. B., Wehlau, W. H., Vogt, S. S., Hatzes, A. P., Tuominen, I., Piskunov, N. E., Hackman, T., & Poutanen, M. 1991, Doppler imaging of hig-latitude SPOT activity on HD 26337. *A&A*, 247, 130.

Washuettl, A., Strassmeier, K. G., & Collier–Cameron, A. Latest Doppler images of the RS CVn Binary EI Eridani (CD-ROM Directory: contribs/washuet). In Garcia Lopez, R. J., Rebolo, R., & Zapaterio Osorio, M. R., editors, *11th Cambridge Workshop on Cool Stars, Stellar Systems and the Sun* 2001,, volume 223 of *Astronomical Society of the Pacific Conference Series*, 1308.

Washuettl, A., Strassmeier, K. G., & Weber, M. 2009, The chromospherically active binary star EI Eridani: II. Long-term Doppler imaging. *Astronomische Nachrichten*, 330(4), 366.

Dynamics of Solar and Stellar Convection Zones and Atmospheres
Proceedings IAU Symposium No. 365, 2024
A. V. Getling & L. L. Kitchatinov, eds.
doi:10.1017/S1743921324000115

Magnetic activity under tidal influences in the 2+2 hierarchical quadruple system V815 Herculis

Zs. Kővári[1,2], K. G. Strassmeier[3], L. Kriskovics[1,2], K. Oláh[1,2], T. Borkovits[1,4,5], B. Seli[1,2], K. Vida[1,2] and Á. Radványi[6]

[1]Konkoly Observatory, HUN-REN Research Centre for Astronomy and Earth Sciences, Budapest, Hungary
email: kovari@konkoly.hu

[2]CSFK, MTA Centre of Excellence, Budapest, Hungary

[3]Leibniz–Institute for Astrophysics, Potsdam, Germany

[4]Baja Astronomical Observatory of University of Szeged, Baja, Hungary

[5]HUN-REN-SZTE Stellar Astrophysics Research Group, Baja, Hungary

[6]Moholy–Nagy University of Art and Design Budapest, Hungary

Abstract. Tidal forces in close binaries and multiple systems that contain magnetically active component are supposed to influence the operation of magnetic dynamo. Through synchronization the tidal effect of a close companion helps maintain fast rotation, thus supporting an efficient dynamo. At the same time, it can also suppress the differential rotation of the convection zone, or even force the formation of active longitudes at certain phases fixed to the orbit. V815 Her is a four-star system consisting of two close binaries orbiting each other, one of which contains an active G-type main-sequence star. Therefore, the system offers an excellent opportunity to investigate the influence of gravitational effects on solar-type magnetic activity using different methods.

Keywords. close binaries, stellar magnetic activity, starspots, Doppler imaging, differential rotation

1. V815 Her: a single-lined multiple system with a chromospherically active primary

V815 Her is a single-lined multiple star system with a G5-6 V type active primary component. It has long been known that the G star features cool spots on its surface. The photometric behavior of the system is dominated by the magnetic activity of the primary component, as shown by the long-term photometric data in Figure 1. For the period analysis we used our Time Frequency Analyzer package TiFrAn (Kolláth and Oláh 2009) with Choi–Williams distribution kernel. Several long-term brightness changes can be attributed to the chromospheric activity of the G star. The century-long archival data from Digital Access to a Sky Century @ Harvard (DASCH) database supplemented with photoelectric data from Jetsu et al. (2000) reveals a slowly increasing cycle with steady amplitude around ~6.5 years, two other cycles of about ~9.1 and ~13 years, while the longest feature points towards a timescale around ~24-26 years with weakening amplitude (see the bottom panel of Fig. 1). The cycle lengths of ~13 and ~26 years are perhaps harmonics of ~6.5 years.

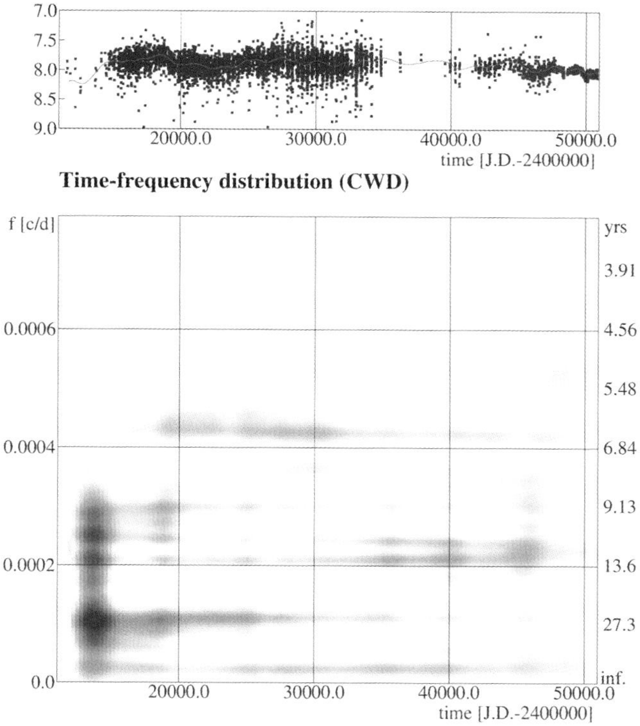

Figure 1. Long-term photometry and time-frequency analysis of V815 Her.

The rapidly rotating ($P_{\rm rot} = 1.8$ d) G star (=V815 Her Aa) together with an unseen, but probably late M-dwarf component (=V815 Her Ab) form the close binary subsystem V815 Her A with $P_{\rm orb} = 1.8$ d, i.e. orbiting synchronously. The binary was also known to have a distant companion with an orbital period of about 5.7 years (Fekel et al. 2005).

2. V815 Her B: an eclipsing binary subsystem formed by two M type dwarfs

Until our discovery, it was not known that the distant third body V815 Her B is actually itself an eclipsing close binary, whose members Ba+Bb are late M dwarfs (most probably M1+M5). In other words, the V815 Her system is essentially a 2+2 hierarchical quadruple. Figure 2 shows the *TESS* light curve of the primary and secondary minima of the V815 Her B eclipsing binary subsystem as an example. The light curve extracted from *TESS* Sector 53 observations is folded with the 0.52-day orbital period of the subsystem. The orbital solution without assuming surface inhomogeneities are drawn with grey line, while the model drawn with red line fits also the light curve variations arising from stellar spots on the surface of one or both M dwarf components.

3. Doppler imaging and differential rotation of V815 Her Aa

We performed a Doppler imaging study for the G star in order to reconstruct its spotted surface. For this we used 545 high-resolution optical spectra taken during a 9-month long observing run in 2018. The spectroscopic observations were carried out with the 1.2-m STELLA-II telescope of the STELLA robotic observatory (Strassmeier et al. 2010) at the Izaña Observatory in Tenerife, Spain. From the spectra we created 19 independent Doppler images in time series (adopted stellar parameters are listed in Table 1).

Table 1. Astrophysical parameters for V815 Her Aa.

Spectral type	G6 V
Gaia distance [pc]	32.087 ± 0.127
V_{br} [mag]	7.56
$(B-V)$ [mag]	0.71
M_{bol} [mag]	4.90 ± 0.03
Luminosity [$L_\odot$]	0.87 ± 0.03
T_{eff} [K]	5582 ± 63
$\log g$ (in cgs)	4.27 ± 0.04
$v \sin i$ [kms^{-1}]	30.0 ± 1.5
$P_{rot} = P_{orb}$ [d]	1.80983433
Inclination [°]	75 ± 5
Radius [$R_\odot$]	1.1 ± 0.1
Mass [$M_\odot$]	$\lesssim 1.0$
Microturbulence [kms^{-1}]	1.14 ± 0.19
Macroturbulence [kms^{-1}]	4.2
Metallicity [Fe/H]	0.06 ± 0.03
Age [Myr]	~ 30

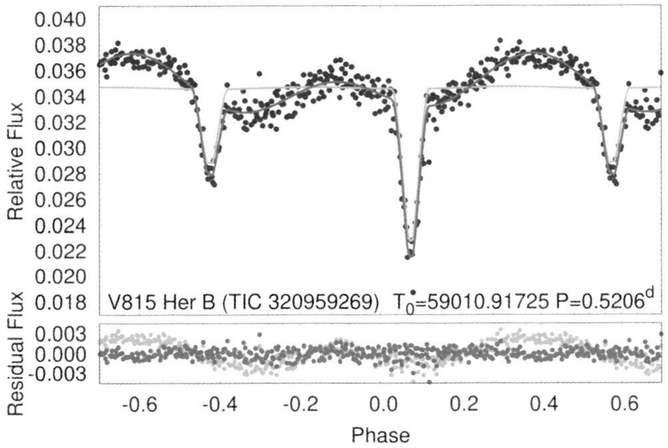

Figure 2. Cleaned and folded *TESS* Sector 53 light curve (blue) for V815 Her B with the orbital solution (grey) and the fitted spot model (red). Below are the residuals.

For the image reconstruction we used the state-of-the-art Doppler imaging code *iMap* (Carroll et al. 2012). The images follow each other at intervals of 11 days on average. As an example, in Figure 3 we show three surface maps, the 8th, the 9th and the 10th from the time series which indicate a constantly changing surface structure on a time scale of a few weeks. The rotational phases are calculated using the orbital period from Fekel et al. (2005) according to the following equation:

$$HJD = 2450204.5802 + 1.80983433 \times E. \tag{1}$$

The latitude dependent surface rotation can usually be inferred from the cross-correlation of two successive Doppler images. Here we use ACCORD (Kővári et al. 2012, 2015), a technique which enables the averaging of latitudinal cross-correlations in the case of a sufficient number of pairs of Doppler images not too distant in time, this way suppressing the effect of randomness and amplifying the correlation pattern attributed to surface differential rotation. In the left panel of Figure 4 we plot the average cross-correlation function map from 18 individual cross-correlations (i.e. for each consecutive image pairs). The figure indicates how much longitude shift occurs at a given latitude due

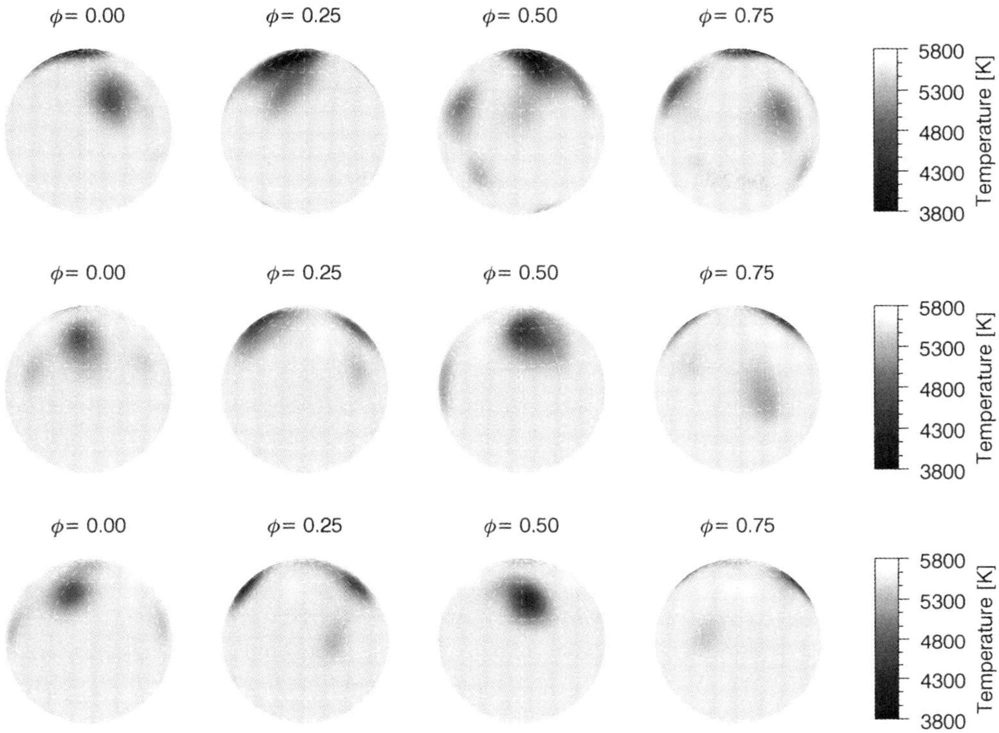

Figure 3. Three time-series Doppler images of V815 Her Aa from 2018.

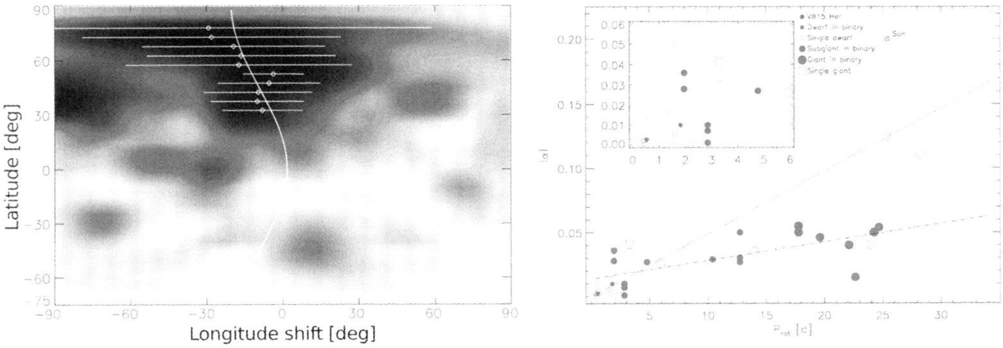

Figure 4. *Left*: The average cross-correlation function map fitted with a solar-type rotation function. *Right*: Surface shear parameter (α) vs. rotation period ($P_{\rm rot}$) graph for active spotted stars.

to surface shear during ~11 days (which is the average time difference between the consecutive Doppler images). The fitted rotation function suggests a weak solar-type surface differential rotation with a dimensionless shear parameter $\alpha = (\Omega_{\rm eq} - \Omega_{\rm pole})/\Omega_{\rm eq} = 0.01$.

In the right panel of Figure 4 we plotted an extended version of the surface shear parameter vs. rotation period graph ($|\alpha|$–$P_{\rm rot}$ graph) for active spotted stars based on Kővári et al. (2017). Open symbols correspond to (effectively) single active stars, grey symbols to active stars in binary (or multiple) systems. Symbol size increases from dwarfs to subgiants to giants. V815 Her Aa is represented by the red dot at $P_{\rm rot}$=1.8 days.

The dotted and dash-dotted lines denote the linear fits for single stars and stars in binary systems, respectively, with slopes of $|\alpha| \propto 0.0049 P_{\rm rot}$[d] and $|\alpha| \propto 0.0014 P_{\rm rot}$[d]. The difference clearly indicates that binarity does play a role in confining differential rotation of the convective envelope.

4. Summary

From the century-long archival photometric data (DASCH) of V815 Her we found a strong signal of a slowly increasing cycle length around 6.5 years on average, close to the 5.73-yr period of the wide (AB) orbit (see Figure 1). At this point we interpret this slowly changing, ∼6.5-year photometric period as the activity cycle of V815 Her Aa, which is perhaps triggered along the eccentric (e=0.77, see Fekel et al. 2005) wide orbit.

We also demonstrated that V815 Her, previously known as a close binary system with a 'third body' in a wide orbit, and later suspected as a *TESS* planet-host candidate, is actually a 2+2 hierarchical quadruple system of two close binaries, one of which (i.e., V815 Her B) is an eclipsing binary (see Figure 2).

We used high-resolution optical spectra from the STELLA robotic observatory taken during the 2018 observing run to create time-series temperature maps (Doppler images) of the spotted surface of the G component V815 Her Aa. Consecutive maps in Figure 3 reflect significant spot activity and spot redistribution on a time scale comparable to the rotation period. From the latitudinal cross-correlation maps we measured a weak solar-type surface differential rotation on the G star. The small shear parameter of α=0.01, equivalent with $\Delta\Omega\approx 2^\circ$/d surface shear, suggests that the differential rotation of the G star is probably suppressed (see Figure 4). This is consistent with our previous finding (Kővári et al. 2017) that tidal forces in close binary systems can indeed suppress the differential rotation of a component with convective envelope.

Acknowledgements Authors from Konkoly Observatory acknowledge the Hungarian National Research, Development and Innovation Office (NKFIH) grants OTKA K-131508 and KKP-143986. LK acknowledges the NKFIH grant OTKA PD-134784. LK and KV are Bolyai János research fellows. KV is supported by the Bolyai+ grant ÚNKP-22-5-ELTE-1093, BS is supported by the ÚNKP-22-3 New National Excellence Program of the Ministry for Culture and Innovation from the source of the National Research, Development and Innovation Fund. STELLA robotic observatory was made possible by funding through the State of Brandenburg (MWFK) and the German Federal Ministry of Education and Research (BMBF). The facility is a collaboration of the AIP in Brandenburg with the IAC in Tenerife.

References

Carroll, T. A., Strassmeier, K. G., Rice, J. B., & Künstler, A. 2012, The magnetic field topology of the weak-lined T Tauri star V410 Tauri. New strategies for Zeeman–Doppler imaging. *A&A*, 548, A95.

Fekel, F. C., Barlow, D. J., Scarfe, C. D., Jancart, S., & Pourbaix, D. 2005, HD 166181 = V815 Herculis, a Single-lined Spectroscopic Multiple System. *AJ*, 129(2), 1001–1007.

Jetsu, L., Hackman, T., Hall, D. S., Henry, G. W., Kokko, M., & You, J. 2000, Time series analysis of V815 Herculis photometry between 1984 and 1998. *A&A*, 362, 223–235.

Kővári, Zs., Korhonen, H., Kriskovics, L., Vida, K., Donati, J. F., Le Coroller, H., Monnier, J. D., Pedretti, E., & Petit, P. 2012, Measuring differential rotation of the K-giant ζ Andromedae. *A&A*, 539, A50.

Kővári, Zs., Kriskovics, L., Künstler, A., Carroll, T. A., Strassmeier, K. G., Vida, K., Oláh, K., Bartus, J., & Weber, M. 2015, Antisolar differential rotation of the K1-giant σ Geminorum revisited. *A&A*, 573, A98.

Kővári, Zs., Oláh, K., Kriskovics, L., Vida, K., Forgács-Dajka, E., & Strassmeier, K. G. 2017, Rotation-differential rotation relationships for late-type single and binary stars from Doppler imaging. *Astronomische Nachrichten*, 338(8), 903–909.

Kolláth, Z. & Oláh, K. 2009, Multiple and changing cycles of active stars. I. Methods of analysis and application to the solar cycles. *A&A*, 501(2), 695–702.

Strassmeier, K. G., Granzer, T., Weber, M., Woche, M., Popow, E., Järvinen, A., Bartus, J., Bauer, S.-M., Dionies, F., Fechner, T., Bittner, W., & Paschke, J. 2010, The STELLA Robotic Observatory on Tenerife. *Advances in Astronomy*, 2010, 970306.

Section 3. Global dynamo. Solar and Stellar Activity Cycles

Invited talks

Dynamics of Solar and Stellar Convection Zones and Atmospheres
Proceedings IAU Symposium No. 365, 2024
A. V. Getling & L. L. Kitchatinov, eds.
doi:10.1017/S1743921323005057

Recent Developments in the Babcock–Leighton Solar Dynamo Theory

Bidya Binay Karak

Department of Physics, Indian Institute of Technology (Banaras Hindu University), Varanasi 221005, India
email: karak.phy@iitbhu.ac.in

Abstract. Babcock–Leighton process, in which the poloidal field is generated through the decay and dispersal of tilted bipolar magnetic regions (BMRs), is observed to be the major process behind the generating poloidal field in the Sun. Based on this process, the Babcock–Leighton dynamo models have been a promising tool for explaining various aspects of solar and stellar magnetic cycles. In recent years, in the toroidal to poloidal part of this dynamo loop, various nonlinear mechanisms, namely the flux loss through the magnetic buoyancy in the formation of BMRs, latitude quenching, tilt quenching, and inflows around BMRs, have been identified. While these nonlinearities tend to produce a stable magnetic cycle, the irregular properties of BMR, mainly the scatter around Joy's law tilt, make a considerable variation in the solar cycle, including grand minima and maxima. After reviewing recent developments in these topics, I end the presentation by discussing the recent progress in making the early prediction of the solar cycle.

Keywords. Solar dynamo, solar cycle, sunspots, solar magnetic field

---

1. Introduction

Many solar-type stars (spectral type: F8V–K2V and effective temperature: 5100–6000 K) show cyclic variations of their magnetic fields (Baliunas et al. 1995; Garg et al. 2019). The (large-scale) magnetic field of our Sun flips every 11 years, and the strength of this field in each cycle varies cycle-to-cycle which is popularly measured by the number of sunspots (Hathaway 2015; Biswas et al. 2023b). While there are variations within the cycle, such as variable rise rate (which is related to the cycle strength and is given by Waldmeier effect; Karak and Choudhuri 2011) and Gnevyshev peaks (Karak et al. 2018), there are variations beyond 11-year periodicity such as the Gnevyshev–Ohl rule, grand minima/maxima and Gleissberg cycle (Usoskin 2023).

It is observed that the magnetic field of the Sun is produced through the dynamo process, which is a cyclic conversion between the toroidal and poloidal fields (Charbonneau 2020). A strong toroidal magnetic field is produced from the poloidal component due to the shearing by the differential rotation in the solar convection zone (SCZ). Due to magnetic buoyancy, this toroidal magnetic field gives rise to bipolar magnetic regions (BMRs) or loosely sunspots. The decay and dispersal of these BMRs give a poloidal magnetic field, which was first proposed by Babcock (1961) and Leighton (1964) and is the heart of surface flux transport (SFT) models (Karak et al. 2014a). While α effect—the lifting and twisting of toroidal field due to helical convection—is another obvious candidate for the generation of poloidal field, the Babcock–Leighton (BL) process is observationally supported and is the major process for generating the poloidal field in the Sun (Cameron and Schüssler 2023). We note that in the BL process, it is the tilt of the BMR which is crucial in generating the poloidal field in the Sun. meridional flow also helps in transporting the field from low to high latitudes (largely by dragging the trailing polarity flux of BMRs), producing the observed pole-ward migration of surface field (Baumann et al. 2004; Mordvinov et al. 2022).

In the basic BL dynamo framework, we expect that the poloidal to toroidal field conversion part is fairly understood due to the following facts. (i) The observed differential rotation in Sun is observed to vary only by a little amount (in the form of torsional oscillation). (ii) Polar field (or its proxy) is linearly correlation with the amplitude of the next cycle (Wang and Sheeley 2009; Kitchatinov & Olemskoy 2011; Muñoz-Jaramillo et al. 2013; Priyal et al. 2014; Kumar et al. 2021b, 2022). (iii) Toroidal flux produced due to shearing of the observed polar flux through the differential rotation matches well with the observation (Cameron and Schüssler 2015). However, the toroidal to poloidal part of the model was less transparent as this process involves some nonlinearity and stochasticity. In recent years, we have made some progress in understanding this part, which I shall highlight in this proceedings.

2. Nonlinearities in BL process

There are several nonlinear processes involved in the BL mechanism for the generation of the poloidal field in the Sun. To explain these, let us begin by observing the butterfly diagram (Figure 9 Hathaway 2015). We find that the strong cycles begin BMRs/sunspots at high latitudes, and thus, the average latitudes of BMR are high for strong cycles and vice versa (Mandal et al. 2017). On the other hand, we find that if a BMR appears at a high latitude, it gives less polar field than the BMR at a low latitude; see Figure 1. Hence, suppose due to fluctuations, if the Sun's magnetic field in a cycle is trying to grow, it will produce BMRs at high latitudes (as observed in the Sun). These high-latitude BMRs will give less polar field. The less polar field will consequently create a less toroidal field for the next cycle, and the indefinite growth of the magnetic field will be halted. This mechanism, so-called latitude quenching as proposed by Petrovay (2020) and Jiang (2020) is shown to be a potential mechanism for stabilizing the magnetic field in the solar dynamo as demonstrated by Karak (2020) in 3D BL dynamo model (Miesch and Dikpati 2014; Miesch and Teweldebirhan 2016; Karak and Miesch 2017).

Strong cycles produce BMRs at higher latitudes, related to a broader feature of the sunspot butterfly diagram, first highlighted by Waldmeier (1955) and recently by Cameron and Schüssler (2016). They show that a strong cycle begins BMRs at higher latitudes; its activity level (sunspot number) rises rapidly (in the activity vs central latitude of the latitudinal distribution of sunspots), and it begins to decline already when the sunspot activity belt is at high latitudes. This feature was explained by Biswas et al. (2022) by including a flux loss due to magnetic buoyancy through the formation of BMRs in their dynamo model (following the idea of Nandy and Choudhuri 2000, including the magnetic buoyancy in axisymmetric dynamo model). They showed that strong cycles produce BMRs at a high rate, and thus, they lose their toroidal flux rapidly. This rapid loss of toroidal flux causes the cycle to decline early when the toroidal belt is at high latitudes.

Another nonlinearity in the BL solar dynamo is the reduction of tilt with the magnetic field, so-called tilt quenching. In the thin flux tube theory of BMR formation, the tilt of a BMR is produced due to the torque induced by the Coriolis force on the diverging flow produced from the apex of the rising flux tube (D'Silva and Choudhuri 1993). Hence, we expect that if the BMR forming flux tube has a strong magnetic field, it will rise quickly, and the Coriolis force will take less time to develop tilt. Thus, we expect the BMR tilt to decrease with the increase of the magnetic field in BMR. Based on this idea, kinematic BL dynamo models use a magnetic field-dependent quenching in the

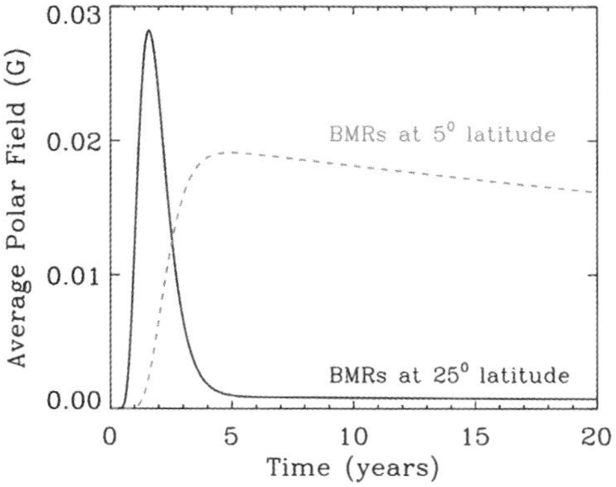

Figure 1. The dashed/red line shows the polar field produced from the decay of two BMRs deposited symmetrically at 5° north and south. Solid/black line represents the same from the same two BMRs but deposited at 5° north and south, and the tilts are assigned according to Joy's law. This plot shows the evidence of latitude quenching—BMRs appearing at high latitude generates less polar field.

tilt (Lemerle & Charbonneau 2017; Karak and Miesch 2017, 2018). However, the observational signature of this tilt quenching is fragile due to limited magnetogram data of strong cycles (Jha et al. 2020); also see Jiao et al. (2021) for the cycle averaged tilt from white-light data.

The final nonlinearity that we shall discuss here is the inflows around BMRs. Observations show converging flows around the BMRs (Gizon et al. 2001; González Hernández et al. 2008). These inflows aggregately generate mean flows around the activity belt (Jiang et al. 2010; Cameron and Schüssler 2012). Due to these flows, the cross-equatorial cancellations of the BMRs are reduced, and the effectivity of the BL process is suppressed. In a strong cycle, this effect is stronger and thus leads to a stabilizing effect in the dynamo (Martin–Belda and Cameron 2017; Nagy et al. 2020; Teweldebirhan et al. 2023).

Besides the above nonlinearities in the toroidal to poloidal part of the dynamo model, the poloidal to toroidal part and the turbulent transport coefficients are also nonlinear. However, due to the fact that the differential rotation is observed to vary only weakly with the solar cycle, and there is a strong correlation between the polar field and the next cycle amplitude, we expect that the poloidal to toroidal part is weakly nonlinear. We refer readers to Section 5 of Karak (2023) for all possible nonlinearities in the solar dynamo.

3. Stochastic effects in BL process

The nonlinearities mentioned above in the BL dynamo model have the tendency to stabilize the magnetic cycle rather than produce irregularity. However, the solar cycle is irregular. Hence, can the BL dynamo models produce the observed modulations in the solar cycle, including grand minima and maxima? The answer is again hidden in the BL process. As mentioned in the introduction, the tilt of BMR plays a vital role in generating a poloidal field in the sun. While this tilt follows Joy's law in a statistical sense, there is a huge scatter around it which can be approximated by a Gaussian of sigma about 19° (Stenflo and Kosovichev 2012; Jha et al. 2020; Sreedevi et al. 2023). Not only the BMR tilt but the BMR eruption rate and flux content also give some variations to the magnetic cycle (Karak and Miesch 2017; Kumar et al. 2023). By including the scatter around Joy's law tilt and variation in the BMR flux and eruption rates Karak and Miesch (2017) showed modulation in the solar cycle, including grand minima and maxima episodes

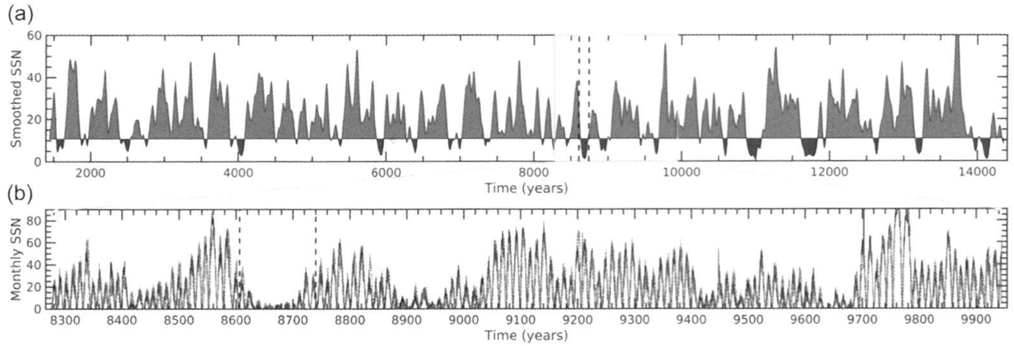

Figure 2. (a) Smoothed (10-year binned and smoothed) BMR number from 3D dynamo model of Karak and Miesch (2018). Blue regions show the locations of grand minima. (b) shows the zoomed-in view of some cycles. A Maunder-like extended grand minimum is highlighted by dashed lines. Note that the model successfully recovers from grand minima without needing any additional source for the poloidal field.

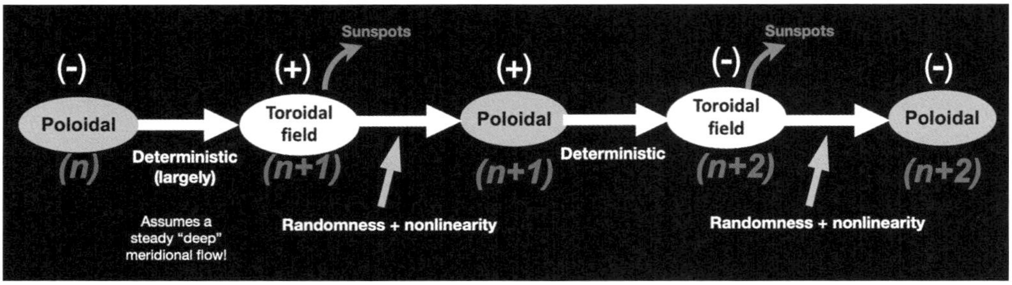

Figure 3. Dynamo chain under the BL framework.

(Figure 2); also see Lemerle & Charbonneau (2017) for a similar work. Several previous studies were also performed to model the grand minima using axisymmetric dynamo models where the BMR scatter was captured by including noise in the BL source term (e.g., Choudhuri and Karak 2009, 2012; Olemskoy and Kitchatinov 2013; Karak and Choudhuri 2013; Passos et al. 2014; Nagy et al. 2017). Double peaks (Gnevyshev peaks) were also modelled using a similar idea (Karak et al. 2018). Recently, Golubeva et al. (2023); Biswas et al. (2023) also showed that the scatter around BMRs, particularly the anti-Hale BMRs, produces considerable variation in the polar field strength and the reversal times.

Another potential source for the irregularity in the solar cycle is the variation in the meridional circulation. Surface observations show a noticeable temporal variation of it over the last few solar cycles (e.g., González Hernández et al. 2008). However, the variation in the deep meridional circulation is uncertain due to the difficulties in the measurement. Karak (2010) have shown that a weak meridional flow makes a cycle long and magnetic field weak (due to getting more time to diffuse the field) in the high diffusivity dynamo model. He showed that a large part of the amplitude is also matched by modelling the cycle period by varying the meridional flow speed. Later, Karak and Choudhuri (2011) showed that a variable meridional circulation helps in modelling the classical Waldmeier effect. A large reduction in the meridional flow also allows the model to push into a grand minimum phase (Karak 2010; Karak and Choudhuri 2012, 2013). SFT models also show that variation in the meridional flow cause variation in the polar field (Upton and Hathaway 2014; Biswas et al. 2023).

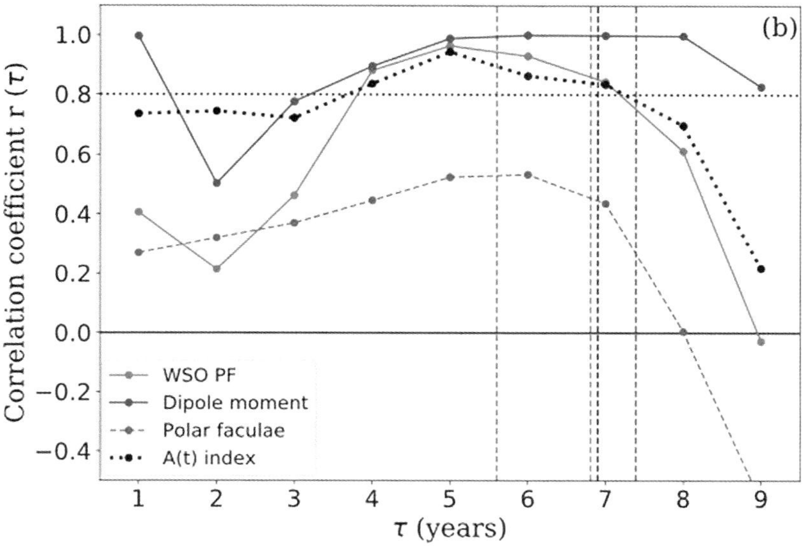

Figure 4. Linear correlation coefficient between the amplitude of the next sunspot cycle and the polar field or its proxy at different times τ measured from the times of reversals. Dashed lines show the time difference between the corresponding time of polar field reversal and the cycle minimum. Figure reproduced from Kumar et al. (2021b).

4. Early prediction of solar cycle

We have seen that the solar dynamo has a stochastic component, and it is nonlinear. So the question is, can we then make a prediction of the future cycle? To answer this question, let us recall the dynamo chain of the solar cycle as shown in Figure 3. We observe that the generation of the poloidal field involves some randomness, and thus, at the end of every cycle, the strength of the poloidal field varies from cycle to cycle. However, the poloidal to toroidal part of the dynamo is deterministic (as the differential rotation in the Sun does not change much with time). The toroidal magnetic field gives sunspots for the next cycle. In fact, this is the basis for producing the observed correlation between the poloidal field (or its proxy) and the next sunspot cycle. Hence, if we can measure the poloidal magnetic field, we can predict the amplitude of the next cycle. Indeed, a part of the poloidal field, namely the radial field on the solar surface, is observed, and it becomes maximum near the solar minimum. Prediction works use the polar field at the solar minimum to predict the amplitude of the next cycle either by using the linear correlation between these two or by feeding the polar field in the dynamo model (Choudhuri et al. 2007; Jiang et al. 2007). As the polar field peaks around the cycle minimum, most of the prediction methods use the polar field at the cycle minimum to predict the amplitude of the next cycle (Bhowmik et al. 2023). However, the question is, 'Do we really need to wait till the solar minimum or a reasonable prediction can be made sometime earlier?' Kumar et al. (2021b) have shown that the correlation between the amplitude of the next cycle and the polar field just after four years becomes strong; see Figure 4. This suggests that a reasonable prediction of the amplitude of the next solar cycle can be made just after four years of the reversal, which is about 2–3 years before the solar minimum (the usual landmark of solar cycle prediction). Later Kumar et al. (2022) have shown that instead of computing the polar field at a fixed time, if we compute the rise rate (slope) of the polar field within the first three years of the polar field after the reversal, then a better and even early prediction can be made. They further showed that the rise rate of the polar field determines its eventual peak (Figure 5a)—it is like the Waldmeier effect

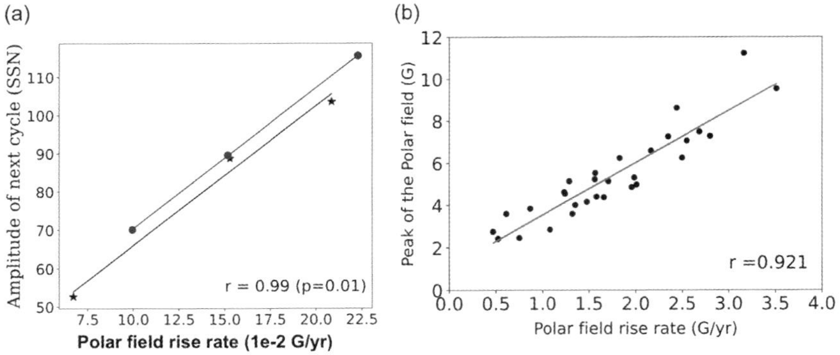

Figure 5. (a) Correlation between the observed rise rate of the polar field with the amplitude of the next sunspot cycle (Kumar et al. 2022). (b) Same as (a) but obtained from an SFT model (Biswas et al. 2023).

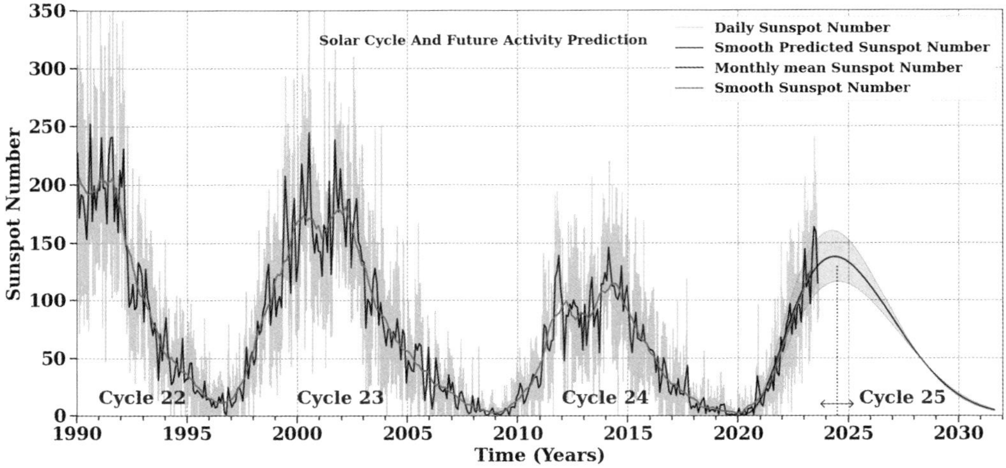

Figure 6. Prediction of the Cycle 25 using the rise rate of the polar field during three years after reversal (Kumar et al. 2022). The red-shaded region shows the prediction window, and the horizontal arrow shows the range of expected peak time.

in the polar field. While their initial inference was based on the limited observed data of the polar field, later Biswas et al. (2023) demonstrated the robustness of this feature using SFT model by feeding the synthetic solar cycle with anti-Hale and non-Joy BMRs; see Figure 5(b). Following this idea, Kumar et al. (2022) forecasted the solar cycle 25 with predicted peak sunspot number: 137 ± 23; see Figure 6.

Finally, the question is, if we can predict the amplitude of the next cycle, can we extend this further to multiple cycles? The simple answer is no. As seen from Figure 3, the toroidal to poloidal part of the solar dynamo diminishes the memory of the polar field due to randomness and nonlinearity involved in this step. Kumar et al. (2021a) have shown that only when the dynamo operates near the critical transition, the nonlinearity is weak, the dynamo growth rate is weak, and the memory of the polar field can be propagated to multiple following cycles. With the increase of supercriticality of the solar dynamo, the memory of the polar field reduces, and in a highly supercritical regime, the memory is completely washed out; see Figure 7 for an illustration. In conclusion, a reliable prediction of the solar cycle amplitude beyond one cycle is impossible. However, a hint of the amplitude of cycle $n+2$ can be obtained from the polar field of cycle n

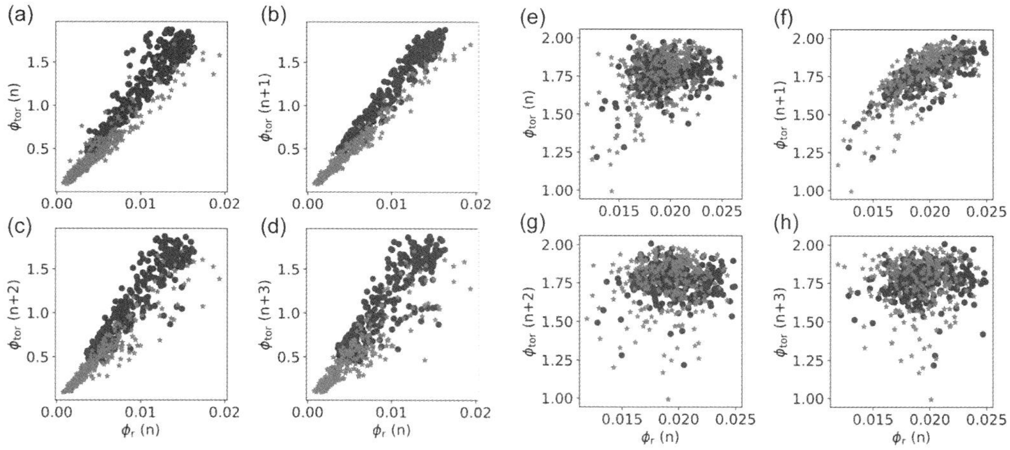

Figure 7. (a–d): Correlations between the polar field of cycle n with the toroidal flux of cycle n, $n+1$, $n+2$, and $n+3$, respectively, when the dynamo operates near-critical regime (two times the critical dynamo number). (e–f) Same as (a–d) but when the model operates at a highly supercritical regime (four times critical); for details, see Kumar et al. (2021a) from where this figure is adopted.

as some studies suggest the solar dynamo is operating in weakly supercritical region (Karak et al. 2015; Kitchatinov and Nepomnyashchikh 2017; Cameron and Schüssler 2017; Vashishth et al. 2021; Tripathi et al. 2021; Albert et al. 2021; Vashishth et al. 2023).

5. Concluding remarks and future outlook

In recent years, we have had a tremendous development in the BL dynamo theory for modelling the various aspects of the solar magnetic cycle. It has been shown that the toroidal to poloidal field part of the BL dynamo model includes some nonlinearities, which at least include the flux loss due to magnetic buoyancy during the formation of BMR, latitude quenching, tilt quenching, and magnetic field dependent inflows around BMRs. These nonlinearities tend to stabilize the dynamo rather than produce a considerable variation in the magnetic cycle. Stochastic properties in the BMRs, namely, the scatter in the BMR tilt around Joy's law, can produce large variations in the solar cycle, including grand minima and maxima. Despite these nonlinearities and stochasticity in the solar dynamo, a prediction of the amplitude of the next solar cycle can be made using the polar field at the solar minimum. We have shown that we indeed do not have to wait till the minimum of the cycle; we can make a prediction of the next cycle's strength just by measuring the rise rate of the polar field of the first three years' data after its reversal. This allows us to make an early prediction of the solar cycle—about three years before the solar minimum, the usual landmark of the forecast. Prediction of multiple cycles is not possible because the memory of the polar field degrades in the poloidal to toroidal part of the dynamo loop. How rapidly the memory degrades beyond one cycle is determined by the supercriticality of the dynamo.

We end the discussion by commenting on the operation of BL dynamo during Maunder-like grand minima. It has been suspected that the generation of the poloidal field through BL process during these episodes is not possible due to lack of sunspots. The obvious candidate for this is the classical α effect (Parker 1955). Some studies have included α effect in the BL dynamo models to recover the models from these quiescent episodes (Karak and Choudhuri 2013; Hazra et al. 2014; Ölçek et al. 2019). However, recent observations show that the Maunder minimum was not as deep as it was thought earlier (e.g., Usoskin et al. 2015; Zolotova and Ponyavin 2016). Also, the small sunspots and the BMRs, which do not appear as sunspots in today's telescope (Jha et al. 2020), were not observed during Maunder minimum. These BMRs also show a systematic tilt and Joy's law (Jha et al. 2020), which can generate some poloidal field during Maunder-like grand minimum as demonstrated by Karak and Miesch (2018). The downward turbulent pumping helps in reducing the diffusion of the poloidal field (Karak and Cameron 2016), and thus, even a weak poloidal field produced through the decay of a few BMRs during the grand minimum can successfully recover the Sun to normal phase. Another comment about the operation of BL dynamo in other solar-type stars. Possibly, the BMRs are also produced in other stars, and they are tilted. With the increase of rotation rate, both the tilt and latitude of the emergence of BMR increase (Schuessler and Solanki 1992). However, the tilt cannot increase beyond 90°. This can cause saturation of magnetic field in rapidly rotating stars (Kitchatinov and Olemskoy 2016). The increase in the latitude of emergence can cause a decrease in the magnetic field (latitude quenching; Jiang 2020; Karak 2020). Karak et al. (2014); Hazra et al. (2019); Karak et al. (2020); Vashishth et al. (2021); Vashishth et al. (2023) have utilized BL dynamo models to explain some aspects of the stellar magnetic cycles, including the saturation of the magnetic field in rapidly rotating stars and the variation of the number of grand minima and cycle variability with the rotation rates of stars. We hope future observations will help to validate the BL dynamo theory in other stars.

6. Acknowledgements

Author thanks the International Astronomical Union for the travel grant to attend the IAU Symposium 365 in Yerevan, Armenia. The partial support from the Ramanujan Fellowship (project no SB/S2/RJN-017/2018) is gratefully acknowledged.

References

Albert, C., Ferriz–Mas, A., Gaia, F., & Ulzega, S. 2021, Can Stochastic Resonance Explain Recurrence of Grand Minima? *Astrophys. J*, 916(2), L9.

Babcock, H. W. 1961, The Topology of the Sun's Magnetic Field and the 22-YEAR Cycle. *Astrophys. J*, 133, 572.

Baliunas, S. L., Donahue, R. A., Soon, W. H., Horne, J. H., Frazer, J., Woodard–Eklund, L., Bradford, M., Rao, L. M., Wilson, O. C., Zhang, Q., Bennett, W., Briggs, J., Carroll, S. M., Duncan, D. K., Figueroa, D., Lanning, H. H., Misch, T., Mueller, J., Noyes, R. W., Poppe, D., Porter, A. C., Robinson, C. R., Russell, J., Shelton, J. C., Soyumer, T., Vaughan, A. H., & Whitney, J. H. 1995, Chromospheric variations in main-sequence stars. *Astrophys. J*, 438, 269–287.

Baumann, I., Schmitt, D., Schüssler, M., & Solanki, S. K. 2004, Evolution of the large-scale magnetic field on the solar surface: A parameter study. *Astrophys. J*, 426, 1075–1091.

Bhowmik, P., Jiang, J., Upton, L., Lemerle, A., & Nandy, D. 2023, Physical Models for Solar Cycle Predictions. *Space Sci. Rev.*, 219(5), 40.

Biswas, A., Karak, B. B., & Cameron, R. 2022, Toroidal Flux Loss due to Flux Emergence Explains why Solar Cycles Rise Differently but Decay in a Similar Way. *Phys. Rev. Lett.*, 129(24), 241102.

Biswas, A., Karak, B. B., & Kumar, P. 2023,a Exploring the reliability of polar field rise rate as a precursor for an early prediction of solar cycle. *MNRAS*, 526a(3), 3994–4003.

Biswas, A., Karak, B. B., Usoskin, I., & Weisshaar, E. 2023,b Long-Term Modulation of Solar Cycles. *Space Sci. Rev.*, 219b(3), 19.

Cameron, R. & Schüssler, M. 2015, The crucial role of surface magnetic fields for the solar dynamo. *Science*, 347, 1333–1335.

Cameron, R. H. & Schüssler, M. 2012, Are the strengths of solar cycles determined by converging flows towards the activity belts? *Astrophys. J*, 548, A57.

Cameron, R. H. & Schüssler, M. 2016, The turbulent diffusion of toroidal magnetic flux as inferred from properties of the sunspot butterfly diagram. *Astrophys. J*, 591, A46.

Cameron, R. H. & Schüssler, M. 2017, Understanding Solar Cycle Variability. *Astrophys. J*, 843(2), 111.

Cameron, R. H. & Schüssler, M. 2023, Observationally Guided Models for the Solar Dynamo and the Role of the Surface Field. *Space Sci. Rev.*, 219(7), 60.

Charbonneau, P. 2020, Dynamo models of the solar cycle. *Living Reviews in Solar Physics*, 17(1), 4.

Choudhuri, A. R., Chatterjee, P., & Jiang, J. 2007, Predicting Solar Cycle 24 With a Solar Dynamo Model. *Physical Review Letters*, 98(13), 131103.

Choudhuri, A. R. & Karak, B. B. 2009, A possible explanation of the Maunder minimum from a flux transport dynamo model. *Res. Astron. Astrophys.*, 9, 953–958.

Choudhuri, A. R. & Karak, B. B. 2012, Origin of Grand Minima in Sunspot Cycles. *Phys. Rev. Lett.*, 109(17), 171103.

D'Silva, S. & Choudhuri, A. R. 1993, A theoretical model for tilts of bipolar magnetic regions. *Astrophys. J*, 272, 621.

Garg, S., Karak, B. B., Egeland, R., Soon, W., & Baliunas, S. 2019, Waldmeier Effect in Stellar Cycles. *Astrophys. J*, 886(2), 132.

Gizon, L., Duvall, T. L., J., & Larsen, R. M. Probing Surface Flows and Magnetic Activity with Time-Distance Helioseismology. In Brekke, P., Fleck, B., & Gurman, J. B., editors, *Recent Insights into the Physics of the Sun and Heliosphere: Highlights from SOHO and Other Space Missions* 2001,, volume 203, 189.

Golubeva, E. M., Biswas, A., Khlystova, A. I., Kumar, P., & Karak, B. B. 2023, Probing the variations in the timing of the Sun's polar magnetic field reversals through observations and surface flux transport simulations. *MNRAS*, 525(2), 1758–1768.

González Hernández, I., Kholikov, S., Hill, F., Howe, R., & Komm, R. 2008, Subsurface Meridional Circulation in the Active Belts. *Sol Phys*, 252(2), 235–245.

Hathaway, D. H. 2015, The Solar Cycle. *Living Reviews in Solar Physics*, 12(1), 4.

Hazra, G., Jiang, J., Karak, B., & Kitchatinov, L. 2019, Stellar Cycles with Dynamo. *Astrophys. J, submitted*, 000, 00.

Hazra, S., Passos, D., & Nandy, D. 2014, A Stochastically Forced Time Delay Solar Dynamo Model: Self-consistent Recovery from a Maunder-like Grand Minimum Necessitates a Mean-field Alpha Effect. *Astrophys. J*, 789, 5.

Jha, B. K., Karak, B. B., Mandal, S., & Banerjee, D. 2020, Magnetic Field Dependence of Bipolar Magnetic Region Tilts on the Sun: Indication of Tilt Quenching. *Astrophys. J*, 889(1), L19.

Jiang, J. 2020, Nonlinear Mechanisms that Regulate the Solar Cycle Amplitude. *Astrophys. J*, 900(1), 19.

Jiang, J., Chatterjee, P., & Choudhuri, A. R. 2007, Solar activity forecast with a dynamo model. *MNRAS*, 381, 1527–1542.

Jiang, J., Işik, E., Cameron, R. H., Schmitt, D., & Schüssler, M. 2010, The Effect of Activity-related Meridional Flow Modulation on the Strength of the Solar Polar Magnetic Field. *Astrophys. J*, 717(1), 597–602.

Jiao, Q., Jiang, J., & Wang, Z.-F. 2021, Sunspot tilt angles revisited: Dependence on the solar cycle strength. *Astrophys. J*, 653, A27.

Karak, B. B. 2010, Importance of Meridional Circulation in Flux Transport Dynamo: The Possibility of a Maunder-like Grand Minimum. *Astrophys. J*, 724, 1021–1029.

Karak, B. B. 2020, Dynamo Saturation through the Latitudinal Variation of Bipolar Magnetic Regions in the Sun. *Astrophys. J*, 901(2), L35.

Karak, B. B. 2023, Models for the long-term variations of solar activity. *Living Reviews in Solar Physics*, 20(1), 3.

Karak, B. B. & Cameron, R. 2016, Babcock–Leighton Solar Dynamo: The Role of Downward Pumping and the Equatorward Propagation of Activity. *Astrophys. J*, 832, 94.
Karak, B. B. & Choudhuri, A. R. 2011, The Waldmeier effect and the flux transport solar dynamo. *MNRAS*, 410, 1503–1512.
Karak, B. B. & Choudhuri, A. R. 2012, Quenching of Meridional Circulation in Flux Transport Dynamo Models. *Sol Phys*, 278, 137–148.
Karak, B. B. & Choudhuri, A. R. 2013, Studies of grand minima in sunspot cycles by using a flux transport solar dynamo model. *Res. Astron. Astrophys.*, 13, 1339–1357.
Karak, B. B., Jiang, J., Miesch, M. S., Charbonneau, P., & Choudhuri, A. R. 2014,a Flux Transport Dynamos: From Kinematics to Dynamics. *Space Sci. Rev.*, 186a, 561–602.
Karak, B. B., Kitchatinov, L. L., & Brandenburg, A. 2015, Hysteresis between Distinct Modes of Turbulent Dynamos. *Astrophys. J*, 803, 95.
Karak, B. B., Kitchatinov, L. L., & Choudhuri, A. R. 2014,b A Dynamo Model of Magnetic Activity in Solar-like Stars with Different Rotational Velocities. *Astrophys. J*, 791b, 59.
Karak, B. B., Mandal, S., & Banerjee, D. 2018, Double Peaks of the Solar Cycle: An Explanation from a Dynamo Model. *Astrophys. J*, 866(1), 17.
Karak, B. B. & Miesch, M. 2017, Solar Cycle Variability Induced by Tilt Angle Scatter in a Babcock–Leighton Solar Dynamo Model. *Astrophys. J*, 847, 69.
Karak, B. B. & Miesch, M. 2018, Recovery from Maunder-like Grand Minima in a Babcock–Leighton Solar Dynamo Model. *Astrophys. J*, 860, L26.
Karak, B. B., Tomar, A., & Vashishth, V. 2020, Stellar dynamos with solar and antisolar differential rotations: Implications to magnetic cycles of slowly rotating stars. *MNRAS*, 491(3), 3155–3164.
Kitchatinov, L. & Nepomnyashchikh, A. 2017, How supercritical are stellar dynamos, or why do old main-sequence dwarfs not obey gyrochronology? *MNRAS*, 470(3), 3124–3130.
Kitchatinov, L. L. & Olemskoy, S. V. 2011, Does the Babcock–Leighton mechanism operate on the Sun? *Astronomy Letters*, 37, 656–658.
Kitchatinov, L. L. & Olemskoy, S. V. 2016, Dynamo model for grand maxima of solar activity: can superflares occur on the Sun? *MNRAS*, 459(4), 4353–4359.
Kumar, P., Biswas, A., & Karak, B. B. 2022, Physical link of the polar field buildup with the Waldmeier effect broadens the scope of early solar cycle prediction: Cycle 25 is likely to be slightly stronger than Cycle 24. *MNRAS*, 513(1), L112–L116.
Kumar, P., Karak, B. B., & Sreedevi, A. 2023, . *MNRAS*, submitted.
Kumar, P., Karak, B. B., & Vashishth, V. 2021,a Supercriticality of the Dynamo Limits the Memory of the Polar Field to One Cycle. *Astrophys. J*, 913a(1), 65.
Kumar, P., Nagy, M., Lemerle, A., Karak, B. B., & Petrovay, K. 2021,b The Polar Precursor Method for Solar Cycle Prediction: Comparison of Predictors and Their Temporal Range. *Astrophys. J*, 909b(1), 87.
Leighton, R. B. 1964, Transport of Magnetic Fields on the Sun. *Astrophys. J*, 140, 1547.
Lemerle, A. & Charbonneau, P. 2017, A Coupled 2 × 2D Babcock–Leighton Solar Dynamo Model. II. Reference Dynamo Solutions. *Astrophys. J*, 834, 133.
Mandal, S., Karak, B. B., & Banerjee, D. 2017, Latitude Distribution of Sunspots: Analysis Using Sunspot Data and a Dynamo Model. *Astrophys. J*, 851, 70.
Martin–Belda, D. & Cameron, R. H. 2017, Inflows towards active regions and the modulation of the solar cycle: A parameter study. *Astrophys. J*, 597, A21.
Miesch, M. S. & Dikpati, M. 2014, A Three-dimensional Babcock–Leighton Solar Dynamo Model. *Astrophys. J*, 785, L8.
Miesch, M. S. & Teweldebirhan, K. 2016, A three-dimensional Babcock–Leighton solar dynamo model: Initial results with axisymmetric flows. *Advances in Space Research*, 58(8), 1571–1588.
Mordvinov, A. V., Karak, B. B., Banerjee, D., Golubeva, E. M., Khlystova, A. I., Zhukova, A. V., & Kumar, P. 2022, Evolution of the Sun's activity and the poleward transport of remnant magnetic flux in Cycles 21-24. *MNRAS*, 510(1), 1331–1339.

Muñoz-Jaramillo, A., Dasi–Espuig, M., Balmaceda, L. A., & DeLuca, E. E. 2013, Solar Cycle Propagation, Memory, and Prediction: Insights from a Century of Magnetic Proxies. *Astrophys. J*, 767, L25.

Nagy, M., Lemerle, A., & Charbonneau, P. 2020, Impact of nonlinear surface inflows into activity belts on the solar dynamo. *Journal of Space Weather and Space Climate*, 10, 62.

Nagy, M., Lemerle, A., Labonville, F., Petrovay, K., & Charbonneau, P. 2017, The Effect of "Rogue" Active Regions on the Solar Cycle. *Sol Phys*, 292, 167.

Nandy, D. & Choudhuri, A. R. 2000, The Role of Magnetic Buoyancy in a Babcock–Leighton Type Solar Dynamo. *Journal of Astrophysics and Astronomy*, 21, 381.

Ölçek, D., Charbonneau, P., Lemerle, A., Longpré, G., & Boileau, F. 2019, Grand Activity Minima and Maxima via Dual Dynamos. *Sol Phys*, 294(7), 99.

Olemskoy, S. V. & Kitchatinov, L. L. 2013, Grand Minima and North-South Asymmetry of Solar Activity. *Astrophys. J*, 777, 71.

Parker, E. N. 1955, Hydromagnetic Dynamo Models. *Astrophys. J*, 122, 293.

Passos, D., Nandy, D., Hazra, S., & Lopes, I. 2014, A solar dynamo model driven by mean-field alpha and Babcock–Leighton sources: fluctuations, grand-minima-maxima, and hemispheric asymmetry in sunspot cycles. *Astrophys. J*, 563, A18.

Petrovay, K. 2020, Solar cycle prediction. *Living Reviews in Solar Physics*, 17(1), 2.

Priyal, M., Banerjee, D., Karak, B. B., Muñoz-Jaramillo, A., Ravindra, B., Choudhuri, A. R., & Singh, J. 2014, Polar Network Index as a Magnetic Proxy for the Solar Cycle Studies. *Astrophys. J*, 793, L4.

Schuessler, M. & Solanki, S. K. 1992, Why rapid rotators have polar spots. *Astrophys. J*, 264, L13–L16.

Sreedevi, A., Jha, B. K., Karak, B. B., & Banerjee, D. 2023, AutoTAB: Automatic Tracking Algorithm for Bipolar Magnetic Regions. *Astrophys. J Supplement Series*, 268(2), 58.

Stenflo, J. O. & Kosovichev, A. G. 2012, Bipolar Magnetic Regions on the Sun: Global Analysis of the SOHO/MDI Data Set. *Astrophys. J*, 745, 129.

Teweldebirhan, K., Miesch, M., & Gibson, S. 2023, Inflows towards Bipolar Magnetic Active Regions and Their Nonlinear Impact on a Three-Dimensional Babcock–Leighton Solar Dynamo Model. *arXiv e-prints*,, arXiv:2310.00738.

Tripathi, B., Nandy, D., & Banerjee, S. 2021, Stellar mid-life crisis: subcritical magnetic dynamos of solar-like stars and the breakdown of gyrochronology. *MNRAS*, 506(1), L50–L54.

Upton, L. & Hathaway, D. H. 2014, Effects of Meridional Flow Variations on Solar Cycles 23 and 24. *Astrophys. J*, 792(2), 142.

Usoskin, I. G. 2023, A history of solar activity over millennia. *Living Reviews in Solar Physics*, 20(1), 2.

Usoskin, I. G., Arlt, R., Asvestari, E., Hawkins, E., Käpylä, M., Kovaltsov, G. A., Krivova, N., Lockwood, M., Mursula, K., O'Reilly, J., Owens, M., Scott, C. J., Sokoloff, D. D., Solanki, S. K., Soon, W., & Vaquero, J. M. 2015, The Maunder minimum (1645-1715) was indeed a grand minimum: A reassessment of multiple datasets. *Astrophys. J*, 581, A95.

Vashishth, V., Karak, B. B., & Kitchatinov, L. 2021, Subcritical dynamo and hysteresis in a Babcock–Leighton type kinematic dynamo model. *Research in Astronomy and Astrophysics*, 21(10), 266.

Vashishth, V., Karak, B. B., & Kitchatinov, L. 2023, Dynamo modelling for cycle variability and occurrence of grand minima in Sun-like stars: rotation rate dependence. *MNRAS*, 522(2), 2601–2610.

Waldmeier, M. 1955, Ergebnisse und Probleme der Sonnenforschung. *Ergebnisse und Probleme der Sonnenforschung (Leipzig: Geest & Portig)*,.

Wang, Y.-M. & Sheeley, N. R. 2009, Understanding the Geomagnetic Precursor of the Solar Cycle. *Astrophys. J*, 694, L11–L15.

Zolotova, N. V. & Ponyavin, D. I. 2016, How Deep Was the Maunder Minimum? *Sol Phys*, 291, 2869–2890.

Dynamics of Solar and Stellar Convection Zones and Atmospheres
Proceedings IAU Symposium No. 365, 2024
A. V. Getling & L. L. Kitchatinov, eds.
doi:10.1017/S1743921324000097

Nonlinear and stochastic mechanisms of the solar cycle and their implications for the cycle prediction

Jie Jiang

School of Space and Environment, Beihang University, Beijing, People's Republic of China,
email: jiejiang@buaa.edu.cn
Key Laboratory of Space Environment Monitoring and Information Processing, Ministry of Industry and Information Technology, Beijing, China

Abstract. Solar activity shows an 11-year (quasi)periodicity with a pronounced, but limited variability of the cycle amplitudes. The properties of active region (AR) emergence play an important role in the modulation of solar cycles and are our central concern in building a model for predicting future cycle(s) in the framework of the Babcock–Leighton (BL)-type dynamo. The emergence of ARs has the property that strong cycles tend to have higher mean latitudes and lower tilt angle coefficients. Their non-linear feedbacks on the solar cycle are referred to as latitudinal quenching and tilt quenching, respectively. Meanwhile, the stochastic properties of AR emergence, e.g., rogue ARs, limit the scope of the solar cycle prediction. For physics-based prediction models of the solar cycle, we suggest that uncertainties in both the observed magnetograms assimilated as the initial field and the properties of the AR emergence should be taken into account.

Keywords. Solar cycle, Solar dynamo, Cycle prediction, Stochastic noise

1. Introduction

Predicting the solar cycle is crucial in our technological society. Many attempts were made to predict cycle 25, including empirical and physical methods to determine its amplitude and cycle profile. There are also several comprehensive reviews about the solar cycle prediction including predictions of cycle 25, e.g., Petrovay (2020); Nandy (2021); Jiang et al. (2023); Bhowmik et al. (2023); Penza et al. (2023). Here no additional cycle 25 predictions or reviews are required. Figure 1 displays the time evolution of the polar field (upper panels) and longitudinally averaged surface field (lower panels) from Wilcox Solar Observatory (WSO, left panels) and SDO/HMI (right panels) data. Both datasets indicate that the polar field reversal began in mid-2023. As of the end of 2023, both the northern and southern polar regions have the magnetic field polarity of cycle 25. The polar reversal indicates that cycle 25 has reached its maximum phase since mid-2023. The maximum phase usually persists about two years. The amplitude of the ongoing cycle 25 is about 125. The result is in agreement with what we once suggested when we investigated the predictability of the solar cycle (Jiang et al. 2018).

Physics-based predictions of the solar cycle provide an effective means of verifying our understanding of it. Numerical weather forecasting provides a good model for physics-based solar cycle predictions. In the 1960s, Lorenz demonstrated the significance of nonlinear dynamics in the context of convecting systems, opening the door to physics-based weather forecasting. (Lorenz 1963, 1965, 1969). Lorenz's discovery regarding the predictability of weather showed that chaotic dynamics result from nonlinearity. It is

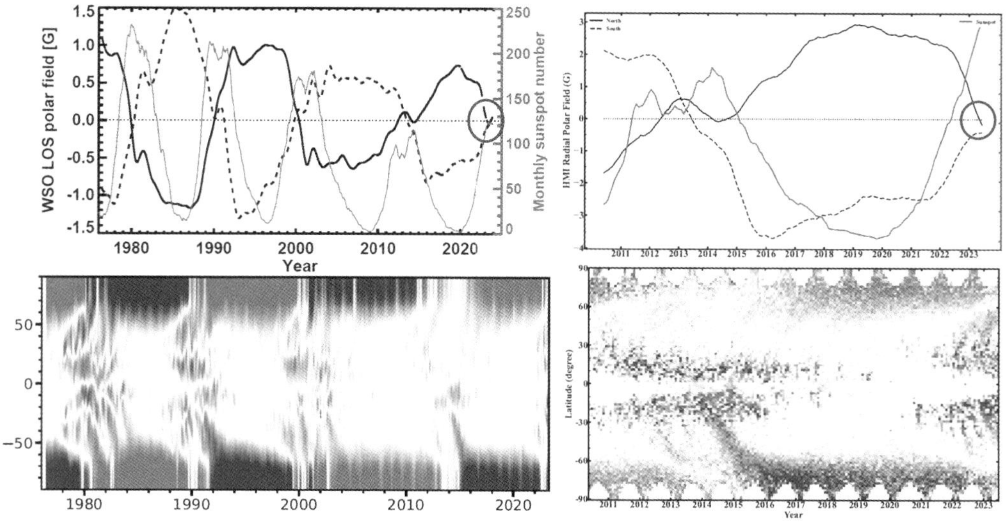

Figure 1. Illustration of the recent polar field reversals with data from WSO (left panels) and from HMI/SDO (right panels). The red curves shown in the top panels are the 13-months smoothed sunspot numbers (version 2.0). Black solid and dashed curves are the northern and southern polar fields, respectively. WSO (left) and HMI (right) polar field is defined as the average of the line-of-sight field between about 55° latitude and the poles and the average of the radial field between 60° and 70° latitudes,respectively. The lower panels show the longitudinal averaged solar surface field during 1976-2023 (left panel) and during 2010-2023 (right panel). Courtesy of the datasets from WSO, HMI/SDO, and SIDC sunspot number.

now known that the atmosphere's chaotic nature limits weather forecasts to about two weeks, even with ideal models and perfect observations.

The sunspot number time series shows the variability of solar cycles, including the odd-even effect, also known as the Gnevyshev–Ohl (G-O) rule (Gnevyshev and Ohl 1948). This pattern is characterized by alternating higher and lower than average solar cycle amplitudes observed in the sunspot number record. However, some cycles do not follow this rule. The variability of the solar cycle indicates that there are nonlinear or stochastic mechanisms that modulate solar cycle amplitudes. However, it is unclear which mechanism is dominant. The investigation of the G-O rule could provide insights into the nature of solar cycle variability. Understanding the nature of solar cycle variability is essential for predicting solar cycles accurately.

2. Understanding the nature of solar cycle variability

There is increasing evidence to support the idea that the solar dynamo is of the Babcock–Leighton (BL) type, as proposed by (Babcock 1961; Leighton 1969). According to the BL-type models, sunspot-generating surface eruptions of the toroidal magnetic field are the source of the poloidal field. The dynamo loop corresponds to the regeneration of the poloidal and toroidal fields. The toroidal field is generated from the poloidal field due to differential rotation. The observed torsional oscillation is approximately 0.5% of the average rotation rate (Howe 2009). Therefore, the process of the toroidal field generation can be regarded as a linear one, which is also supported by the correlation between the polar field at cycle minimum and the subsequent cycle strength (Jiang et al. 2007; Wang and Sheeley 2009). The generation of the poloidal field from the toroidal field by the BL processes is expected to include nonlinear and stochastic mechanisms. The emergence and evolution of ARs can be observed. Given that the generation of the

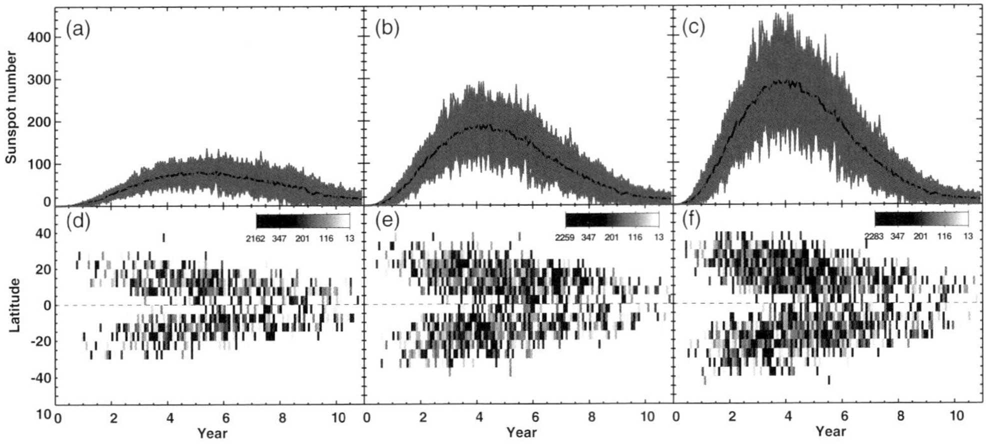

Figure 2. Examples of synthetic solar cycles of different amplitudes S_n. Top: Time evolution of the synthetic sunspot number. Red curves show the 100 simulated random realizations. Black indicates the expected sunspot number value at each time. The blue curve marks one of the random realizations. Bottom: Time evolution of the sunspot emergence latitude for the cases shown using blue curves. The color displays the average area in unit of MSH in bins of one month by 5 degrees in latitude. From left to right S_n=70, S_n=180, and S_n=280, respectively. Image reproduced with permission from Jiang (2020), copyright by the American Astronomical Society.

toroidal field is a linear process, understanding solar variability may only require processes over the solar surface with well-measured empirical relations for AR emergence and evolution. An interior dynamo model is not particularly useful in understanding the nature of solar cycle variability at present since the process of flux emergence remains an outstanding question.

ARs are characterized by the tilt of the line connecting their two polarities, which typically deviates from the east-west direction. This tilt angle introduces the poloidal component. The poloidal field contribution, usually measured by the quality axial dipole moment, is roughly proportional to the flux and tilt of the sunspot upon emergence (Wang and Sheeley 1991). However, it is important to note that surface flux transport processes that occur after the emergence of sunspot groups have the potential to significantly alter the dipole moment (Jiang et al. 2014; Petrovay et al. 2020; Wang et al. 2021). These processes can be simulated using surface flux transport models (Jiang et al. 2014; Yeates et al. 2023). This raises the question of whether observable properties of sunspot emergence are subject to nonlinear and stochastic mechanisms that modulate solar cycles.

Now let's first see the systematic property of the sunspot emergence. Past studies shows a good correlation between the mean latitude of sunspot emergence during a cycle and the cycle strength (Li et al. 2003; Solanki et al. 2008; Jiang et al. 2011). Strong cycles tend to have higher latitude emergence. Now let's see the effect of the systematic property on the solar cycle. For each sunspot group, net flux transported across the equator determines the final contribution to the dipole field. Lower latitude emergence tends to finally contribute more to the axial dipole field. Together with the statistical relation that stronger cycles tend to higher mean latitudes of ARs, the systematic properties of AR emergence in latitude work as a nonlinearity to modulate the dipole field generation. We refer to it as latitudinal quenching (Jiang 2020; Petrovay 2020; Karak 2020).

Besides the latitude, the tilt angle is another key parameter of an active region. The anti-correlation between the tilt angle and the cycle strength found by Dasi–Espuig et al.

(2010) initiated the revival of the Babcock–Leighton type dynamo during the past decade. However, the standard errors of the averaged tilt angle for each cycle are significant, and the anti-correlation is mainly due to cycle 19. Recently, Jiao et al. (2021) discovered a strong anti-correlation between the tilt coefficient and the cycle strength by improving the method of dealing with historical data. This anti-correlation can modulate the generation of the dipole field, as the dipole field is proportional to the tilt angle. The phenomenon is referred to as tilt quenching.

Apart from the systematic property of the tilt angle, there is a large scatter in tilts resulting from randomness due to convective flows (Weber et al. 2011). The dependence of the tilt scatter on the umbra area can be measured using historical datasets. Any variation in the properties of the spots, such as emergence latitude or tilt angle, introduces a corresponding variation in the resulting poloidal field and, consequently, in the amplitude of the subsequent cycle.

In summary, both stochastic and systematic properties are involved in sunspot emergence in their tilt angle, latitude, area, and number. Jiang et al. (2011, 2018) gave empirical formula describing relations between sunspot emergence and cycle strength. Hence, with a given cycle strength, we may synthesize the sunspot emergence.

Figure 2 adapted from Jiang (2020) displays three examples of synthetic solar cycles with weak, medium, and strong amplitudes, respectively. For each cycle amplitude, we generate 100 random realizations including the number, latitude, longitude, tilt, and area of each AR using formerly mentioned empirical relations. Thus we may quantify the effects of randomness and non-linearity resulting from stochastic and systematic properties of sunspot emergence using surface flux transport models.

For each solar cycle, sunspots initially reverse the dipole field of the previous cycle and then establish the opposite polarity of the new cycle. The total dipole moment generated by the emerged sunspots during cycle n is calculated by the sum of the absolute values of the dipole moments at the beginning and the end of cycle n. This allows us to compare the final total dipole moments for cycles with different amplitudes. Despite the large variations in the averaged amplitudes of the three columns in Figure 2, the expected value of the final total dipole moment remains similar. The final total dipole moment is weakly determined by cycle amplitudes due to the effect of nonlinearities. Additionally, the final dipole moments have a large uncertainty range due to the stochastic properties of sunspot emergence, which corresponds to the effects of randomness.

Figure 3 from Jiang (2020) shows the dependence of the final total dipole moment on different cycle strengths. The black solid curve indicates the expected values from 100 surface flux transport simulations using random sunspot group realizations including latitudinal and tilt quenching. They lead to a saturation state in which normal and strong cycles tend to produce similar amounts of the total dipole moment regardless of their amplitudes, whereas weak cycles become even more effective in generating a total dipole moment. The black dashed curve is the curve fit to the black solid curve. The error bars correspond to the 1σ standard deviation, caused by the randomness in the properties of sunspot groups. The shaded region shows the range of the standard deviation. This result paves the way for understanding the nature of solar cycle variability, chaotic or stochastic.

Let's first analyze the effects of nonlinearities without considering the stochastic mechanisms. Based on the fitted relation between the total final dipole moment and the cycle strength along with the calibrated relation between the dipole moment and the subsequent cycle strength (Jiang et al. 2018), with a given cycle n strength S_n, we may determine the strength of the subsequent cycle $n+1$, S_{n+1}. Thus the recursion function of the solar cycle can be derived, which is shown by the black curves of Figure 4. The left panel of the cobweb diagram shows results with different initial conditions in different

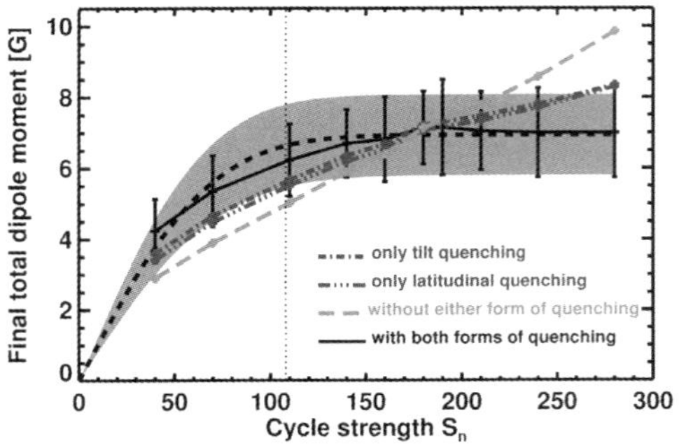

Figure 3. Effects of randomness and non-linearity of sunspot emergence on the solar cycle. The solid black curve shows the expected values from 100 SFT simulations with random sunspot group realisations, including latitude and tilt quenching. The error bars correspond to the 1σ standard deviation caused by the randomness in the sunspot group properties. The shaded area corresponds to the standard deviations. Green dashed triple-dot and purple dashed dot curves show the expected values for SFT simulations with only latitude and tilt quenching, respectively. Orange long dashes show the expected value of SFT simulations without either latitudinal or tilt quenching. Image reproduced with permission from Jiang (2020), copyright by the American Astronomical Society.

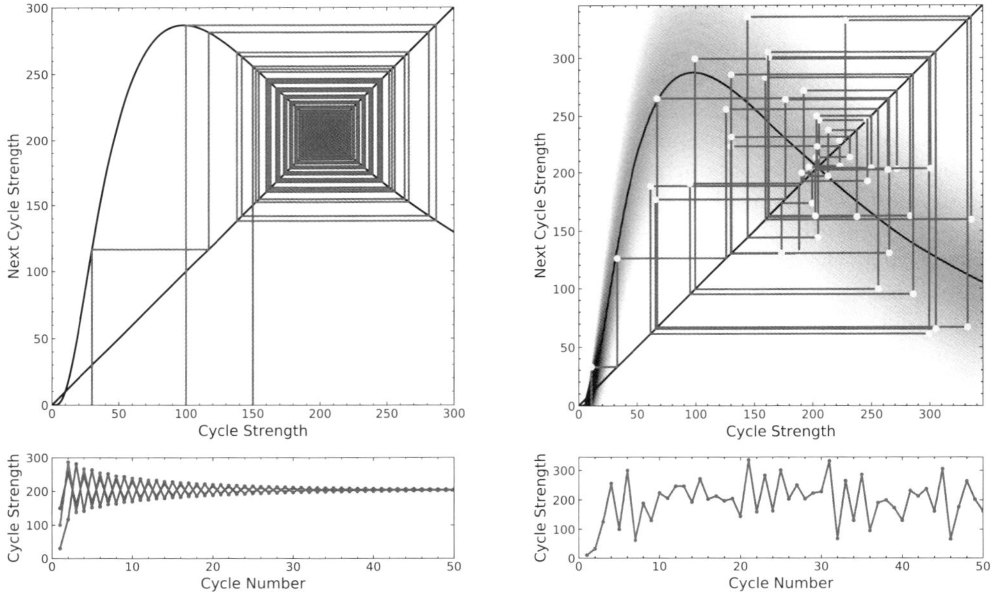

Figure 4. Cobweb diagrams for solar cycle recursion. Left panels: with only nonlinearity. The curves in 3 different colors correspond to different initial conditions. Right panels: with both randomness and nonlinearity. The over-plotted red curve is one random realization.

colors. The lower panel of Figure 4 gives the corresponding solar cycle series. We may see two distinct properties. One is the regular G-O effect at the beginning phase. The other is that the cycle strength eventually reaches a stable fixed point. This implies that the observed nonlinearities lead to a stable limit cycle, rather than a chaotic system (Wang et al. 2024).

Now we include the effect of the randomness in the properties of sunspot groups on the generation of the final total dipole moment. Thus we may generate a large number of cycles series for analyzing solar cycle variability, especially the G-O effect. The right panel of Figure 4 shows an example of cycle series of 50 cycles. This allows for the analysis of solar cycle variability, including the odd-even effect, through the generation of numerous cycle series. The analysis based on the 1 million synthesized solar cycles indicates that the normalized occurrence frequency of various lengths of G-O blocks follows an exponential distribution, which suggests a Poisson random process. The reconstructed solar cycles of the past 100 cycles by Usoskin et al. (2021), present roughly the same occurrence frequency of the G-O blocks. The nonlinearity contributes to the G-O rule, while the stochastic nature of sunspot emergence causes the probability of obtaining the same even-odd relationship to decrease exponentially as the cycle number increases.

3. Impacts on solar cycle predictions

The aforementioned attempt indicates that the variability of the solar cycle could be attributed to a weakly nonlinear limit cycle affected by stochastic noise. The result is consistent with Cameron and Schüssler (2017), who used a generic normal-form model with parameters constrained by observations. The nature of the solar cycle variability has significant impacts on solar cycle predictions. The dominant effects of randomness put constraints on the scope of the solar cycle prediction. No sensible prediction of the strength of the next cycle can be made before the solar minimum. The uncertain range increases with the scope of the cycle prediction. On the other hand, people need a cycle prediction as early as possible. In this situation, people should keep in mind the uncertain range of dipole moments. Randomness and nonlinearities should be realistically incorporated or considered to get the uncertain range of the dipole moment.

Jiang et al. (2018) developed a scheme to investigate the predictability of the solar cycle over one cycle. There are three main steps. The first is to predict the sunspot emergence of the current cycle using the empirical formulae on the stochastic and systematic properties of sunspot emergence from Jiang et al. (2011). Please note here that the stochastic properties have usually been ignored by prediction models. Ignoring the effect of stochastic properties in sunspot numbers also explains the variable predictions of solar cycle 25 given at different times by Hathaway and Upton †. In January 2022, they once gave the prediction of the maximum amplitude of cycle 25 as less than 100. And the prediction jumped to about 150 in April 2022. From January 2024, their prediction dropped back to about 135. As shown in Figure 3 of Jiang et al. (2018), the prediction of an ongoing cycle using the curve fit to the observed sunspot number has large relative errors, which can be over 50% when it is within 3 years into a cycle. In addition to the uncertainty in the sunspot number, there are also uncertainties in latitude, tilt angle, and area of sunspot emergence, which can significantly affect the sunspot contribution to the polar field.

The second step in investigating the predictability of the solar cycle over one cycle is to utilize the surface flux transport model, with the predicted sunspot group emergence as the flux source, to get the predicted polar field or dipole moment at the cycle minimum. The predicted result includes both the expected value and the uncertain range at cycle minimum. During this step, the observed magnetograms are required as the initial condition. As noted by Cameron et al. (2016); Jiang and Cao (2018), the uncertainty due to measurement error in the magnetogram should also be considered. Upton and Hathaway (2018) once predicted that cycle 25 would be about 95% smaller in strength than cycle 24, and that cycle 25 would be the weakest solar cycle in the last hundred years. Solar behavior has shown that the prediction by Upton and Hathaway (2018) is not the real

† `http://solarcyclescience.com/forecasts.html`

case. As analyzed by Jiang et al. (2023), the weak prediction is mainly due to the synchronic magnetogram at CR2172 that they used as the initial field. As shown in Figures 4 and 5 of Jiang et al. (2023), the map has some differences from the synoptic magnetogram at CR2172 especially at low latitudes. The flux at low latitudes has a significant effect on the predicted dipole moment at cycle minimum (Wang and Sheeley 1991; Jiang et al. 2014; Petrovay et al. 2020), leading to the weak dipole moment at cycle minimum and hence a weak prediction of cycle 25.

The final step is to obtain a predicted subsequent cycle profile, including its uncertainty, by using a carefully calibrated linear relation between the dipole moment at cycle minimum and the subsequent cycle strength. This step involves using the relation between cycle strength and the averaged cycle profile. It is expected that each real solar cycle will exhibit some deviations from the averaged profile within a certain range.

4. Conclusions

In this paper we argue that a prerequisite for meaningful solar cycle predictions is to understand the nature of solar cycle variability. Unlike atmospheric dynamics for weather forecasting, solar cycle variability could result from a weakly nonlinear limit cycle dominated by stochastic noise. Systematic and stochastic properties of sunspot emergence contribute to the nonlinear and stochastic mechanisms of the solar cycle, respectively. The stochastic properties constrain the scope of the solar cycle prediction. Uncertainty increases with the scope of the prediction. The uncertainties in observed magnetograms and sunspot emergence should be well evaluated and included in the physics-based solar cycle predictions. Otherwise, the models could make misleading predictions as the article mentioned in the last section.

Physics-based solar cycle prediction usually directly or indirectly use the polar field (dipole moment) at cycle minimum as precursor. As shown in Figure 4 of Labonville et al. (2019), some of such predictions gave a weaker cycle 25 than cycle 24. This is mainly due to the fact that the stochastic noise in magnetograms and sunspot formation, or the systematic properties of sunspot formation, have not been well incorporated into the prediction models. Fortunately, Jiang et al. (2018); Pesnell and Schatten (2018); Svalgaard (2020), who also used the precursor, gave different results. The behavior of cycle 25 provides further evidence that the polar field precursor method is reliable.

Acknowledgements. The research is supported by the National Natural Science Foundation of China (Nos. 12173005 and 12350004) and the National Key R&D Program of China No. 2022YFF0503800. The SDO/HMI data are courtesy of NASA and the SDO/HMI team. Wilcox Solar Observatory data used in this study was obtained via the web site `http://wso.stanford.edu` courtesy of J.T. Hoeksema. The sunspot records are courtesy of WDC-SILSO, Royal Observatory of Belgium, Brussels.

References

Babcock, H. W. 1961, The Topology of the Sun's Magnetic Field and the 22-YEAR Cycle. *apj*, 133, 572.

Bhowmik, P., Jiang, J., Upton, L., Lemerle, A., & Nandy, D. 2023, Physical Models for Solar Cycle Predictions. *Space Sci. Rev.*, 219(5), 40.

Cameron, R. H., Jiang, J., & Schüssler, M. 2016, Solar Cycle 25: Another Moderate Cycle? *Astrophys. J. Lett.*, 823(2), L22.

Cameron, R. H. & Schüssler, M. 2017, Understanding Solar Cycle Variability. *Astrophys. J.*, 843(2), 111.

Dasi–Espuig, M., Solanki, S. K., Krivova, N. A., Cameron, R., & Peñuela, T. 2010, Sunspot group tilt angles and the strength of the solar cycle. *Astron. Astrophys.*, 518, A7.

Gnevyshev, M. N. & Ohl, A. I. 1948, On the 22-year cycle of solar activity. *Living Reviews in Solar Physics*, 25, 18.

Howe, R. 2009, Solar Interior Rotation and its Variation. *Living Reviews in Solar Physics*, 6(1), 1.

Jiang, J. 2020, Nonlinear Mechanisms that Regulate the Solar Cycle Amplitude. *Astrophys. J.*, 900(1), 19.

Jiang, J., Cameron, R. H., Schmitt, D., & Schüssler, M. 2011, The solar magnetic field since 1700. I. Characteristics of sunspot group emergence and reconstruction of the butterfly diagram. *Astron. Astrophys.*, 528, A82.

Jiang, J., Cameron, R. H., & Schüssler, M. 2014, Effects of the scatter in sunspot group tilt angles on the large-scale magnetic field at the solar surface. *Astrophysical Journal*, 791(1).

Jiang, J. & Cao, J. 2018, Predicting solar surface large-scale magnetic field of Cycle 24. *Journal of Atmospheric and Solar-Terrestrial Physics*, 176, 34–41.

Jiang, J., Chatterjee, P., & Choudhuri, A. R. 2007, Solar activity forecast with a dynamo model. *Mon. Not. Roy. Astron. Soc.*, 381(4), 1527–1542.

Jiang, J., Hathaway, D. H., Cameron, R. H., Solanki, S. K., Gizon, L., & Upton, L. 2014, Magnetic Flux Transport at the Solar Surface. *Space Sci. Rev.*,.

Jiang, J., Wang, J.-X., Jiao, Q.-R., & Cao, J.-B. 2018, Predictability of the Solar Cycle Over One Cycle. *Astrophys. J.*, 863(2), 159.

Jiang, J., Zhang, Z., & Petrovay, K. 2023, Comparison of physics-based prediction models of solar cycle 25. *Journal of Atmospheric and Solar-Terrestrial Physics*, 243, 106018.

Jiao, Q., Jiang, J., & Wang, Z.-F. 2021, Sunspot tilt angles revisited: Dependence on the solar cycle strength. *Astron. Astrophys.*, 653, A27.

Karak, B. B. 2020, Dynamo Saturation through the Latitudinal Variation of Bipolar Magnetic Regions in the Sun. *Astrophys. J. Lett.*, 901(2), L35.

Labonville, F., Charbonneau, P., & Lemerle, A. 2019, A Dynamo-based Forecast of Solar Cycle 25. *Solar Phys.*, 294(6), 82.

Leighton, R. B. 1969, A Magneto-Kinematic Model of the Solar Cycle. *Astrophys. J.*, 156, 1.

Li, K. J., Wang, J. X., Zhan, L. S., Yun, H. S., Liang, H. F., Zhao, H. J., & Gu, X. M. 2003, On the Latitudinal Distribution of Sunspot Groups over a Solar Cycle. *Solar Phys.*, 215(1), 99–109.

Lorenz, E. N. 1963, Deterministic Nonperiodic Flow. *Journal of Atmospheric Sciences*, 20(2), 130–148.

Lorenz, E. N. 1965, A study of the predictability of a 28-variable atmospheric model. *Tellus*, 17(3), 321–333.

Lorenz, E. N. 1969, The predictability of a flow which possesses many scales of motion. *Tellus*, 21(3), 289–307.

Nandy, D. 2021, Progress in Solar Cycle Predictions: Sunspot Cycles 24-25 in Perspective. *Solar Phys.*, 296(3), 54.

Penza, V., Bertello, L., Cantoresi, M., Criscuoli, S., & Berrilli, F. 2023, Prediction of solar cycle 25: applications and comparison. *Rendiconti Lincei. Scienze Fisiche e Naturali*, 34(3), 663–670.

Pesnell, W. D. & Schatten, K. H. 2018, An Early Prediction of the Amplitude of Solar Cycle 25. *Solar Phys.*, 293(7), 112.

Petrovay, K. 2020, Solar cycle prediction. *Living Reviews in Solar Physics*, 17(1), 2.

Petrovay, K., Nagy, M., & Yeates, A. R. 2020, Towards an algebraic method of solar cycle prediction. I. Calculating the ultimate dipole contributions of individual active regions. *Journal of Space Weather and Space Climate*, 10, 50.

Solanki, S. K., Wenzler, T., & Schmitt, D. 2008, Moments of the latitudinal dependence of the sunspot cycle: a new diagnostic of dynamo models. *Astron. Astrophys.*, 483(2), 623–632.

Svalgaard, L. 2020, Prediction of Solar Cycle 25. *arXiv e-prints*,, arXiv:2010.02370.

Upton, L. A. & Hathaway, D. H. 2018, An Updated Solar Cycle 25 Prediction With AFT: The Modern Minimum. *Geophys. Res. Lett.*, 45(16), 8091–8095.

Usoskin, I. G., Solanki, S. K., Krivova, N. A., Hofer, B., Kovaltsov, G. A., Wacker, L., Brehm, N., & Kromer, B. 2021, Solar cyclic activity over the last millennium reconstructed from annual ^{14}C data. *Astron. Astrophys.*, 649, A141.

Wang, Y. M. & Sheeley, N. R., J. 1991, Magnetic Flux Transport and the Sun's Dipole Moment: New Twists to the Babcock–Leighton Model. *Astrophys. J.*, 375, 761.
Wang, Y. M. & Sheeley, N. R. 2009, Understanding the Geomagnetic Precursor of the Solar Cycle. *Astrophys. J. Lett.*, 694(1), L11–L15.
Wang, Z.-F., Jiang, J., & Wang, J.-X. 2021, Algebraic quantification of an active region contribution to the solar cycle. *Astron. Astrophys.*, 650, A87.
Wang, Z.-F., Jiang, J., & Wang, J.-X. 2024, Algebraic quantification of an active region contribution to the solar cycle. *in prep.*,.
Weber, M. A., Fan, Y., & Miesch, M. S. 2011, The Rise of Active Region Flux Tubes in the Turbulent Solar Convective Envelope. *Astrophys. J.*, 741(1), 11.
Yeates, A. R., Cheung, M. C. M., Jiang, J., Petrovay, K., & Wang, Y.-M. 2023, Surface Flux Transport on the Sun. *Space Sci. Rev.*, 219(4), 31.

Contributed papers

Dynamics of Solar and Stellar Convection Zones and Atmospheres
Proceedings IAU Symposium No. 365, 2024
A. V. Getling & L. L. Kitchatinov, eds.
doi:10.1017/S1743921324000140

A new generation of solar dynamo model and its application to explore the stellar magnetic cycle

Zebin Zhang[1,2]

[1]School of Space and Environment, Beihang University, Beijing, China
email:zebinzhang@buaa.edu.cn

[2]Key Laboratory of Space Environment Monitoring and Information Processing of MIIT, Beijing, China

Abstract. In most of the Babcock–Leighton (BL) type solar dynamo models, the toroidal magnetic field is assumed to be generated in the tachocline. However, magnetic activity of fully convective stars and MHD simulations of global stellar convection have recently raised serious doubts about the importance of the tachocline in the generation of the toroidal field. We have developed a BL-type dynamo model operating in the bulk of the convection zone, and are extending this model to solar-type stars. In this study, we aim at exploring how the starspot properties affect the stellar magnetic cycle. Observations show that faster rotating stars tend to have stronger magnetic activity and shorter magnetic cycles. By considering the higher latitudes and larger tilt angles of starspots for faster rotators, our simulations reproduce observations that faster rotating stars have shorter magnetic cycle and stronger activity.

Keywords. Dynamo, Stars: activity, Stars: starspots

1. Introduction

Solar large-scale magnetic fields exhibit quasi-11-year cyclic variations, as evidenced by sunspots. Starspots and magnetic cycles are common among cool stars. The progress in solar magnetism is paving the way for understanding stellar activity. Observations of stellar magnetic activity indicate general relations of P_{rot}–P_{cyc} and P_{rot}–A_{cyc} (Wilson 1978). Specifically, younger and faster rotating stars tend to have longer magnetic cycle periods (P_{cyc}) and higher magnetic activity amplitudes (A_{cyc}).

Solar observations indicate that stronger solar cycles are associated with smaller tilt angles and sunspot emergence at higher latitudes (Jiang et al. 2011). Systematic studies have demonstrated the effect of tilt and emergence latitude variations on solar cycle (Jiang 2020). This raises an interesting question of how the starspots' emergence latitudes and tilt angles affect stellar magnetic cycles. Sunspots appear at latitudes lower than about 40°. While for starspots, the emergence latitudes could be distributed over the whole stellar disk (Strassmeier 2009). Observations also indicate that stars with faster rotation rates tend to have starspots located at higher latitudes, and even in polar regions (Vogt and Penrod 1983).

Zhang and Jiang (2022) have developed a new generation of the BL-type dynamo model that operates in the bulk of the convective zone rather than in the tachocline. The tachocline is not a necessary ingredient in their model, which is supported by stellar observations and MHD simulations that the dynamo process is more likely to operate in the bulk of the convection zone (Nelson et al. 2013; Wright and Drake 2016). This work aims at extending the solar dynamo model of Zhang and Jiang (2022) to stars with a solar mass $M_\odot$, ranging the rotation periods from 10 to 25 days, and explore the effect of starspots emergence properties on magnetic cycles.

2. Model

In this study, we adopt the kinematics dynamo model developed by Zhang and Jiang (2022). The flow fields and BL-type source term S_{BL} will be prescribed in the next subsection. The other ingredients are almost the same as Zhang and Jiang (2022). In this study, the computation domain is $0.72R_\odot \leq r \leq R_\odot, 0 \leq \theta \leq \pi$, which excludes the region of the tachocline.

2.1. *Differential rotation and meridional flow*

We will study the dynamo process of solar-mass stars with rotation periods ranging from 10 to 25 days. The flow fields will be obtained from a mean-field hydrodynamical model (Kitchatinov and Nepomnyashchikh 2017b).

We have calculated the latitudinal distribution of the latitudinal shear and the average return flow for various rotators. For stars with varied rotation periods, the strength of latitudinal shear all reaches their maxima at about 55 °, which means that the generation efficiency of the toroidal fields is the strongest there. This latitude dependence of the latitudinal shear also has a contribution to the butterfly diagram (Zhang et al. 2022). The strength of return flow decreases from 3.15 ms^{-1} to 2.4 ms^{-1} with the decreasing of rotation period.

2.2. *Babcock–Leighton source term*

The BL-type source term is defined as

$$S_{BL}(r,\theta,t) = \frac{\alpha_0 f(\theta)}{2}\left[1 + \text{erf}\left(\frac{\text{r} - 0.95\text{R}_\odot}{0.01\text{R}_\odot}\right)\right]\bar{B}(\theta,t). \tag{1}$$

Observations show that faster rotators tend to have starspots at higher latitudes. One explanation is the larger Coriolis force for faster rotators, which pushes the toroidal flux to higher latitudes (Schuessler et al. 1996). In this study, to explore the influence of appearing latitudes of starspots on the dynamo process, we consider

$$f(\theta) = \frac{\cos\theta \sin^n \theta}{Max[\cos\theta \sin^n \theta, \theta \in (0,\pi)]}, \tag{2}$$

in Eq. (1). A simple linear relation between rotation period and n that $n = 1 + (P_{rot} - 10) * 0.55$ is considered. With the increasing of rotation rate, n will decrease, allowing the faster-rotating stars to form starspots at higher latitudes.

A key parameter in the dynamo process is the dynamo number, $\alpha_0 \Delta\Omega R^3/\eta^2$. When α_0 is greater than or equal to its critical value α_c, a self-sustaining dynamo process occurs. Most previous dynamo models set α_0 arbitrarily. Kitchatinov and Nepomnyashchikh (2017a) (hereafter KN17) suggests that α_0 could be constrained on the basis of observations. Motivated by KN17, we give α_0 for a star with rotation period P_{rot} based on

$$\alpha_0 = (1 + m\frac{P_{max}}{P_{rot}})\alpha_c, \tag{3}$$

where m is a free parameter that constrains the amount of supercriticality. For the solar-mass stars we study, we take P_{max} to be 28 days. For the solar case, the saturated unsigned toroidal flux in the convection zone is about 10^{23} Mx. The flux puts a constraint on the value of m, which is taken to be $m = 0.2$.

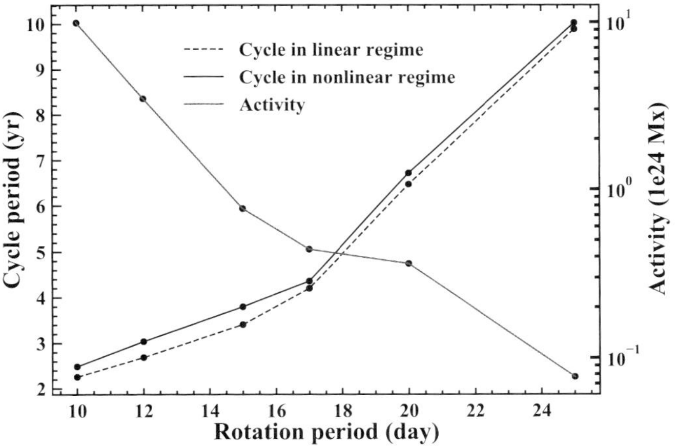

Figure 1. P_{rot}–P_{cyc} and P_{rot}–A_{cyc} relations. The solid (dashed) line represents the solution operating in the nonlinear regime (linear regime). Image reproduced with permission from Zhang et al. (2024).

3. Results

Figure 1 shows magnetic activity and cycle period versus rotation period. We explore the P_{rot}–P_{cyc} relation in both linear critical and nonlinear regime, and the cycle period of nonlinear solutions is slightly longer than that of linear marginal mode. The cycle period increases as rotation period, close to observations. When the meridional flow is faster, the magnetic cycle is longer. This indicates that our model is different from the flux transport dynamo (FTD) models. The strength of magnetic activity increases with rotation rate, as shown by observations. The reason for this P_{rot}–A_{cyc} relation is that larger tilt angles of faster rotators make dynamo operate in a stronger supercritical regime, leading to stronger saturated magnetic fields.

To further explore why faster rotators have a shorter magnetic cycle period, Figure 2 shows the time-latitude diagram of subsurface toroidal flux and surface radial fields of rotators of 10 and 25 d. Figure 2 (b) shows that faster rotators have starspots closer to $\pm 55^\circ$ latitudes, where the strongest latitudinal shear exists. The newly generated poloidal flux could be sheared efficiently then generating the toroidal field for the subsequent cycle quickly. High-latitude starspots emergence shorten the dynamo circle and hence the magnetic cycle. Slower rotators have starspots farther from the $\pm 55^\circ$ latitude (see Figure 2 (d)), so that additional time is required for the transport and winding of the poloidal flux, resulting in longer cycle periods. The emergence features of starspots in latitudes plays a crucial role in creating the observed relation between P_{rot} and P_{cyc}.

4. Conclusion

We have extended the BL mechanism to solar-mass stars and investigated the effect of starspot emergence properties on stellar magnetic cycles. The rotation-dependent emergence latitude and tilt angle of starspots are considered in a BL-type dynamo operating in the bulk of the convection zone developed by Zhang and Jiang (2022).

In our BL-type dynamo model, the emergence latitudes of starspots play a crucial role in reproducing the P_{rot}–P_{cyc} and P_{rot}–A_{cyc} relations. Faster rotators have poloidal flux appearing closer to about $\pm 55^\circ$ latitudes, where the toroidal field generation efficiency is the strongest. It takes a shorter time for faster rotators to transport the surface poloidal field from their emergence latitude to the $\pm 55^\circ$ latitudes of efficient latitudinal shear thus shortening magnetic cycles. The faster rotators operate in a more supercritical regime

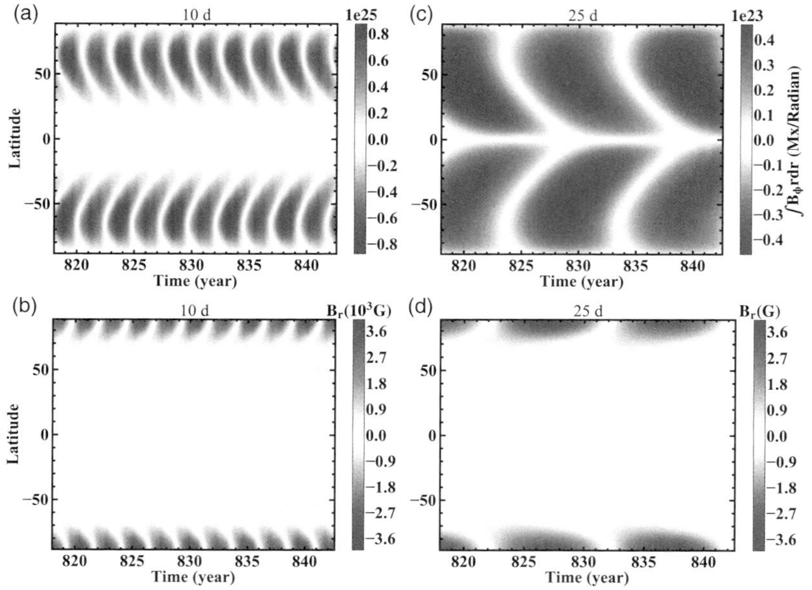

Figure 2. Time-latitude diagram of subsurface toroidal flux (top) and surface radial fields (bottom). Image reproduced with permission from Zhang et al. (2024).

due to a stronger BL α effect with respect to the tilt angles, leading to stronger saturated magnetic fields.

Acknowledgements. The research is supported through the National Natural Science Foundation of China (grant Nos. 12173005 and 12350004), National Key R&D Program of China No. 2022YFF0503800.

References

Jiang, J. 2020, Nonlinear Mechanisms that Regulate the Solar Cycle Amplitude. *Astrophys. J.*, 900(1), 19.

Jiang, J., Cameron, R. H., Schmitt, D., & Schüssler, M. 2011, The solar magnetic field since 1700. I. Characteristics of sunspot group emergence and reconstruction of the butterfly diagram. *Astron. Astrophys.*, 528, A82.

Kitchatinov, L. & Nepomnyashchikh, A. 2017,a How supercritical are stellar dynamos, or why do old main-sequence dwarfs not obey gyrochronology? *Mon. Not. Roy. Astron. Soc.*, 470a(3), 3124–3130.

Kitchatinov, L. L. & Nepomnyashchikh, A. A. 2017,b A joined model for solar dynamo and differential rotation. *Astronomy Letters*, 43b(5), 332–343.

Nelson, N. J., Brown, B. P., Brun, A. S., Miesch, M. S., & Toomre, J. 2013, Magnetic Wreaths and Cycles in Convective Dynamos. *Astrophys. J.*, 762(2), 73.

Schuessler, M., Caligari, P., Ferriz–Mas, A., Solanki, S. K., & Stix, M. 1996, Distribution of starspots on cool stars. I. Young and main sequence stars of $1M_{sun-}$. *Astron. Astrophys.*, 314, 503–512.

Strassmeier, K. G. 2009, Starspots. *Astron. Astrophys. Rev.*, 17(3), 251–308.

Vogt, S. S. & Penrod, G. D. 1983, Doppler imaging of spotted stars : application to the RS Canum Venaticorum star HR 1099. *Pub. Astron. Soc. Pac.*, 95, 565–576.

Wilson, O. C. 1978, Chromospheric variations in main-sequence stars. *Astrophys. J.*, 226, 379–396.

Wright, N. J. & Drake, J. J. 2016, Solar-type dynamo behaviour in fully convective stars without a tachocline. *Nature*, 535(7613), 526–528.

Zhang, Z. & Jiang, J. 2022, A Babcock–Leighton-type Solar Dynamo Operating in the Bulk of the Convection Zone. *Astrophys. J.*, 930(1), 30.

Zhang, Z., Jiang, J., & Kitchatinov, L. L. 2024, Modeling effects of starspots on stellar magnetic cycles. *Astron. Astrophys.*, under review.

Zhang, Z., Jiang, J., & Zhang, H. 2022, A Potential New Mechanism for the Butterfly Diagram of the Solar Cycle: Latitude-dependent Radial Flux Transport. *Astrophys. J. Lett.*, 941(1), L3.

Dynamics of Solar and Stellar Convection Zones and Atmospheres
Proceedings IAU Symposium No. 365, 2024
A. V. Getling & L. L. Kitchatinov, eds.
doi:10.1017/S1743921323003861

Resonance and stellar dynamos

D. Sokoloff[1,2], E. Yushkov[1,3] and A. Serenkova[1]

[1]Department of Physics, Moscow State University
email: sokoloff.dd@gmail.com

[2]IZMIRAN

[3]Space Research Institute, Moscow

Abstract. Planetary influence on a stellar convective shell can result in a periodic modulation of stellar dynamo drivers. Similar modulation can arise in stellar binary systems. Using the Parker low-mode dynamo model we investigate the properties of nonlinear parametric resonance. This model is a system of four ordinary differential equations and, in the first approximation, describes the processes of generation and oscillation of large-scale magnetic fields in stellar systems. In the absence of nonlinear suppression effects, the problem, by analogy with a system of harmonic oscillations, allows an asymptotic selection of multiple resonant frequencies. Despite the fact that at first glance at these frequencies it is reasonable to expect an increase in the amplitude, the behavior of the system can be just the opposite. All this stuff deserves a systematic analysis of swing excitation in the dynamo sistems in comparison with classical swing excitation in the framework of the Mathieu equation.

Keywords. Solar activity cycle, stellar cycles, stellar dynamo

1. Introduction

Solar activity cycle is a (quasi-)periodic phenomenon. Activity cycles similiar to the solar one are known for many stars similar to the Sun. The activity cycles are believed to be supported by a process in the stellar convective zone known as stellar dynamo. On the other hand, stellar convective zones may be affected by various periodic processes, say, tidal forces arising by gravitational forces from planetary systems and/or members of the binary system. The idea of swing excitation participating in the stellar activity cycles looks natural. The idea was addressed in some papers for quite a long time (see, e.g. Gilman, Dikpati 2011, Moss, Sokoloff 2013). Indeed, numerical modelling provides particular dynamo models which demonstrate behavior similar to the swing excitation (e.g. Moss, Piskunov, Sokoloff 2002) however there are examples of more complicated behavior (e.g. Moss, Sokoloff 2017) rather than just a parametric resonance.The situation becomes dramatic because Jupiter orbital period (about 11 yrs) is closed to the nominal solar cycle length (about 11 yrs) and the idea that this is something more rather a coincidence is discussed in astronomical community (e.g. Klevs, Stefani, Jouve 2023) while search of similar cases in exoplanetary systems with known activity of central star (Obridko, Katsova, Sokoloff 2022) do not support this expectation. It is why a systematic comparison and confrontation of dynamo parametric resonance and classical swing excitation undertaken recently by Kitchatinov, Nepomnyashchikh 2015 and Serenkova, Sokoloff, Yushkov 2023 from different viewpoints becomes important. Here we concentrate on the last approach.

2. Comparing dynamo resonance and classical swing excitation

Comparing dynamo resonance and classical swing excitation we have to take into account that the last one is considered in terms of ordinary differential equations (Mathieu equation), i.e. for system with finite degrees of freedom, while standard solar and stellar dynamo models contain partially differential equations, i.e. consider systems with infinite degrees of freedom (field equations). Of course, any particular kind of swing excitation can be developed to include field equations, however more practical is to reduce a dynamo model to a dynamical system. This can be done by presenting corresponding partial differential equations by Fourier series with further truncation to keep as few variables (degrees of freedom) to get a cyclic behaviour. Nefedov, Sokoloff 2010 performed such truncation for the classical Parker migratory dynamo with an algebraic α-quenching to get a four-dimensional dynamical system which we exploit below. In contrast, the second-order Mathieu equation corresponds to a second order dynamical system. There is no problem to perform such truncation for, say, much more realistic contemporary flux-transport dynamo models, however the exploited over-simplified model is sufficient to isolate substantial difference between dynamo resonance and classical swing excitation.

2.1. *Mathieu resonance*

The classical parametric resonance is described by the Mathieu equation (e.g. McLachlan 1947):

$$\ddot{f} + \omega_0^2(1 + \sigma \sin(\omega t))f = 0. \tag{1}$$

Resonant solutions for the Mathieu equation are sought in the form of a harmonic function with small exponential growth:

$$f(t) = f_1 \exp(st + i(\omega_0 + \varepsilon/2)t) + f_1^* \exp(st - i(\omega_0 + \varepsilon/2)t). \tag{2}$$

Exponential growth is observed in a small corridor near the doubled frequency of the system. In the nonlinear case of parametric resonance for the system, with an increase in the amplitude, the parameter ω_0^2 decreases, due to which the system gradually decouples, and this, in turn, leads to stabilization of exponential growth and quasistationary output. Stationary is understood in the sense that the solution oscillates, but the energy of the oscillations remains constant. The parameter ω_0^2 reduction is set as follows:

$$\omega_{eff}^2 = \frac{\omega_0^2}{1 + \langle f^2 \rangle}. \tag{3}$$

2.2. *Dynamo resonance*

Let us consider the case of dynamo resonance on the example of the Parker system. The key simplification proposed by Parker in the transition to a spherical coordinate system consists in dividing the axially symmetric magnetic field into poloidal and toroidal components (A and B respectively):

$$\begin{aligned} \dot{A} &= R_\alpha B + A_{\theta\theta} - \mu^2 A, \\ \dot{B} &= R_\omega (A \sin(\theta))_\theta + B_{\theta\theta} - \mu^2 B. \end{aligned} \tag{4}$$

The migration waves generated in such a system (they are often called dynamo waves) are most sensitive to the dynamo number D, i.e. the product of dimensionless parameters responsible for differential rotation (R_ω) and helicity (R_α). The number μ determines the radial part of the diffusion and is proportional to the ratio of the radius to the thickness of the convective shell. Hypothetically, some periodic influence on the system (for example, binary stars on each other or exoplanets on their star) can lead to periodic changes in R_α or R_ω, which, in turn, can cause parametric resonance. To analyze such a situation, the following change is made to the system:

$$R_\omega \rightarrow R_\omega(1+\sigma \sin(\omega t)), \tag{5}$$

here, σ is the amplitude of the parametric effect. A more convenient approach for mathematical interpretation of the solution is the so-called low-mode approximation, which has already been used in some works (e.g. Kalinin, Sokoloff 2018, Tarbeeva, Semikoz, Sokoloff 2011). In the case of expansion by four principal modes of the Fourier series, it is possible to obtain a system of Mathieu-type equations:

$$\begin{aligned} &\ddot{f}_1 - (R_\alpha R_\omega/2)(1+\sigma \sin(\omega t))(f_1 - f_2) = 0, \\ &\ddot{f}_2 - (R_\alpha R_\omega/2)(1+\sigma \sin(\omega t))(f_1 + f_2) = 0. \end{aligned} \tag{6}$$

In this case, the solution has the form of four complex-conjugate exponentials, since now the frequencies have a complex additive responsible for generating outside of a narrow corridor about twice the frequency (see for details Serenkova et al.):

$$\begin{aligned} f_{1,2} = \bar{f}_{1,2}\exp((\gamma_0+s+i\beta)t + i(\omega_0+\varepsilon/2)t) + \bar{f}^*_{1,2}\exp((\gamma_0+s-i\beta)t - i(\omega_0+\varepsilon/2)t) + \\ + \bar{\bar{f}}_{1,2}\exp((\gamma_0+s-i\beta)t + i(\omega_0+\varepsilon/2)t) + \bar{\bar{f}}^*_{1,2}\exp((\gamma_0+s+i\beta)t - i(\omega_0+\varepsilon/2)t). \end{aligned} \tag{7}$$

In the case, when the field is non-linearly stabilized, the growth of the magnetic field leads to a decrease in hydrodynamic helicity, the oscillations have a constant amplitude, and with resonance it can become larger than in the absence of exposure. To ensure stabilization, we will nonlinearly suppress the parameter $R\alpha$:

$$R_\alpha \rightarrow \frac{R_\alpha}{1+\max|B^2|}. \tag{8}$$

Some differences between the dynamo resonances and classical swing excitation are clear from advance. Indeed, dynamo systems include self-excitation, i.e. initial exponential growth, and a nonlinear saturation. Natural oscillators with swing excitation may contain similar features however they are not included in the classical Mathieu equation. It is however far to be clear which difference and to what extent is responsible for specific properties of dynamo resonances known from available numerics. In order to clarify this point we mimic the classical derivation of the Mathieu equation to get a similar equation for the dynamical system under consideration. Below we refer to this equation as a generalised Mathieu equation. Because parametric resonance can be interesting for another physical problems with self-excitation and its further nonlinear suppression and in order to mimic somehow more complicated dynamo models we consider numerical coefficients in the generalised Mathieu equation as free parameters and play with numbers nearby their nominal quantities obtained for the Parker migratory dynamo.

3. Specific features of dynamo resonances

The generalised Mathieu equation is quite similar to the classical one however is more bulky. It admits numerical as well to some extent analytical investigation (see for details Serenkova, Sokoloff, Yushkov 2023). Performing the analysis we arrive at the following conclusions.

First of all, periodic modulation of dynamo equations may result in two types of effects, i.e. effects associated with a specific frequency as well as effects presented for a wide range of frequencies. We describe both effects separately.

3.1. *Wide-range effect*

Stellar dynamo excitation is a threshold effect, i.e. intensity of dynamo drivers presented in dynamo equations as dimensionless dynamo numer D should exceed a marginal value D_0 to get self-excitation. Periodic modulation can slightly diminish D_0. In other words, periodic modulation can transform a slightly subcritical dynamo into a slightly supercritical one (Sokoloff, Serenkova, Yushkov 2022). It looks quite improbable that the dynamo number for a particular stellar dynamo is so close to the corresponding marginal value however if it happens the resulting effect on the dynamo supported magnetic field occurs to be crucial.

This effect does not require a fine tuning of the modulation frequency. Because standard swings do not presume self-excitation without modulation, the effect is impossible for this case. We do not insist that the effect have to be referred to as a resonance however it looks interesting by itself.

3.2. *True resonances*

The generalized Mathieu equation contains growing solutions associated with a particular value of the perturbation frequency. In order to separate them from the wide-range effect we refer to them as true resonances. Depending on coefficients in the dynamical system, i.e. on details of dynamo system, we can obtain a resonance peak at a particular modulation frequency which may coincide with the eigenfrequency of the system or do not coincide with it or do not obtain a peak at any frequency. In some cases resonance effects diminish the amplitude of the stellar cycle rather than enlarge it. In other words, a resonant absorption instead of resonant excitation is possible. In any case, resonance effects are not very strong, they modify the stellar cycle amplitude of about 10% at most.

4. Conclusions

We conclude that varity of the effects associated with periodic modulation of dynamo drivers is much more wide rather that one for the standard swing excitation. From the other hand, the effects are not very strong in comparison with the initial dynamo self-excitation and hardly can be very important in many stellar systems. From the other hand the effects can be important for some quite rare stellar systems.

A.S. and D.S. are grateful to the Basis Foundation for financial support under grant 21-1-1-4-1.

References

Gilman P.A., Dikpati M., 2011, *ApJ*, 738, 108
Kalinin, A. O., Sokoloff, D. D., 2018, Astron. Rep., 62, 689
Kitchatinov L., Nepomnyashchikh A., 2015, Astron. Lett., 41, 374
Klevs M., Stefani F., Jouve L., 2023, *Solar Phys.*, 298, 83
McLachlan, N. W., Theory and Application of Mathieu Function, 1947, Clarendon Press. Oxford
Moss D., Sokoloff D., 2013, *A&A*, 553, A37
Moss D., Piskunov N., Sokoloff D., 2002, *A&A*, 396, 885
Moss D., Sokoloff D., 2017, Astron. Rep., 61, 878
Nefedov S.N., Sokoloff D.D., 2010, Astron. Rep., 54, 247
Obridko V.N., Katsova M. M., Sokoloff D.D., 2022, *MNRAS*, 516, 2151
Serenkova A.Yu., Sokoloff D.D., Yushkov E.V., 2023, JETP, 136, 456
Sokoloff D.D., Serenkova A.Yu., Yushkov E.V., 2022, *Comm. Byurakan Astrophys. Obs.*, 69, 231
Tarbeeva, S. M., Semikoz, V. B., Sokoloff, D. D., 2011, Astron. Rep., 55, 456

Dynamics of Solar and Stellar Convection Zones and Atmospheres
Proceedings IAU Symposium No. 365, 2024
A. V. Getling & L. L. Kitchatinov, eds.
doi:10.1017/S1743921323005033

The role of nonlinear toroidal flux loss due to flux emergence in the long-term evolution of the solar cycle

Akash Biswas

Department of Physics, Indian Institute of Technology (Banaras Hindu University), Varanasi 221005, India
email: akashbiswas.rs.phy20@itbhu.ac.in

Abstract. A striking feature of the solar cycle is that at the beginning, sunspots appear around mid-latitudes, and over time the latitudes of emergences migrate towards the equator. The maximum level of activity varies from cycle to cycle. For strong cycles, the activity begins early and at higher latitudes with wider sunspot distributions than for weak cycles. The activity and the width of sunspot belts increase rapidly and begin to decline when the belts are still at high latitudes. However, in the late stages of the cycles, the level of activity, and properties of the butterfly wings all have the same statistical properties independent of the peak strength of the cycles. We have modelled these features using Babcock–Leighton type dynamo model and shown that the toroidal flux loss from the solar interior due to magnetic buoyancy is an essential nonlinearity that leads to all the cycles decline in the same way.

Keywords. Solar Dynamo, Solar Cycles, Solar Magnetism

1. Introduction

The magnetic activity of the Sun, as generally measured by the presence of sunspots (or more generally bipolar magnetic regions) on the solar surface evolves in a cyclic fashion with a period of about 11 years, commonly known as the 'Solar Cycle' (Hathaway 2015). The long term study of the solar magnetic activity reveals that the 11 year solar cycles possess significant variations in their characteristics which makes the otherwise similar looking cycles unique and widely different from each other (Karak 2023; Usoskin 2013; Biswas et al. 2023b).

The strong cycles rise faster and take less time to reach peak, whereas the weaker ones rise slowly and take more time to attain their peak strength. This phenomena is known as the Waldmeier effect (Waldmeier 1935; Karak and Choudhuri 2011) which is also detected in the magnetic cycles of other solar-type stars (Garg et al. 2019). A more detailed look at the rising phases of the solar cycles reveals that the stronger cycles start early, and shows sunspot activities on higher latitudes, with wider latitudinal distribution in the initial phase, on the other hand, the weak cycles exhibit a delayed start of their activity and from lower latitude regions with narrower latitudinal bands of activity. Hence the rising phases of the cycles are quite different from each other depending on their eventual strength.

In 1955, Waldmeier reported about an even more intriguing aspect of the evolution of the solar cycles is that, during their decline phases of the cycles, they evolve with similar statistical properties irrespective of their strength (Waldmeier 1955). Later, Cameron and Schüssler (2016) analysed the century-scale sunspot observations and found out that when the cycles are represented in terms of the average latitude and width of their annual activity belt and the annual strength of their activity, the rising phases of the cycles show different trajectories of evolution however, soon after the cycles reach their peak, they evolve with a common trajectory, i.e. all the cycles decay in the same way with similar statistical properties irrespective of their peak strength.

The phenomena of solar cycles evolving in different manners through their rising phases, but possessing similar properties while declining indicates towards the presence of a nonlinearity in solar dynamo process. In this work, we perform numerical simulations of the solar dynamo to explore the role of this nonlinearity behind the aforementioned traits of the solar cycle.

2. Model description:

The underlying reason behind the sustained cyclic variation of strength of the solar magnetic fields is believed to be the dynamo mechanism working in the solar convection zone. In recent times, the Babcock–Leighton (Babcock 1961; Leighton 1969) type solar dynamo models have made significant progress in explaining various observed features of magnetic activity of the Sun and Sun like stars (Charbonneau 2020; Karak 2010; Karak and Miesch 2018; Karak et al. 2018; Vashishth et al. 2021, 2023) and have been extensively used for forecasting solar cycle strength (Choudhuri et al. 2007; Bhowmik and Nandy 2018; Kumar et al. 2021b; Kumar et al. 2022; Biswas et al. 2023).

In this work, we are utilizing a Babcock–Leighton type, axisymmetric, kinematic solar dynamo code 'SURYA' (Nandy and Choudhuri 2002; Chatterjee et al. 2004; Karak and Choudhuri 2012) which solves the following two coupled equations to capture the evolution of the poloidal and toroidal component of the solar magnetic field:

$$\frac{\partial A}{\partial t} + \frac{1}{s}(\boldsymbol{v}_p \cdot \nabla)(sA) = \eta_p \left(\nabla^2 - \frac{1}{s^2}\right) A + \alpha B, \tag{1}$$

$$\frac{\partial B}{\partial t} + \frac{1}{r}\left[\frac{\partial (r v_r B)}{\partial r} + \frac{\partial (v_\theta B)}{\partial \theta}\right] = \eta_t \left(\nabla^2 - \frac{1}{s^2}\right) B + s(\boldsymbol{B}_p \cdot \nabla)\Omega + \frac{1}{r}\frac{d\eta_t}{dr}\frac{\partial (rB)}{\partial r}, \tag{2}$$

Where, $s = r\sin\theta$, A is the magnetic potential of the poloidal magnetic field, B is the toroidal magnetic field. A detailed discussion about the model parameters for this particular work can be found in Biswas et al. (2022).

Here we describe the source of nonlinearity in this dynamo model which is related to the process of buoyant rise of the toroidal flux, i.e. the formation of sunspots. In every certain time interval it is checked whether the strength of the toroidal magnetic field in the convection zone has crossed a critical value B_c. If at any grid point the field is found to have crossed B_c, it is assumed that flux tube will buoyantly emerge to the surface. The value of the magnetic field is locally reduced by half and the other half is added to the surface. This is how the formation of sunspot due to the buoyant rise of the toroidal flux tubes are captured in the model in a simple way to ensure that with each sunspot eruption, a part of the toroidal flux is lost from the solar convection zone. The value of the critical field is taken as $B_c = 0.8 \times 10^4\ G$ to keep the surface radial field strength within the observed range (Mordvinov et al. 2022; Golubeva et al. 2023), and to keep the total amount of toroidal flux loss consistent with observational estimates (Cameron and Schüssler 2020). It is worth mentioning that, just by constraining the model parameters from observations, we find that the value of the critical field B_c is similar to the equipartiiton field strength at the base of solar convection zone, as calculated from the mixing length approximation.

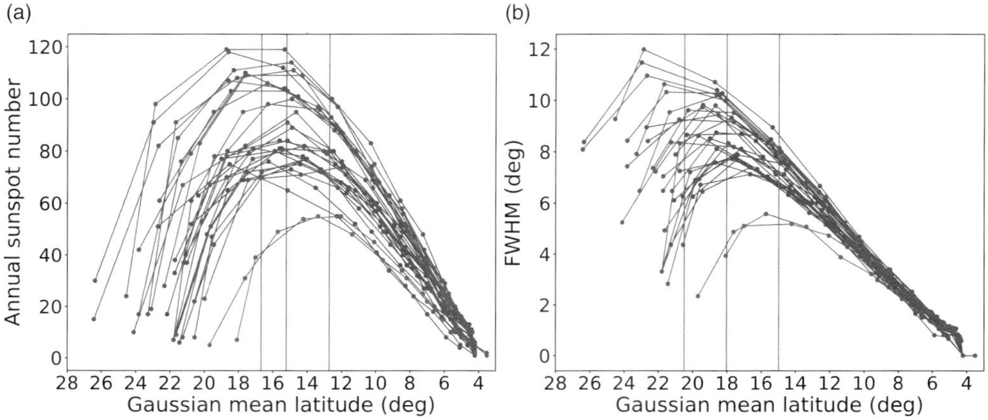

Figure 1. (a) The model sunspot number for each year plotted as the function of the mean latitude of the annual sunspot distribution. Each curve correspond to one cycle. (b) Same as (a) but with the width of the annual latitudinal distribution. The cycles begin from the left (at higher latitudes) and progress towards the right with time. The vertical lines correspond to the average latitude where the strong (red), moderate (blue), and weak (green) cycles reach their peak.

3. Results and Discussions

Following the model setup described in the previous section, we perform simulation spanning over 40 cycles. The (pseudo) sunspots for each of the cycles from the simulation are tracked by the latitude of their emergence. The annual latitudinal distribution of the sunspots approximately follow a Gaussian distribution similar to what is seen in observation (Cameron and Schüssler 2016). From these distributions, the central latitude of the activity belts and the width of their distribution are measured. Here in the Figure 1, the trajectories of the cycles are shown as the sunspot activity belts migrate from high latitude regions towards the equator with the progress of the cycle. It can be clearly seen that during the beginning phases of the cycles, when the sunspot activities are centred on higher latitude regions, the cycles evolve very differently from each other, the strong cycles start early at higher latitudes, with wider activity belts, they rise rapidly and attain their peak at higher latitudes, on the other hand, the weaker cycles, start lately, from somewhat lower latitude regions, rise slowly and eventually attain their maxima at even lower latitudes. On the other hand, during the decline phases of the cycles, the trajectories merge quickly after obtaining the peaks and the cycles evolve with similar properties. Hence, the observed features of the latitudinal distribution of the solar magnetic activity throughout the phases of solar cycle as described earlier is very well reproduced in this model.

Now, let us discuss the role of the nonlinear flux loss in the evolution of solar cycles with different strengths. In a strong cycle, the poloidal field strength during the minima of the cycle is strong, as a result the rate of production of the toroidal field due to the action of differential rotation is high, hence the toroidal flux emergence condition is fulfilled very early in the cycle in a wider latitudinal region. As a result, the sunspot activity for these cycles begin very early with a wider latitudinal distribution and at higher latitudes. Due to the high production rate of the toroidal field, the emergence rate of the sunspots are also high here, however, with each sunpsot emergence, a significant portion of the toroidal flux is lost from the solar interior. As a result, the toroidal field strength, attains saturation very quickly at high latitudes, when the subsurface toroidal field strength becomes comparable to B_c, marking the maxima of the solar cycle. During the declining phase of the cycles, any further production of the toroidal flux is taken care of by the further loss of flux due to emergences as the activity belts get advected towards the equator by the meridional circulation.

In the case of a weak cycle, the poloidal field during the minima is weak and as a result, the toroidal flux is generated at a low pace. Hence, it takes more time for the toroidal field to meet the threshold for flux emergence. By the time the toroidal field becomes strong enough to produce sunspots, the merdiional circulation transports the activity belt to lower latitudes, this explains why the weak cycles exhibit a delayed start and begin their journey from a low latitude. Also, due to the low rate of toroidal field production, lead to the flux eruption condition getting satisfied only at a narrow latitudinal regions and the activity of the cycle rise slower than the stronger cycles. The weak cycles lose less amount of toroidal flux due to their low rise rate, hence the toroidal field takes more time to be saturated and in the meanwhile, the meridional circulation advects the activity belts to further lower latitudes making the weaker cycles reach their peak at lower latitudes in comparison to the stronger ones. However, during the decline phase, the weaker cycles evolve exactly the same way as the stronger ones, the value of the toroidal field stays comparable to the critical value B_c and any further build-up of the toroidal flux is taken care of by the further flux lost due to emergencies.

4. Conclusion

In conclusion, we find that a strong nonlinear mechanism of toroidal flux loss due to flux emergence is essential to reproduce the observed features of the latitudinal distributions of solar cycle activity. This mechanism significantly enhances the amount and rate of flux lost during the rising phases of the strong cycles in comparison to weak cycles ensuring that the morphology of the toroidal magnetic field inside the Sun's convection zone becomes similar for all cycles during their decline phases, hence all the cycles decline in the same way having similar statistical properties irrespective of their peak strength. By comparing the model outputs with observations, it has been found that, the critical field strength (B_c) for the threshold of flux emergence is similar to the equipartition field strength at solar convection zone. Our study also provides an explanation of the so-called latitudinal quenching—strong cycles produce BMR at high latitudes (Mandal et al. 2017), which in turn generates a weak polar field and stabilizes the solar dynamo (Jiang 2020; Karak 2020)

5. Acknowledgements

I thank the International Astronomical Union for the award of the generous travel grant to attend the in-person IAU Symposium 365, in Yerevan Armenia. The financial support from the University Grants Commission, Govt. of India is gratefully acknowledged.

References

Babcock, H. W. 1961, The Topology of the Sun's Magnetic Field and the 22-YEAR Cycle. *Astrophys. J.*, 133, 572.

Bhowmik, P. & Nandy, D. 2018, Prediction of the strength and timing of sunspot cycle 25 reveal decadal-scale space environmental conditions. *Nature Communications*, 9, 5209.

Biswas, A., Karak, B. B., & Cameron, R. 2022, Toroidal Flux Loss due to Flux Emergence Explains why Solar Cycles Rise Differently but Decay in a Similar Way. *Phys. Rev. Lett.*, 129(24), 241102.

Biswas, A., Karak, B. B., & Kumar, P. 2023,a Exploring the reliability of polar field rise rate as a precursor for an early prediction of solar cycle. *Mon. Not. R. Astron. Soc.*, 526a(3), 3994–4003.

Biswas, A., Karak, B. B., Usoskin, I., & Weisshaar, E. 2023,b Long-Term Modulation of Solar Cycles. *Space Sci. Rev.*, 219b(3), 19.

Cameron, R. H. & Schüssler, M. 2016, The turbulent diffusion of toroidal magnetic flux as inferred from properties of the sunspot butterfly diagram. *Astron. Astrophys.*, 591, A46.

Cameron, R. H. & Schüssler, M. 2020, Loss of toroidal magnetic flux by emergence of bipolar magnetic regions. *Astron. Astrophys.*, 636, A7.

Charbonneau, P. 2020, Dynamo models of the solar cycle. *Living Reviews in Solar Physics*, 17(1), 4.

Chatterjee, P., Nandy, D., & Choudhuri, A. R. 2004, Full-sphere simulations of a circulation-dominated solar dynamo: Exploring the parity issue. *Astron. Astrophys.*, 427, 1019–1030.

Choudhuri, A. R., Chatterjee, P., & Jiang, J. 2007, Predicting Solar Cycle 24 With a Solar Dynamo Model. *Physical Review Letters*, 98(13), 131103.

Garg, S., Karak, B. B., Egeland, R., Soon, W., & Baliunas, S. 2019, Waldmeier Effect in Stellar Cycles. *arXiv e-prints*,, arXiv:1909.12148.

Golubeva, E. M., Biswas, A., Khlystova, A. I., Kumar, P., & Karak, B. B. 2023, Probing the variations in the timing of the Sun's polar magnetic field reversals through observations and surface flux transport simulations. *Mon. Not. R. Astron. Soc.*, 525(2), 1758–1768.

Hathaway, D. H. 2015, The Solar Cycle. *Living Reviews in Solar Physics*, 12(1), 4.

Jiang, J. 2020, Nonlinear Mechanisms that Regulate the Solar Cycle Amplitude. *Astrophys. J.*, 900(1), 19.

Karak, B. B. 2010, Importance of Meridional Circulation in Flux Transport Dynamo: The Possibility of a Maunder-like Grand Minimum. *Astrophys. J.*, 724, 1021–1029.

Karak, B. B. 2020, Dynamo Saturation through the Latitudinal Variation of Bipolar Magnetic Regions in the Sun. *Astrophys. J. Lett.*, 901(2), L35.

Karak, B. B. 2023, Models for the long-term variations of solar activity. *Living Reviews in Solar Physics*, 20(1), 3.

Karak, B. B. & Choudhuri, A. R. 2011, The Waldmeier effect and the flux transport solar dynamo. *Mon. Not. R. Astron. Soc.*, 410, 1503–1512.

Karak, B. B. & Choudhuri, A. R. 2012, Quenching of Meridional Circulation in Flux Transport Dynamo Models. *Sol. Phys.*, 278, 137–148.

Karak, B. B., Mandal, S., & Banerjee, D. 2018, Double Peaks of the Solar Cycle: An Explanation from a Dynamo Model. *Astrophys. J.*, 866(1), 17.

Karak, B. B. & Miesch, M. 2018, Recovery from Maunder-like Grand Minima in a Babcock–Leighton Solar Dynamo Model. *Astrophys. J. Lett.*, 860, L26.

Kumar, P., Biswas, A., & Karak, B. B. 2022, Physical link of the polar field buildup with the Waldmeier effect broadens the scope of early solar cycle prediction: Cycle 25 is likely to be slightly stronger than Cycle 24. *Mon. Not. R. Astron. Soc.*, 513(1), L112–L116.

Kumar, P., Nagy, M., Lemerle, A., Karak, B. B., & Petrovay, K. 2021, The Polar Precursor Method for Solar Cycle Prediction: Comparison of Predictors and Their Temporal Range. *Astrophys. J.*, 909(1), 87.

Leighton, R. B. 1969, A Magneto-Kinematic Model of the Solar Cycle. *Astrophys. J.*, 156, 1.

Mandal, S., Karak, B. B., & Banerjee, D. 2017, Latitude Distribution of Sunspots: Analysis Using Sunspot Data and a Dynamo Model. *Astrophys. J.*, 851, 70.

Mordvinov, A. V., Karak, B. B., Banerjee, D., Golubeva, E. M., Khlystova, A. I., Zhukova, A. V., & Kumar, P. 2022, Evolution of the Sun's activity and the poleward transport of remnant magnetic flux in Cycles 21-24. *Mon. Not. R. Astron. Soc.*, 510(1), 1331–1339.

Nandy, D. & Choudhuri, A. R. 2002, Explaining the Latitudinal Distribution of Sunspots with Deep Meridional Flow. *Science*, 296(5573), 1671–1673.

Usoskin, I. G. 2013, A History of Solar Activity over Millennia. *Living Reviews in Solar Physics*, 10, 1.

Vashishth, V., Karak, B. B., & Kitchatinov, L. 2021, Subcritical dynamo and hysteresis in a Babcock–Leighton type kinematic dynamo model. *Research in Astronomy and Astrophysics*, 21(10), 266.

Vashishth, V., Karak, B. B., & Kitchatinov, L. 2023, Dynamo modelling for cycle variability and occurrence of grand minima in Sun-like stars: rotation rate dependence. *Mon. Not. R. Astron. Soc.*, 522(2), 2601–2610.

Waldmeier, M. 1935, Neue Eigenschaften der Sonnenfleckenkurve. *Astronomische Mitteilungen der Eidgenössischen Sternwarte Zurich*, 14, 105–136.

Waldmeier, M. 1955, Ergebnisse und Probleme der Sonnenforschung. *Ergebnisse und Probleme der Sonnenforschung (Leipzig: Geest & Portig).*

Dynamics of Solar and Stellar Convection Zones and Atmospheres
Proceedings IAU Symposium No. 365, 2024
A. V. Getling & L. L. Kitchatinov, eds.
doi:10.1017/S1743921323004982

Dominating active regions in the minima of solar activity

R. A. Suleymanova and V. I. Abramenko

Crimean Astrophysical Observatory of Russian Academy of Sciences, Nauchny, Russia
email: bictr97@gmail.com

Abstract. During minima of solar activity, it is possible to estimate the influence of convection zone turbulence on the magnetic flux tubes forming active regions (ARs), because the toroidal field of the old cycle weakens, and the new toroidal field is still weak. We analyzed ARs of solar minima between 23-24 and 24-25 solar cycles. ARs were classified as regular, irregular and unipolar spots. Regular ARs follow the empirical laws consistent with the Babcock–Leighton dynamo theory. We found that regular ARs dominate by flux and by number during the solar minima. Irregular ARs are mainly represented by bipolar structures of deformed orientation and contribute only one-third in the total flux and one-third in the total number. Very complex multipolar ARs are extremely rare. So, during solar minima the global dynamo still guides the formation of ARs, whereas the turbulence only slightly affects the toroidal flux tubes orientation.

Keywords. Sun:magnetic fields, Sun:active regions, Sun:solar cycle

1. Introduction

According to Babcock–Leighton dynamo theory Babcock (1961), Leighton (1969) the solar activity is cyclic: in the maximum phase there are strong toroidal filed, increased number of active regions (ARs) and flares, whereas in the minimum there is strong poloidal field, the toroidal field of the previous cycle ceased and the new toroidal field is still weak. At the same time the turbulence in the convection zone works with the same intensity through all of these periods, so at the minima of solar activity the manifestation of turbulence on the AR formation might be the most pronounced. The aim of our work is to clear up what component of solar dynamo is the most effective in the minima of solar activity: global dynamo or turbulent component of dynamo.

2. Data and methods

We considered two periods: from January, 2008 till June, 2009 (the deep minimum between cycles 23rd-24th) and from April, 2018 till April, 2020 (the deep minimum between cycles 24th-25th). These intervals were chosen according to the curve of the total sunspot area variations presented at Abramenko *et al.* (2023) and obtained from the monthly average sunspot area data (`http://solarcyclescience.com/AR_Database/sunspot_area.txt`). The curve is marked with dash-dotted line and intervals are marked with vertical dotted lines in Fig. 1 and 2.

For 23rd cycle we used line-of-sight (LOS) magnetograms in Ni I 6768 Å line with a spatial resolution of 4 arcsec acquired by Michelson Doppler Imager (MDI) on board the Solar and Heliospheric Observatory (SOHO) Scherrer *et al.* (1995). For 24th cycle we used LOS magnetograms (Space-weather HMI Active Region Patches, SHARP, sharp_cea_720s) in Fe I 6173.3 Å line with the spatial resolution of 1 arcsec acquired by Helioseismic and Magnetic Imager (HMI) on board the Solar Dynamics Observatory (SDO) Scherrer *et al.*

Table 1. The ARs during solar minima 23rd-24th, 24th-25th (beginning of the table).

NOAA	Date of observation	Coordinates (latitude; longitude)	Magnetic flux, 10^{21} Mx	HJL	MMC class	MW class
10982	2008.01.31	-8.39; -35.37	11.0	YYN	B1	β
10983	2008.02.26	-5.29; 42.41	2.4	YYN	B1	α
10987	2008.03.27	-7.60; -3.88	14.3	YYY	A1	β
10988	2008.03.271	-7.61; -28.10	15.1	YYY	A1	β
10989	2008.03.28	-11.00; -46.44	9.8	Y0P	U	β
10992	2008.04.23	12.25; 5.72	2.4	YNY	B1	β

(2012), Schou *et al.* (2012). To retrieve the data we used the website of the Joint Science Operations Center (JSOC, `http://jsoc.stanford.edu/`).

Each AR located closer than 60° from the disc center and having the unsigned magnetic flux above 10^{21} Mx was counted once at the moment of its maximum development. This way, we found 21 ARs during the first minimum and 41 ARs during the second minimum.

To distinguish between the global dynamo action and the turbulent component of the dynamo, we need to classify each AR in the framework of the magneto-morphological classification (MMC), described in Abramenko *et al.* (2018), Abramenko (2021), Abramenko *et al.* (2023). In brief, the regular ARs produced by the global dynamo are classified as A-class. They follow the empirical lows of the mean-field dynamo theory, in particular, the Hale polarity law, the Joy's law, the prevalence of the leading spot Babcock (1961), van Driel–Gesztelyi & Green (2015). All the rest, except unipolar spots, are the ARs violating the empirical laws and therefore might be considered as disturbed by turbulence in the convection zone. The degree of disturbance is reflected in the complexity of the ARs which increases from B1, to B2, to B3 class.

In this manner, we classified all 62 ARs. Results are collected in the table, the beginning of which is shown in Table 1. Except for the common parameters in the four columns (NOAA number, date of observation, coordinates, total magnetic flux), the column 5 represents the perfection of the empirical laws: YES (Y) or NO (N) in the following order: Hale law, Joy's law, the leading sunspot prevalence. The next two columns show the AR class: according to MMC and to the Mount Wilson classification Hale et al. (1919). Out of all 62 ARs, 52 turned to belong to the β class by Hale classification. So, we don't get much for our purposes of the dynamo diagnostics, therefore below we considered only the MMC class.

3. The ARs distributions during the minima of solar activity

Fig. 1 represents how ARs are distributed in the time-flux diagram during to minima 23rd-24th (panel a) and 24th-25th (panel b). The dash-dotted line shows the average monthly sunspot areas smoothed over an interval of 13 months, and dashed vertical lines show the accepted boundaries of the deep cycle minimum. Regular ARs are represented by blue color: asterics for A1 class, straight crosses for A2 class. Red symbols show irregular ARs: filled circles for B1 class, open circles for B2 class. Green rhombuses show the unipolar sunspots.

The comparison of distribution of regular ARs (top panels) with that for irregular ARs (bottom panels) is shown in Fig. 2. During each minimum the number of regular ARs more than twice exceeds the number of irregular. So, the toroidal field of the global dynamo steel prevails the turbulent convection in generation of ARs. Out of the total number of 18 irregular ARs during both minima 15 of them are wrong bipolar structures of B1 class, so the influence of the turbulence is rather mild during the minimum. It is also interesting to note that there are only 3 ARs of B2 class and none of ARs of the most complex B3 class.

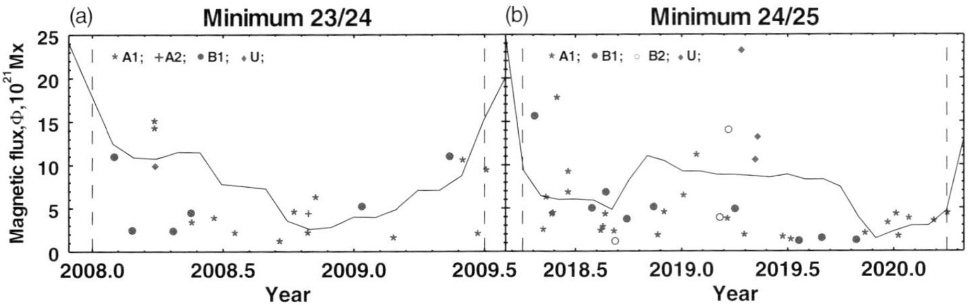

Figure 1. Occurrance of ARs during 2 solar minima: blue symbols represent regular ARs (asterics for A1 class, straight crosses for A2 class), red symbols represent irregular ARs (filled circles for B1 class, open circles for B2 class). Filled green rhombuses represent the unipolar sunspots. The dash-dotted line and dashed vertical lines show the average mounthly sunspot areas and selected time interval respectively.

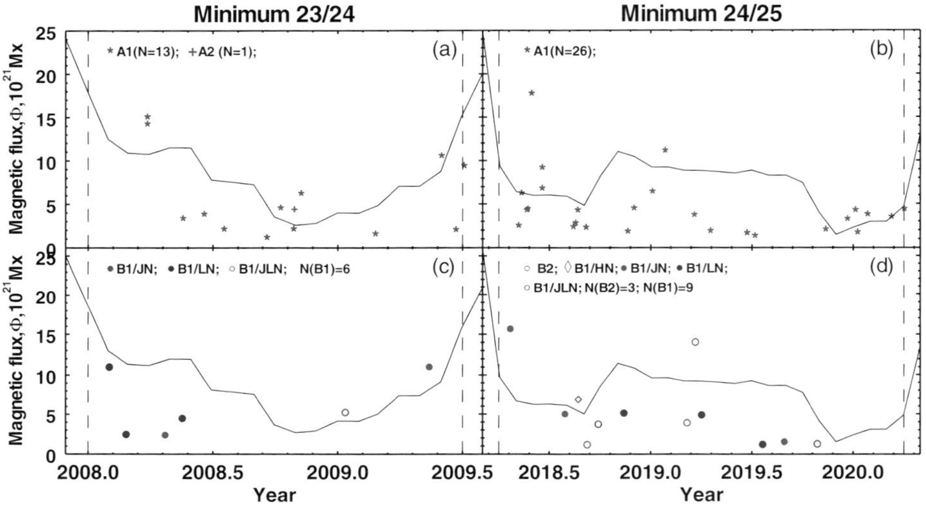

Figure 2. A separate representation of regular and irregular ARs occurrance during 2 solar minima. The designations of A1, A2 and B1 classes coinside with designations in Fig. 1. Red rhombuses represent ARs of B1 class with Hale's law violation (B1/HN). ARs with wrong Joy's law violation are represented by filled red circles (B1/JN). Purple filled circles are represent ARs with violation the rule of dominance of the leader spot (B1/LN). Open purple circles show the ARs that not obey the Joy's law and the rule of dominance of the lider spot (B1/JLN). Other notations are the same as in Fig. 1.

It can be noticed that for the entire considered intervals, regular ARs dominate over irregular ones. Consequently, both the previous toroidal field of old cycle and the new toroidal field of the oncoming cycle work against the background of weak fluctuations caused by turbulence. It could be concluded that turbulence by itself cannot cause a large flow, since it can only distort the existing flow.

The composition of the summed flux from all ARs (gray bars) during minima (top panels) and maxima (bottom panels) is presented in Fig. 3. We see that the total flux from regular ARs dominates twice the total flux of irregular ARs. However, during the maxima, the total flux of regular ARs is approximately equal to the summed flux from all irregular ARs. In other words, during minima the global dynamo dominates in ARs production whereas during the maxima about half of total ARs flux is distorted by turbulence.

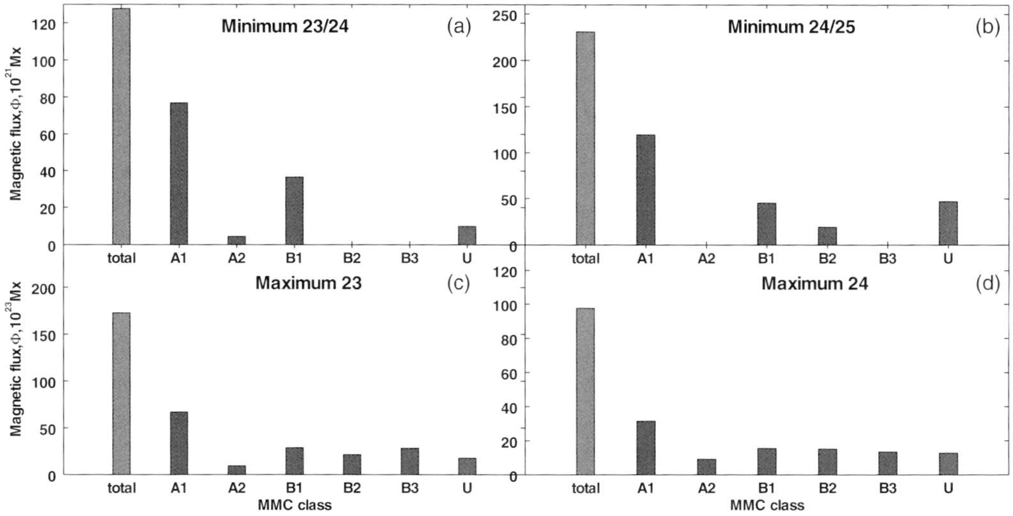

Figure 3. A histogram of ARs over different classes. Upper row, data on two minima studied in this work; bottom row - data for the 23rd cycle maximum, and for the 24th cycle maximum. The vertical scales differ by two orders of magnitude.

4. Conclusions

The following conclusions can be drawn:

- Regular ARs were observed twice more frequently than irregular ARs.
- The summed flux from regular ARs was twice higher than that from irregular ARs.
- The simplest bipolar structures of A1 and B1 classes consist of 87% from all observed ARs.
- During solar minima the magnetic flux from ARs is still dominated by the toroidal flux of global dynamo with very mild influence from the turbulent component of the dynamo.

Acknowledgments

SDO/HMI data is available through the SDO/HMI and SDO/AIA Science Groups. SDO is NASA's Living With a Star (LWS) mission. SOHO/MDI data is available through the SOHO/MDI and SOHO/EIT Science Groups. SOHO is an international collaboration between ESA and NASA. The SDO/HMI and SOHO/MDI data were provided by the Joint Science Operation Center (JSOC).

References

Abramenko, V.I. 2021, *MNRAS*, 507, 3698.
Abramenko, V.I., Suleymanova, R.A., Zhukova, A.V. 2023, *MNRAS*,518,, 4746.
Abramenko, V.I., Zhukova, A.V., Kutsenko, A.S. 2018, *G&A*,58, 1159.
Babcock, H.W. 1961, *A&A*, 133, 572.
Hale, G.E., Ellerman, F., Nicholson, S.B., *et. al* 1919, *ApJ*,49, 153.
Leighton, R.B. 1969, *ApJ*, 156, 1.
Scherrer, P.H., Bogart, R.S., Bush, R.I.,, *et al.* 1995, *SoPh*,162, 129.
Scherrer, P.H., Schou, J., Bush, R.I.,, *et al.* 2012, *SoPh*,275, 207.
Schou, J., Scherrer, P.H., Bush, R.I.,, *et al.* 2012, *SoPh*,275, 229.
van Driel–Gesztelyi, L., Green, L.M. 2015, *LRSP*,12, 1.

Dynamics of Solar and Stellar Convection Zones and Atmospheres
Proceedings IAU Symposium No. 365, 2024
A. V. Getling & L. L. Kitchatinov, eds.
doi:10.1017/S1743921324000905

Understanding Grand Minima in Solar Activity: Confronting Observations with Dynamo Simulations

Chitradeep Saha[1] and **Dibyendu Nandy**[1,2]

[1]Center of Excellence in Space Sciences India, Indian Institute of Science Education and Research Kolkata, Mohanpur 741246, West Bengal, India
email: cs20rs018@iiserkol.ac.in

[2]Department of Physical Sciences, Indian Institute of Science Education and Research Kolkata, Mohanpur 741246, West Bengal, India

Abstract. The grand minimum in the Sun's activity is a distinctive mode characterized by a magnetic lull that almost completely lacks the emergence of sunspots on the solar surface for an extended duration. The factors driving this transition of an otherwise magnetically active star into a quiescent phase, the processes occurring within the solar interior and across the heliosphere during this period, and the mechanisms leading to the eventual resurgence of surface magnetic activity remain enigmatic. However, there have been sustained efforts in the past few decades to unravel these mysteries by employing a combination of observation, reconstruction and simulation of solar magnetic variability. Here, we summarize recent research on the solar grand minimum and highlight some outstanding challenges – both intellectual and practical – that necessitate further investigations.

Keywords. Solar magnetic activity, Solar grand minimum, Solar dynamo, Sunspot cycle

1. Introduction

The Sun is a magnetized star that exhibits variability in its magnetic activity across various timescales ranging from minutes to millennia and beyond (Nandy et al. 2021; Pevtsov et al. 2023; Biswas et al. 2023). The characteristic decadal variation in solar magnetic activity, i.e., the ~11-yr Schwabe cycle is well observed and recorded for multiple centuries (Hathaway 2015). Besides, there have been several instances of intermittent periods characterized by a magnetically quiescent state of the Sun persisting over decades and even centuries – known as the solar grand minimum – as evidenced in reconstructed solar activity using indirect proxies such as cosmogenic isotopes and auroral activity records (Solanki et al. 2004; Steinhilber et al. 2009; Usoskin 2023).

From an observational perspective, solar grand minima are marked by a significant decrease in sunspot numbers (see, Fig. 1) and a reduction of large-scale solar magnetic field strength and open flux for a prolonged period (Vaquero et al. 2015; Carrasco et al. 2019; Usoskin et al. 2021b). Historically, solar grand minima have been associated with periods of cooler temperatures on Earth, as evidenced by the temporal proximity of the Maunder Minimum (1645–1715) and the Little Ice Age, hinting towards a potential link between solar grand minima and terrestrial cooling (Eddy 1976). However, whether the latter is causally concomitant with the former is debatable (Owens et al. 2017). Nevertheless, the Sun being the primary driver of heliospheric space environmental conditions, variabilities in solar open flux eventually influence the space weather and space

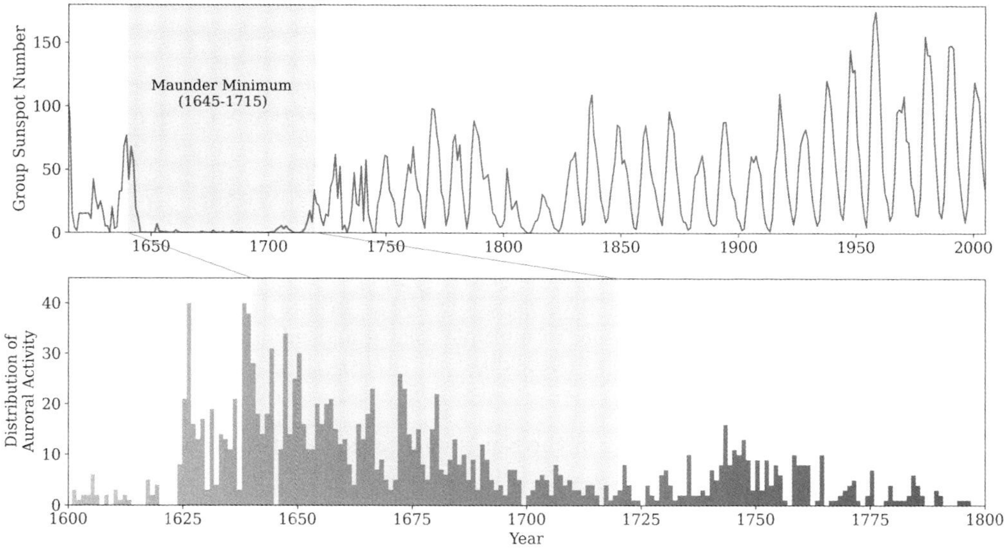

Figure 1. Top panel: group sunspot number time series (top panel) highlighting one of the most recent solar grand minima during 1645-1715 – known as the Maunder minimum (Eddy 1983). Bottom panel: Temporal distribution of auroral activity records (Wang et al. 2021) suggesting a persistent solar activity modulation during Maunder minimum.

climate over short and long timescales, respectively (Schrijver et al. 2015; Nandy et al. 2023).

On the other hand, the precision in modeling regular solar-like activity cycles and predicting the amplitude and progression of upcoming cycles has significantly advanced in recent decades (Bhowmik and Nandy 2018; Petrovay 2020; Nandy 2021). In contrast, the ability to predict the onset of solar grand minimum and its duration has yet to achieve substantial success. In fact, the unusually long minimum of sunspot cycle 23 (Nandy et al. 2011) prompted debate and speculation among researchers about the potential occurrence of an impending solar grand minimum (Solanki and Krivova 2011; Zolotova and Ponyavin 2014).

It is now widely acknowledged that there operates a dynamo mechanism in the solar convection zone (SCZ), and it is responsible for the periodic generation and recycling of large-scale solar magnetic fields (Charbonneau 2020). Owing to the extreme physical conditions in the SCZ, such as density and temperature stratification, turbulent plasma motions, magnetic stresses and back-reactions, among others, the dynamo mechanism exhibits irregular behavior. This irregularity is believed to account for the observed variability in the strength and timing of sunspot cycles. As a theoretical limit, it is hypothesized that this irregularity may occasionally intensify to the extent that it can drive the dynamo operation below a critical threshold, thereby causing an intermittent behavior in the solar magnetic activity – the solar grand minimum (Passos et al. 2014). A more detailed discussion on various approaches towards modeling solar activity extrema follows in section 3.

The primary challenge in understanding solar grand minima is posed due to insufficient observational constraint on dynamo models, unlike regular solar activity cycles (Muñoz-Jaramillo and Vaquero 2019). Thanks to various terrestrial archives – such as tree rings and ice cores – that preserve the millennial timescale solar activity signatures and give significant insights into solar magnetic behavior during grand minima (see, Fig. 1 & 2). However, numerical simulations remain indispensable for gaining knowledge

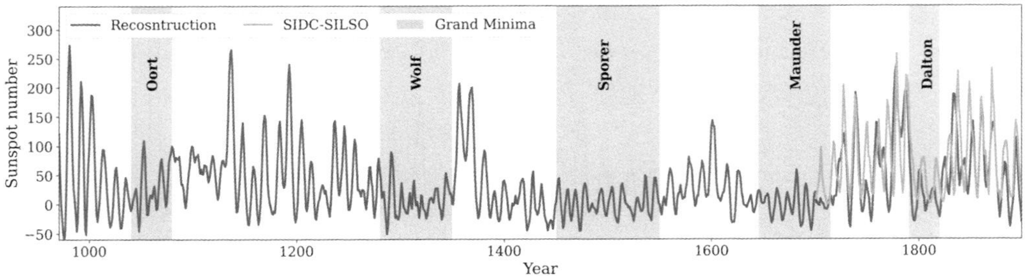

Figure 2. Direct sunspot records complemented by solar activity reconstructions using ^{14}C cosmogenic isotope put forward evidence of multiple solar grand minima (shaded in gray) that have occurred over the past millennium (Usoskin et al. 2021a).

of the dynamics during such phases in the Sun's uncharted territories, such as its interior and polar regions.

In the subsequent sections, we aim to revisit our current understanding of the solar grand minimum in terms of both numerical simulations and observations, emphasize the complementarity between reconstruction, observation and physics-based numerical simulations of extreme solar activity, discuss critical insights gleaned from recent results and their implications on our space environment. We also list down some of the outstanding aspects of solar/stellar grand minima that are required to be explored further.

2. Solar grand minimum: Fact or Artifact?

Before investigating deeper into solar grand minima, a common skepticism arises whether these episodes of reduced sunspot numbers genuinely exist or they are merely artifacts resulting from limited and sparse observational data points (Hoyt and Schatten 1996). Based on historical records, Zolotova and Ponyavin (2015) argued that the Maunder Minimum could be an ordinary secular minimum with diminished decadal variability. (Feynman and Ruzmaikin 2011) claimed the Maunder Minimum to be the trough of the centennial Gleissberg cycle. Even if grand minima do exist, there remains considerable uncertainty and a lack of consensus regarding their impact on terrestrial cooling and climatology (Owens et al. 2017).

Nevertheless, sophisticated methods of solar magnetic activity reconstruction based on various proxies like the terrestrial abundance of cosmogenic isotopes reveal that there were indeed multiple episodes of critically low magnetic flux output from the Sun for prolonged duration over the past few millennia (Steinhilber et al. 2009; Usoskin et al. 2015; Usoskin 2023). Data gleaned from these sources could be calibrated well with the modern-day sunspot record, suggesting the reliability of the reconstruction techniques. Recently, Hayakawa et al. (2024) explored ancient sunspot records at the onset of the Maunder minimum that corroborate well with the proxy records. Evidence of declining auroral activity during the deep phase of the Maunder Minimum put independent support to the reduced solar activity during this time (Wang et al. 2021) (see, Fig. 1).

Moreover, multiple works have theorized the solar grand minimum as an outcome of the irregularity in the dynamo mechanism and have successfully reproduced many of the observational signatures and statistics of solar grand minimum (Passos et al. 2014; Tripathi et al. 2021; Saha et al. 2022). This strengthens the scenario in favour of the solar grand minimum being a reality and prompts that there may be possibilities of impending intermittency again in the solar variability.

Muñoz-Jaramillo and Vaquero (2019) presented an exhaustive summary of various observational datasets covering the past few centuries and their collective reliability.

While it is a fact that there exist not many coherent observations of the Sun during the Maunder Minimum phase, a period of prolonged quiescence emerges in all the datasets, suggesting that one cannot completely rule out the occurrence of solar grand minima Usoskin et al. (2015).

3. Numerical modeling of intermittent solar activity: Current understanding

In a cursory view, solar grand minima episodes are mainly devoid of surface eruptions of sunspots. This phenomenon can theoretically be modelled as a reduced abundance of bipolar magnetic sources in the solar surface flux transport models (Mackay 2003; Wang and Sheeley Jr 2003). Such models regulate the production rate of magnetic bipoles at different phases of sunspot cycles to mimic observed statistics of the depth and duration of grand minima (Usoskin et al. 2007). However, the underlying reasons for such non-uniform regulation are not imbibed into these models.

One of the proposed avenues to numerically model grand minima like extreme solar variability is to introduce self-consistent irregularities, i.e., stochastic forcing on the dynamo processes (Moss et al. 2008; Usoskin et al. 2009). These irregularities predominantly manifest during the induction of large-scale poloidal fields. Therefore, fluctuation in poloidal source intensity across the solar hemispheres can drive the solar dynamo into a sub-critical regime producing intermittent grand minimum-like episodes (Charbonneau et al. 2004; Olemskoy et al. 2013; Brandenburg and Spiegel 2008; Passos et al. 2014; Saha et al. 2022). These models incorporate random fluctuations in the poloidal source, in the Babcock–Leighton framework or the electromotive force in the Parker-type mean-field dynamo framework as a signature of stochastic forcing. Spatially reduced simpler time-delay dynamo models can also simulate the entry and recovery from grand minima episodes when subjected to stochastic driving (Wilmot–Smith et al. 2005, 2006; Hazra et al. 2014; Tripathi et al. 2021).

Magnetically buoyant flux tubes produced deep in the SCZ rise upward and are subject to turbulent buffeting by the convective plasma flows. This eventually introduces a dispersion around the mean tilt angle distribution of sunspots (Dasi–Espuig, M. et al. 2010; Nagy et al. 2017; Pal et al. 2023). Migration and diffusion of randomly tilted bipolar magnetic regions on the solar surface eventually influence the polar field production rate and, in turn, the strength of the next sunspot cycle (Yeates et al. 2008). Large anomaly in the bipoles can potentially simulate grand minima episodes (Karak and Miesch 2018).

In axisymmetric flux transport dynamo models, a reduction in meridional circulation flow speed either in seclusion or in conjunction with fluctuating poloidal sources can replicate grand minimum-like activity phases (Karak 2010; Choudhuri and Karak 2012). However, the observed variation in this weak flow is relatively small. More importantly, due to active region inflows, the meridional circulation is observed to be faster during solar activity minimum and vice-versa as shown by Hathaway and Rightmire (2010). They also pointed out a fundamental limitation of flux transport dynamo models in order to explain this discrepancy. Some more recent observational evidence indicating accelerated dynamics in solar activity during grand minima (Yan et al. 2023; Herrera et al. 2024) challenges the proposition of weak meridional flow as a cause of this intermittency.

The nonlinear back-reaction of magnetic fields on plasma flows in the convection zone is a plausible mechanism for generating magnetically quiescent phases (Inceoglu et al. 2017). Mathematically, this phenomenon can be modeled by incorporating either Lorentz feedback, algebraic α-quenching, Λ-quenching, or a combination of these effects (Tobias 1996; Küker et al. 1999; Ossendrijver 2000; Simard and Charbonneau 2020). However,

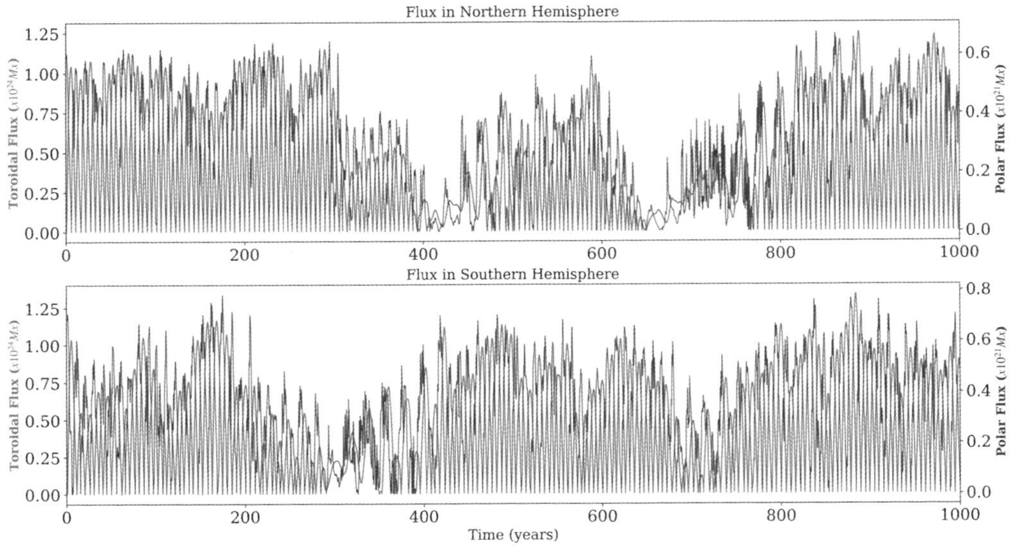

Figure 3. Results from stochastic numerical dynamo simulation by Saha et al. (2022) showing the evolution of solar poloidal and toroidal magnetic flux. Multiple episodes of grand minima with significantly low magnetic output are captured independently in the solar northern (top panel) and southern (bottom panel) hemispheres.

the solar dynamo is shown to be only weakly nonlinear. Additionally, the random fluctuations are shown to play a more crucial role in explaining long-term solar variabilities (Cameron and Schüssler 2019).

Only a few studies have been conducted to investigate extreme solar activity using MHD simulation framework. Augustson et al. (2015) demonstrated episodes resembling grand minima in a convection-driven stellar dynamo model. Achieving a self-consistent onset and recovery of a Sun-like star into a grand minimum has yet to be accomplished in direct numerical simulations – underscoring that our theoretical comprehension of grand minima and the overall solar dynamo mechanism is not yet complete.

4. Persistence of dynamo activity during solar grand minimum: Observational and numerical evidence

Does the underlying dynamo mechanism cease when the solar surface magnetic activity slips to a slumber? For the first time, a reasonable clue was put forward by Beer et al. (1998) in the form of observational evidence suggesting a persistent decadal timescale modulation in reconstructed solar activity, even during grand minima. Later, several other reconstructions have corroborated this finding (Miyahara et al. 2004; Usoskin et al. 2015; Usoskin 2023).

Numerical simulations like Charbonneau et al. (2004); Saha et al. (2022) have shown persistence in phase and cyclicity during intermittent episodes (see, Fig. 3). An intriguing aspect of simulating this persistence of magnetic activity and eventual recovery from a solar grand minimum in flux transport dynamo models is to introduce an additional polar field generation mechanism slightly deeper in the convection zone in addition to the Babcock–Leighton source term near the solar surface (Passos et al. 2014; Saha et al. 2022). This is to compensate for the deficit of polar flux near the solar surface due to the absence of sunspots therein. A downward magnetic pumping in the solar convection zone also can help recover the Sun from grand minima (Karak and Miesch 2018).

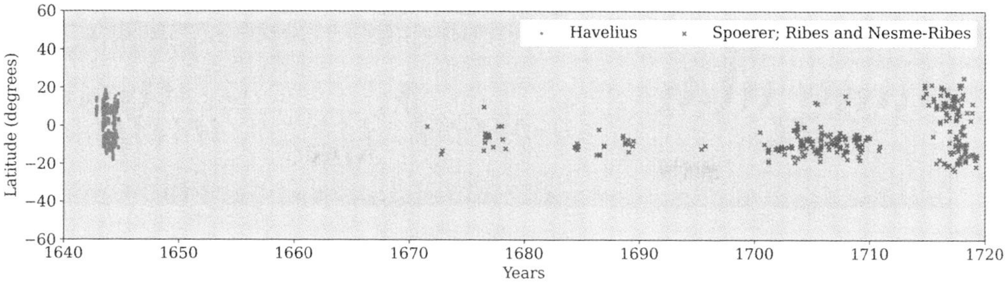

Figure 4. Sunspot butterfly diagram constructed from historical observational datasets (Ribes and Nesme–Ribes 1993; Vaquero et al. 2015; Carrasco et al. 2019), depicting strong hemispheric asymmetry in the solar activity during Maunder minimum with a prolonged preferential bias in the latitudes of sunspot emergence – predominantly confined to the southern solar hemisphere.

However, the 11-yr sunspot cycle disrupts to some extent, and power distribution in different periodicities considerably re-organizes during grand minima. Several shorter and longer periodic phenomena become prominent, as elucidated by Saha et al. (2022). We will elaborately discuss this in the upcoming sections.

5. Hemispheric asymmetry in solar activity during grand minimum

Evidence from helioseismic measurements hints towards a nearly zero meridional plasma flow at the solar equator, putting an important constraint on solar dynamo models (Muñoz-Jaramillo et al. 2009). This implies that the solar hemispheres are weakly coupled, predominant contribution of which comes from the cross-equatorial diffusion of magnetic fields over a longer timescale. Moreover, the meridional circulation itself demonstrates a hemispheric asymmetry as it evolves, plausibly causing an asymmetry in solar activity across the two hemispheres (Lekshmi et al. 2018). Such hemispheric asymmetry causes a deviation from the dipolar parity in large-scale solar magnetic fields. This effect is more pronounced during solar grand minima as confirmed by both simulations (Hazra and Nandy 2019) and observations (see, Fig. 4).

Olemskoy and Kitchatinov (2013) simulated asymmetric solar activity during a grand minimum by randomizing the preferred latitude of sunspot emergence in the active latitudinal belts across the hemispheres. Another justifiable approach is to introduce stochastic fluctuations independently into the poloidal sources in the two hemispheres, resulting in the asymmetric occurrence of hemispheric grand minima (Passos et al. 2014; Saha et al. 2022).

6. Probing the solar interior and polar dynamics

Direct probing of the solar interior is not possible. On the other hand, existing high latitude observations of the Sun also suffer from large projection effects (Nandy et al. 2023, and references therein). These make the solar interior and polar regions almost uncharted territories. Knowledge about the solar interior and poles is crucial because, in the absence of sunspots in the active latitudes during grand minima, the polar and deep-seated toroidal magnetic fields determine the dynamics of solar activity. Numerical simulations of solar dynamo are essential in this context. Multiple solar dynamo simulations have independently reported instances of a temporary halt in the polar field reversal during deep solar grand minima (Mackay 2003; Saha et al. 2022) (see, Fig. 5). The reversal can eventually resume due to gradual accumulation of magnetic fluxes from ephemeral regions and pores (Švanda et al. 2016) transported by the poleward branch of

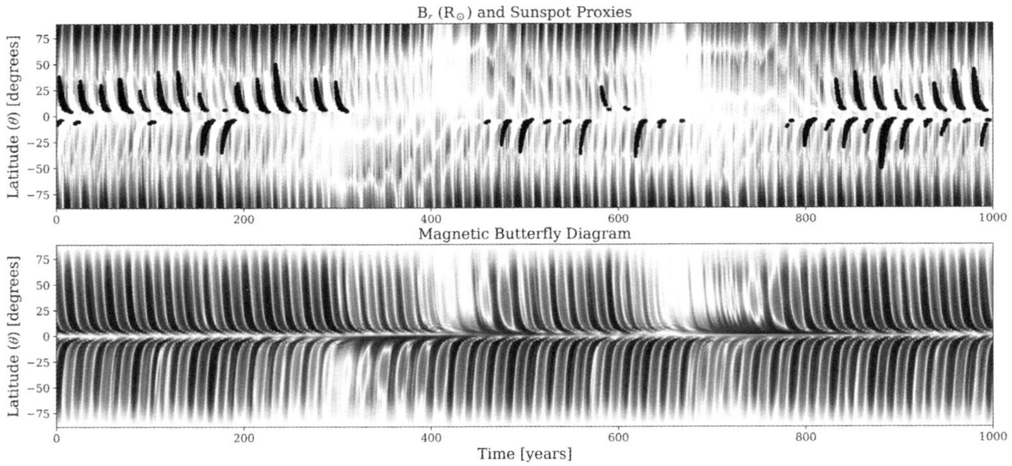

Figure 5. Stochastically forced numerical solar dynamo simulations can potentially provide novel insights into the dynamics on the Sun's surface including the polar regions (top panel) and at the base of the convection zone (bottom panel) during regular and extreme activity phases, direct observations of which are not available otherwise (ref. Saha et al. 2022).

meridional circulation (Saha et al. 2022), until sufficient polar magnetic flux is built up to kick start regular sunspot cycles.

7. Grand minima through a mathematical prism: Spectral components of solar grand minima

Spectral analysis techniques, when applied to reconstructed and simulated solar open flux time series, provide valuable insights into the dominant periodicities in solar magnetic activity, including that of during grand minima (Miyahara et al. 2004; Inceoglu et al. 2015; Usoskin et al. 2021b). While the spectral power stored in the 11-year periodicity diminishes during grand minima, several other shorter and longer periodicities are enhanced (Saha et al. 2022). Recent studies on auroral activity records indicate towards a shortened solar cycle period during the Maunder minimum (Yan et al. 2023; Herrera et al. 2024). This can be explained by faster meridional flows during grand minima, resulting in an increased rate of magnetic flux transportation, as discussed in section 3.

Previous studies showed that a deep meridional flow acts as a clock and regulates the solar cycle timescales (Nandy and Choudhuri 2002; Hathaway et al. 2003). Recently, Saha et al. (2022) speculated the ceaseless meridional plasma flows in tandem with mean-field α-effect to be responsible for the dynamo operation to recuperate from a grand minimum. In fact, the signature of meridional plasma motion is captured in the spectral domain in the form of a $\sim$5-yr component (see, Fig. 6), which is also the characteristic timescale to dredge up magnetic fields from the tachocline to the active latitude on the solar surface solely by the meridional circulation in the absence of any buoyancy mechanism (Saha et al. 2022). Interestingly, a subsequent study by Inceoglu (2024) has discovered a robust presence of a similar spectral component in a reconstructed solar activity dataset with high temporal resolution.

8. Influence of solar grand minima on the space environmental conditions

The large-scale magnetic field in the solar corona acts as a bridge between the dynamo-generated fields in the solar interior and the interplanetary magnetic fields pervading

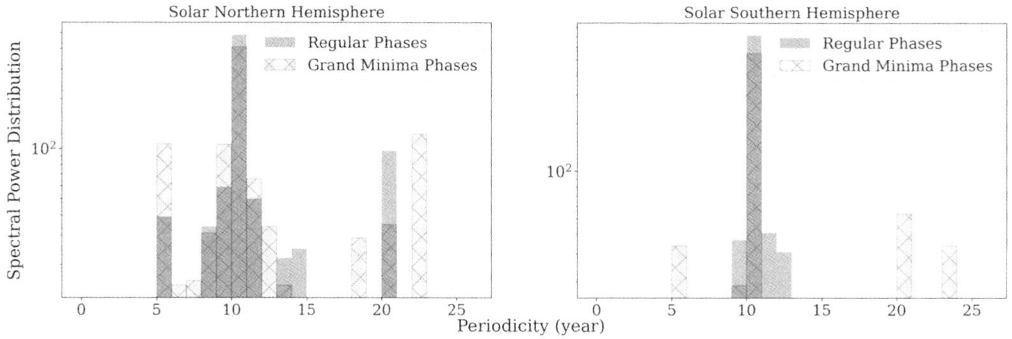

Figure 6. Spectral power distribution indicates towards a relative suppression of 11-year cyclicity in solar activity and enhancement of other shorter and longer periodicities during simulated solar grand minima. see, Saha et al. (2022) for further details.

all through the heliosphere, thereby establishing a causal connection between the solar interior and the state of the heliosphere, as elucidated by Nandy et al. (2023). Therefore, it goes without saying that a solar grand minimum can profoundly influence the space environmental conditions (Owens et al. 2012; Riley et al. 2015; Hayakawa et al. 2021).

With the aid of potential field source surface extrapolations Dash et al. (2023) have illuminated that during these periods of reduced solar magnetic activity, the heliosphere may experience significant changes – the weakening of the solar open magnetic flux, leading to decreased solar wind pressure and interplanetary magnetic field strength. This can alter the Earth's magnetosphere, ionosphere, and thermosphere structure and dynamics. Furthermore, the reduced solar activity during grand minima is associated with decreased occurrence and intensity of solar flares and coronal mass ejections (Richardson and Cane 2012).

The topology of the large-scale solar coronal magnetic field during a grand minimum (see, Fig. 7, top panel) can be strikingly different from that of the regular solar cycle activity minimum (see, Fig. 7, middle panel, Year 2296 and 2308). The closed magnetic loop-like structures in the solar corona predominantly manifest throughout the heliographic latitudes, including the polar regions, unlike a dipolar configuration during solar minimum. This reasonably explains the unusually large number of auroral events during the Maunder minimum (Wang et al. 2021).

9. Outstanding questions and concluding remarks

There are still multiple unresolved inquiries about solar grand minima, leading to continued research endeavours to enhance our comprehension of these events. A few of these inquiries consist of:

1. What are the primary mechanisms responsible for initiating grand minima events in the solar cycle? Understanding the triggers behind these prolonged periods of reduced solar activity is essential for predicting their occurrence and assessing their potential impacts.

2. How do solar dynamo processes behave during grand minima events? Investigating the behavior of the solar dynamo during periods of reduced magnetic activity can provide insights into the underlying mechanisms governing the solar cycle.

3. What factors determine the duration and frequency of grand minima events? Investigating the variability in the length and recurrence of grand minima can provide insights into the underlying processes driving solar cycle dynamics.

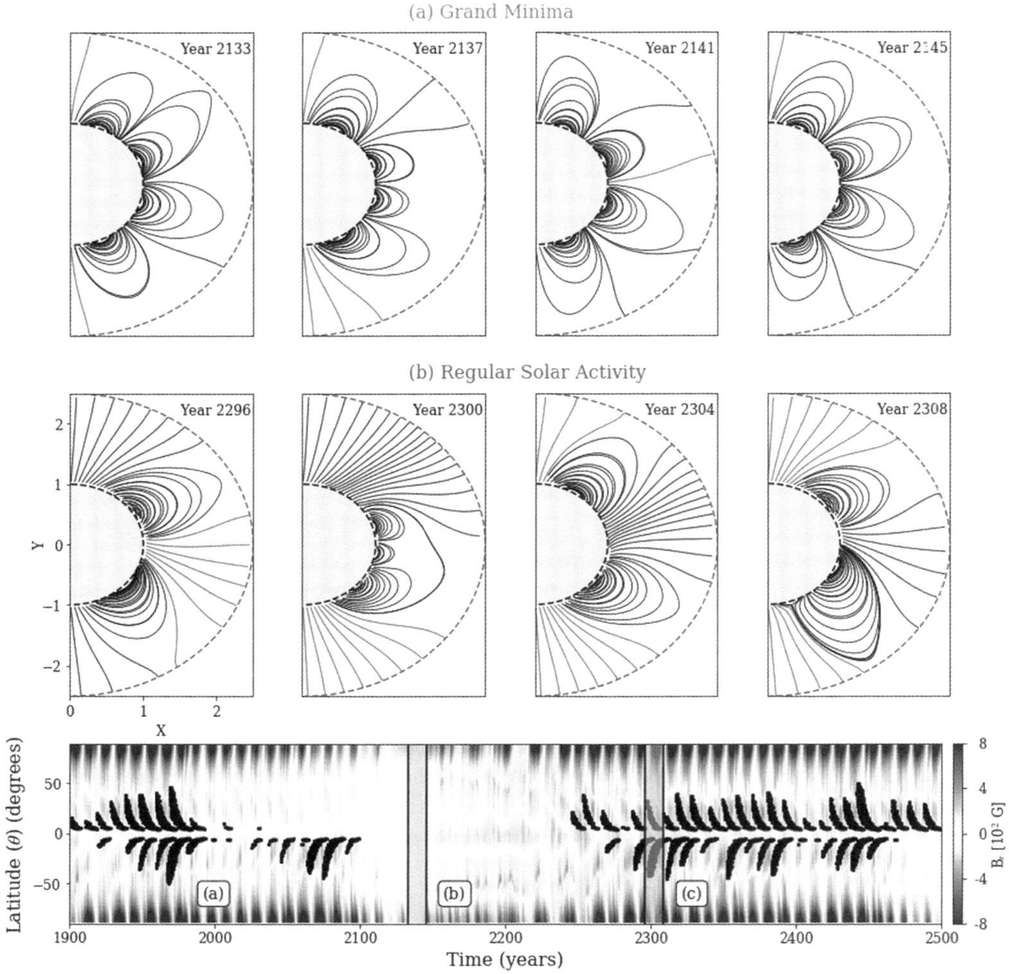

Figure 7. Evolution of large-scale magnetic field configuration in the solar corona during different levels of solar activity including grand minimum (ref. Dash et al. 2023).

4. What is the relationship between solar grand minima and terrestrial climate variability? Understanding the linkages between prolonged periods of reduced solar activity and climate changes on Earth can shed light on the mechanisms driving long-term climate trends.

Addressing these outstanding questions is crucial for advancing our understanding of solar grand minima and their implications for space weather, climate variability, and solar dynamo processes.

10. Acknowledgements

C.S. is grateful to have received a travel grant from the International Astronomical Union to attend the IAU Symposium 365 in Yerevan, Armenia. C.S. also acknowledges the financial support from CSIR through grant no. 09/921(0334)/2020-EMR-I. CESSI is funded by IISER Kolkata, Ministry of Education, Government of India.

References

Herrera, V. M. V., Soon, W., Babynets, N., Muraközy, J., Tlatov, A. G., Nagovitsyn, Y. A., Qiu, S., Švanda, M., & Velasco Herrera, P. A. 2024, Reconstructing daily group sunspot numbers since the maunder minimum with objective inter-calibration algorithms. *Advances in Space Research*, 73(5), 2788–2815.

Augustson, K., Brun, A. S., Miesch, M., & Toomre, J. 2015, Grand minima and equatorward propagation in a cycling stellar convective dynamo. *The Astrophysical Journal*, 809(2), 149.

Beer, J., Tobias, S., & Weiss, N. 1998, An active sun throughout the maunder minimum. *Solar Physics*, 181, 237–249.

Bhowmik, P. & Nandy, D. 2018, Prediction of the strength and timing of sunspot cycle 25 reveal decadal-scale space environmental conditions. *Nature Communications*, 9, 5209.

Biswas, A., Karak, B. B., Usoskin, I., & Weisshaar, E. 2023, Long-term modulation of solar cycles. *Space Science Reviews*, 219(3), 19.

Brandenburg, A. & Spiegel, E. A. 2008, Modeling a maunder minimum. *Astronomische Nachrichten: Astronomical Notes*, 329(4), 351–358.

Cameron, R. H. & Schüssler, M. 2019, Solar activity: periodicities beyond 11 years are consistent with random forcing. *A&A*, 625, A28.

Carrasco, V. M. S., Vaquero, J. M., Gallego, M. C., Muñoz-Jaramillo, A., de Toma, G., Galaviz, P., Arlt, R., Pavai, V. S., Sánchez-Bajo, F., Álvarez, J. V., & Gómez, J. M. 2019, Sunspot characteristics at the onset of the maunder minimum based on the observations of hevelius. *The Astrophysical Journal*, 886(1), 18.

Charbonneau, P. 2020, Dynamo models of the solar cycle. *Living Reviews in Solar Physics*, 17(1), 4.

Charbonneau, P., Blais–Laurier, G., & St-Jean, C. 2004, Intermittency and phase persistence in a babcock-leighton model of the solar cycle. *The Astrophysical Journal*, 616(2), L183.

Choudhuri, A. R. & Karak, B. B. 2012, Origin of grand minima in sunspot cycles. *Physical Review Letters*, 109(17), 171103.

Dash, S., Nandy, D., & Usoskin, I. 2023, Long-term forcing of the Sun's coronal field, open flux, and cosmic ray modulation potential during grand minima, maxima, and regular activity phases by the solar dynamo mechanism. *Monthly Notices of the Royal Astronomical Society*, 525(4), 4801–4814.

Dasi–Espuig, M., Solanki, S. K., Krivova, N. A., Cameron, R., & Peñuela, T. 2010, Sunspot group tilt angles and the strength of the solar cycle. *A&A*, 518, A7.

Eddy, J. A. 1976, The maunder minimum. *Science*, 192(4245), 1189–1202.

Eddy, J. A. 1983, The maunder minimum: a reappraisal. *Solar Physics*, 89, 195–207.

Feynman, J. & Ruzmaikin, A. 2011, The sun's strange behavior: Maunder minimum or gleissberg cycle? *Solar Physics*, 272(2), 351.

Hathaway, D. H. 2015, The solar cycle. *Living Reviews in Solar Physics*, 12(1), 4.

Hathaway, D. H., Nandy, D., Wilson, R. M., & Reichmann, E. J. 2003, Evidence that a deep meridional flow sets the sunspot cycle period. *The Astrophysical Journal*, 589(1), 665.

Hathaway, D. H. & Rightmire, L. 2010, Variations in the sun's meridional flow over a solar cycle. *Science*, 327(5971), 1350–1352.

Hayakawa, H., Carrasco, V. M., Aparicio, A. J., Villalba Álvarez, J., & Vaquero, J. M. 2024, An overview of sunspot observations in the early maunder minimum: 1645–1659. *Monthly Notices of the Royal Astronomical Society*, 528(4), 6280–6291.

Hayakawa, H., Lockwood, Mike, Owens, Matthew J., Sôma, Mitsuru, Besser, Bruno P., & van Driel–Gesztelyi, Lidia 2021, Graphical evidence for the solar coronal structure during the maunder minimum: comparative study of the total eclipse drawings in 1706 and 1715. *J. Space Weather Space Clim.*, 11, 1.

Hazra, S. & Nandy, D. 2019, The origin of parity changes in the solar cycle. *Mon. Not. R. Astron. Soc.*, 489(3), 4329–4337.

Hazra, S., Passos, D., & Nandy, D. 2014, A Stochastically Forced Time Delay Solar Dynamo Model: Self-consistent Recovery from a Maunder-like Grand Minimum Necessitates a Mean-field Alpha Effect. *Astrophys. J.*, 789(1), 5.

Hoyt, D. V. & Schatten, K. H. 1996, How well was the sun observed during the maunder minimum? *Solar Physics*, 165, 181–192.

Inceoglu, F. 2024, Exploring solar dynamo behavior using an annually resolved carbon-14 compilation during multiple grand solar minima. *Scientific Reports*, 14(1), 5617.

Inceoglu, F., Arlt, R., & Rempel, M. 2017, The nature of grand minima and maxima from fully nonlinear flux transport dynamos. *The Astrophysical Journal*, 848(2), 93.

Inceoglu, F., Simoniello, R., Knudsen, M. F., Karoff, C., Olsen, J., Turck–Chiéze, S., & Jacobsen, B. H. 2015, Grand solar minima and maxima deduced from 10be and 14c: magnetic dynamo configuration and polarity reversal. *A&A*, 577, A20.

Karak, B. B. 2010, Importance of meridional circulation in flux transport dynamo: The possibility of a maunder-like grand minimum. *The Astrophysical Journal*, 724(2), 1021.

Karak, B. B. & Miesch, M. 2018, Recovery from maunder-like grand minima in a babcock–leighton solar dynamo model. *The Astrophysical Journal Letters*, 860(2), L26.

Küker, M., Arlt, R., & Rüdiger, G. 1999, The maunder minimum as due to magnetic lambda-quenching. *Astronomy and Astrophysics, v. 343, p. 977-982 (1999)*, 343, 977–982.

Lekshmi, B., Nandy, D., & Antia, H. M. 2018, Asymmetry in Solar Torsional Oscillation and the Sunspot Cycle. *Astrophys. J.*, 861(2), 121.

Mackay, D. H. 2003, Magnetic flux transport simulations of solar surface magnetic distributions during a grand minimum. *Solar Physics*, 213, 173–193.

Miyahara, H., Masuda, K., Muraki, Y., Furuzawa, H., Menjo, H., & Nakamura, T. 2004, Cyclicity of solar activity during the maunder minimum deduced from radiocarbon content. *Solar Physics*, 224, 317–322.

Moss, D., Sokoloff, D., Usoskin, I., & Tutubalin, V. 2008, Solar grand minima and random fluctuations in dynamo parameters. *Solar Physics*, 250, 221–234.

Muñoz-Jaramillo, A., Nandy, D., & Martens, P. C. H. 2009, Helioseismic Data Inclusion in Solar Dynamo Models. *Astrophys. J.*, 698(1), 461–478.

Muñoz-Jaramillo, A. & Vaquero, J. M. 2019, Visualization of the challenges and limitations of the long-term sunspot number record. *Nature Astronomy*, 3(3), 205–211.

Nagy, M., Lemerle, A., Labonville, F., Petrovay, K., & Charbonneau, P. 2017, The effect of "rogue" active regions on the solar cycle. *Solar Physics*, 292, 1–22.

Nandy, D. 2021, Progress in Solar Cycle Predictions: Sunspot Cycles 24-25 in Perspective. *Sol. Phys.*, 296(3), 54.

Nandy, D., Banerjee, D., Bhowmik, P., BRUN, A. S., Cameron, R. H., Gibson, S. E., Hanasoge, S., Harra, L., Hassler, D. M., Jain, R., Jiang, J., Jouve, L., Mackay, D. H., Mahajan, S. S., Mandrini, C. H., Owens, M., Pal, S., Pinto, R. F., Saha, C., Sun, X., Tripathi, D., & Usoskin, I. G. Exploring the Solar Poles: The Last Great Frontier of the Sun. In *Bulletin of the American Astronomical Society* 2023,, volume 55, 287.

Nandy, D., Baruah, Y., Bhowmik, P., Dash, S., Gupta, S., Hazra, S., Lekshmi, B., Pal, S., Pal, S., Roy, S., Saha, C., & Sinha, S. 2023, Causality in heliophysics: Magnetic fields as a bridge between the sun's interior and the earth's space environment. *Journal of Atmospheric and Solar-Terrestrial Physics*, 248, 106081.

Nandy, D. & Choudhuri, A. R. 2002, Explaining the Latitudinal Distribution of Sunspots with Deep Meridional Flow. *Science*, 296(5573), 1671–1673.

Nandy, D., Martens, P. C. H., Obridko, V., Dash, S., & Georgieva, K. 2021, Solar evolution and extrema: current state of understanding of long-term solar variability and its planetary impacts. *Progress in Earth and Planetary Science*, 8(1), 40.

Nandy, D., Muñoz-Jaramillo, A., & Martens, P. C. H. 2011, The unusual minimum of sunspot cycle 23 caused by meridional plasma flow variations. *Nature*, 471(7336), 80–82.

Olemskoy, S., Choudhuri, A., & Kitchatinov, L. 2013, Fluctuations in the alpha-effect and grand solar minima. *Astronomy Reports*, 57, 458–468.

Olemskoy, S. & Kitchatinov, L. 2013, Grand minima and north–south asymmetry of solar activity. *The Astrophysical Journal*, 777(1), 71.

Ossendrijver, M. 2000, Grand minima in a buoyancy-driven solar dynamo. *Astronomy and Astrophysics, v. 359, p. 364-372 (2000)*, 359, 364–372.

Owens, M. J., Lockwood, M., Hawkins, E., Usoskin, I., Jones, G. S., Barnard, L., Schurer, A., & Fasullo, J. 2017, The maunder minimum and the little ice age: an update from recent reconstructions and climate simulations. *Journal of Space Weather and Space Climate*, 7, A33.

Owens, M. J., Usoskin, I., & Lockwood, M. 2012, Heliospheric modulation of galactic cosmic rays during grand solar minima: Past and future variations. *Geophysical Research Letters*, 39(19).

Pal, S., Bhowmik, P., Mahajan, S. S., & Nandy, D. 2023, Impact of anomalous active regions on the large-scale magnetic field of the sun. *The Astrophysical Journal*, 953(1), 51.

Passos, D., Nandy, D., Hazra, S., & Lopes, I. 2014, A solar dynamo model driven by mean-field alpha and Babcock–Leighton sources: fluctuations, grand-minima-maxima, and hemispheric asymmetry in sunspot cycles. *Astron. Astrophys.*, 563, A18.

Petrovay, K. 2020, Solar cycle prediction. *Living Reviews in Solar Physics*, 17(1), 2.

Pevtsov, A. A., Nandy, D., Usoskin, I., Pevtsov, A. A., Corti, C., Lefèvre, L., Owens, M., Li, G., Krivova, N., Saha, C., Perri, B., Brun, A. S., Strugarek, A., Dayeh, M. A., Nagovitsyn, Y. A., & Erdélyi, R. 2023, Long-term solar variability: Iswat s1 cluster review for cospar space weather roadmap. *Advances in Space Research*,.

Ribes, J. C. & Nesme–Ribes, E. 1993, The solar sunspot cycle in the Maunder minimum AD1645 to AD1715. *Astron. Astrophys.*, 276, 549.

Richardson, I. G. & Cane, H. V. 2012, Near-earth solar wind flows and related geomagnetic activity during more than four solar cycles (1963–2011). *J. Space Weather Space Clim.*, 2, A02.

Riley, P., Lionello, R., Linker, J. A., Cliver, E., Balogh, A., Beer, J., Charbonneau, P., Crooker, N., DeRosa, M., Lockwood, M., Owens, M., McCracken, K., Usoskin, I., & Koutchmy, S. 2015, Inferring the structure of the solar corona and inner heliosphere during the maunder minimum using global thermodynamic magnetohydrodynamic simulations. *The Astrophysical Journal*, 802(2), 105.

Saha, C., Chandra, S., & Nandy, D. 2022, Evidence of persistence of weak magnetic cycles driven by meridional plasma flows during solar grand minima phases. *Mon. Not. R. Astron. Soc.*, 517(1), L36–L40.

Schrijver, C. J., Kauristie, K., Aylward, A. D., Denardini, C. M., Gibson, S. E., Glover, A., Gopalswamy, N., Grande, M., Hapgood, M., Heynderickx, D., Jakowski, N., Kalegaev, V. V., Lapenta, G., Linker, J. A., Liu, S., Mandrini, C. H., Mann, I. R., Nagatsuma, T., Nandy, D., Obara, T., Paul O'Brien, T., Onsager, T., Opgenoorth, H. J., Terkildsen, M., Valladares, C. E., & Vilmer, N. 2015, Understanding space weather to shield society: A global road map for 2015-2025 commissioned by COSPAR and ILWS. *Advances in Space Research*, 55(12), 2745–2807.

Simard, C. & Charbonneau, P. 2020, Grand minima in a spherical non-kinematic $\alpha 2\omega$ mean-field dynamo model. *Journal of Space Weather and Space Climate*, 10, 9.

Solanki, S. K. & Krivova, N. A. 2011, Analyzing solar cycles. *Science*, 334(6058), 916–917.

Solanki, S. K., Usoskin, I. G., Kromer, B., Schüssler, M., & Beer, J. 2004, Unusual activity of the sun during recent decades compared to the previous 11,000 years. *Nature*, 431(7012), 1084–1087.

Steinhilber, F., Beer, J., & Fröhlich, C. 2009, Total solar irradiance during the holocene. *Geophysical Research Letters*, 36(19).

Švanda, M., Brun, A. S., Roudier, T., & Jouve, L. 2016, Polar cap magnetic field reversals during solar grand minima: could pores play a role? *Astronomy & Astrophysics*, 586, A123.

Tobias, S. M. 1996, Grand minimia in nonlinear dynamos. *Astron. Astrophys.*, 307, L21.

Tripathi, B., Nandy, D., & Banerjee, S. 2021, Stellar mid-life crisis: subcritical magnetic dynamos of solar-like stars and the breakdown of gyrochronology. *Mon. Not. R. Astron. Soc.*, 506(1), L50–L54.

Usoskin, I., Sokoloff, D., & Moss, D. 2009, Grand minima of solar activity and the mean-field dynamo. *Solar Physics*, 254, 345–355.

Usoskin, I. G. 2023, A history of solar activity over millennia. *Living Reviews in Solar Physics*, 20(1), 2.

Usoskin, I. G., Arlt, Rainer, Asvestari, Eleanna, Hawkins, Ed, Käpylä, Maarit, Kovaltsov, Gennady A., Krivova, Natalie, Lockwood, Michael, Mursula, Kalevi, O'Reilly, Jezebel, Owens, Matthew, Scott, Chris J., Sokoloff, Dmitry D., Solanki, Sami K., Soon, Willie, & Vaquero, José M. 2015, The maunder minimum (1645–1715) was indeed a grand minimum: A reassessment of multiple datasets. *A&A*, 581, A95.

Usoskin, I. G., Solanki, S. K., Krivova, N., Hofer, B., Kovaltsov, G. A., Wacker, L., Brehm, N., & Kromer, B. 2021,a VizieR Online Data Catalog: 1000-year sunspot series (Usoskin+, 2021). *VizieR Online Data Catalog*,, J/A+A/649/A141.

Usoskin, I. G., Solanki, S. K., & Kovaltsov, G. A. 2007, Grand minima and maxima of solar activity: new observational constraints. *A&A*, 471(1), 301–309.

Usoskin, I. G., Solanki, S. K., Krivova, N. A., Hofer, B., Kovaltsov, G. A., Wacker, L., Brehm, N., & Kromer, B. 2021,b Solar cyclic activity over the last millennium reconstructed from annual 14c data. *A&A*, 649b, A141.

Vaquero, J. M., Nogales, J. M., & Sánchez-Bajo, F. 2015, Sunspot latitudes during the maunder minimum: A machine-readable catalogue from previous studies. *Advances in Space Research*, 55(6), 1546–1552.

Wang, Y., Chen, S., Xu, K., Yan, L., Yue, X., He, F., & Wei, Y. 2021, Ancient auroral records compiled from korean historical books. *Journal of Geophysical Research: Space Physics*, 126(1), e2020JA028763. e2020JA028763 2020JA028763.

Wang, Y.-M. & Sheeley Jr, N. 2003, Modeling the sun's large-scale magnetic field during the maunder minimum. *The Astrophysical Journal*, 591(2), 1248.

Wilmot–Smith, A. L., Martens, P. C. H., Nandy, D., Priest, E. R., & Tobias, S. M. 2005, Low-order stellar dynamo models. *Mon. Not. R. Astron. Soc.*, 363(4), 1167–1172.

Wilmot–Smith, A. L., Nandy, D., Hornig, G., & Martens, P. C. H. 2006, A Time Delay Model for Solar and Stellar Dynamos. *Astrophys. J.*, 652(1), 696–708.

Yan, L., He, F., Yue, X., Wei, Y., Wang, Y., Chen, S., Fan, K., Tian, H., He, J., Zong, Q., & Xia, L. 2023, The 8-year solar cycle during the maunder minimum. *AGU Advances*, 4(5), e2023AV000964. e2023AV000964 2023AV000964.

Yeates, A. R., Nandy, D., & Mackay, D. H. 2008, Exploring the Physical Basis of Solar Cycle Predictions: Flux Transport Dynamics and Persistence of Memory in Advection- versus Diffusion-dominated Solar Convection Zones. *Astrophys. J.*, 673(1), 544–556.

Zolotova, N. V. & Ponyavin, D. I. 2014, Is the new grand minimum in progress? *Journal of Geophysical Research: Space Physics*, 119(5), 3281–3285.

Zolotova, N. V. & Ponyavin, D. I. 2015, The maunder minimum is not as grand as it seemed to be. *The Astrophysical Journal*, 800(1), 42.

Dynamics of Solar and Stellar Convection Zones and Atmospheres
Proceedings IAU Symposium No. 365, 2024
A. V. Getling & L. L. Kitchatinov, eds.
doi:10.1017/S1743921323005264

Solar cycle variability induced by stochastic fluctuations of BMR properties and at different amounts of dynamo supercriticality

Pawan Kumar

Department of Physics, Indian Institute of Technology (Banaras Hindu University), Varanasi 221005, India
email: pawan.kumar.rs.phy18@itbhu.ac.in

Abstract. Understanding the irregular variation of the solar cycle is crucial due to its significant impact on global climates and the heliosphere. Since the polar magnetic field determines the amplitude of the next solar cycle, variations in the polar field can lead to fluctuations in the solar cycle. We have explored the variability of the solar cycle at different levels of dynamo supercriticality. We observe that the variability depends on the dynamo operation regime, with the near-critical regime exhibiting more variability than the supercritical regime. Furthermore, we have explored the effects of the irregular BMR properties (emergence rate, latitude, tilt, and flux) on the polar field and the solar cycle. We find that they all produce considerable variation in the solar cycle; however, the variation due to the tilt scatter is the largest.

Keywords. Bipolar Magnetic Regions, Polar Field, Solar Cycle, Solar Dynamo

1. Introduction

The irregular variation of the solar magnetic cycle is the characteristic feature of the solar magnetic field with a periodicity of about 11 years. In addition to this, short-term variations, often referred to as the "seasons of the sun," are also observed in the cycle (Rieger et al. 1984; Gurgenashvili et al. 2016; Biswas et al. 2023b). These variations manifest in phenomena such as double peaks (Gnevyshev peaks) in the solar cycle, timing of polar field reversal, duration of polar field and solar cycle, active regions, extended periods of low and high solar activity, and total solar irradiance (Karak et al. 2018; Mordvinov et al. 2020; Mordvinov et al. 2022). Therefore, understanding the variable nature of the solar magnetic cycle is essential, as it has direct and indirect impacts on human society. (Petrovay 2020).

The underlying dynamo process in the sun can explain the variability of the cycle (Karak et al. 2014a; Karak 2023). It is believed that solar dynamo is $\alpha\Omega$ type, and the magnetic field grows when the dynamo number exceeds a critical value which is defined by,

$$D = \frac{\alpha_0 \Delta\Omega R_\odot^3}{\eta_0^2}, \tag{1}$$

where α_0 is the measure of the α effect, η_0 is the turbulent magnetic diffusivity, $\Delta\Omega$ is angular velocity variation, and $R_\odot$ is the solar radius. There is a huge amount of literature on the understanding of solar cycle variability based on the dynamo process (Charbonneau 2020; Karak 2023) and, which can be broadly classified as 1) nonlinear dynamics and deterministic chaos 2) stochastic forcing of the dynamo. In the

Babcock–Leighton dynamo scenario, the nonlineary primarily comes from the flux loss due to magnetic buoyancy (Biswas et al. 2022), latitude quenching (Karak 2020), and tilt quenching (Jha et al. 2020). The variability in the magnetic cycle comes from the fluctuating part of the Babcock–Leighton dynamo, i.e., from the Babcock–Leighton mechanism of polar field generation, which involves randomness. This extensive randomness arises in the Babcock–Leighton mechanism from the irregular properties of Bipolar magnetic Regions (BMRs), mainly from tilt scatter (Olemskoy and Kitchatinov 2013; Jha et al. 2020). The other sources of irregularity in the solar dynamo of observed BMR properties are time delay in BMR emergence, variation in flux, and BMR emergence latitude. The time delay of successive BMR emergence depends on the strength of the toroidal magnetic field in the convection zone (Jouve et al. 2010) and may cause solar cycle variation. Observations suggest that the mean latitude of BMR emergence on the solar surface is not uniform from cycle to cycle and depends on the strength of the cycle (Mandal et al. 2017), which can lead to solar cycle variability. The flux of BMRs is also not constant and varies irregularly throughout the cycle in the range 10^{21} to 10^{23} Mx (Sreedevi et al. 2023) and can produce variation in the solar cycle.

2. Model and Method

To see the effect of irregular properties of BMR and supercriticality on the solar cycle variation, we used 2D and 3D dynamo models. We used the Surface Flux Transport And Babcock–Leighton (STABLE) dynamo model to incorporate the observed irregular properties of BMR in the model. For a detailed description of the model, see (Karak and Miesch 2017). The STABLE dynamo model can be used as the Surface Flux Transport (SFT) model and the dynamo model. To use it as the SFT, we feed the synthetic BMR data into STABLE and see the evolution of the polar field. We generated the synthetic BMR through synthetic BMR generation code following the Jiang et al. (2018); also see (Kumar et al. 2023). The BMR produced by the STABLE model and synthetic BMR generation code follows the following properties. 1) The time delay of BMR emergence roughly follows the lognormal distribution. 2) The flux variation is also lognormally distributed. However, to see the solar cycle variation at different level of supercriticality, we used the Babcock–Leighton type flux transport dynamo model (Karak et al. 2014a; Charbonneau 2020). For a detailed understanding of the models used in the study, see Kumar et al. (2021a).

3. Results

In this section, we will demonstrate the solar cycle and polar field variability by utilizing 2D and 3D dynamo simulations. Initially, we will explain dynamo supercriticality's influence on the solar cycle's variation. Subsequently, we will elucidate the effects of irregular BMR properties on the polar field and its consequential impact on cycle variability.

3.1. *Variability in the different operation regime of the dynamo*

We shall first examine the solar cycle variability across different levels of supercriticality in the dynamo, as illustrated in Figure 1. Figure 1 (a and c) depict the variation in the solar cycle in the near-critical or weakly supercritical regime, while (b) and (d) illustrate the supercritical regime. The analysis of Figure 1 reveals that when the dynamo operates in the weakly supercritical regime, it induces large variability in the cycle and extended episodes of grand minima (Cameron and Schüssler 2017; Kumar et al. 2021a; Vashishth et al. 2023). On the other hand, when the dynamo operates near the supercritical regime, the variability in the cycle is notably reduced. The cause of this is that when the dynamo operates near the critical regime, the inherent nonlinearity of the

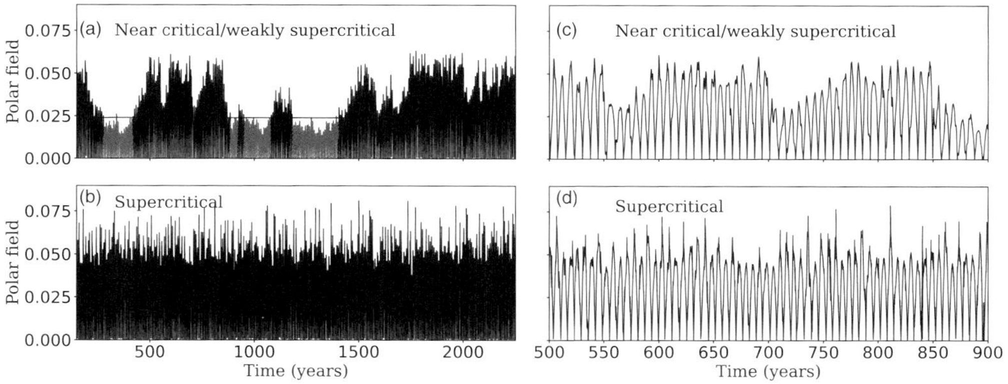

Figure 1. The plot shows the variation in the solar cycle from a dynamo simulation which operates in the weakly supercritical region (a and c) and in the supercritical region (b and d). **c** and **d** show 400 years of data from the long Run **a** and **b**. The red color shows the extended episodes of grand minima in weakly supercritical regime. The plot are produced from the model used in Kumar et al. (2021a).

dynamo weakens. Consequently, even a slight change in the dynamo number (D) results in small magnetic field growth, as the impact is linear. Thus, after some time, when D is increased due to fluctuations, the magnetic field experiences linear growth. Therefore, a large variability is observed in the magnetic field when the dynamo operates near the critical regime. Conversely, in the supercritical regime, the dynamo's nonlinearity is strong. In this scenario, significant changes in D do not translate into substantial growth in the magnetic field, as the nonlinearity suppresses such effects. Thus, in the supercritical regime, despite the large dynamo number, the variation in the magnetic field is observed less than the critical regime; also see (Karak et al. 2015; Vashishth et al. 2021; Karak 2023).

3.2. *Variation in the polar field*

Now we shall discuss the variation in the polar field and in the solar cycle due to irregular properties of BMR. Using the STABLE dynamo model as SFT, we see the variation in the polar due to different irregular properties of BMR. We see variations due to time delay in successive BMR emergence for different random realizations. Due to time delay, we see a negligible variation in the polar field; see Figure 1. The reason for the negligible variation in the polar field due to time delays is that the maximum delay in BMR emergence is much smaller than the transport time of the flux towards the pole from lower latitudes. Consequently, the variation in the polar field due to time delay is smoothed out by the time required for the flux to reach the pole. As a result, the net change in the deposited flux in the polar region is negligible due to the smoothing effect of this transport time.

Figure 3 shows the variation in the polar field due to fluctuation in BMR flux, latitude, scatter in tilt, and due to overall fluctuation, (a), (b), (c), and (d), respectively. We see that the maximum variation in the polar field is produced due to scatter in the BMR tilt. However, fluctuation in mean BMR latitude and flux variation also produces a significant amount of variability in the polar field; see Figure 3. The cause of the variation due to these parameters is the following in brief. The anticipated variation in the polar field due to latitude variation stems from the fact that when BMRs emerge at low latitudes, the cross-equatorial flux cancellation is notably more effective compared to those at

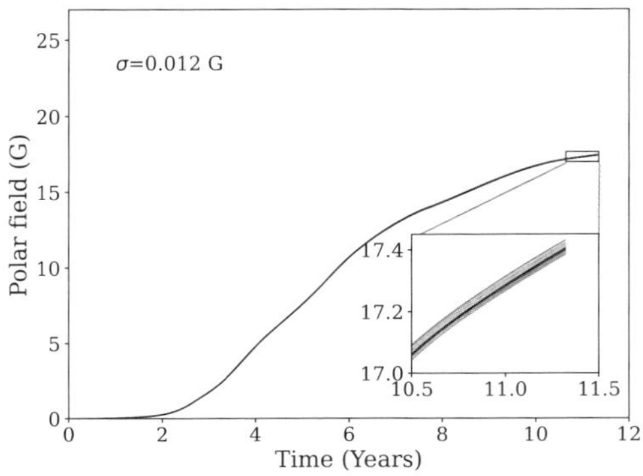

Figure 2. The plot illustrates the impact of time delays in successive BMR eruptions on the polar field. The black curve shows a reference polar field in the plot, with a zoomed-in view provided in the inset.

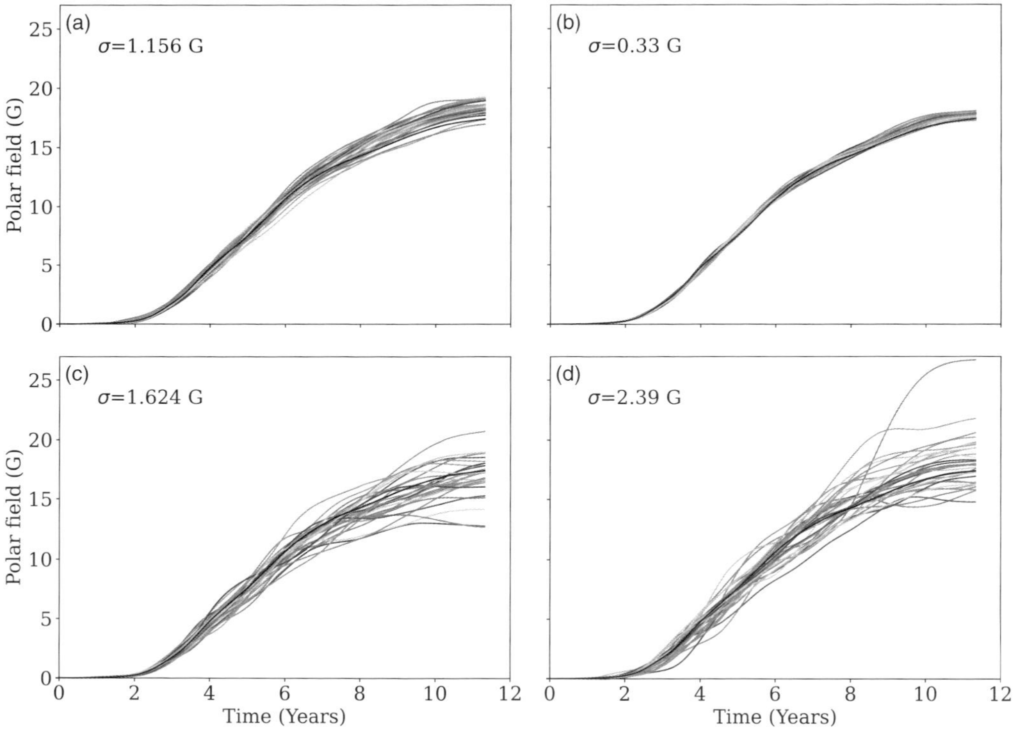

Figure 3. Format of subplots are the same as in Figure 1, but shows the evolution of the polar field from simulations with (a) variation in the BMRs flux, (b) variation in BMRs emergence latitude, (c) scatter in BMR tilt, and (d) combined fluctuations.

higher latitudes (Karak and Miesch 2018; Karak 2020). Moreover, it is important to note that the flux contribution to polar field formation in a given hemisphere is directly proportional to the total active region flux for that hemisphere (Kitchatinov & Olemskoy 2011). Therefore, any changes in the flux can induce variations in the polar field. In addition, the scatter in the tilt arises from wrongly (anti-Joy and anti-Hale) tilted BMRs (Stenflo and Kosovichev 2012). These wrongly tilted BMRs engage in annihilation with

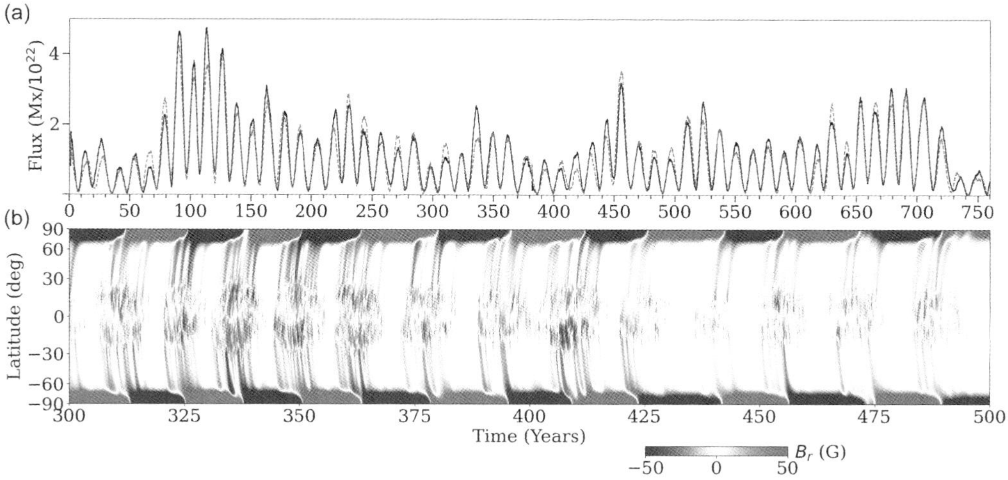

Figure 4. The plot shows the variation in the solar cycle from simulation, including all fluctuations (time delay, scatter in tilt, variation in latitude, and flux). Time delay is considered magnetic field dependent in this simulation.

the normal BMR field, leading to a significant variation in the polar field (Nagy et al. 2017; Golubeva et al. 2023).

3.3. *Variability in the solar cycle*

The variability introduced in the solar cycle due to the irregular properties of BMRs is similar to that observed in the polar field. This result is evident because the polar magnetic field or its proxy strongly correlates with the next cycle's activity level (Schatten et al. 1978; Choudhuri et al. 2007; Kumar et al. 2021b, 2022; Biswas et al. 2023). Similar to the polar field, the impact of time delay on solar cycle variation is found to be negligible. The maximum variability is produced due to the scatter in tilt, wherein BMRs with erroneously oriented Joy's and Hale's laws contribute significantly to fluctuations in the solar cycle. Furthermore, the fluctuations in flux and emergence latitude also contribute substantially to the observed variation in the solar cycle. When all these fluctuations are considered collectively, the resulting variability in the solar cycle aligns consistently with the observed patterns, as depicted in Figure 4.

4. Conclusion

In our investigation, we utilized a 2D and 3D flux transport dynamo model to explore the variability in both the solar cycle and the polar field, considering the observed irregular properties of BMRs and different levels of supercriticality. The simulations revealed that, for the same level of fluctuation, if the dynamo operates near the critical regime, the resulting variation in the solar cycle is large. Conversely, when the dynamo operates near the supercritical regime, the observed variation in the solar cycle is comparatively small. Furthermore, when we consider the irregular properties of BMR, the maximum variation produced in the polar field and the solar cycle is due to scatter in BMR tilt. The variation produced by the time delay is minimal; however, fluctuation in emergence latitude and the BMR flux produces significant variability in the polar field and the solar cycle. Moreover, tilt scatter in BMR tilt, flux variation, latitude variation, and fluctuation in time delay produce variability in the polar field and the solar cycle in the descending order, respectively. These results also suggest that the irregular BMR properties sufficiently produce the variation in the polar field and the solar magnetic cycle, which is consistent with the observations.

5. Acknowledgements

The author thanks the International Astronomical Union (IAU) for providing the travel grant to attend the IAU Symposium 365 in Yerevan, Armenia. The author also acknowledges CSIR Symposia and Travel Grant Unit and IIT (BHU) for providing partial travel support to attend the IAU Symposium.

References

Biswas, A., Karak, B. B., Usoskin, I., & Weisshaar, E. 2023,b Long-Term Modulation of Solar Cycles. *Space Sci. Rev.*, 219b(3), 19.

Biswas, A., Karak, B. B., & Cameron, R. 2022, Toroidal Flux Loss due to Flux Emergence Explains why Solar Cycles Rise Differently but Decay in a Similar Way. *Phys. Rev. Lett.*, 129(24), 241102.

Biswas, A., Karak, B. B., & Kumar, P. 2023, Exploring the reliability of polar field rise rate as a precursor for an early prediction of solar cycle. *MNRAS*, 526(3), 3994–4003.

Charbonneau, P. 2020, Dynamo models of the solar cycle. *Living Reviews in Solar Physics*, 17(1), 4.

Cameron, R. H. & Schüssler, M. 2017, Understanding Solar Cycle Variability. *ApJ*, 843(2), 111.

Choudhuri, A. R., Chatterjee, P., & Jiang, J. 2007, Predicting Solar Cycle 24 With a Solar Dynamo Model. *Physical Review Letters*, 98(13), 131103.

Gurgenashvili, E. and 8 colleagues 2016. Rieger-type Periodicity during Solar Cycles 14-24: Estimation of Dynamo Magnetic Field Strength in the Solar Interior. The Astrophysical Journal 826. doi:10.3847/0004-637X/826/1/55

Golubeva, E. M., Biswas, A., Khlystova, A. I., Kumar, P., & Karak, B. B. 2023, Probing the variations in the timing of the Sun's polar magnetic field reversals through observations and surface flux transport simulations. *MNRAS*, 525(2), 1758–1768.

Jha, B. K., Karak, B. B., Mandal, S., & Banerjee, D. 2020, Magnetic Field Dependence of Bipolar Magnetic Region Tilts on the Sun: Indication of Tilt Quenching. *ApJ Letters*, 889(1), L19.

Jouve, L., Proctor, M. R. E., Lesur, G. 2010. Buoyancy-induced time delays in Babcock–Leighton flux-transport dynamo models. Astronomy and Astrophysics 519. doi:10.1051/0004-6361/201014455

Jiang, J., Wang, J.-X., Jiao, Q.-R., Cao, J.-B. 2018. Predictability of the Solar Cycle Over One Cycle. The Astrophysical Journal 863. doi:10.3847/1538-4357/aad197

Karak, B. B. 2020, Dynamo Saturation through the Latitudinal Variation of Bipolar Magnetic Regions in the Sun. *ApJ Letters*, 901(2), L35.

Karak, B. B. 2023, Models for the long-term variations of solar activity. *Living Reviews in Solar Physics*, 20(1), 3.

Karak, B. B., Jiang, J., Miesch, M. S., Charbonneau, P., & Choudhuri, A. R. 2014,a Flux Transport Dynamos: From Kinematics to Dynamics. *Space Sci. Rev.*, 186a, 561–602.

Karak, B. B., Mandal, S., & Banerjee, D. 2018, Double Peaks of the Solar Cycle: An Explanation from a Dynamo Model. *ApJ*, 866(1), 17.

Karak, B. B., Kitchatinov, L. L., & Brandenburg, A. 2015, Hysteresis between Distinct Modes of Turbulent Dynamos. *ApJ*, 803, 95.

Karak, B. B. & Miesch, M. 2017, Solar Cycle Variability Induced by Tilt Angle Scatter in a Babcock–Leighton Solar Dynamo Model. *ApJ*, 847, 69.

Karak, B. B. & Miesch, M. 2018, Recovery from Maunder-like Grand Minima in a Babcock–Leighton Solar Dynamo Model. *ApJ Letters*, 860, L26.

Kitchatinov, L. L. & Olemskoy, S. V. 2011, Does the Babcock–Leighton mechanism operate on the Sun? *Astronomy Letters*, 37, 656–658.

Kumar, P., Karak, B. B., & Vashishth, V. 2021,a Supercriticality of the Dynamo Limits the Memory of the Polar Field to One Cycle. *ApJ*, 913a(1), 65.

Kumar, P., Nagy, M., Lemerle, A., Karak, B. B., & Petrovay, K. 2021,b The Polar Precursor Method for Solar Cycle Prediction: Comparison of Predictors and Their Temporal Range. *ApJ*, 909b(1), 87.

Kumar, P., Biswas, A., & Karak, B. B. 2022, Physical link of the polar field buildup with the Waldmeier effect broadens the scope of early solar cycle prediction: Cycle 25 is likely to be slightly stronger than Cycle 24. *MNRAS*, 513(1), L112–L116.

Kumar, P., Karak, B. B., & Sreedevi, A. 2023, . *MNRAS*, submitted.

Mandal, S., Karak, B. B., & Banerjee, D. 2017, Latitude Distribution of Sunspots: Analysis Using Sunspot Data and a Dynamo Model. *ApJ*, 851, 70.

Mordvinov, A. V., Karak, B. B., Banerjee, D., Chatterjee, S., Golubeva, E. M., Khlystova, A. I. 2020. Long-term Evolution of the Sun's Magnetic Field during Cycles 15-19 Based on Their Proxies from Kodaikanal Solar Observatory. The Astrophysical Journal 902. doi:10.3847/2041-8213/abba80

Mordvinov, A. V., Karak, B. B., Banerjee, D., Golubeva, E. M., Khlystova, A. I., Zhukova, A. V.& Kumar, P. 2022, Evolution of the Sun's activity and the poleward transport of remnant magnetic flux in Cycles 21-24. *MNRAS*, 510, 1.

Nagy, M., Lemerle, A., Labonville, F., Petrovay, K., Charbonneau, P. 2017. The Effect of "Rogue" Active Regions on the Solar Cycle. Solar Physics 292. doi:10.1007/s11207-017-1194-0

Olemskoy, S. V. & Kitchatinov, L. L. 2013, Grand Minima and North-South Asymmetry of Solar Activity. *ApJ*, 777, 71.

Petrovay, K. 2020, Solar cycle prediction. *Living Reviews in Solar Physics*, 17(1), 2.

Rieger, E., Share, G. H., Forrest, D. J., Kanbach, G., Reppin, C., Chupp, E. L. 1984. A 154-day periodicity in the occurrence of hard solar flares?. Nature 312, 623–625. doi:10.1038/312623a0

Stenflo, J. O. & Kosovichev, A. G. 2012, Bipolar Magnetic Regions on the Sun: Global Analysis of the SOHO/MDI Data Set. *ApJ*, 745(2), 129.

Schatten, K. H., Scherrer, P. H., Svalgaard, L., Wilcox, J. M. 1978. Using Dynamo Theory to predict the sunspot number during Solar Cycle 21. Geophysical Research Letters 5, 411–414. doi:10.1029/GL005i005p00411

Sreedevi, A., Jha, B. K., Karak, B. B., & Banerjee, D. 2023, AutoTAB: Automatic Tracking Algorithm for Bipolar Magnetic Regions. *ApJ Supplement*, 268(2), 58.

Vashishth, V., Karak, B. B., & Kitchatinov, L. 2021, Subcritical dynamo and hysteresis in a Babcock–Leighton type kinematic dynamo model. *Research in Astronomy and Astrophysics*, 21(10), 266.

Vashishth, V., Karak, B. B., & Kitchatinov, L. 2023, Dynamo modelling for cycle variability and occurrence of grand minima in Sun-like stars: rotation rate dependence. *MNRAS*, 522(2), 2601–2610.

Dynamics of Solar and Stellar Convection Zones and Atmospheres
Proceedings IAU Symposium No. 365, 2024
A. V. Getling & L. L. Kitchatinov, eds.
doi:10.1017/S1743921323005045

Exploring the predictability of the solar cycle from the polar field rise rate: Results from observations and simulations

Akash Biswas

Department of Physics, Indian Institute of Technology (Banaras Hindu University), Varanasi 221005, India
email: akashbiswas.rs.phy20@itbhu.ac.in

Abstract. The inherent stochastic and nonlinear nature of the solar dynamo makes the strength of the solar cycles vary in a wide range, making it difficult to predict the strength of an upcoming solar cycle. Recently, our work has shown that by using the observed correlation of the polar field rise rate with the peak of polar field at cycle minimum and amplitude of following cycle, an early prediction can be made. In a follow-up study, we perform SFT simulations to explore the robustness of this correlation against variation of meridional flow speed, and against stochastic fluctuations of BMR tilt properties that give rise to anti-Joy and anti-Hale type anomalous BMRs. The results suggest that the observed correlation is a robust feature of the solar cycle and can be utilized for a reliable prediction of peak strength of a cycle at least 2 to 3 years earlier than the minimum.

Keywords. Solar Magnetism, Solar Cycle Prediction, Space Weather

1. Introduction

The Sun's magnetic fields dominates and determines the space weather conditions of the heliosphere. The interaction of the magnetic fields embedded in the coronal mass ejections from the Sun with the Earth's magnetosphere gives rise to various phenomena in near-Earth space that can have serious consequences on our lives. The geomagnetic storms that occur through these interactions can damage our space and ground-based communication systems, impact the power grids and can pose a challenge to our space exploration capabilities (Gopalswamy 2022). Hence, the study of space weather and an adequate understanding of the evolution of solar magnetic fields is crucial for taking necessary measures to safeguard assets from the impact of these adverse conditions.

In this context, the forecast of space weather is incomplete without the capability of reliably predicting the strength of an upcoming solar cycle well in advance. The solar cycles are driven by the action of the dynamo operations that facilitate the cyclic conversion of strength from the poloidal component of the magnetic field to the toroidal field and back (Karak et al. 2014a; Charbonneau 2020; Hathaway 2015; Karak 2023). The conversion of the toroidal field strength into the poloidal field exhibits some nonlinear mechanisms (Biswas et al. 2022; Jha et al. 2020; Jiang 2020; Karak 2020) and stochastic fluctuations due to the randomness in the properties of the Bipolar Magnetic Regions (BMRs) (Biswas et al. 2023b; Karak and Miesch 2017; Jha et al. 2020; Sreedevi et al. 2023). Due to the combined impact of these processes, the strength and other features of the solar cycles vary widely from one cycle to the other, which makes the job of predicting an upcoming solar cycle incredibly difficult (Petrovay 2020; Kumar et al. 2021a; Bhowmik et al. 2023).

Among the various methodologies used for the prediction of solar cycle strength, the polar precursor method has gained consensus in the recent times owing to its success in providing successful prediction of Cycle 24 and convergence in the predicted strength of Cycle 25 (Choudhuri et al. 2007; Bhowmik and Nandy 2018; Nandy 2021; Upton and Hathaway 2018; Jiang et al. 2023). However, one major drawback of this method is that, the polar field strength at around the solar cycle minimum is used as an input for a reliable prediction of the upcoming solar cycle, this significantly limits the time window of the forecast.

Kumar et al. (2021b) reported that the direct correlation of the polar precursors with the strength of the following cycle can be used for cycle prediction after 4 to 5 years of the polar field reversal. Recently, Kumar et al. (2022) analysed the observed polar field data and showed that, there exists a strong correlation between the rise rate of the polar field and the peak of polar field at the cycle minimum as well as the strength of the upcoming cycle. As the sample size of the observed data is limited to only last three solar cycles, this correlation is not yet statistically robust. However, if this correlation stays high for a larger set of cycles, it can enable us to make reliable prediction of the upcoming solar cycle much ahead of the cycle minimum.

In this work, we explore the reliability of the correlation of the polar field rise rate with the peak of the polar field and the peak strength of the upcoming cycle under the impact of the stochastic properties of the BMRs, especially in the presence of 'rogue' (or anti-Hale and anti-Joy) BMRs through the Surface Flux Transport (SFT) simulations.

2. Surface Flux Transport Model

The primary objective of the SFT model is to simulate the build-up of the Sun's polar field through the Babcock–Leighton mechanism (Babcock 1961; Leighton 1969) for the evolution of radial component of the surface magnetic field from the decaying BMRs under the action of meridional circulation, differential rotation, and horizontal diffusion by numerically solving the following equation:

$$\begin{aligned}\frac{\partial B_r}{\partial t} &= -\Omega(\lambda)\frac{\partial B_r}{\partial \phi} - \frac{1}{R_\odot \cos\lambda}\frac{\partial}{\partial \lambda}\left[v(\lambda)B_r\cos\lambda\right] + \\ &\eta_H\left[\frac{1}{R_\odot^2\cos\lambda}\frac{\partial}{\partial\lambda}\left(\cos\lambda\frac{\partial B_r}{\partial\lambda}\right) + \frac{1}{R_\odot^2\cos^2\lambda}\frac{\partial^2 B_r}{\partial\phi^2}\right] + \\ &D(\eta_r) + S(\lambda,\phi,t). \qquad (1)\end{aligned}$$

For this study, we use the SFT code that has been used in multiple previous studies like Baumann et al. (2004); Cameron et al. (2010) to name a few. We use synthetic spatio-temporal profile of the solar cycles and the BMRs as prescribed by Jiang et al. (2018) and Hathaway et al. (1994). For the calculation of the toroidal flux to be produced by the action of the differential rotation on the polar field developed in the SFT simulations, we use the following equation as prescribed by Cameron and Schüssler (2015):

$$\frac{\mathrm{d}\Phi_{\mathrm{tor}}^{\mathrm{N}}}{\mathrm{d}t} = \int_0^1 (\Omega - \Omega_{eq})B_r R_\odot^2\,\mathrm{d}(\cos\theta) - \frac{\Phi_{tor}^N}{\tau} \qquad (2)$$

More details about the SFT model parameters, properties of the BMRs, and the methodology of analysis of the simulated data can be found in Biswas et al. (2023).

3. Results and Discussion

In the build-up of the poloidal field of the Sun through the Babcock–Leighton mechanism, the meridional circulation plays a major role in transporting the remnant radial surface fields from the decaying BMRs towards the polar regions. Hence any variation

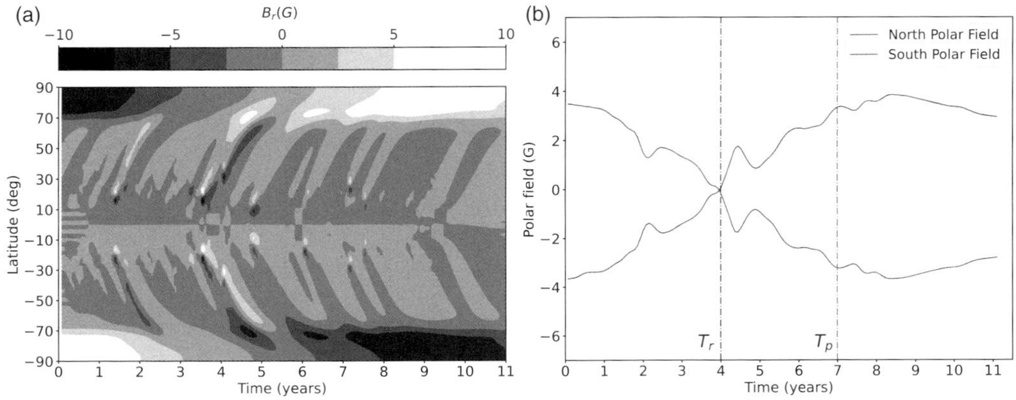

Figure 1. (a) The evolution of the surface radial field under the impact of the surface flows and horizontal diffusion. (b) Evolution of the polar field under the impact of the stochastic properties of the BMRs. The black and green lines represent the time of polar field reversal (T_r) and the time of prediction (T_p) of the upcoming cycle's strength respectively.

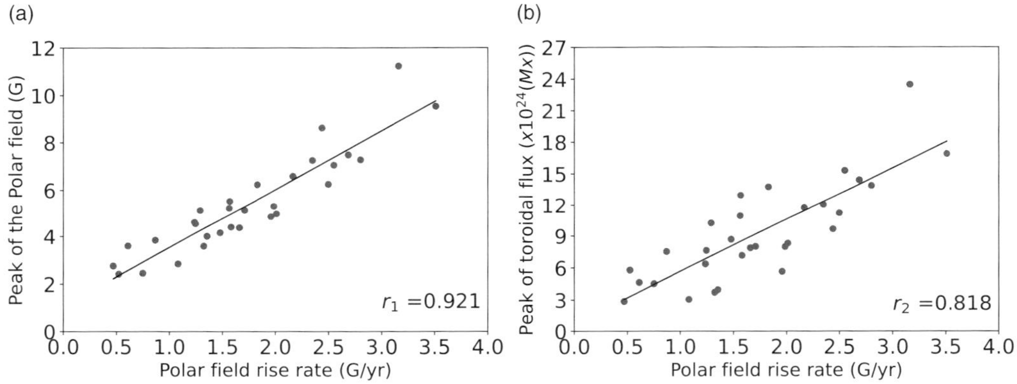

Figure 2. The scatter plots of the polar field rise rate with (a) the peak of the polar field, and (b) the peak strength of the toroidal field to be produced in the following cycle.

in the meridional circulation can change the rate of the production of the polar field and other features of the solar cycle (Karak 2010; Karak and Choudhuri 2011, 2012). On the other hand, the properties of the BMRs, especially their tilts are a major factor in determining their contribution towards the polar field build-up. Hence, the observed large scatter in the tilts (γ) of the BMRs around the famous Joy's law (Hale et al. 1919) can be a major factor in producing fluctuation in polar field strength. In extreme cases, the tilt of some BMRs is opposite of that of the convention for that particular cycle, these BMRs with unusual properties are named as the 'rogue' BMRs which are mainly of two types depending on the value of their tilt, the anti-Joy ($-90° < \gamma < 0°$) and anti-Hale BMRs ($-180° < \gamma < -90°$) (McClintock et al. 2014; Muñoz-Jaramillo et al. 2021). Earlier studies have shown that these 'rogue' BMRs can severely impact the build-up of the polar field in a negative sense and can create a challenge for the predictability of solar cycle strength (Nagy et al. 2017; Golubeva et al. 2023; Pal et al. 2023).

Considering the above-mentioned factors, we perform our simulations for the following cases to test our method of solar cycle prediction against different stochastic processes present throughout different phases of the cycles:

I) Variation in cycle amplitudes with BMR tilts strictly following Joy's law.

Table 1. The correlation coefficients similar to the Figure 2 for all the six cases mentioned.

Cases	r_1	r_2
I)	0.933	0.883
II)	0.929	0.867
III)	0.921	0.818
IV)	0.924	0.830
V)	0.837	0.774
VI)	0.903	0.851

II) Same as Case I, but with varying meridional flow speed in each cycle.
III) Variation in the BMR tilt scatter prperties with anti-Hale BMRs present in all the phases of the cycles.
IV) Same as Case III, but with all anti-Hale BMRs in the rising phases of the cycles.
V) Same as Case III, but with all anti-Hale BMRs in the declining phases of the cycles.
VI) Same as Case III, but with the BMR tilt scatter dependent on the area of the BMRs.

We perform simulations for 30 cycles in each of these cases. The amount of the anti-Hale, and anti-Joy (present throughout all the phases of the cycles) BMRs has been taken to be within around 3%-7% and 10%-30% respectively for the last four cases.

For the limitation of space, here we present results for only the case-III and we mention the correlation coefficients for the rest of the cases in the Table 1.

Here, in the Figure 1, the results from a typical simulation of a cycle from case III is presented. Panel (a) shows the evolution of the surface radial field from the diffused BMRs that get transported towards the polar regions under the action of meridional circulation. Pane (b) shows the polar field strength calculated from the average field strength of the polar cap regions ($55^\circ < |\lambda| < 90^\circ$). The simulated results reproduce the observed polar field evolution reasonably well. The black and the green vertical lines in panel (b) show the times of the polar field reversal (T_r) and the time of prediction ($T_p = T_r + 3$) in our method. We calculate the average rise rate of the polar field build-up within the time range between T_r and T_p and analyse its relation with the peak of the polar field and the strength of the toroidal field in the next cycle as computed from (2) for the above mentioned six cases. It can be easily seen that the time of prediction T_p is a few years earlier than the cycle minimum, which significantly extends the time window of cycle prediction.

The Figure 2 shows the scatter plots of the polar field rise rate with the (a) peak of the polar field and the (b) peak of the toroidal flux for the cycles of case III. The significantly high values of correlation coefficients (r_1 and r_2) between these quantities indicate that a reliable prediction of the strength of the solar cycle can be possible from the polar field rise rate even in the presence of significant amount of stochastic fluctuations throughout the cycles.

Here in Table 1 the correlation coefficients for all the six cases are mentioned. It is clear that the correlation stays high for all the cases, except case V, where all the anti-Hale BMRs are deposited in the declining phases of the cycles and that significantly disturbs the polar field build-up and eventually it hampers the correlation between these quantities. However, even in this case, the correlation is reasonable enough to get an indication of the strength of the upcoming solar cycle.

4. Conclusion

The results of this work suggest that the observed strong correlation of the polar field rise rate with the peak of the polar field and the peak strength of the upcoming cycle is a robust feature of the solar cycles. The extensive simulations performed in this work showed that the correlation stays high even under significant impact of stochastic fluctuations of the cycle parameters, meridional flow strength and BMR tilt properties. Hence, this feature can be utilized for an early prediction of the solar cycle much before the cycle minimum. Based on this method, the predicted amplitude of the Cycle 25 is 137 ± 23 (Kumar et al. 2022).

5. Acknowledgements

I thank Dr. Robert Cameron for kindly providing the SFT code and for fruitful discussions regarding the work. I also thank the International Astronomical Union for the award of the generous travel grant to attend the in-person IAU Symposium 365, in Yerevan Armenia. The financial support from the University Grants Commission, Govt. of India is gratefully acknowledged.

References

Babcock, H. W. 1961, The Topology of the Sun's Magnetic Field and the 22-YEAR Cycle. *Astrophys. J.*, 133, 572.

Baumann, I., Schmitt, D., Schüssler, M., & Solanki, S. K. 2004, Evolution of the large-scale magnetic field on the solar surface: A parameter study. *Astron. Astrophys.*, 426, 1075–1091.

Bhowmik, P., Jiang, J., Upton, L., Lemerle, A., & Nandy, D. 2023, Physical Models for Solar Cycle Predictions. *arXiv e-prints*,, arXiv:2303.12648.

Bhowmik, P. & Nandy, D. 2018, Prediction of the strength and timing of sunspot cycle 25 reveal decadal-scale space environmental conditions. *Nature Communications*, 9, 5209.

Biswas, A., Karak, B. B., & Cameron, R. 2022, Toroidal Flux Loss due to Flux Emergence Explains why Solar Cycles Rise Differently but Decay in a Similar Way. *Phys. Rev. Lett.*, 129(24), 241102.

Biswas, A., Karak, B. B., & Kumar, P. 2023,a Exploring the reliability of polar field rise rate as a precursor for an early prediction of solar cycle. *Mon. Not. R. Astron. Soc.*, 526a(3), 3994–4003.

Biswas, A., Karak, B. B., Usoskin, I., & Weisshaar, E. 2023,b Long-Term Modulation of Solar Cycles. *Space Sci. Rev.*, 219b(3), 19.

Cameron, R. & Schüssler, M. 2015, The crucial role of surface magnetic fields for the solar dynamo. *Science*, 347, 1333–1335.

Cameron, R. H., Jiang, J., Schmitt, D., & Schüssler, M. 2010, Surface Flux Transport Modeling for Solar Cycles 15-21: Effects of Cycle-Dependent Tilt Angles of Sunspot Groups. *Astrophys. J.*, 719(1), 264–270.

Charbonneau, P. 2020, Dynamo models of the solar cycle. *Living Reviews in Solar Physics*, 17(1), 4.

Choudhuri, A. R., Chatterjee, P., & Jiang, J. 2007, Predicting Solar Cycle 24 With a Solar Dynamo Model. *Physical Review Letters*, 98(13), 131103.

Golubeva, E. M., Biswas, A., Khlystova, A. I., Kumar, P., & Karak, B. B. 2023, Probing the variations in the timing of the Sun's polar magnetic field reversals through observations and surface flux transport simulations. *Mon. Not. R. Astron. Soc.*, 525(2), 1758–1768.

Gopalswamy, N. 2022, The Sun and Space Weather. *Atmosphere*, 13(11), 1781.

Hale, G. E., Ellerman, F., Nicholson, S. B., & Joy, A. H. 1919, The Magnetic Polarity of Sun-Spots. *Astrophys. J.*, 49, 153.

Hathaway, D. H. 2015, The Solar Cycle. *Living Reviews in Solar Physics*, 12(1), 4.

Hathaway, D. H., Wilson, R. M., & Reichmann, E. J. 1994, The Shape of the Sunspot Cycle. *Sol. Phys.*, 151(1), 177–190.

Jha, B. K., Karak, B. B., Mandal, S., & Banerjee, D. 2020, Magnetic Field Dependence of Bipolar Magnetic Region Tilts on the Sun: Indication of Tilt Quenching. *Astrophys. J. Lett.*, 889(1), L19.

Jiang, J. 2020, Nonlinear Mechanisms that Regulate the Solar Cycle Amplitude. *Astrophys. J.*, 900(1), 19.

Jiang, J., Wang, J.-X., Jiao, Q.-R., & Cao, J.-B. 2018, Predictability of the Solar Cycle Over One Cycle. *Astrophys. J.*, 863(2), 159.

Jiang, J., Zhang, Z., & Petrovay, K. 2023, Comparison of physics-based prediction models of solar cycle 25. *Journal of Atmospheric and Solar-Terrestrial Physics*, 243, 106018.

Karak, B. B. 2010, Importance of Meridional Circulation in Flux Transport Dynamo: The Possibility of a Maunder-like Grand Minimum. *Astrophys. J.*, 724, 1021–1029.

Karak, B. B. 2020, Dynamo Saturation through the Latitudinal Variation of Bipolar Magnetic Regions in the Sun. *Astrophys. J. Lett.*, 901(2), L35.

Karak, B. B. 2023, Models for the long-term variations of solar activity. *Living Reviews in Solar Physics*, 20(1), 3.

Karak, B. B. & Choudhuri, A. R. 2011, The Waldmeier effect and the flux transport solar dynamo. *Mon. Not. R. Astron. Soc.*, 410, 1503–1512.

Karak, B. B. & Choudhuri, A. R. 2012, Quenching of Meridional Circulation in Flux Transport Dynamo Models. *Sol. Phys.*, 278, 137–148.

Karak, B. B., Jiang, J., Miesch, M. S., Charbonneau, P., & Choudhuri, A. R. 2014, Flux Transport Dynamos: From Kinematics to Dynamics. *Space Sci. Rev.*, 186, 561–602.

Karak, B. B. & Miesch, M. 2017, Solar Cycle Variability Induced by Tilt Angle Scatter in a Babcock–Leighton Solar Dynamo Model. *Astrophys. J.*, 847, 69.

Kumar, P., Biswas, A., & Karak, B. B. 2022, Physical link of the polar field buildup with the Waldmeier effect broadens the scope of early solar cycle prediction: Cycle 25 is likely to be slightly stronger than Cycle 24. *Mon. Not. R. Astron. Soc.*, 513(1), L112–L116.

Kumar, P., Karak, B. B., & Vashishth, V. 2021,a Supercriticality of the Dynamo Limits the Memory of the Polar Field to One Cycle. *Astrophys. J.*, 913a(1), 65.

Kumar, P., Nagy, M., Lemerle, A., Karak, B. B., & Petrovay, K. 2021,b The Polar Precursor Method for Solar Cycle Prediction: Comparison of Predictors and Their Temporal Range. *Astrophys. J.*, 909b(1), 87.

Leighton, R. B. 1969, A Magneto-Kinematic Model of the Solar Cycle. *Astrophys. J.*, 156, 1.

McClintock, B. H., Norton, A. A., & Li, J. 2014, Re-examining Sunspot Tilt Angle to Include Anti-Hale Statistics. *Astrophys. J.*, 797, 130.

Muñoz-Jaramillo, A., Navarrete, B., & Campusano, L. E. 2021, Solar Anti-Hale Bipolar Magnetic Regions: A Distinct Population with Systematic Properties. *Astrophys. J.*, 920(1), 31.

Nagy, M., Lemerle, A., Labonville, F., Petrovay, K., & Charbonneau, P. 2017, The Effect of "Rogue" Active Regions on the Solar Cycle. *Sol. Phys.*, 292, 167.

Nandy, D. 2021, Progress in Solar Cycle Predictions: Sunspot Cycles 24-25 in Perspective. *Sol. Phys.*, 296(3), 54.

Pal, S., Bhowmik, P., Mahajan, S. S., & Nandy, D. 2023, Impact of Anomalous Active Regions on the Large-scale Magnetic Field of the Sun. *arXiv e-prints*,, arXiv:2305.13145.

Petrovay, K. 2020, Solar cycle prediction. *Living Reviews in Solar Physics*, 17(1), 2.

Sreedevi, A., Jha, B. K., Karak, B. B., & Banerjee, D. 2023, AutoTAB: Automatic Tracking Algorithm for Bipolar Magnetic Regions. *Astrophys. J. Suppl.*, 268(2), 58.

Upton, L. A. & Hathaway, D. H. 2018, An Updated Solar Cycle 25 Prediction With AFT: The Modern Minimum. *Geophys. Res. Lett.*, 45(16), 8091–8095.

Dynamics of Solar and Stellar Convection Zones and Atmospheres
Proceedings IAU Symposium No. 365, 2024
A. V. Getling & L. L. Kitchatinov, eds.
doi:10.1017/S1743921323005124

The North–South asymmetry of the number and magnetic fluxes of active regions of different magneto-morphological types in cycles 23 and 24

Anastasiya Zhukova

Crimean Astrophysical Observatory of RAS, Nauchny, Crimea, Russia
email: anastasiya.v.zhukova@gmail.com

Abstract. Statistical study of 3047 active regions (ARs) from 1996 to 2021 was performed using the catalog of the magneto-morphological classes (MMC) of ARs CrAO. According to the magneto-morphological classification of ARs proposed earlier, all ARs, except for unipolar spots, were sorted out between two categories: regular (bipolar groups obeying the Hale's polarity law, the Joy's law, etc.) and irregular ARs (all the rest). We analyzed the number and fluxes of ARs depending on their location in different (relative to the equator) hemispheres. We found that the trends for fluxes are more pronounced. For ARs of both MMC types, they demonstrate signs of both a multi-peak and double-peak structure. Some peaks coincide with the main maxima of the cycle. The second main maximum is mainly formed by the irregular ARs in the S-hemisphere. This might be due to the interaction of the dipole and quadrupole components of the global magnetic field.

Keywords. Sun:magnetic fields, Sun:active regions, Sun:solar cycle

1. Introduction

The North–South (N-S) asymmetry of the solar activity is confirmed by observations in different global indices (sunspot number and area, magnetic fields, flares, prominences, etc.). This phenomenon is crucial in terms of the global mean-field dynamo theory (Moffatt 1978; Krause and Rädler 1980). According to this concept, breaking of the symmetry of turbulence is a necessary condition for the solar dynamo action (this ensures overcoming the limitations of anti-dynamo theorems) (see, e.g., review by Charbonneau and Sokoloff 2023). In addition, during time intervals of a critical decrease in the solar activity, the strong N-S asymmetry is observed (for example, in the late Maunder minimum, sunspots occurred only in the S-hemisphere) (see, e.g., Vokhmyanin and Zolotova 2023). Unfortunately, the mechanism of the N-S asymmetry is not fully understood. The study of this phenomenon is of interest for the development of dynamo theory, as well as for predicting critical changes in the level of solar activity.

2. Data and method

The updated version of the catalog of the magneto-morphological classes (MMC) of solar active regions (ARs) CrAO (https://sun.crao.ru/databases/catalog-mmc-ars) was used to perform a statistical study of 3047 ARs from May 1996 to December 2021. According to the classification of ARs proposed earlier (Abramenko et al. 2018; Abramenko 2021), all ARs, except for unipolar spots, were sorted out between two

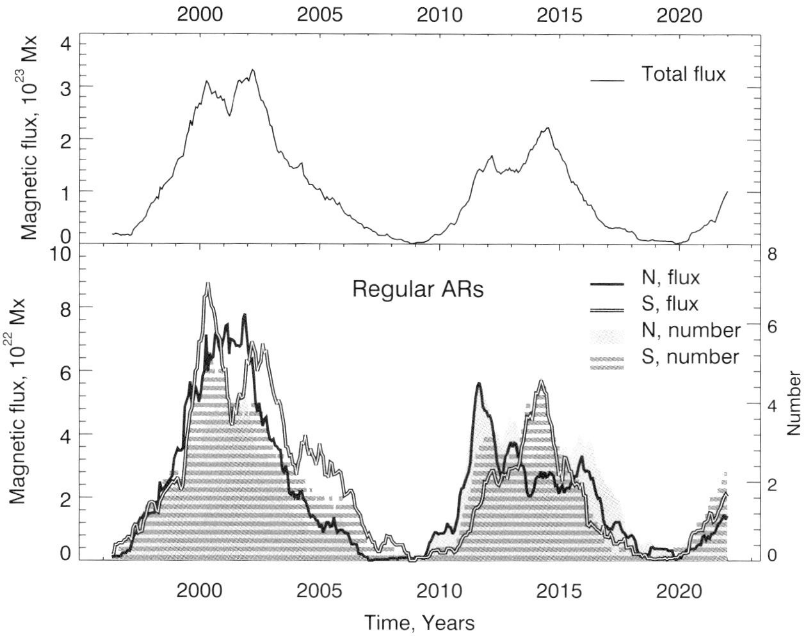

Figure 1. Temporal variations of the number and fluxes of regular ARs.

categories: regular bipolar sunspot groups obeying empirical rules for sunspots (the Hale polarity law, the Joy law, the leading sunspot dominance rule) (Hale et al. 1919; van Driel–Gesztelyi and Green 2015) and the irregular ARs (all the rest). In this study, we analyze the number and magnetic fluxes of different-type ARs depending on their location in different (relative to the equator) hemispheres. Note that magnetic fluxes can be defined as a "generative" activity indices (Nagovitsyn et al. 2021) and allow a more complete conclusions of the details of the dynamo process (Abramenko et al. 2018; Nagovitsyn et al. 2021).

3. Results

3.1. *Temporal variations of the regular ARs*

Temporal variations of the regular ARs number in the N-hemisphere (solid fill) and S-hemisphere (striped fill) are presented in Figure 1. In each hemisphere, for ARs of this MMC type, the profiles demonstrate a multi-peak structure. The peaks occur sometimes in-phase and sometimes out-of-phase. The flux data for the N- and S-hemispheres are overplotted as solid and double lines, respectively. The trends for the fluxes are the same. However, the relative height of the peaks for number and fluxes is different for different maxima. The mutual position of the number and flux profiles depends on the strength of the cycle and the level of activity in each of the cycle maximum.

3.2. *Temporal variations of the irregular ARs*

Temporal variations of the irregular ARs number in the N-hemisphere (solid fill) and S-hemisphere (striped fill) are presented in Figure 2. Time profiles for the irregular groups also have several peaks, but differ from the profiles for the regular groups. Some of the peaks coincide with one of the main maxima of the cycle. Obviously, the double-peak structure of a cycle (discovered by Gnevyshev 1963) as a whole is provided by the totality of ARs of different MMC types in both hemispheres.

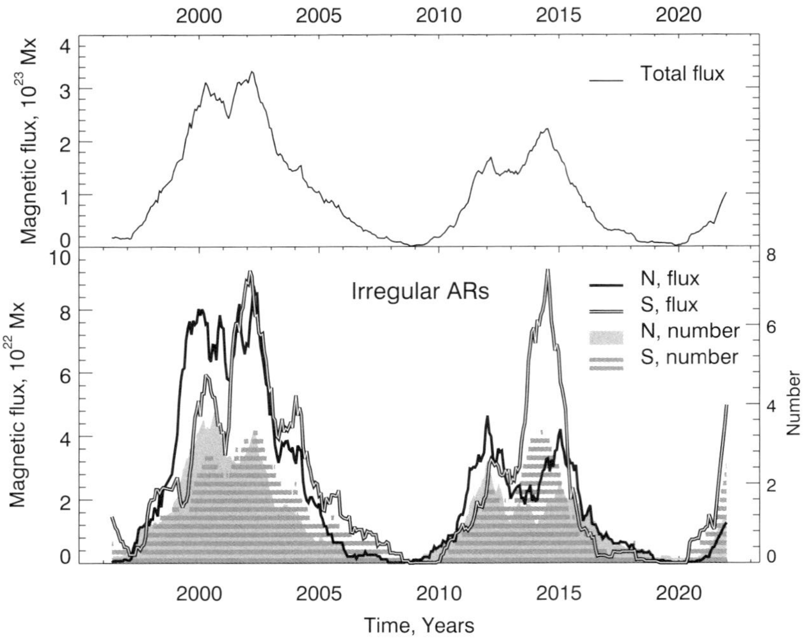

Figure 2. Temporal variations of the number and fluxes of irregular ARs.

The flux data for the N- and S-hemispheres are shown as solid and double lines, respectively. Note that the scales in Figures 1, 2 are the same. Although the number of the irregular ARs is less than ones of the regular ARs (Abramenko et al. 2023), the profiles for the number of irregular ARs are located lower than the profiles of the regular groups. However, the fluxes of different-type ARs are comparable. The trends for fluxes are more pronounced. The strongest irregular ARs fluxes are observed in the second maximum of the cycle in the S-hemisphere (Figure 2). The occurrence of the irregular ARs during this time interval is consistent with previous studies and may be due to the influence of the turbulent component of the dynamo on the magnetic flux tubes of ARs in the convection zone (Zhukova et al. 2022; Abramenko et al. 2023). The predominance of the irregular ARs fluxes in one of the hemispheres might be due to the interaction of the dipole and quadrupole components of the global magnetic field (Zhukova et al. 2023).

3.3. *The interaction between the dipole and quadrupole components of the magnetic field*

The idea of the interaction between the dipole and quadrupole componens of the global magnetic field is as follows. The orientation of the dipole component of the field is changed once per cycle. Note that the observed change in the polarity of sunspots (accepted as the beginning of the 11-year cycle) occurs later than the global magnetic field reversal (in the previous cycle maximum). The quadrupole component may always be oriented in two ways. Depending on the mutual orientation of the components, in different hemispheres, the magnetic field lines of the components have the same/opposite direction. In the respective hemisphere, the total field is getting stronger/weaker.

The phenomenon also concerns the toroidal part of the field, which is the source of the ARs occurrence. The point is that the differential rotation (one of the main triggers of the dynamo process) stretches magnetic field lines (frozen in plasma) along the equator. And this is true for both the dipole and quadrupole components. Depending on the mutual orientation of the components, the magnetic field lines of the toroidal field have the same/opposite direction in different hemispheres. Accordingly, the toroidal field is strengthening/weakening in the respective hemisphere. As the result, the stronger/weaker the field in a given hemisphere, the weaker/stronger the influence of turbulence on the arising flux tubes in the convection zone, the less/more irregular ARs appear in the given hemisphere.

4. Conclusions

A comparison of cyclic variations of the number and magnetic fluxes of the regular and irregular ARs showed the following.

i) The trends for ARs magnetic fluxes are more pronounced than that for the ARs number;

ii) The second main maximum of the cycle is weakly pronounced for the regular ARs and is mainly formed by the irregular ARs in the S-hemisphere;

iii) The predominance of the irregular ARs in one of the hemispheres might be due to the interaction of the dipole and quadrupole components of the global magnetic field.

5. Acknowledgements

The author is grateful to V.I.Abramenko for her valuable comments and to R.A.Suleymanova for the SC 23 data provided for the MMC ARs CrAO catalog.

Supplementary material

To view supplementary material for this article, please visit http://dx.doi.org/10.1017/S1743921323005124

References

Abramenko, V. I. 2021, *MNRAS*, 507(3), 3698.
Abramenko, V. I., Suleymanova, R. A., & Zhukova, A. V. 2023, *MNRAS*, 518(3), 4746.
Abramenko, V. I., Zhukova, A. V., & Kutsenko, A. S. 2018, *Geomag. Aeron.*, 58(8), 1159.
Charbonneau, P. & Sokoloff, D. 2023, *Space Sci. Rev.*, 219(5), 35.
Gnevyshev, M. N. 1963, *Soviet Astron.*, 7, 311.
Hale, G. E., Ellerman, F., Nicholson, S. B., & Joy, A. H. 1919, *Ap.J.*, 49, 153.
Krause, F. & Rädler, K. H. 1980,. *Mean-field magnetohydrodynamics and dynamo theory.*
Moffatt, H. K. 1978,. *Magnetic Field Generation in Electrically Conducting Fluids.* Cambridge Univ. Press, Cambridge, UK.
Nagovitsyn, Y. A., Osipova, A. A., & Nagovitsyna, E. Y. 2021, *Solar Phys.*, 296(2), 32.
van Driel–Gesztelyi, L. & Green, L. M. 2015, *Liv. Rev. Solar Phys.*, 12(1), 1.
Vokhmyanin, M. & Zolotova, N. 2023, *Solar Phys.*, 298(10), 113.
Zhukova, A., Sokoloff, D., Abramenko, V., & Khlystova, A. 2023, *Adv. Space Res.*, 71(4), 1984.
Zhukova, A. V., Khlystova, A. I., Abramenko, V. I., & Sokoloff, D. D. 2022, *Geomag. Aeron.*, 62(7), 823.

Dynamics of Solar and Stellar Convection Zones and Atmospheres
Proceedings IAU Symposium No. 365, 2024
A. V. Getling & L. L. Kitchatinov, eds.
doi:10.1017/S1743921323005306

Prognostic opportunity of the shifted correlation between Wolf numbers and their time derivatives

S. V. Starchenko and S. V. Yakovleva

IZMIRAN, Moscow, Troitsk, 108840 Russia
email: sstarchenko@mail.ru

Abstract. We correlate the annual Wolf numbers W and their time derivatives W' by shifting time fragments of W and W' relative to each other. The most significant (up to 0.874) correlation is with 3 years shifts for fragments covering 14 years. For longer and shorter periods, the correlation coefficients 0.771–0.855 with 2–3 years shift. The most significant 9 years shift corresponds to -0.852/-0.824 anti-correlation coefficient for 14/11 years period. The other periods are less significant. To evaluate predictive estimates, we use the times series fragments of W shifted back into the past. A forecast can be made using the leading graphs based upon the derived calibration factor. Test calculations show that the most effective is the calibration factor calculated for changing the phase of the cycle. The best linear pairwise correlation coefficient of the approximation is 0.94.

Keywords. Solar activity cycles, sunspots, Wolf numbers, physical foundations of the solar dynamo, forecast

1. Introduction

Despite numerous attempts, forecasting solar-stellar magnetic activity has remained an outstanding challenge (Pesnell (2012), Obridko & Nagovitsyn (2017), Nandy (2021)). In a detailed review, Nandy (2021) shows that forecasts for Solar Cycle XXV based on different techniques still diverge widely, with a majority of the forecasts indicating that Solar Cycle XXV would be stronger than Solar Cycle XXIV. The mean of the various forecasts of the maximum amplitude in solar cycle XXV is 136.2 ± 41.6. Some of the most frequently used forecasting methods are statistical methods, including the study of the correlation properties of Wolf numbers W (Vitinsky (1973), Bondar *et al.* (1995), Ishkov & Shibaev (2006), Pishkalo (2008), Abdel–Rahman & Marzouk (2018), Petrovay (2020), McIntosh *et al.* (2020)). Therefore, correlations, and in fact autocorrelations, are primarily sought between original or transformed fragments of W series, separated by time intervals. In (Starchenko & Yakovleva (2022)), the basic correlation dependencies between the series of Wolf numbers shifted in time relative to each other and their derivatives (W and W') were determined. Annual means of Wolf numbers were used from 1700 to 2022 in version v2 http://sidc.oma.be/silso/datafiles. Time derivatives $dW/dt = W'$ were obtained at the same middle of the year as W, by taking the average of the derivative on the left and right. This technique allows us to smooth some errors originated from the observational and contractual definition of W. Both the Wolf numbers and their derivatives in the considered interval was modified by extracting the corresponding average values. We use such deviations for more clearly demonstration of variability,

Table 1. Correlation coefficients between temporal rows of W and W' of different lengths. K is the resulting correlation coefficient, I is the length of the time series fragments in years, Z is the step (in years) by which the time fragments are shifted (the sign "−" means an anti-correlation).

Z	K(I=11)	K(I=14)	K(I=22)	K(I=33)	K(I=44)	K(I=66)	K(I=88)	K(I=176)	K(I=322)
0	0.015	-0.076	0.023	0.012	0.011	0.004	0.002	0.006	0.000
1	0.584	0.521	0.531	0.551	0.553	0.554	0.548	0.542	0.531
2	0.771	0.783	0.793	**0.807**	**0.812**	**0.814**	**0.812**	**0.793**	**0.761**
3	**0.855**	**0.874**	**0.836**	**0.808**	0.805	0.790	0.792	0.769	0.714
4	0.672	0.737	0.639	0.545	0.533	0.518	0.528	0.515	0.463
5	0.314	0.401	0.281	0.123	0.116	0.117	0.131	0.143	0.115
6	0.120	0.077	-0.126	-0.300	-0.282	-0.284	-0.289	-0.244	-0.244
7	-0.229	-0.303	-0.522	-0.637	-0.588	-0.586	-0.611	-0.548	-0.518
8	-0.624	-0.681	**-0.725**	**-0.753**	**-0.719**	**-0.719**	**-0.746**	**-0.699**	**-0.623**
9	**-0.824**	**-0.852**	-0.697	-0.616	-0.611	-0.611	-0.621	-0.637	-0.509
10	-0.781	-0.786	-0.433	-0.217	-0.273	-0.241	-0.237	-0.330	-0.209

leading to more significant correlations, essentially corresponding with the classic Pearson correlation.

In the next section we show how the correlation between the shifted fragments of the W and W' series was determined, and we also try to substantiate the physical meaning and the corresponding predictive potential of these shifts.

2. Correlation of shifted series and the physical meaning of the maximal shifts

Unlike a previously published study by Starchenko & Yakovleva (2022), in the present work we determine a correlation between the complete series W and W', which seems more reasonable in terms of future forecasting. The correlation coefficient K is calculated using the Pearson equation where corresponding values W are used together with their mean values $\overline{W}$ and $\overline{W'}$:

$$\mathrm{K} = \sum_{i=1}^{\mathrm{I}} (W_i - \overline{W})(W'_i - \overline{W'}) / \sqrt{\sum_{i=1}^{\mathrm{I}} (W_i - \overline{W})^2 \sum_{i=1}^{\mathrm{I}} (W'_i - \overline{W'})^2} \qquad (1)$$

Here i counted (from the past to the present) from the I-th value in the past for the complementary series and from the (I+$|Z|$)-th for the leading series.

All obtained correlation coefficients K are presented in Table 1 for fragments of series with lengths I = 11, 14, 22, 33, 44, 66, 88, 176 (years) and the entire available series with I = 322 (years). The most significant correlations/anti-correlations are highlighted in bold.

A comparison of correlation coefficients K for fragments of different lengths showed, that the most significant (0.85–0.87) correlation coefficients are associated with the shifts of 2 or 3 years for fragments covering the last 11–14 years. For longer fragments, the coefficients remain significant (0.76–0.84) at the same shifts. Therefore, the main phase shift between W and W' is approximately a quarter of a solar cycle, which physically corresponds to the preferential connection of spots with magnetic energy in accordance with the simplest dynamo model (see next paragraph). There is also a significant shift by 8-9 years, characterizes by anti-correlation coefficients from -0.70 to -0.85. Summarizing the resulting 2–3-year (one quarter) and 8–9-year (the remaining three quarters) shifts, we obtain the observed variety of dynamo magnetic energy periodicity from 10 to 12 years. Figure 1 shows the evolution of the Wolf numbers and their derivatives for the interval of 1976.5–2021.5 (magnetic solar cycles XXI–XXIV), supplemented with fragments of W shifted to 3 and 9 years ago. In particular, these plots allow us to estimate the predictive potential of the approach proposed here.

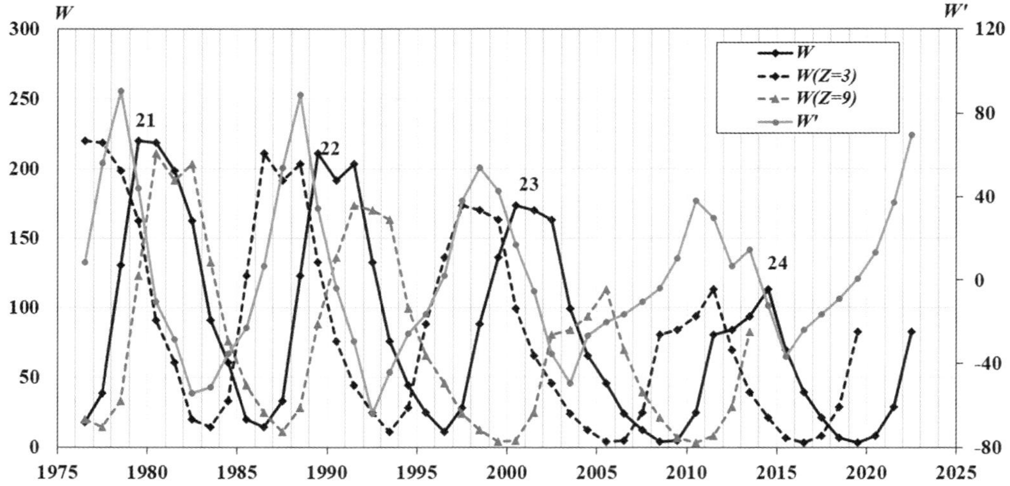

Figure 1. Displacement plots of W with respect to W' for Z = 3 and Z = 9 based on calculations for the magnetic solar cycles XXI-XXIV.

We use the idea of our previous study (Starchenko & Yakovleva (2022)) that the magnetic field variation is a sinusoidal: $\sin(\pi t/T_c)$, where T_c is the length of the solar activity cycle (about 11 years). Then, taking into account that the energy is quadratic to the field, the dimensionless normalized functions of the corresponding energy (essentially Wolf numbers) and derivative (essentially power) minus the average are $-\cos(2x)$ and $\sin(2x)$ respectively, where $x = \pi t/T_c$ is a phase variable. Obviously, the phase shifts (up to complete correlation or anti-correlation) correspond to 1/4 and 3/4 of the T_c. This conclusion is consistent with the maximum correlations/anti-correlations mentioned above. It should be noted that such a physical reason for anti-correlation was not emphasized in (Starchenko & Yakovleva (2022)) and we are filling this gap here.

3. Application of correlation of shifted series for Wolf numbers forecasting

For potential prognostic estimates based on the W series, we chose the correlation coefficient from Table 1 for K(I=11) and K(I=14). Based on four values of W corresponding to the beginning of the cycle XXV (2019.5-2022.5) and using the least squares, we calculate the coefficients of the equation $y = A + Bx$, where y are the values of W shifted by 3 years ago, and x are the values of unshifted W'. So we get $W = A + BW'$. Using coefficients A and B, we can test the accuracy of the approximation and make a forecast for 3 years based on the known values of the derivative W' (see Figure 2). To estimate the quality of approximation of the reconstructed series, the determination coefficient R^2 was calculated. In our case of linear regression, the coefficient of determination is equal to the square of the Pearson correlation coefficient similar to one shown in (1).

As an example, we present the equations calculated for several options for four W:

(a) $W = 70.524 + 3.228\ W'$; R^2=0.89 or K=0.94 for the interval 2019.5-2022.5;

(b) $W = 75.829 + 3.680\ W'$; R^2=0.87 or K=0.92 for 2 pairs of four W for the intervals 2008.5–2011.5 2019.5–2022.5, respectively;

Tests have shown that the calibration coefficient calculated for the same phase of the previous cycles (XXI–XXIII) is the most effective. The maximum accuracy of pairwise correlation is 0.94 for option (a).

Table 2. Example of approximations using four W for the interval 2019.5–2022.5 (option (a)).

Years	Initial fragment W	Derivative W'	Shifted fragment W	Approximated W	Predicted W
2016.5	39.8	-24.05	3.6	7.12	
2017.5	21.7	-16.4	8.8	17.58	
2018.5	7.0	-9.05	29.7	41.31	
2019.5	3.6	0.9	83.1	73.43	7.12
2020.5	8.8	13.05		**112.65**	17.58
2021.5	29.7	37.15		**190.46**	41.31
2022.5	83.1	69.65		**295.38**	73.43
2023.5					**112.65**
2024.5					**190.46**
2025.5					**295.38**

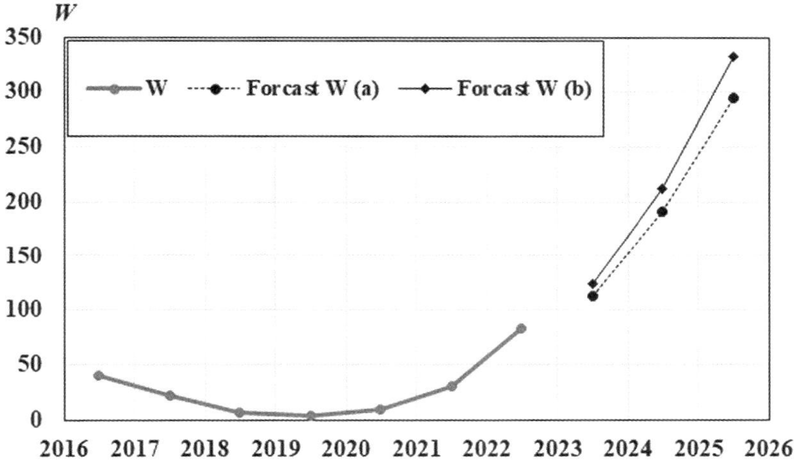

Figure 2. An example of a forecast for three years ahead for four W at the beginning of cycle XXV(a) and for 2 pairs of four W corresponding to the beginning of the cycles XXIV and XXV(b).

We presented an example in Table 2 to illustrates the calculation techniques. First, a fragment of W of the required length was shifted back 3 years, then the linear regression coefficients A and B were calculated using the least squares method. Then, the approximated series W was calculated using the known values of W', and finally, the resulting series was shifted forward to its place, thereby making a forecast for three years.

So, we are expecting a relatively long growing phase with sufficiently high maximum and extended duration of the current cycle. Apparently, this forecasting method is better suited for forecasting intermediate and average annual (or monthly average) values, rather than for determination of maximum amplitude in solar cycle and the time of it's occurrence. We are currently working on this problem. At the same time, the predicted values of W (option (a)) for the coming years confirm the results of the forecast of a number of researchers (Bhowmik & Nandy (2018), Du & Du (2006), Okoh *et al.* (2018)). This includes the work that particularly interested us (Pishkalo (2008)), which used the correlation between cycle parameters to predict maximum and minimum Wolf numbers for the solar cycle XXV. According to the study by Pishkalo (2008), the maximum amplitude in solar cycle XXV will be 112.37 ± 33.4 in mid-2023. At the same time, the predicted time of occurrence of the upcoming maximum covers a fairly large range from mid-2023 to end-2025 (Nandy (2021)).

4. Conclusions

The correlations of the annual means of W and their time derivatives W' were studied with shifts of the series fragments relative to each other. The most significant (up to 0.87 and -0.85) correlation and anti-correlation coefficients are obtained with shifts of 3 years and 9 years for fragments of 11 and 14 years. Fragments of 22, 33, 44, 66, 88, 176, and 322 (the entire series) years have also been studied. Their coefficients are at the levels of -0.8 for shifts of 2–3 years and 8–9 years.

The corresponding phase shifts between the Wolf numbers and their derivatives are approximately 1/4 and 3/4 of the solar cycle, which is consistent with the predominant relationship between the dynamics of solar spots and the variability of magnetic energy.

Based on the high correlation coefficients between the shifted fragments of the W relative to W', the forecast of Wolf numbers for 2023.5–2025.5 is made with the maximum accuracy of pairwise correlation is 0.94. Moreover, this accuracy remains approximately at the same level with the joint participation of similar fragments of neighboring cycles.

References

Abdel–Rahman, H.I., & Marzouk, B.A. 2018, NRIAG J. Astron. Geophys., 7, 175
Bhowmik, P., & Nandy, D. 2018, Nat. Commun., 9, 5209.
Bondar, T.N., Rotanova, N.M., & Obridko, V.N. 1995, Astron. Rep., 39, 130
Du, Z., & Du, S. 2006, Solar Phys., 238, 431
Ishkov, V.N., & Shibaev, I.G. 2006, Bull. Russ. Acad. Sci., (Physics), 70, 1643
McIntosh, S.W., Chapman, S., Leamon, R.J., Egeland, R., & Watkins, N.W. 2020, Solar Phys., 295, 163
Nandy, D. 2021, Solar Phys., 296, 54
Obridko, V.N., & Nagovitsyn, Yu.A. 2017, Solar activity, cyclicity and forecasting methods, SPb.: VVM Publishing House, 466pp. (in Russian)
Okoh, D.I., Seemala, G.K., Rabiu, A.B., Uwamahoro, J., Habarulema, J.B., Aggarwal, M. 2018, Space Weather, 16, 1424.
Pesnell, W.D. 2012, Solar Phys., 281, 507
Petrovay, K. 2020, Living Rev. Sol. Phys., 17, 2
Pishkalo, M.I. 2008, Kinematics and Physics of Celestial Bodies, 24, 242.
Starchenko S.V. & Yakovleva S.V. 2022, G&A, 62, 685.
Vitinsky, Yu.I. 1973, Cyclicity and forecasts of solar activity, L.: Science, 258 pp. (in Russian)

Dynamics of Solar and Stellar Convection Zones and Atmospheres
Proceedings IAU Symposium No. 365, 2024
A. V. Getling & L. L. Kitchatinov, eds.
doi:10.1017/S1743921323004295

Chromosphere activity: relations with Solar cycles (SC)

S. Koutchmy[1] **and E. Tavabi**[2]

[1]Institut d'Astrophysique de Paris, Sorbonne Universités, UMR 7095, CNRS and UPMC, 75014 Paris, France

[2]Physics Department, Payame Noor University, Tehran, Iran, 19395-3697
email: e_tavabi@pnu.ac.ir

Abstract. The study's focus on the modulation of geomagnetism by low latitude solar magnetically activity, including coronal mass ejections (CMEs), solar flares, and solar energetic particles (SEPs). It mentions the Babcock–Leighton (B-L) dynamo model used to predict sunspot numbers in Solar Cycle 25 (SC25) and highlights the challenges in understanding aspects such as the regeneration of the poloidal field and the occurrence of magnetic regions, active longitudes, and coronal holes. The abstract introduces the study's concentration on the activity of polar regions using chromosphere jets activity proxies and other parameters like polar faculae density and cool ejection events. It also mentions the observation of chromospheric prolateness during the minimum solar activity periods.

Keywords. chromosphere, solar cycle, activity, sunspot

1. Introduction

The solar activity through the sunspot number (SN) modulates geomagnetism, CMEs, flares, SEPs and associated disturbances. The prediction of the height of the forthcoming SC (expressed by the SN) is since 2019 the subject of many studies. They are mainly based on statistical and/or mathematical and/or Heuristic methods, taking parameters from the analysis of past cycles; they leaded to a predicted SC 25 similar to the low height preceding SC 24, sometimes significantly lower, indeed. Some predictions uses solar activity parameters justified by the solid belief that the solar activity is fully governed by a dynamo mechanism occurring inside the Sun (Filippov and Koutchmy 2002; Filippov et al. 2007; Koutchmy et al. 2022b,a; Siebert et al. 2021; Kosovichev et al. 2020).

2. data analysis

The images were obtained from 2010 to 2022, for 13 years data from the area above the coronal holes in the north pole of the Sun, from the release of 304 A due to the Helium 2 resonance line formed at about 50,000 K, taken by the EE telescope (Solar Atmospheric Imaging Collection)(approx. 2340 data). The cadence between frames is 12 seconds (images every minute), in FITS protocol. After obtaining and summing the images on the disk, we attended them using MATLAB software, then we compressed the areas above the coronal holes in the South and North Poles regions to the extent of one column (see Figure 1). The graphs related to these data were plotted (15 data on the 15th day of every month) and measured the width at half the height of the maximum exactly at the first maximum. We recorded them as the width of the chromosphere at this point of the poles of the Sun. This thickness actually shows the width of the area above the

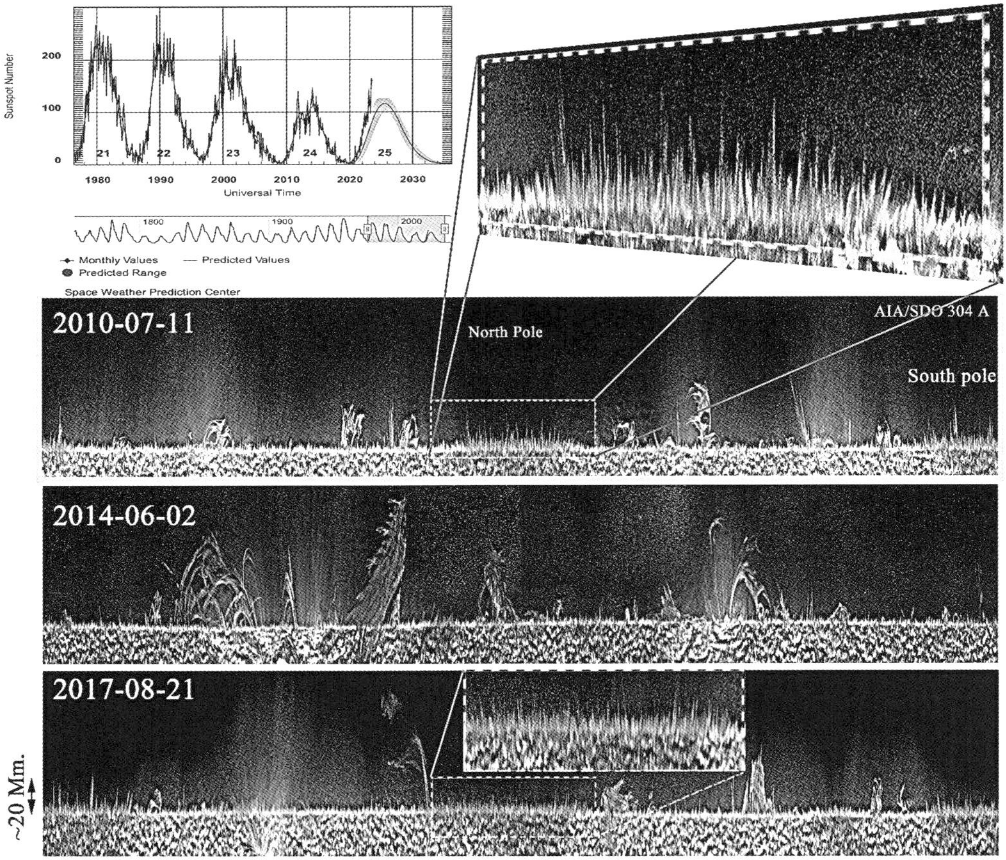

Figure 1. Partial frames in R-Theta coordinate, extracted image of a typical three polar regions at time of minimum and maximum activity (2010 minimum, 2014 maximum and 2017 minimum) to demonstrate the "abnormal" thickness of the 304 emissions (due to the HeII resonance line formed around 80 000K). Such extensions were measured during more than a full solar cycle (24th) in both the South and the North regions (fig. 2). Extensions are believed to be due to many ejection events in nearly radial directions called macro-spicules, in contrast to spicules seen everywhere, including regions outside the coronal hole regions. AIA filtergrams of the NSO NASA mission were used after summing original frames for 10 min include 50 successive frames.

coronal cavity where the magnetic field lines and as a result plasma and coronal mass escape into the space, and by measuring and studying it, we can find out the extent of the sun's magnetic activity in that area.

3. Result

The study examined various proxies to assess the activity of polar regions in relation to Solar Cycle 25 (SC25). The density of polar faculae, evaluated from WL filtergrams of the SDO mission, showed a definite increase before SC25, contradicting the low amplitude predicted by the NASA Solar Cycle 25 Prediction Panel. Similarly, the number of cool ejection events observed in Halpha observations suggested that SC25 could reach higher levels compared to the panel's moderate prediction. These off-disk parameters provided more convincing evidence than classical on-disk observations. Additionally, the study noted the observation of chromospheric prolateness (ovalisation) in the years preceding the minimum solar activity, as seen in cool spectral lines like Halpha and the H and K

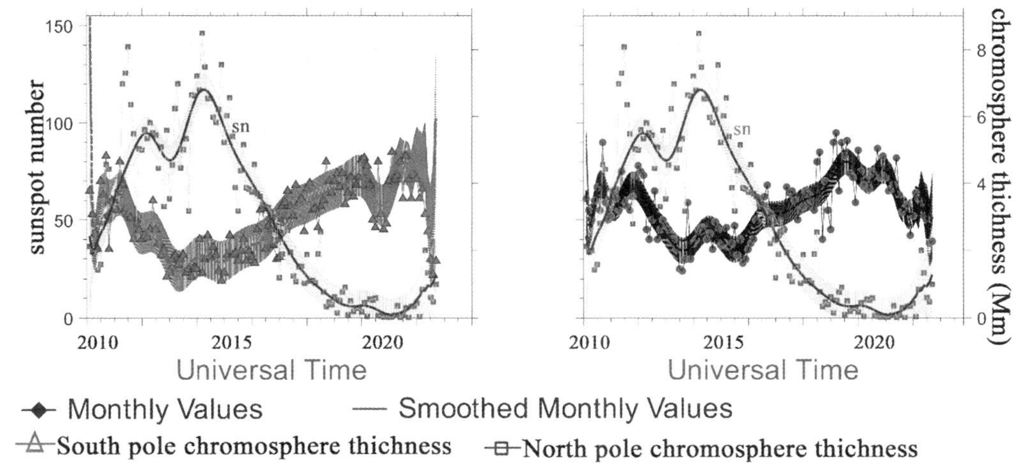

Figure 2. The results of measuring the FWHM of the 304 A shell for the last SC24. Again in both hemisphere we recorded a more important activity reflected by the FWHM thickness of the 304 A shell before the SC25 compared to the level recorded before the SC24. Apparent "thicknesses" Variations of the transition region polar regions shells as measured using the 304 emissions (He II) from AIA/SDO. Both the North (N) in red and the South (S) in black poles are showing an extended shell of macro-spicules activity at Years of the solar minima but during the last period around 2019, an enhanced activity is recorded suggesting that the next SC 25 will be high. In black line the "sn" during the same period of time.

lines of CaII. The prolateness was observed in full disk images and indicated a north-south elongation of the solar chromosphere during the minimum periods (figure 2).

4. Discussion

The study highlights the discrepancies between the predictions of the NASA Solar Cycle 25 Prediction Panel and the observations based on various proxies. The increase in the density of polar faculae and the number of cool ejection events suggest a higher level of solar activity for SC25 than initially predicted. The authors note that off-disk observations provide more reliable indications of solar activity than on-disk observations. The observation of chromospheric prolateness during the minimum solar activity periods raises interesting questions about the behavior of the solar chromosphere and its connection to solar cycles. The study suggests further investigation of this topic and plans to revisit the chromospheric prolateness near the time of solar minimum. Overall, the results and discussion highlight the complexity of predicting solar cycles and the importance of considering various proxies and observational techniques to gain a comprehensive understanding of solar activity.

References

Filippov, B. & Koutchmy, S. L. 2002, About the prominence heating mechanisms during its eruptive phase. *Solar Physics*, 208, 283–295.

Filippov, B., Koutchmy, S. L., & Vilinga, J. 2007, On the dynamic nature of the prolate solar chromosphere: jet formation. *Astronomy and Astrophysics*, 464, 1119–1125.

Kosovichev, A., Strassmeier, S., & Jardine, M. 2020, Solar and stellar magnetic fields: Origins and manifestations. *Solar and Stellar Magnetic Fields: Origins and Manifestations*, 354.

Koutchmy, S., Filippov, B., Tavabi, E., Noëns, J., & Wurmser, O. 2022,a Acta astrophysica taurica. *Acta Astrophysica Taurica*,.

Koutchmy, S. L., Filippov, B., Tavabi, E., Noens, J.-C., & Wurmser, O. Polar regions activity and the prediction of the height of the solar cycle 25 2022,b.

Siebert, A., Baillie, K., Lagadec, E., Lagarde, N., Malzac, J., Marquette, J., N'Diaye, M., Richard, J., & Venot, O. 2021, Sf2a-2021: Proceedings of the annual meeting of the french society of astronomy and astrophysics.

Dynamics of Solar and Stellar Convection Zones and Atmospheres
Proceedings IAU Symposium No. 365, 2024
A. V. Getling & L. L. Kitchatinov, eds.
doi:10.1017/S1743921323005173

Modelling the rotation dependence of cycle variability in sun-like stars: Answering why only slowly rotating stars produce grand minima

Vindya Vashishth

Department of Physics, Indian Institute of Technology (Banaras Hindu University)
Varanasi 221005 India
email: vindyavashishth.rs.phy19@itbhu.ac.in

Abstract. The Sun and solar-type stars exhibit irregular cyclic variations in their magnetic activity over long time scales. To understand this irregularity, we employed the flux transport dynamo models to investigate the behavior of one solar mass star at various rotation rates. To achieve this, we have utilized a mean-field hydrodynamic model to specify differential rotation and meridional circulation, and we have incorporated stochastic fluctuations in the Babcock–Leighton source of the poloidal field to capture inherent fluctuations in the stellar convection. Our simulations successfully demonstrated consistency with the observational data, revealing that rapidly rotating stars exhibit highly irregular cycles with strong magnetic fields and no Maunder-like grand minima. On the other hand, slow rotators produce smoother cycles with weaker magnetic fields, long-term amplitude modulation, and occasional extended grand minima. We observed that the frequency and duration of grand minima increase with the decreasing rotation rate. These results can be understood as the tendency of a less supercritical dynamo in slow rotators to be more prone to produce extended grand minima. We further explore the possible existence of the dynamo in the subcritical regime in a Babcock–Leighton-type framework and in the presence of a small-scale dynamo.

Keywords. Stars: magnetic field, stars: rotation, dynamo: sun-like stars

1. Introduction

Like our Sun, many other sun-like stars have magnetic fields and cycles as unveiled by various observations (Donati et al. 1992; Baliunas et al. 1995; Wright et al. 2011; Wright & Drake 2016; Metcalfe, Egeland, and van Saders 2016). According to these observations, a star's rotation rate plays an important role in determining its magnetic activity. Rapidly rotating (young) Sun-like stars exhibit a high activity level with no Maunder-like grand minimum (flat activity) and rarely display smooth regular activity cycles. On the other hand, slowly rotating old stars like the Sun and older have lower activity levels and smooth cycles with occasional grand minima (Skumanich 1972; Rengarajan 1984; Baliunas et al. 1995; Oláh et al. 2016; Boro Saikia et al. 2018; Garg et al. 2019). Recently, Shah et al. (2018) observed the decreasing magnetic activity of HD 4915, which might indicate it as the Maunder minimum candidate. Later Baum et al. (1995) confirmed that HD 166620 is entering into a grand minimum phase. Interestingly, these stars (including Sun) are slow rotators.

Magnetic cycles in the Sun and other sun-like stars are maintained by dynamo action powered by helical convection and differential rotation in their convection zones

(Karak et al. 2014a; Charbonneau 2020; Karak 2023). This is because the toroidal field is generated through the stretching of the poloidal field by the differential rotation, which is known as the Ω effect. There is strong evidence that the poloidal field is generated through a mechanism, so-called the Babcock–Leighton process (Dasi–Espuig et al. 2010; Kitchatinov & Olemskoy 2011; Muñoz-Jaramillo et al. 2013; Priyal et al. 2014; Mordvinov et al. 2020, 2022; Kumar et al. 2021). In this process, tilted sunspots (more generally bipolar magnetic regions; Sreedevi et al. 2023) decay and disperse to produce a poloidal field through turbulent diffusion, meridional flow, and differential rotation. While the systematic tilt in the BMR is crucial to generate the poloidal field, the scatter around Joy's law tilt (e.g, McClintock et al. 2014; Jha et al. 2020) produces a variation in the solar cycle (Lemerle & Charbonneau 2017; Karak and Miesch 2017, 2018; Karak et al. 2018; Karak 2020; Biswas et al. 2023).

Many observations suggest that as the stars rotate faster, the magnetic activity becomes stronger, but the relation between the activity cycle period and the rotation rate does not seem a straightforward trend. The cycle period tends to decrease with the rotation rate for the slowly rotating stars, whereas the trend is quite complicated for the fast rotators. Previous studies have explored the trend of magnetic field strength and the cycle period with the rotation rate of the stars (Karak et al. 2014; Hazra et al. 2019). Karak et al. (2020); Noraz et al. (2022) have also explored the possibility of magnetic cycle and reversals in slowly rotating stars possibly having anti-solar differential rotations (e.g., Karak et al. 2015, 2018). Here we aim to understand these observational trends of stellar magnetic activity using dynamo modeling. We shall extract the dependency of the rotation rate of the sun-like stars on its cycle variability and the occurrence of the grand minima using the dynamo models of Karak et al. (2014); Hazra et al. (2019) in which the regular behavior of the stellar cycle was simulated. As the stellar cycles are irregular, it is natural to explore the irregular features of the stellar cycles using these models. For this, we have included stochastic noise to capture the inherent fluctuations in the stellar convection, as seen in the form of variations in the flux emergence rates and the tilts of BMRs around Joys law. To do so, we have included the stochastic fluctuations in the Babcock–Leighton source for the poloidal field in the dynamo.

2. Model

In our work, we have developed three kinematic mean-field dynamo models, namely, Models I-III, by assuming the axisymmetry. Thus, the evolution equation of the poloidal ($\nabla \times [A(r,\theta)\mathbf{e}_\phi]$) and toroidal ($B(r,\theta)\mathbf{e}_\phi$) fields are followings.

$$\frac{\partial A}{\partial t} + \frac{1}{s}(\boldsymbol{v}_p \cdot \boldsymbol{\nabla})(sA) = \eta\left(\nabla^2 - \frac{1}{s^2}\right)A + S(r,\theta;B), \tag{1}$$

$$\frac{\partial B}{\partial t} + \frac{1}{r}\left[\frac{\partial}{\partial r}(rv_rB) + \frac{\partial}{\partial \theta}(v_\theta B)\right] = \eta\left(\nabla^2 - \frac{1}{s^2}\right)B + s(\boldsymbol{B}_p \cdot \boldsymbol{\nabla})\Omega + \frac{1}{r}\frac{d\eta}{dr}\frac{\partial(rB)}{\partial r}, \tag{2}$$

where $s = r\sin\theta$, $\boldsymbol{v}_p = v_r\hat{r} + v_\theta\hat{\theta}$ is the meridional flow and the Ω is the angular velocity whose profile is obtained from the mean-field hydrodynamic model of Kitchatinov & Olemskoy (2011), η is the turbulent magnetic diffusivity which is written as the function of r alone and take the following form,

$$\eta(r) = \eta_{\rm RZ} + \frac{\eta_{\rm SCZ}}{2}\left[1 + {\rm erf}\left(\frac{r - r_{\rm BCZ}}{d_t}\right)\right] + \frac{\eta_{\rm surf}}{2}\left[1 + {\rm erf}\left(\frac{r - r_{\rm surf}}{d_2}\right)\right] \tag{3}$$

with $r_{\rm BCZ} = 0.7R_s$ (R_s being the stellar radius), $d_t = 0.015R_s$, $d_2 = 0.05R_s$, $r_{\rm surf} = 0.95R_s$, $\eta_{\rm RZ} = 5\times10^8$ cm^2 s^{-1}, $\eta_{\rm SCZ} = 5\times10^{10}$ cm^2 s^{-1}, and $\eta_{\rm surf} = 2\times10^{12}$ cm^2 s^{-1}.S

is the source for the poloidal field and its parameterized form is written as

$$S(r,\theta;B)=\frac{\alpha_0\alpha_{\rm BL}(r,\theta)}{1+\left(\overline{B}(r_t,\theta)/B_0\right)^2}\overline{B}(r_t,\theta), \tag{4}$$

where $\overline{B}(r_t,\theta)$ is the toroidal field at latitude θ averaged over the whole tachocline from $r=0.685R_s$ to $r=0.715R_s$, α_0 is the measure of the strength of the Babcock–Leighton process, which is expressed as the dependence on the rotation in the following way,

$$\alpha_0=\alpha_{0,s}\frac{T_s}{T}, \tag{5}$$

where $\alpha_{0,s}$ is the value of α_0 for the solar case, which is taken as 0.7 cm s^{-1} in Model I-II and T_s and T are the rotation period of Sun and the star, respectively. And finally, $\alpha_{\rm BL}$ is the parameter for Babcock–Leighton process which is written in Model I-II as,

$$\alpha_{\rm BL}(r,\theta)=\frac{1}{4}\left[1+{\rm erf}\left(\frac{r-r_4}{d_4}\right)\right]\left[1-{\rm erf}\left(\frac{r-r_5}{d_5}\right)\right]\times\sin\theta\cos\theta \tag{6}$$

where, $r_4=0.95R_s$, $r_5=R_s$, $d_4=0.05R_s$, $d_5=0.01R_s$, and for Model III, the α profile used is given by,

$$\alpha_{\rm BL}(r,\theta)=\frac{1}{2}\left[1+{\rm erf}\left(\frac{r-r_{\rm surf}}{d}\right)\right]\sin^2\theta\cos\theta, \tag{7}$$

where $d=0.01R_s$.

Thereafter, to study large-scale magnetic fields, we included fluctuations in the source of the poloidal field. The main reason for including the randomness was that since the star cycle's amplitude is not equal, it varies from time to time. This happens due to the fluctuating nature of the stellar convection. The dynamo parameters fluctuate around their mean. In Babcock–Leighton, the fluctuations are due to the scatter in the bipolar active region tilts around the Joy's law. This randomness changes the poloidal field and makes irregular magnetic cycles as observed in Sun and Sun-like stars. In order to include randomness in our Babcock–Leighton α, we include fluctuations in the α appearing in Eq. (5) as, $\alpha_{0,s}=\alpha_{0,s}r$, where r is the Gaussian random number with mean unity and standard deviation (σ) as 2.67. We keep the value of σ the same for all the stars. In our models, the value of α_0 is updated randomly after a certain time, which we take to be one month.

The α_0 in all three models have the same form (Eq. (5)), except in Model III, $\alpha_{0,s}=$ 4 cm s^{-1} and fluctuations in this model are included separately in the two hemispheres. We note that above α in Eq. (7) has a $\sin^2\theta\cos\theta$ dependence instead of $\sin\theta\cos\theta$ as used in Models I-II to make the α effect strong (weaker) in low (high) latitudes. Also, the radial extent of this α is a bit wider than that used in Models I-II.

Finally, In Model III, we have added radial magnetic pumping. This inclusion of pumping is inspired by Hazra et al. (2019), who found some agreement of the cycle period vs rotation trend with observations. It was realized that a downward magnetic pumping helps to make the magnetic field radial near the surface and reduce the toroidal flux loss through the surface, making the dynamo model in accordance with the surface flux transport models and observations (Cameron et al. 2012; Karak and Cameron 2016). The pumping has the following form:

$$\gamma=-\gamma_0\left[1+{\rm erf}\left(\frac{r-0.9R_s}{0.02R_s}\right)\right], \tag{8}$$

where the amplitude of the radial magnetic pumping is given by γ_0 which is 24 m s^{-1} in all the stars.

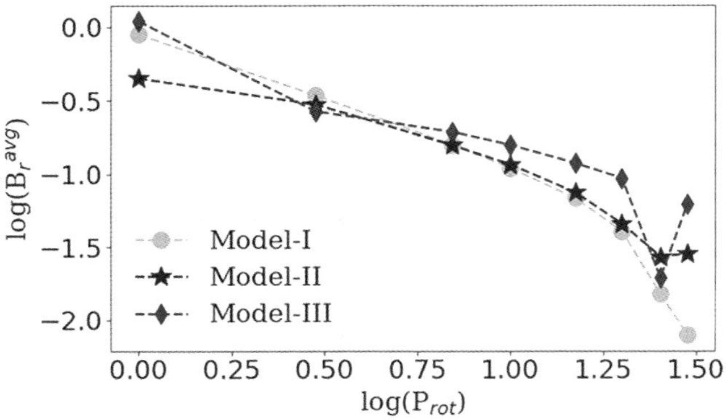

Figure 1. Variation of the magnetic activity with the rotation period of stars for all three models. Figure adopted from Vashishth et al. (2023).

3. Results & Discussion

The simulations were done for $M_{\odot}$ mass stars having rotation periods of 1, 3, 7, 10, 15, 20, 25.30 (Sun), and 30 days, respectively. Here we discuss the various aspects of magnetic cycles obtained from all the stars and from all the three models.

Firstly, the regular polarity reversals in the toroidal field were noted. Model II presents an irregular magnetic cycle with significant hemispheric asymmetry. In contrast, Model III shows regular polarity reversals for fast rotators and the Sun but irregularity for slower ones. Depending on the rotation period, stars exhibit varying magnetic field configurations: slow rotators mainly have dipolar fields, while those rotating in 7 days or less can have quadrupolar configurations. The pictorial representation and detailed analysis are in Vashishth et al. (2023).

One obvious feature in these simulations, as seen in Fig. 1, is that the magnetic field becomes strong in fast-rotating stars. This is because the strength of α increases with the rotation rate of the star (the shear, however, remains more or less unchanged in different stars). If a star rotates faster, the tilt of the BMR associated with the Babcock–Leighton α is expected to increase. Therefore, with age, as the rotation rate decreases, the dynamo process becomes weaker, and the dynamo number also decreases. This implies that the rapidly rotating stars will likely have stronger magnetic cycles. This result agrees with the Karak et al. (2014) and the observations (Noyes et al. 1984; Wright et al. 2011).

We also computed the cycle periods for all three models. To achieve this, we analyzed the Fourier power spectrum peaks of the toroidal field time series within the tachocline. This analysis was conducted separately for the northern and southern hemispheres, distinguishing between symmetric and anti-symmetric cycles. The variations of the cycle duration in each case with the rotation rate are shown in Fig. 2. Notably, Models I and II exhibit an increasing trend in the cycle period with higher stellar rotation rates. This behavior is due to the weakening of the meridional flow with the decrease in the rotation period which leads to intensifying the flow speed in the thin layers near the boundaries.

Although these two models reproduced various stellar observations, they failed to reproduce the magnetic cycle period vs. rotation trend correctly for the slowly-rotating stars. One way to resolve this discrepancy was to include radial magnetic pumping in the stellar CZs as done by Hazra et al. (2019). As a result, in Model III, we got the cycle-rotation period trend closer to the observations.

When strong downward magnetic pumping is included in this model, the diffusion of the magnetic field across the surface becomes negligible, and then the dynamo allows it to operate at a low α (Karak and Cameron 2016). The lower the α, the longer the

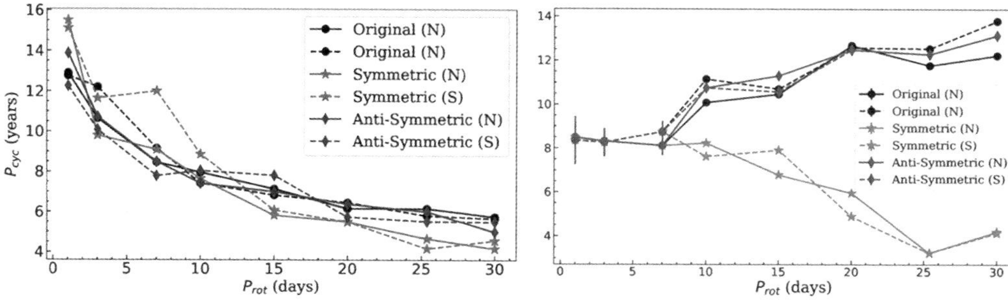

Figure 2. Change in the stellar cycle duration with the rotation period of stars for (a) Models I-II and (b) Model III. From Vashishth et al. (2023).

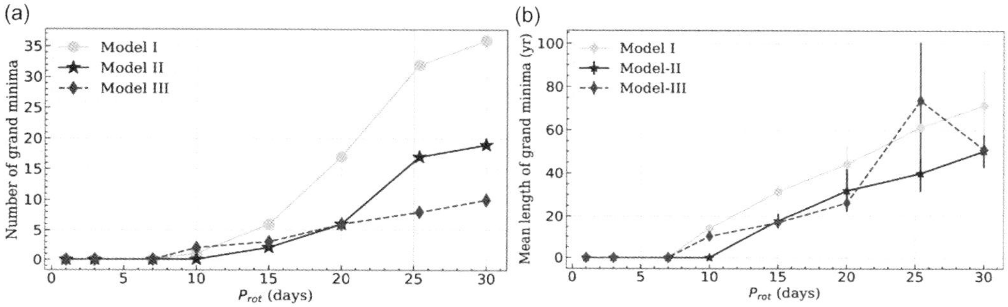

Figure 3. Change of (a) the number and (b) the average duration of grand minima with the rotation period of stars. Yellow circles, blue asterisks, and red diamonds depict Models I, II, and III trends, respectively. In (b), the error bars are computed from the standard deviation of the durations of the grand minima in each case. From Vashishth et al. (2023).

cycle period. We can see from Fig. 2 that at 30 days rotation period, while Models I-II were producing a cycle period of 6 years, Model III produced a much longer period of 13 years. Then with the decrease of the rotation period, the α becomes stronger and thus the poloidal field generation process becomes more efficient. This makes the reversal of the field faster. This effect in the pumping-dominated regime overpowers the increase of the cycle period due to a decrease in meridional flow speed.

The focal point of our study was to understand how the long-term variability of stellar cycles is influenced by their rotation rates. We found that fast rotators manifest irregular cycles with less pronounced long-term variability, while slower rotators exhibit longer modulations in their cycles, interspersed with periods of weaker magnetic fields. We then identified extended periods of low magnetic activity, known as grand minima, using a method adapted from solar studies by Usoskin et al. (2007). We examined the number of grand minima observed in each case with the help of a time-series plot of the toroidal field at the base of CZ and radial magnetic field from simulations for 11,000 years. We infer that the number of grand minima observed in all the models shows an increasing trend with the rotation period. We saw that the rapidly rotating stars hardly produce any grand minima, whereas the slowly rotating stars produce some grand minima, and also, as the rotation period increases, the number of grand minima is seen to increase (see Fig. 3). This is because, with the increase of rotation period, the supercriticality of the dynamo decreases, and the dynamo is more prone to produce extended grand minima in this regime. This result is as per Vashishth et al. (2021) where we observed that as the supercriticality increases (i.e., as the dynamo number increases), the frequency of occurrence of grand minima decreases. This is also supported by the observation that the detected grand/Maunder minima candidates are the slow rotators. Notably, the average duration of these grand minima also increases with rotation duration, with most lasting under 150 years and the majority even below 70 years.

4. Conclusion

Based on the kinematic dynamo simulations of one solar mass at different rotation rates with stochastically forced Babcock–Leighton source, we make the following conclusions. In slowly rotating stars, the cycles are smooth and show long-term variation with occasional grand minima. Whereas the magnetic field is strong for rapidly rotating stars, cycles are more irregular, and no grand minima are detected. The number of grand minima increases with the decrease in the star's rotation rate. Details of this work have been presented in Vashishth et al. (2023). We further explore the possible existence of the dynamo in the subcritical regime in a Babcock–Leighton-type framework and in the presence of a small-scale dynamo, whose details have been presented in Vashishth et al. (2021) and Vashishth et al. 2023 (under preparation).

References

Baliunas, S. L. et al. 1995 *ApJ*, 438, 269
Boro Saikia, S. et al. 2018 *A&A*, 616, A108
Baum, Anna C., Wright, Jason T., Luhn, Jacob K. and Isaacson, Howard, 2022 *ApJ*, 163, 183
Biswas, A., Karak, B. B., & Kumar, P. 2023,a. *MNRAS*, 526a(3), 3994–4003.
Cameron, R. H., Schmitt, D., Jiang, J., Işık, E. 2012. Astronomy and Astrophysics 542.
Charbonneau, P. 2020, Dynamo models of the solar cycle. *Living Reviews in Solar Physics*, 17(1), 4.
Dasi–Espuig, M., Solanki, S. K., Krivova, N. A., Cameron, R. and Peñuela, T. 2010 *A&A*, 518, A7
Donati, J. -F., Semel, M. and Rees, D. E. 1992 *A&A*, 26, 353
Garg, Suyog, Karak, Bidya Binay, Egeland, Ricky and Soon, Willie and Baliunas, Sallie 2019 *ApJ*, 886, 2
Hazra, G., Jiang, J., Karak, B. B., Kitchatinov, L. 2019. *ApJ*, 884.
Jeffers, S. V. et al. 2022 *A&A*, 661, A152
Jha, Bibhuti Kumar, Karak, Bidya Binay, Mandal, Sudip and Banerjee, Dipankar 2020 *ApJ*, 889, 1
Karak, B. B. 2023, *Living Reviews in Solar Physics*, 20(1), 3.
Karak, B. B. 2020, *Astrophys. J Lett.*, 901(2), L35.
Karak, B. B. & Cameron, R. 2016, *Astrophys. J*, 832, 94.
Karak, B. B., Jiang, J., Miesch, M. S., Charbonneau, P., & Choudhuri, A. R. 2014,a *Space Sci. Rev.*, 186a, 561–602.
Karak, B. B., Kitchatinov, L. L. and Choudhuri, A. R. 2014, *ApJ*, 791, 59
Karak, B. B., Käpylä, P. J., Käpylä, M. J., Brandenburg, A., Olspert, N., Pelt, J. 2015. Astronomy and Astrophysics 576.
Karak, B. B., Mandal, S., & Banerjee, D. 2018, *Astrophys. J*, 866(1), 17.
Karak, B. B. and Miesch, M. 2017, *ApJ*, 847, 69
Karak, B. B. and Miesch, M. 2018, *ApJ*, 860, L26
Karak, B. B., Miesch, M., Bekki, Y. 2018. Physics of Fluids 30.
Karak, B. B., Tomar, A., Vashishth, V. 2020. *MNRAS*491, 31553164.
Kitchatinov, L. L. and Olemskoy, S. V. 2011, *MNRAS*, 411, 1059
Kumar, Pawan, Karak, Bidya Binay and Vashishth, Vindya 2021 *ApJ*, 913, 1
Kumar, Pawan, Biswas, Akash and Karak, Bidya Binay 2022 *MNRAS*, 513, 1
Lemerle, A. and Charbonneau, P. 2017 *ApJ*, 834, 133
McClintock, B. H., Norton, A. A. and Li, J. 2014 *ApJ*, 797, 130

Mordvinov, A. V., Karak, B. B., Banerjee, D., Golubeva, E. M., Khlystova, A. I., Zhukova, A. V., & Kumar, P. 2022, *MNRAS*, 510(1), 1331–1339.

Mordvinov AV, Karak BB, Banerjee D, Chatterjee S, Golubeva EM, Khlystova AI (2020) *ApJ*, 902(1):L15

Muñoz-Jaramillo, A., Dasi–Espuig, M., Balmaceda, L. A. and DeLuca, E. E. 2013 *ApJ*, 767, L25

Noraz, Q., Brun, A. S., Strugarek, A., Depambour, G. 2022. Astronomy and Astrophysics 658.

Noyes, R. W., Hartmann, L. W., Baliunas, S. L., Duncan, D. K. and Vaughan, A. H. 1984 *ApJ*, 279, 763

Oláh, K., Kővári, Zs., Petrovay, K., Soon, W., Baliunas, S.,Kolláth, Z. and Vida, K. 2016 *A&A*, 590, A133

Priyal, M. et al. 2014 *ApJ*, 793, L4

Rengarajan, T. N. 1984 *ApJ*, 283, L63

Shah, Shivani P. et al. 2018 *ApJ*, 863, L26

Skumanich, A. 1972 *ApJ*, 171, 565

Sreedevi, A., Jha, B. K., Karak, B. B., & Banerjee, D. 2023, *ApJS*, 268(2), 58.

Metcalfe, T.S., Egeland, R., and van Saders, J.: 2016, *The Astrophysical Journal* **826**, L2.

Usoskin, I. G., Solanki, S. K., Kovaltsov, G. A. 2007. Astronomy and Astrophysics 471, 301309.

Vashishth, Vindya, Karak, Bidya Binay and Kitchatinov, Leonid 2021 *RAA*, 21, 10

Vashishth, V., Karak, B. B., Kitchatinov, L. 2023. *MNRAS* 522, 26012610.

Wright, N. J., Drake, J. J., Mamajek, E. E. and Henry, G. W. 2011, *ApJ*, 743, 48

Wright, N. J., Drake, J. J., Mamajek, E. E. and Henry, G. W. 2016, *Nature*, 535, 526

Dynamics of Solar and Stellar Convection Zones and Atmospheres
Proceedings IAU Symposium No. 365, 2024
A. V. Getling & L. L. Kitchatinov, eds.
doi:10.1017/S1743921323005343

On origin of active/inactive branches on moderate rotating solar analogs

Valery V. Pipin

Institutte solar-terrestrial physics, Irkutsk, 6604033, Russia
email: pip@iszf.irk.ru

Abstract. The fast rotating solar analogs show a decrease of the dynamo period with an increase of the rotation rate for the moderate stellar rotation periods in the range between 10 and 25 days. Simultaneously, observations indicate two branches: the "in-active" branch stars shows short dynamo cycles and the active branch stars show the relatively long magnetic cycles. We suggest that this phenomenon can be produced by effect of the doubling frequency of the dynamo waves, which is due to excitation of the second harmonic. It is generated because of the nonlinear B^2 effects in the large-scale dynamo.

Keywords. stellar dynamo,magnetic activity cycles

1. Introduction

For our understanding of origin of solar activity, it is extremely important that magnetic cycles can be seen on other solar-type stars (Soon et al. 1994). Analysis of Lehtinen et al. (2016); Brandenburg et al. (2017) showed the multiple activity branches on diagram chromoshpheric activity parameter and ratio of rotation and dynamo cycle period. Some results of their analysis are illustrated in Figure 1. It is noteworthy that these surveys deal with the solar type stars with the effective temperature in between 5000 and 6000 K. This set includes G-dwarfs and some amount of the early K and late F-type stars. The effect of rotation on the turbulent convection inside the stars is quantified by the Coriolis number $Co = 2\Omega\tau_c$, where Ω is the global rotation rate and τ_c is the typical convective tunover time. On the given Figure we can identify saturation branch (green line) for very fast rotators ($Co \gg 1$) and two branches for moderate rotators, $Co \leq 20$. They represent the so-called active (high chromospheric activity, blue color) and inactive stars (low chromospheric activity, red color). The last one shows very short cycle periods.

The active branch stars on diagram Fig. 1 can be is reproduced using the Parker–Yoshimura dynamo waves (Yoshimura 1975; Parker 1979). In this case the dynamo period is determined by the wave type of the dynamo solution with the wave frequency (Stix 1976):

$$\omega_{cyc} = \left| \frac{\alpha_{\phi\phi} k_\theta}{2} r \cos\theta \frac{\partial\Omega}{\partial r} \right|^{1/2}, \tag{1}$$

where k_θ is the latitudinal dynamo wave number, $\alpha_{\phi\phi}$ is the hydrodynamic α effect, and Ω is the angular velocity profile. It is noteworthy that turbulent diffusion controls the dynamo wave length. For condition of maximum of the dynamo wave length we can get (Brandenburg et al. 2017),

$$\omega_{cyc} \propto (\alpha_{\phi\phi} r \frac{\partial\Omega}{\partial r})^{2/3}. \tag{2}$$

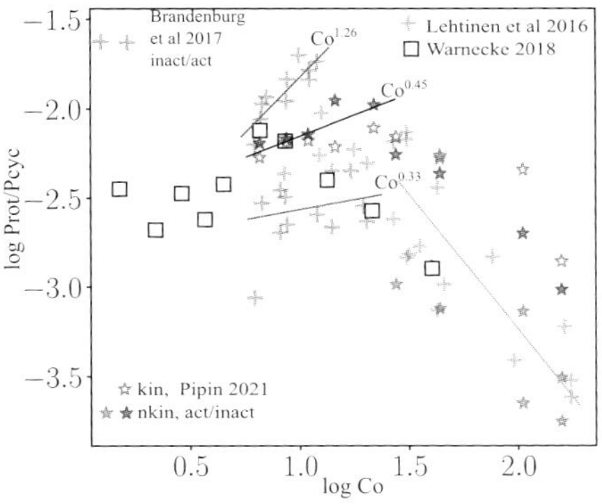

Figure 1. Relation of the dynamo period with stellar rotation period. The red (active branch) and black (quiet branch) crosses show the results of Brandenburg et al. (2017) for F- and G-type stars; the green crosses show the results of Lehtinen et al. (2016) for young solar-type stars; the black hollow squares show the results of Warnecke (2018); the white stars show the kinematic dynamo models of Pipin (2021); the black and red stars show the non-kinematic models from that paper, where the red stars mark the long dynamo period. Updated from Pipin (2021).

Therefore, if $\alpha_{\phi\phi} \propto \Omega$ then $\omega_{cyc} \propto \Omega^{4/3}$ and $P_{rot}/P_{cyc} \propto \Omega^{1/3}$. Saturation branch on Fig. 1 corresponds to marginal modes which are determined by the typical diffusive time. Stars symbols show results for our axisymmetric mean-field dynamo model (Pipin 2021). The results of the global convective simulation (GCD) of Warnecke (2018) reproduce roughly the dynamo cycle variations on the saturated branch. The GCD of Warnecke (2018) show sign of the active branch as well. None of the dynamo simulations reproduce the inactive branch of stars for the moderate rotators, $Co \leq 20$. It is noteworthy that the power law $Co^{0.45}$ for the non-kinematic dynamo model runs of (Pipin 2021) is only partly due to the Parker's dynamo wave law $Co^{0.33}$. An additional power increment is due to an increase of the magnetic flux loss because of the magnetic buoyancy. This effect was anticipated from the standard mean-field theory as well, see, Noyes et al. (1984).

2. Basic ideas and dynamo model

We consider the nonlinear dynamo model with the non-local turbulent electromotive force developed recently by Pipin (2023). The mean magnetic field follows the induction equation,

$$\partial_t \bar{\mathbf{B}} = \nabla \times \left(\bar{\mathcal{E}} + \bar{\mathbf{U}} \times \bar{\mathbf{B}}\right), \tag{3}$$

where the mean electromotive force, $\bar{\mathcal{E}} = \overline{\mathbf{u} \times \mathbf{b}}$ expresses the effects of the turbulence on the mean magnetic field evolution. Here, we assume the large-scale flow, $\bar{\mathbf{U}}$, which includes effects of the differential rotation and meridional circulation, as given (see, Pipin 2023), and we neglect the magnetic feedback on $\bar{\mathbf{U}}$. To derive the turbulent electromotive force either analytically or numerically, the scale separation approximation is applied. Such an approximation is hardly satisfied from observations of the solar/stellar dynamos. Following suggestion of Rheinhardt and Brandenburg (2012) we approximate the integro-differential equation for the mean electromotive force by the reaction–diffusion type equation,

$$\left(1 + \tau\frac{\partial}{\partial t} + a_{\mathcal{E}}\eta_T\nabla^2\right)\overline{\boldsymbol{\mathcal{E}}} = \overline{\boldsymbol{\mathcal{E}}}^{(0)}, \tag{4}$$

where, $a_{\mathcal{E}} \approx 0 - 1$ is the spatial non-locality parameter, the RHS of the Eq (4) corresponds to the local expression of the mean electromotive force obtained either numerically, e.g.,

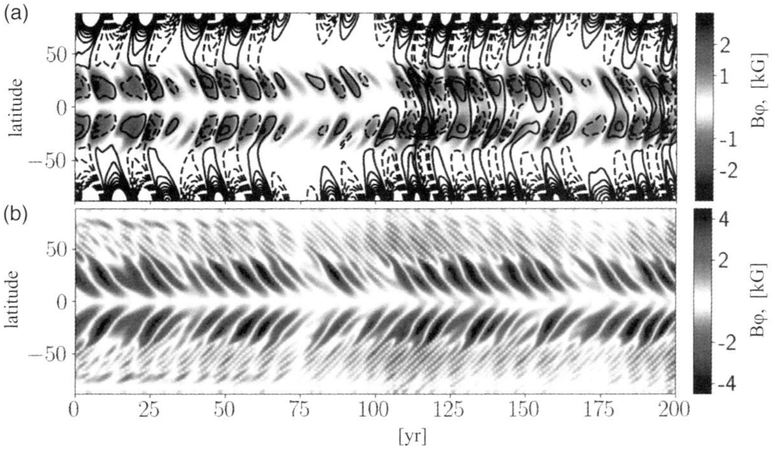

Figure 2. Time-latitude diagrams for the solar analog rotating with the period of 17 days, the radial magnetic field at the surface is shown by contours in range of $\pm 10G$, and color image shows the toroidal magnetic field at r= 0.9R; b) show the toroidal magnetic field at the bottom of the convection zone.

by the test-field method or analytically. It can be written as follows,

$$\overline{\mathcal{E}}_i^{(0)} = (\alpha_{ij} + \gamma_{ij})\,\overline{B}_j - \eta_{ijk}\nabla_j\overline{B}_k, \tag{5}$$

here, α_{ij} describes the turbulent generation of the magnetic field by helical motions (the α-effect), γ_{ij} describes the turbulent pumping, and η_{ijk} is the eddy magnetic diffusivity tensor. The details of the tensors profile in the solar convection zone can be found in the above cited paper (Pipin 2023). The α-effect tensor includes effects of the magnetic helicity. The other nonlinear effects are due the mean-field magnetic buoyancy and the "algebraic" α-quenching. The given generalization of the dynamo evolution equation results to a number of interesting consequences, such as the decrease of the dynamo instability threshold, quenching the turbulent dynamo effects in depth of the convection zone by means of the nonlocal $\overline{\mathcal{E}}$, excitation of the different dynamo modes with different localization inside of the convection zone etc (see, Rheinhardt and Brandenburg 2012; Pipin 2023). The last two effects present a particular interest if we assume that the active/inactive magnetic cycles correspond to the distinct dynamo modes, which can co-exist in the nonlinear supercritical dynamo regime. Following the results of Pipin (2021) we expect the "inactive" magnetic cycle can results from the nonlinear generation of the sub-harmonic B^2 dynamo modes. It is noteworthy that in the solar observations sub-harmonics of the dynamo cycle is weak. The similar results are demonstrated by the dynamo model. We search the dynamo solutions on the solar analogs with the rotation period shorter that for the Sun.

In our previous runs, which were discussed in Pipin (2021), we found the transition of the main dynamo mode to the mode with the double harmonic frequency happens for the rotation period of about 15 days. That transition ends with the stationary dynamo evolution pattern in the asymptotic state. To avoid the steady asymptotic state we add the small, 20 percents magnitude, random fluctuations of the α effect with the typical renovation time about 5 years. Below we consider some results for the solar analog which rotate with the period of 17 days. Figure 2 shows the time latitude diagrams for the large-scale magnetic field evolution in the upper part of convection zone Fig. 2(a), and at the bottom, Fig. 2(b). The two dynamo periods are clearly manifest themselves in the run. Similar to our eigen problem analysis, (Pipin 2021), the nonlocal $\overline{\mathcal{E}}$ results to excitation

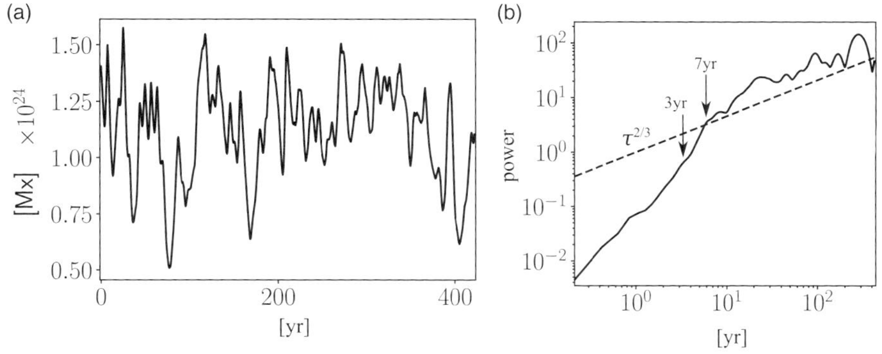

Figure 3. a) Evolution of the total flux of the unsigned toroidal magnetic field in the bulk of the convection zone; b) The integral wavelet spectrum (Morlet) for the total toroidal magnetic field flux, power in relative units.

of two dynamo modes. One mode operate at low latitudes in the bulk of the convection zone. It has the dynamo period about 7.5 years. A weaker dynamo mode is excited at high latitudes near the bottom of the convection zone. Its period is as twice as small in compare with the main dynamo mode. It is noteworthy that the magnetic buoyancy as well as the turbulent generation by the α effect are suppressed near the bottom of the convection zone due to the nonlocality effects of the mean electromotive force, which are caused by the turbulent diffusion of $\overline{\mathcal{E}}$, see, the Eq (4). The main dynamo mode is from time to time, because of the magnetic flux loss in the upper part of convection zone. These periods are characterized by the low activity which is defined by the polar dynamo modes which has a short dynamo period of about 3. years, and $\log P_{rot}/P_{cyc} \sim -1.8$, which corresponds to the inactive branch, the main dynamo mode of this star $\log P_{rot}/P_{cyc} \sim -2.2$ lies on the branch, which has power law $Co^{0.45}$ (see, Fig. 1).

The Figure 3(a) shows evolution of the total flux of the unsigned toroidal magnetic field in the bulk of the convection zone. The time series is characterized by deep minims. As we have seen from Fig. 2(a), these minims are attributed to the dynamo mode which is concentrated to the bottom of the convection zone and it has the short dynamo period. On the longtime scales the spectrum of the dynamo activity is close to $\tau^{2/3}$, here, τ is the time scale of the magnetic field evolution.

3. Conclusions

Our results suggest that mean-field stellar dynamo model can support coexistence of two distinct cycle periods for interval of rotation period less than 20 days. It can explain upper part of inactive branch on the diagram of Fig. 1. Further investigation are needed to understand the position of the Sun on that diagram and the nature of active/inactive stellar activity branches for the moderately rotating solar analogs.

Acknowledgments The author thanks the financial support of the Ministry of Science and Higher Education of the Russian Federation (Subsidy No.075-GZ/C3569/278).

References

Brandenburg, A., Mathur, S., & Metcalfe, T. S. 2017, Evolution of co-existing long and short period stellar activity cycles. *ApJ*, 845(1), 79.

Lehtinen, J., Jetsu, L., Hackman, T., Kajatkari, P., & Henry, G. W. 2016, Activity trends in young solar-type stars. *A&A*, 588, A38.

Noyes, R. W., Weiss, N. O., & Vaughan, A. H. 1984, The relation between stellar rotation rate and activity cycle periods. *ApJ*, 287, 769–773.

Parker, E. N. 1979,. *Cosmical magnetic fields: Their origin and their activity*. Oxford: Clarendon Press.

Pipin, V. V. 2021, Solar dynamo cycle variations with a rotational period. *MNRAS*, 502(2), 2565–2581.

Pipin, V. V. 2023, Spatio-temporal non-localities in a solar-like mean-field dynamo. *MNRAS*, 522(2), 2919–2927.

Rheinhardt, M. & Brandenburg, A. 2012, Modeling spatio-temporal nonlocality in mean-field dynamos. *Astronomische Nachrichten*, 333, 71–77.

Soon, W. H., Baliunas, S. L., & Zhang, Q. 1994, Variations in surface activity of the sun and solar-type stars. *Solar Phys.*, 154, 385–391.

Stix, M. Dynamo Theory and the Solar Cycle. In Bumba, V. & Kleczek, J., editors, *Basic Mechanisms of Solar Activity* 1976,, volume 71, 367.

Warnecke, J. 2018, Dynamo cycles in global convection simulations of solar-like stars. *A&A*, 616, A72.

Yoshimura, H. 1975, Solar-cycle dynamo wave propagation. *ApJ*, 201, 740–748.

Section 4. Helioseismology and Asteroseismology

Invited talks

Dynamics of Solar and Stellar Convection Zones and Atmospheres
Proceedings IAU Symposium No. 365, 2024
A. V. Getling & L. L. Kitchatinov, eds.
doi:10.1017/S1743921324000176

Recent Progress in Time–Distance Helioseismology: Meridional Circulation, Far-side Imaging, and Sunquakes

Ruizhu Chen and **Junwei Zhao**

W. W. Hansen Experimental Physics Laboratory, Stanford University, Stanford, CA, USA
email: rzchen@stanford.edu

Abstract. Time–distance helioseismology uses solar surface Doppler observations to measure areas that are not directly observable, such as solar interior, far side, and sunquake sources. In this work, we briefly review recent advancements in time–distance helioseismology, focusing on meridional circulation measurements, far-side imaging, and sunquakes. Solar deep meridional flows are crucial for understanding the dynamics of the solar interior, but precise measurements of these flows are challenging. This review explores recent developments in this area, particularly highlighting new findings related to systematic effects that have long challenged meridional circulation determination. We also review recent progress in solar far-side imaging, which is useful in improving space weather forecasting. Recent developments in far-side imaging using time–distance techniques and Deep Learning are introduced. Additionally, we review a new approach in sunquake reconstruction by incorporating observation-based Green's functions constructed by time–distance helioseismology.

Keywords. Sun: helioseismology, Sun: interior, Sun: magnetic fields, Sun: oscillations

1. Introduction

Helioseismology utilizes solar surface Doppler observations to investigate regions of the Sun that are beyond direct observation, such as the solar interior and the far side. Time–distance helioseismology, a local-scale method within this field, studies these regions by tracking the propagation of acoustic waves and analyzing their travel times over specific distances. The subtle changes in wave travel times reflect the physical conditions along their paths, allowing for the inferences of various physical quantities. This technique enables researchers to measure solar subsurface structures, examine variations in wave-speed perturbations and flows in sunspots, and physical condition down to the base of the convection zone. In this paper, we will review recent developments using time–distance helioseismology, focusing on three topics: solar meridional circulation, far-side imaging, and sunquakes.

The Sun's meridional circulation is a global-scale interior flow of solar material from the solar equator toward the poles and back in the Sun's meridional plane. It is a crucial component for understanding the solar dynamo and interior dynamics, facilitating the transport of magnetic flux and the redistribution of angular momentum (e.g., Wang *et al.* 1989; Miesch 2005; Upton & Hathaway 2014a,b; Featherstone & Miesch 2015). Additionally, it has been reported to be associated with the length and strength of solar cycles (Hathaway & Rightmire 2010; Dikpati *et al.* 2010). Surface meridional flows have been extensively measured in various studies, consistently reporting a poleward flow on the order of $10-20$ m s^{-1}. This has been achieved through methods such as surface

Doppler measurements (e.g. Duvall 1979; Hathaway 1996), surface feature tracking (e.g. Howard & Gilman 1986; Komm *et al.* 1993; Švanda *et al.* 2007), time–distance helioseismology (e.g. Giles *et al.* 1997; Zhao & Kosovichev 2004), ring-diagram analysis (e.g. González Hernández *et al.* 1999; Haber *et al.* 2000), and Fourier–Legendre analysis (e.g. Braun & Fan 1998; Roth *et al.* 2016). However, determining deep meridional flow proves to be notoriously difficult, resulting in inconsistencies across previous studies. Questions regarding the number of circulation cells and the depth of the return flow still remain open. Most measurements of deep meridional circulation employ the time–distance helioseismology method. In Section 2, we'll provide a brief review of these studies, with an emphasis on recent developments about the associated systematic effects that have posed challenges in the meridional-flow determination.

The Solar far side is crucial for effective space weather forecasting. Given the Sun's rotation period of about 27 days, phenomena developing on the far side take time to rotate into view on the Earth side. Accurate solar wind modeling requires boundary conditions of full solar surface spanning 360°, and missing half of the surface diminishes predictive capabilities. For a few years, direct observations of the far side in extreme ultraviolet (EUV) were available through the Solar Terrestrial Relations Observatory (*STEREO*; Howard *et al.* 2008), with two spacecraft orbiting the Sun ahead and behind Earth. With *STEREO* currently positioned back on the Earth side, the solar far side is beyond direct observation. The recent launch of the *Solar Orbiter* (*SO*; Müller *et al.* 2013) has reintroduced direct far-side observation, albeit sporadically on a shorter time scale. On the other hand, through a technique known as helioseismic far-side imaging, helioseismologists can utilize near-side Doppler observations to map the far side's surface and identify large active regions (ARs). This method offers valuable insights into the far side in near-real-time when direct observation is unavailable. Section 3 will provide a brief review of developments in helioseismic far-side imaging, with a particular focus on recent improvements with time–distance methods and their integration with deep learning.

Sunquakes manifest as enhancements in photospheric oscillatory power triggered by solar flares, with photospheric helioseismic waves becoming visible approximately 10-30 minutes after the flare onset in the surrounding area. Proposed mechanisms to initiate sunquakes are diverse, with suggestions that they may be excited by sudden photospheric pressure perturbations caused by shock waves (Kosovichev & Zharkova 1998; Kosovichev 2007; Zharkova & Zharkov 2007), photospheric heating (Donea *et al.* 2006; Lindsey & Donea 2008), or Alfvén-wave heating (Russell & Fletcher 2013; Reep & Russell 2016). Other studies also propose that the Lorentz force drives sunquakes (Hudson *et al.* 2008; Fisher *et al.* 2012). However, these mechanisms fall short in explaining why not all flares result in sunquakes. Despite the tendency for sunquakes to occur during strong flares, only a fraction of such flares induce sunquakes, and intriguingly, even weak flares can occasionally cause sunquakes (Buitrago–Casas *et al.* 2015; Sharykin & Kosovichev 2020). In Section 4, we introduce a recent study that aims to address this question. This work utilizes time–distance methods to derive observation-based Green's functions, which are then incorporated into sunquake reconstructions.

2. Meridional Circulation

2.1. *Inconsistent Measurements of Meridional Circulation*

An early measurement of solar deep meridional flow was conducted by Giles (2000) using time–distance method, employing data from the *Solar and Heliospheric Observatory* / Michelson Doppler Imager (*SOHO*/MDI; (Scherrer *et al.* 1995)). Poleward flows were identified spanning nearly the entire convection zone, and an equator-ward flow was inferred near the base of the convection zone after applying a mass-conservation

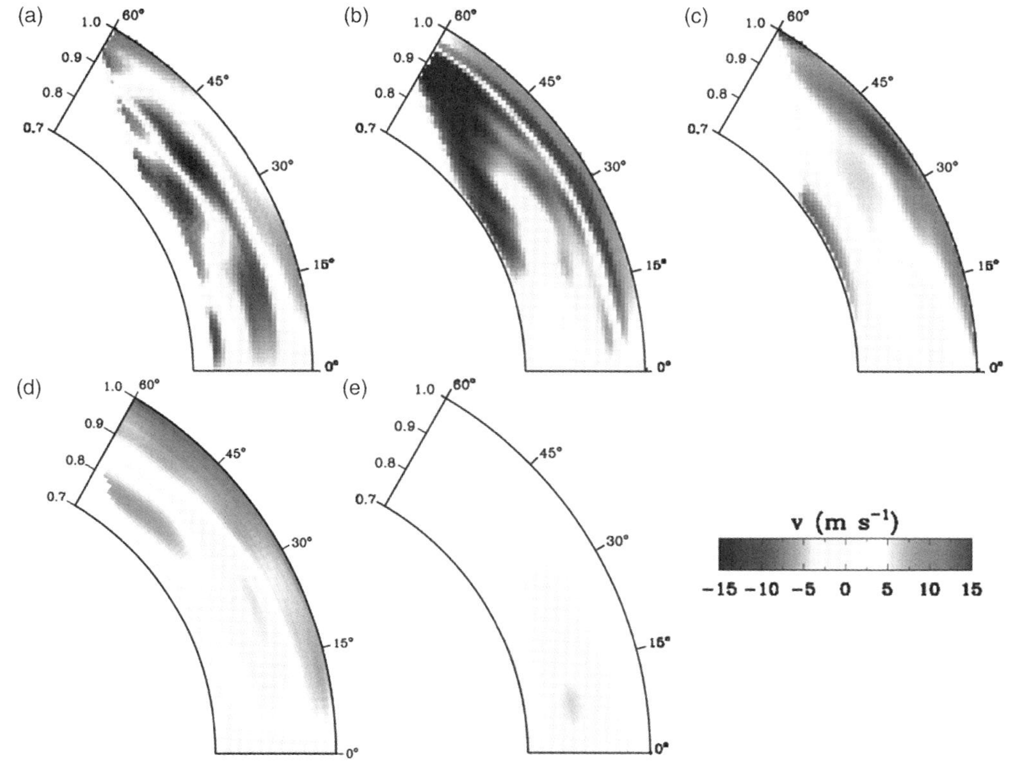

Figure 1. Comparison of hemispherically symmetrized meridional circulation for selected results by time–distance method, from (a) Zhao *et al.* (2013), (b) Kholikov *et al.* (2014); Jackiewicz *et al.* (2015), (c) Rajaguru & Antia (2015), (d) Chen & Zhao (2017), and (e) Lin & Chou (2018). Adapted from Figure 1.8 of Chen (2019).

constraint, forming an single-cell circulation. This result was achieved long before the recognition of a systematic effect, identified later as the center-to-limb (CtoL) effect, which significantly affects deep flow inferences by introducing an additional systematic signal (Zhao *et al.* 2012). It is found that the CtoL effect increases when going deeper and from the disk center to the limb, reaching magnitudes as large as 5–10 times of the flow-caused signal. By empirically removing this effect, Zhao *et al.* (2013) successfully detected the equator-ward flow using first two years of data from *Solar Dynamics Observatory* / Helioseismic and Magnetic Imager (*SDO*/HMI; Scherrer *et al.* 2012; Schou *et al.* 2012), revealing a shallow return flow between 0.82 $R_\odot$ and 0.91 $R_\odot$, below which the meridional flow exhibites poleward again (Figure 1(a)). The resulting flow profile indicates a double-cell circulation pattern.

Subsequent studies affirmed the presence of the center-to-limb (CtoL) effect across various instruments. Employing the same effect-removal strategy and similar measurement techniques, these investigations also identified the return flow but at varying depths, resulting in different meridional flow profiles. Utilizing GONG data, Kholikov *et al.* (2014) and Jackiewicz *et al.* (2015) observed a shallow return flow at a depth similar to Zhao *et al.* (2013), yet reported no additional poleward flow in deeper layers (Figure 1(b)). Analyzing 4 years of HMI observations, Rajaguru & Antia (2015) found the equatorward flow only beneath 0.77 $R_\odot$, suggesting a single-cell circulation (Figure 1(c)). Utilizing MDI data, Liang & Chou (2015b) and Lin & Chou (2018) reported a 3-layer meridional flow profile during two solar minima, with a return flow in the middle of the convection zone,

Table 1. Comparison of procedures in different meridional circulation measurements with time–distance method

Authors	Instruments	Observation Time	Filtering	Magnetic Masking	CtoL Removal Method	Sensitivity Kernels	Inversion method	Circulation
Zhao *et al.* (2012)	HMI	2010 May - 2012 April	running difference	None	EW measurement proxy	Ray path approximation	Rregularized Least Squares (RLS)	double-cell
Kholikov *et al.* (2014) Jackiewicz *et al.* (2015)	GONG	652 selected days in 2004–2012	phase-speed filter	None	EW measurement proxy	Ray path approximation	Subtractive Optimally Localized Average (SOLA)	single-cell
Böning *et al.* (2017)						Born approximation	refined SOLA	single- or multiple-cell
Rajaguru & Antia (2015)	HMI	2010 May - 2014 April	running difference	None	EW measurement proxy	Ray path approximation	stream function RLS with built-in mass conservation	single-cell
Rajaguru & Antia (2020)		2010 May - 2021 April						single-cell (3 mHz) double-cell (4 mHz)
Mandal *et al.* (2018)		2010 May - 2016 April				Born approximation		single-cell
Chen & Zhao (2017)	HMI	2010 May - 2017 April	running difference	Yes	Least-square disentangle	Ray path approximation	RLS	double-cell
Liang & Chou (2015b) Lin & Chou (2018)	MDI	1996 May - 2010 November	High-pass > 1.5mHz	Yes	EW measurement proxy	Ray path approximation	SOLA	double-cell at solar minima
Gizon *et al.* (2020)	MDI & GONG	1996 May - 2011 April & 2001 August - 2019 April		Yes	EW measurement proxy	Born approximation	stream function RLS with mass conservation	single-cell

Adapted and Updated from Table 1 of Chen (2019)

indicating a double-cell circulation (Figure 1(e)). However, during the solar maximum, the circulation pattern became more complex. Gizon *et al.* (2020) conducted extensive measurements using HMI, MDI, and GONG data. They had to exclude HMI results due to an unclear systematic issue causing severe North–South asymmetry in their HMI measurements. Their results, using MDI and GONG data, reported a single-cell meridional circulation throughout Solar Cycles 23 and 24.

In all these investigations, a uniform strategy for removing the center-to-limb (CtoL) effect was employed. This approach uese East–West measurements along the equator as a proxy for the CtoL effect, which is then subtracted from North–South measurements along the central meridian. The assumption made was that the CtoL effect is consistent along both the equator and the central meridian. To examine this assumption, Chen & Zhao (2017) devised a comprehensive method for measurements along all disk-radial directions, subsequently solving the CtoL effect through a least-square method. Their study confirmed that the CtoL effect is isotropic relative to the azimuthal angle (within certain limitations). Moreover, they were able to disentangle the CtoL effect from the meridional-flow-induced travel-time shifts in a more robust manner. The outcomes of their investigation revealed a double-cell circulation profile (Figure 1(d)).

The exact cause of the inconsistencies among these studies remains elusive. Multiple differences exist in the detailed measurement procedures, as summarized in Table 1. Due to space constraints, this review will not go through all of them. However, it is believed that the discrepancies likely stem from variations in the flow inversion process and the CtoL effect removal procedures. Notably, Rajaguru & Antia (2015), Mandal *et al.* (2018), and Gizon *et al.* (2020) incorporated a strict mass conservation in their inversion using regularized least squares (RLS) methods. All three reported a deep return flow, indicating a single-cell circulation pattern. The incorporation of the mass conservation constraint may contribute to the systematic differences in the inverted circulation profile. On the other hand, challenges arise in effectively removing the systematic CtoL effect. In deep regions, the CtoL effect could be 5-10 times larger than the flow-induced signals. Therefore, a small error in the CtoL effect removal can lead to a substantial change in the inverted flow profile. However, thoroughly evaluating whether the CtoL effect has been effectively removed by these empirical methods poses a significant challenge.

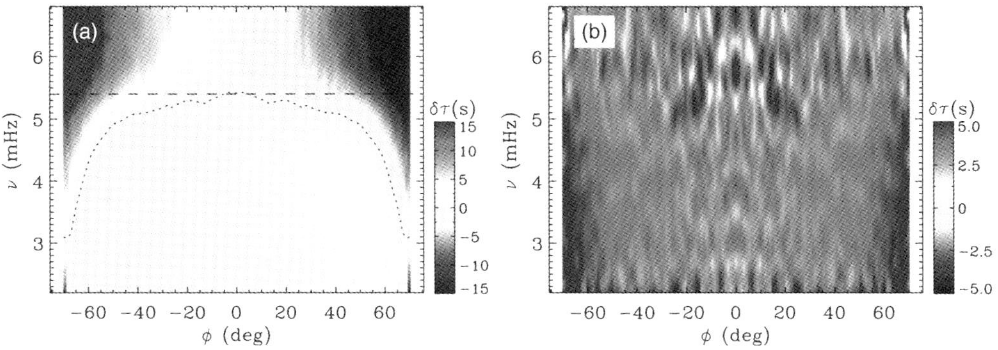

Figure 2. (a)Profile of the CtoL effect for an effective travel distance of 21°.6. The dotted line is a smoothed contour at value of 0, and the dashed line represents the frequency of 5.4 mHz. Adapted from Figure 7(a) of Chen & Zhao (2018). (b)Rotation-induced travel-time shifts, displayed as a function of disk-centric distance and frequency for a travel-distance of 14°.4. Adapted from Figure 11(a) of Chen & Zhao (2018).

A recent discovery by Chen & Zhao (2018) offers a potential means to assess the efficacy of systematic effect removal and enhances our understanding of the physics behind the CtoL effect. Notably, the CtoL effect exhibits a significant frequency dependence, changing signs at around 5.4 mHz. Furthermore, the frequency-dependent behavior varies markedly with disk-centric distance but not much with travel distance (Figure 2(a)). In contrast, travel-time shifts induced by flows show little frequency dependence (Figure 2(b)). This pronounced difference provides an opportunity to enhance the disentanglement of the CtoL effect from meridional-flow measurements. The findings of this study introduce a new dimension of information to the meridional flow problem. Ideally, flows measured and inverted across different frequencies should be invariant, allowing the examination of CtoL effect removal by comparing results at multiple frequencies. Later, Rajaguru & Antia (2020) conducted a frequency-dependent meridional flow measurement by applying frequency filters to their measurements. Their results revealed a single-cell circulation profile at 3 mHz and a double-cell profile at 4 mHz, suggesting unresolved issues in the measurement process.

Similar frequency-dependent travel-time aymmetries were also observed in measuring travel times between different atmospheric heights using HMI line-core and continuum intensities (Zhao & Chen 2020). In fact, the Center-to-Limb (CtoL) effect occurs between different disk-centric locations, suggesting that the cause of the CtoL effect likely originates from the different line-formation heights..

To investigate this scenario, Zhao *et al.* (2022) employed a set of high-spectral-resolution, high-cadence observational data from the *Interferometric BI-dimensional Spectrometer (IBIS)* installed at the *Dunn Solar Telescope*. This instrument observes the same spectral line as HMI (Fe I λ6173.3 Å). Dopplergrams at different atmospheric heights were constructed using the bisector method. The evanescent waves above the photosphere were found carrying nonnegligible phase shifts that cannot be explained by flows. Moreover, these phase shifts also exhibit significant frequency dependence. Figure 3 compares the frequency dependence of the CtoL effect in Panel (a) with that of the non-zero phase shifts between multiple heights in Panel (b). Although not quantitatively close, they exhibit qualitatively similar dependencies on frequency. Both change from positive to negative at a certain frequency, and their magnitudes increase with the atmospheric height difference. The cause of the phase shifts between different heights is not immediately clear, but it is likely related to the non-adiabatic nature of

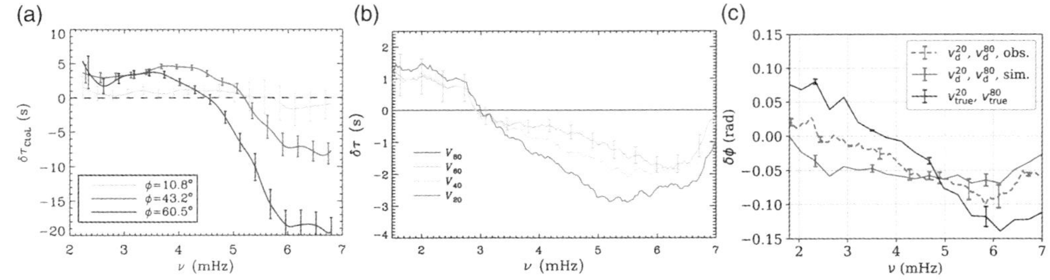

Figure 3. (a)The CtoL effect as functions of frequency at selected disk-centric locations. Adapted from Figure 9(a) of Chen & Zhao (2018). (b) Time shifts measured between IBIS Depplergrams derived at 20%, 40%, 60%, and 80% intensity level (higher to lower atmospheric heights) relative to that of 80% intensity level (lowest height derived), displayed as functions of frequency. Adapted from Figure 5(b) of Zhao *et al.* (2022). (c) Phase differences at the disk center as functions of frequency for Doppler velocities derived from the bisector method for IBIS data (dashed curves), derived from simulation data (solid), and true velocities (black solid curves). Adapted from Figure 8(a) of Waidele *et al.* (2023).

the atmosphere and complicated radiative transfer above the photosphere. This finding also alerts helioseismologists about systematic effects in helioseismic measurements when using oscillatory signals observed at different atmospheric heights, not limited to meridional flow measurements by time–distance helioseismology.

To further investigate the physics behind these unexpected phase shifts, Waidele *et al.* (2023) conducted similar measurements of phase shifts on a simulated, fully radiative, and convective atmosphere from which the Fe I $\lambda 6173.3\,\mathring{A}$ line is synthesized. Doppler velocities at different atmospheric heights were calculated using the bisector of the spectral line. Subsequently, phase shifts were calculated between these Doppler velocities and also between the true velocities at corresponding heights in the simulation. Nonnegligible phase shifts that vary with frequency were measured in both cases. However, their magnitudes are not close, and both deviate from the observations (Figure 3(c)). All these frequency-dependent phase shifts pose challenges to the interpretation of helioseismic measurements but also provide more observational constraints on understanding the physics behind various systematics.

In summary, current studies on solar meridional circulation involve ongoing efforts to refine measurements, address systematic effects, and explore new approaches for understanding the complex systematics. Challenges persist, but these studies contribute valuable insights and constraints, pushing forward the boundaries of our knowledge about meridional flow.

3. Far-Side Imaging

3.1. *Helioseismic Far-Side Imaging*

The first successful far-side active region detection was made by Lindsey & Braun (2000a) using the Doppler-velocity data from MDI by employing helioseismic holography method Lindsey & Braun (2000b)). The method was later expanded to achieve entire far-side coverage (Braun & Lindsey 2001; Lindsey & Braun 2017; Yang *et al.* 2023a). Time–distance methods were also developed to detect far-side ARs (Duvall *et al.* 2000) by tracking waves propagating from the near side to the far side and then back to the near side. These waves get reflected multiple times on the solar surfaces, carrying information that can be used to map the far-side surface conditions. Zhao (2007) developed this method by designing and measuring different wave-propagation schemes, achieving better

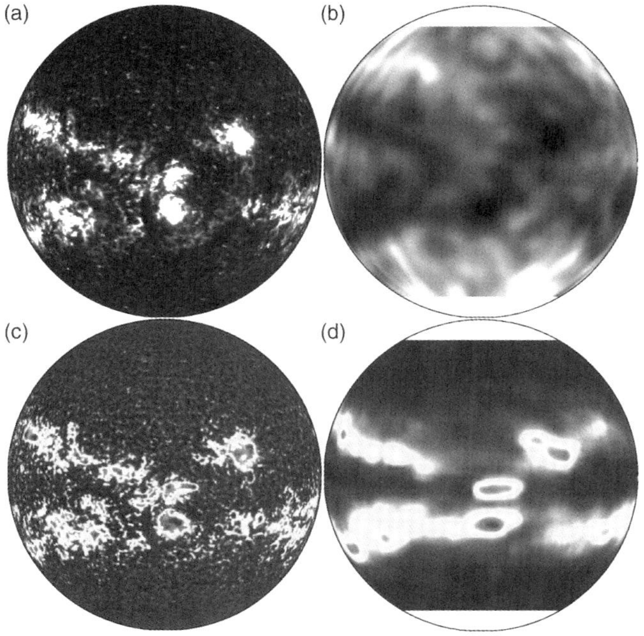

Figure 4. (a) Observed EUV 304 Å map merged from *STEREO* A & B spacecraft data; (b) Helioseismic far-side image obtained from the new timedistance far-side imaging codes (Zhao *et al.* 2019); (c) Far-side unsigned magnetic-flux map generated using *STEREO* observations and the machine-learned EUV-to-magnetic mapping relation (Chen *et al.* 2022); (d) Far-side magnetic-flux map generated using helioseismic far-side images and the machine-learned acoustic-to-magnetic mapping relation (Chen *et al.* 2022).

performance by superimposing the results of multiple schemes. Initially, Zhao (2007) developed codes for calculating 4-skip and 5-skip schemes, and Ilonidis *et al.* (2009) added 3-skip ones for improvements.

With *STEREO*'s direct far-side observations, far-side active regions (ARs) can be monitored in EUV channels, enabling a systematic evaluation of the performance of helioseismic far-side imaging techniques. In a series of articles, Liewer *et al.* (2012, 2014, 2017) compared *STEREO* 304 Å images with helioseismic holography far-side images and found that 95% of the helioseismically-detected far-side ARs correspond to an EUV brightening area.

Recently, Zhao *et al.* (2019) improved the time–distance far-side imaging technique by introducing additional wave propagating schemes, resulting in a total of 14 sets of measurements. This significantlly elevated the sensitivities and reliability of their time–distance far-side imaging tool. Through an examination of helioseismic maps generated using near-side HMI Dopplergrams and *STEREO*'s direct far-side EUV observation, the study demonstrated that 97.3% of helioseismically detected far-side ARs larger than a certain size were found to correlate with an observed region exhibiting strong EUV brightening. Furthermore, 85.7% of ARs observed by *STEREO*, larger than a certain size, proved detectable in the helioseismic images. Such remarkable performance holds great potential to improve space weather forecasting. Figure 4 showcases the *STEREO*-observed EUV image (Panel a) and the helioseismic far-side map (Panel b) of the Sun's far side on 2014 Mar 13. A pipeline has been established to generate such helioseismic far-side maps in near-real time.

3.2. *Incorporating Machine Learning techniques.*

While significant progress has been achieved in the field of helioseismic far-fide imaging, there remains one particular gap to fill. The helioseismic far-side maps provide information in terms of travel-time shifts, not magnetic fields. However, solar wind modeling demands global magnetic field data for a high quality in space weather forecasting (Nitta & DeRosa 2008; Schrijver & Title 2011). To fulfill this need, modelers have developed flux-transport models (e.g., Schrijver & DeRosa 2003; Upton & Hathaway 2014b) to approximate the far-side magnetic field by migrating the magnetic field observed on the near side towards the far side with decay. However, such methods are incapable of predicting the growth or new emergence of active regions.

A recent study by Chen *et al.* (2022) aimed to address this problem using machine learning. This work developed an approach to calibrate far-side acoustic images into far-side unsigned magnetic-flux maps using *STEREO* EUV observations as a bridge. Two machine-learning models were trained to achieve this goal. The first model converts EUV 304 Å maps to unsigned magnetic flux maps, trained using 9 years of near-side observations by *SDO*/AIA and *SDO*/HMI. This model transforms the *STEREO* EUV 304 Å images into far-side unsigned magnetic-flux proxies (with certain calibrations between *STEREO* and AIA EUV observations). Using these (*STEREO*-based) far-side magnetic-flux proxies as targets, the second model was trained to convert helioseismic far-side images into (acoustic-based) magnetic-flux maps. Once trained, the second model can produce future acoustic-based unsigned magnetic-flux maps in near-real-time solely based on near-side Doppler observations. Figure 4(c) shows their *STEREO*-based far-side magnetic flux on this date (as an intermediate data product), and Figure 4(d) shows their final acoustic-based unsigned magnetic-flux map. Although being low-resolution, these acoustic-based far-side magnetic-flux maps can provide useful information about the far side's magnetic fields when direct observation is unavailable.

This is not the only work leveraging the power of machine learning to improve helioseismic products. In fact, machine learning has been applied in multiple aspects to enhance the detection of far-side ARs and estimating magnetic fields. Felipe & Asensio Ramos (2019) and Broock *et al.* (2021) trained a machine-learning model that compared the far-side EUV Images and the helioseismic-holography images before these helioseismic images were routinely overlapped for 5 days to enhance the signal-to-noise ratio. This approach resulted in a higher confidence in detecting smaller far-side ARs that might otherwise be neglected. Additionally, signed magnetic fields were obtained by authors using deep-learning methods from *SDO*/AIA EUV maps (e.g., Kim *et al.* 2019), and in following studies, from *STEREO* EUV maps as well (Jeong *et al.* 2020; Alshehhi 2020). However, considering the low resolution and large uncertainty of helioseismic far-side imaging, generating signed magnetic fields is challenging, as acoustic signals exhibit no sensitivity to the polarity of the magnetic field.

3.3. *Direct Far-side Magnetogram Observation from SO.*

The recent launch of the *Solar Orbiter* (*SO*; (Müller *et al.* 2013)) has reintroduced direct far-side observation. In particular, the Polarimetric and Helioseismic Imager (PHI) on board *SO* provides line-of-sight magnetograms of the far side during the cruise and the early science phase of the mission (Solanki *et al.* 2020). This marks the first direct observation of far-side magnetic fields, offering valuable data for calibrating far-side magnetic field proxies. With access to a limited sample of 6 ARs on 3 magnetograms, (Yang *et al.* 2023b) established an empirical relation between helioseismic far-side measurements (Yang *et al.* 2023a) and directly observed magnetic flux. Future observations will enable calibration on a larger sample, holding the potential to significantly enhance solar far-side imaging.

4. Sunquakes

Sunquakes often occur in active regions following strong flares. Although the sunquake regions are usually observed, the velocity data are often corrupted due to irregular spectrum lines in high temperature and violent dynamics following the flares. Sunquakes' velocity fields are often reconstructed and measured by the helioseismic holography method (Lindsey & Braun 1997, 2000a). This approach essentially reverses the expanding sunquake ripples backward to the sunquake initiation time and location to reconstruct the sunquake source (Donea *et al.* 2000), measure the sunquake energy (Zharkov *et al.* 2013; Sharykin *et al.* 2017), or probe into the submerged sources (Lindsey *et al.* 2020). The reversal of waves is achieved by deconvolution with a Green's function, typically calculated theoretically using a standard quiet-Sun model. However, albeit being noise-free, theoretical Green's functions carry biases in determining the sunquake initiation time, as sunquakes often occur in active regions rather than the quiet regions where the Green's function is calculated.

Recently, Chen & Zhao (2021) employed time–distance methods to derive observation-based Green's functions for waves propagating outside sunspot umbra, penumbra, and surrounding quiet regions. These observation-based Green's functions were then integrated into the helioseismic holography method to reconstruct the sunquake velocity fields. This approach is expected to provide more precise time determinations. With this enhancement, Chen & Zhao (2021) investigated the relationship between flares and sunquakes, aiming to answer the question of why some flares trigger sunquakes while others do not.

Their hypothesis introduces the consideration of photospheric background oscillation as a contributing factor for sunquakes, a factor often overlooked in previous studies. They propose that during the impulsive phase of a flare, when the flare's impulse influences the photosphere through shock waves, energetic particles, or downward Lorentz Force, a sunquake is more likely to occur if the background oscillation at the flare footpoint oscillates downward in the same direction as the impulse from above. This alignment enhances the background oscillation to a detectable level against the ambient background oscillations. To test this hypothesis, they analyzed 60 strong flares in Solar Cycle 24, examining the background oscillatory velocity at the sunquake sources during the flares' impulsive phases. Given the short time periods of solar oscillations, precise time determination of the sunquake's velocity field, as mentioned earlier, is crucial for this investigation. Their findings indicate that out of the 60 studied flares, 24 are helioseismically active, resulting in a total of 41 sunquake events (in either frequency range). In the $3 - 5$ mHz frequency band, 80.6% of the 31 sunquakes exhibit average downward velocities during the flares' impulsive phases, and in the $5 - 7$ mHz frequency band, 86.8% of the 38 sunquakes have net downward velocities. This consistency supports their hypothesis that sunquakes are more likely to occur when the downward flare impact strengthens a pre-existing downward background oscillation. Despite supporting statistics, it is acknowledged that the proposed background oscillation serves as a selection rule that restricts the chance for a sunquake to occur, but does not explain how or when a flare triggers a sunquake in the first place. Nevertheless, this work may provide new insights to understand the puzzling sunquake triggering mechanisms.

5. Conclusion

In conclusion, recent advancements in time–distance helioseismology have significantly improved our understanding of various solar phenomena. These developments include refining measurements, addressing systematic effects, and exploring novel approaches to overcome challenges in the field. Studies on solar meridional circulation have revealed discrepancies in flow profiles, prompting investigations into the removal of systematic effects like the CtoL effect. Frequency-dependent observations have provided insights into the nature of the CtoL effect, introducing new dimensions to meridional flow studies. Machine learning applications have enhanced the detection of far-side active regions and the estimation of magnetic fields, opening avenues for improved space weather forecasting. Additionally, the integration of time–distance-derived Green's functions has enhanced the precision of time determination for sunquakes using holography methods. This improvement contributes to our understanding of the potential influence of photospheric background oscillations on sunquake triggering during solar flares. Despite these advancements, challenges and gaps persist, underscoring the need for continued research to unravel the complexities of solar dynamics and improve the accuracy of helioseismic measurements.

Acknowledgements. *SDO* is a NASA mission, and HMI project is supported by NASA contract NAS5-02139 to Stanford University. R.C. and J.Z. are partly supported by NASA DRIVE Center COFFIES grant 80NSSC22M0162.

References

Alshehhi, R. 2020, *Proceedings of the IEEE/CVF Conference on Computer Vision and Pattern Recognition Workshops*, 204–205
Böning, V. G. A., Roth, M., Jackiewicz, J., & Kholikov, S. 2017, *Astrophys. J.*, 845, 2
Braun, D. C., & Fan, Y. 1998, *Astrophys. J. Letters*, 508, L105
Braun, D. C. & Lindsey, C. 2001, *Astrophys. J. Lett.*, 560, L189.
Broock, E. G., Felipe, T., Asensio Ramos, A. 2021, *A&A*, 652, A132
Buitrago–Casas, J. C., Martínez Oliveros, J. C., Lindsey, C., *et al.* 2015, *Solar Phys.*, 290, 3151
Chen, R., 2019, PhD dissertation, Stanford University
Chen, R., & Zhao, J. 2017, *Astrophys. J.*, 849, 144
Chen, R., & Zhao, J. 2018, *Astrophys. J.*, 853(2), 161
Chen, R. & Zhao, J. 2021, *Astrophys. J.*, 908, 182
Chen, R., Zhao, J., Hess Webber, S., *et al.* 2022, *Astrophys. J.*, 941, 197
Dikpati, M., Gilman, P. A., de Toma, G., & Ulrich, R. K. 2010, *Geophys. Res. Lett.*, 37, L14107
Donea, A.-C., Besliu–Ionescu, D., Cally, P. S., Lindsey, C., & Zharkova, V. V. 2006, *Solar Phys.*, 239, 113
Donea, A.-C., Lindsey, C., & Braun, D. C. 2000, *Solar Phys.*, 192, 321
Duvall, T. L., Jr. 1979, *Solar Phys.*, 63, 3
Duvall, T. L., Kosovichev, A. G., & Scherrer, P. H. 2000, *AAS/Solar Physics Division Meeting* #31
Featherstone, N. A., & Miesch, M. S. 2015, *Astrophys. J.*, 804, 67
Felipe, T., Asensio Ramos, A. 2019, *A&A*, 632, A82
Fisher, G. H., Bercik, D. J., Welsch, B. T., & Hudson, H. S. 2012, *Solar Phys.*, 277, 59
Giles, P. M. 2000, PhD dissertation, Stanford University
Giles, P. M., *et al.* 1997, *Nature*,
Gizon, L., Cameron, R. H., Pourabdian, M., *et al.* 2020, *Science*, 368, 1469
González Hernández, I., *et al.* 1999, *Astrophys. J. Lett.*, 510, L153
Haber, D. A., Hindman, B. W., Toomre, J., *et al.* 2000, *Solar Phys.*, 192, 335
Hathaway, D. H. 1996, *Astrophys. J.*, 460, 1027
Hathaway, D. H., & Rightmire, L. 2010, *Science*, 327, 1350
Howard, R., & Gilman, P. A. 1986, *Astrophys. J.*, 307, 389
Howard, R. A., Moses, J. D., Vourlidas, A., *et al.* 2008, *Space Sci. Rev.*, 136, 67
Hudson, H. S., Fisher, G. H., & Welsch, B. T. 2008, in ASP Conf. Ser. 383, Subsurface and Atmospheric Influences on Solar Activity, ed. R. Howe,
Ilonidis, S., Zhao J., & Hartlep, T. 2009, *Solar Phys..*, 258, 181
Jackiewicz, J., Serebryanskiy, A., & Kholikov, S. 2015, *Astrophys. J.*, 805, 133
Jeong, H.-J., Moon, Y.-J., Park, E., *et al.* 2020, *Astrophys. J. Lett.*, 903, L25

Kim, T., Park, E., Lee, H., *et al.* 2019, *Nature Astron*, 3, 397
Komm, R. W., Howard, R. F., & Harvey, J. W. 1993, *Solar Physics*, 147, 207
Kholikov, S., Serebryanskiy, A., & Jackiewicz, J. 2014, *Astrophys. J.*, 784, 145
Kosovichev, A. G. 2007, *Astrophys. J. Lett.*, 670, L65
Kosovichev, A. G., & Zharkova, V. V. 1998, *Nature*, 393, 317
Liang, Z.-C., & Chou, D.-Y. 2015, *Astrophys. J.*, 809, 150
Liewer, P. C., González Hernández, I., Hall, J. R., Lindsey, C., & Lin, X. 2014, *Solar Phys.*., 289, 3617
Liewer, P. C., González Hernández, I., Hall, J. R., Thompson, W. T., & Misrak, A. 2012, *Solar Phys.*., 281, 3
Liewer, P. C., Qiu, J., & Lindsey, C. 2017, *Solar Phys.*., 292, 146
Lindsey, C., & Braun, D. C. 1997, *Astrophys. J. Lett.*, 485, 895
Lin, C.-H., & Chou, D.-Y. 2018, *Astrophys. J.*, 860, 48
Lindsey, C., & Braun, D. C. 2000a, *Science*, 287, 5459
Lindsey, C., & Braun, D. C. 2000b, *Solar Phys.*., 192, 261
Lindsey, C., & Braun, D. C. 2017, *Space Weather*, 15, 761
Lindsey, C., & Donea, A.-C. 2008, *Solar Phys.*, 251, 627
Lindsey, C., Buitrago–Casas, J. C., Martínez Oliveros, J. C., *et al.* 2020, *Astrophys. J. Lett.*, 901, L9
Mandal, K., Hanasoge, S. M., Rajaguru, S. P., & Antia, H. M. 2018, *Astrophys. J.*, 863, 39
Miesch, M. S. 2005, Living Rev. *Solar Phys.*., 2, 1
Müller, D., Marsden, R. G., St. Cyr, O. C., *et al.* 2013, *Astrophys. J.*, 285, 25
Nitta, N. V., & DeRosa, M. L. 2008, *Astrophys. J. Lett.*, 673, L207
Rajaguru, S. P., & Antia, H. M. 2015, *Astrophys. J.*, 813, 114
Rajaguru, S. P. & Antia, H. M. 2020, *Dynamics of the Sun and Stars; Honoring the Life and Work of Michael J. Thompson*, 57, 107.
Reep, J. W., & Russell, A. J. B. 2016, *Astrophys. J. Lett.*, 818, L20
Roth, M., Doerr, H.-P., & Hartlep, T. 2016, *A&A*, 592, A106
Russell, A. J. B., & Fletcher, L. 2013, *Astrophys. J.*, 765, 81
Scherrer, P. H., Bogart, R. S., Bush, R. I., *et al.* 1995, *Solar Phys.*, 162, 129
Scherrer, P. H., Schou, J., Bush, R. I., et al. 2012, *Solar Phys.*, 275, 207
Schou, J., Scherrer, P. H., Bush, R. I., et al. 2012, *Solar Phys.*, 275, 229
Schrijver, C. J., & DeRosa, M. L. 2003, *Solar Phys.*, 212, 165
Schrijver, C. J., & Title, A. M. 2011, *J. Geophys. Res.*, 116, A04108
Sharykin, I. N. & Kosovichev, A. G. 2020, *Astrophys. J.*, 895, 76
Sharykin, I. N., Kosovichev, A. G., Sadykov, V. M., Zimovets, I. V., & Myshyakov, I. I. 2017, *Astrophys. J.*, 843, 67
Solanki, S. K., del Toro Iniesta, J. C., Woch, J., *et al.* 2020, *A&A*, 642, A11
Švanda, M., Zhao, J., & Kosovichev, A. G. 2007, *Solar Phys.*, 241, 27
Upton, L., & Hathaway, D. H. 2014a, *Astrophys. J.*, 780, 5
Upton, L., & Hathaway, D. H. 2014b, *Astrophys. J.*, 792, 142
Waidele, M., Zhao, J., & Kitiashvili, I. N. 2023, *Astrophys. J.*, 949, 99.
Wang, Y.-M., Nash, A. G., & Sheeley, N. R., Jr. 1989, *Science*, 245, 712
Yang, D., Gizon, L., Barucq, H., *et al.* 2023, *A&A*, 674, A183
Yang, D., Gizon, L., & Barucq, H. 2023a, *A&A*, 669, A89
Yang, D., Gizon, L., Barucq, H., *et al.* 2023b, *A&A*, 674, A183
Zhao, J., & Kosovichev, A. G. 2004, *Astrophys. J.*, 603, 776
Zhao, J., Bogart, R. S., Kosovichev, A. G., Duvall, T. L., Jr. 2013, *Astrophys. J. Lett.*, 774, L29
Zhao, J. & Chen, R. 2020, *Dynamics of the Sun and Stars; Honoring the Life and Work of Michael J. Thompson*, 57, 123.
Zhao, J. 2007, *Astrophys. J. Lett.*, 664, L139
Zhao, J., Hing, D., Chen, R., & Hess Webber, S. 2019, *Astrophys. J.*, 887, 216

Zhao, J., Nagashima, K., Bogart, R. S., Kosovichev, A. G., Duvall, T. L., Jr. 2012, *Astrophys. J. Lett.*, 749, L5

Zhao, J., Rajaguru, S. P., & Chen, R. 2022, *Astrophys. J.*, 933, 109

Zharkov, S., Green, L. M., Matthews, S. A., & Zharkova, V. V. 2013, *Solar Phys.*, 284, 315

Zharkova, V. V., & Zharkov, S. I. 2007, *Astrophys. J.*, 664, 573

Dynamics of Solar and Stellar Convection Zones and Atmospheres
Proceedings IAU Symposium No. 365, 2024
A. V. Getling & L. L. Kitchatinov, eds.
doi:10.1017/S1743921323005471

Convection, rotation, and magnetic activity of solar-like stars from asteroseismology

Savita Mathur

Instituto de Astrofísica de Canarias (IAC), E-38205 La Laguna, Tenerife, Spain
Universidad de La Laguna (ULL), Departamento de Astrofísica, E-38206 La Laguna, Tenerife, Spain
email: smathur@iac.es

Abstract. During the last decade, our understanding of stellar physics and evolution has undergone a tremendous revolution thanks to asteroseismology. Space missions such as CoRoT, *Kepler*, K2, and TESS have already been observing millions of stars providing high-precision photometric data. With these data, it is possible to study the convection of stars through the convective background in the power spectrum density of the light curves. The properties of the convective background or granulation has been shown to be correlated to the surface gravity of the stars. In addition, when we have enough resolution (so long enough observations) and a high signal-to-noise ratio (SNR), the individual modes can be characterized in particular to study the internal rotational splittings and magnetic field of stars. Finally, the surface magnetic activity also impacts the amplitude and hence detection of the acoustic modes. This effect can be seen as a double-edged sword. Indeed, modes can be studied to look for magnetic activity changes. However, this also means that for stars too magnetically active, modes can be suppressed, preventing us from detecting them.

In this talk, I will present some highlights on what asteroseismology has allowed us to better understand the convection, rotation, and magnetism of solar-like stars while opening doors to many more questions.

Keywords. Asteroseismology, Convection, Rotation, Magnetic activity, Solar-type stars

---

1. Introduction

While the study of the Sun with both local and global helioseismology has been going on for many decades (e.g. Claverie et al. 1979), the possibility to study other stars like the Sun has been possible on a very large scale thanks to the ESA and NASA space missions such as CoRoT (Convection, Rotation, and planetary Transits, Baglin et al. 2006), *Kepler* (Borucki et al. 2010), K2 (Howell et al. 2014), and TESS (Transiting Exoplanet Survey Satellite, Ricker et al. 2015). Most of these missions focus on the search for exoplanets but the importance of characterizing the star at the center of planetary system has been part of the main objectives of CoRoT and the more recent missions. Indeed the high quality of the photometric data collected by such missions allowed us to do asteroseismic analyses with many new discoveries that are revolutionizing our understanding of stellar evolution structure and dynamics.

The ESA M3 mission, PLATO (PLAnetary Transits and Oscillations of stars, Rauer et al. 2014), which launched is scheduled for the end of 2026 will continue in this direction with the goal of characterizing stars like the Sun fo spectral types F, G, and K on the main sequence and increase by more than a factor of 100 the current *Kepler* Legacy catalog (e.g. Campante et al. 2011; Mathur et al. 2011a; Metcalfe et al. 2012; Silva Aguirre et al. 2015, 2017).

Finally, the ESA Roman Space telescope (Wilson et al. 2023) with a launch scheduled in May 2027 and whose main objectives will be to study dark energy and dark matter, as well as image exoplanets will provide data that will be well suited for asteroseismic analyses (Huber et al. 2023).

Up to now, *Kepler* is the mission that provided the largest sample of solar-like main-sequence and subgiant stars with detected solar-like oscillations (Chaplin et al. 2014; Davies et al. 2016; Lund et al. 2017; Mathur et al. 2022) with more than 620 stars. The K2 mission that followed the nominal *Kepler* mission after the second reaction wheel failed also contributed to increase the sample of solar-like stars with seismic detections (e.g. Lund et al. 2016; Ong et al. 2021; González-Cuesta et al. 2023). Finally, while the TESS mission has been observing millions of stars since its launch in 2018, it's observing feature and an instrumental noise level higher the one of *Kepler* did not lead as many detections of solar-like oscillations in main sequence and subgiant stars (e.g. Huber et al. 2019). The new 20-second cadence improved the signal-to-noise ratio (Huber et al. 2022; Hatt et al. 2023) and the yield of detections should increase with the new cycles.

The path for asteroseismic studies has been well paved and is very bright for the near future.

When computing the power spectrum density (PSD) computed from photometric observations contains information on several physical processes in stars: surface rotation, convection, and oscillations (García and Ballot 2019). In the next sections, we will go through the different highlights on convection, rotation and magnetism of stars thanks to asteroseismic analyses.

2. Convection

The continuum in the PSD of a solar-like stars results from the convection (or granulation) occurring in the outer layers of solar-like stars (García and Ballot 2019). That convective background is generally modeled with several components by many of the asteroseismic pipelines. For the Sun, Harvey (1985) proposed the following function to model the granulation:

$$B = \frac{4\sigma^2 \tau_{\rm gran}}{1 + (2\pi\nu\tau_{\rm gran})^{\alpha}}, \tag{1}$$

where σ is the characteristic amplitude of the granulation, $\tau_{\rm gran}$ the characteristic time scale of granulation. For the Sun $\alpha = 2$ but (Kallinger et al. 2014) showed that for other stars where we model the background with two Harvey-like functions – corresponding to different scales of granulations – a slope $\alpha = 4$ was favored. By studying the granulation of several hundreds of red giants and main-sequence solar-like stars observed by *Kepler*, a correlation between the surface gravity and σ and τ was found (Mathur et al. 2011b; Kallinger et al. 2014). The analysis also showed that the granulation timescale increases with the evolution of the star.

The comparison of the *Kepler* observations of granulation with 3D simulations of convection in a box by Trampedach and Stein (2011)emphasized that the simulations underestimated the timescale by a factor of 2 while they overestimated the granulation power by an order of magnitude. These differences could be explained by the fact that the grid of 3D simulations was sparse and extrapolations to the stellar parameters of the observations were performed and also by the fact that the simulations were done for solar metallicity, which can impact the stellar convection. More studies are being done where metallicity from spectroscopic surveys are used (Perillo et al. in prep.).

The strong correlation between surface gravity and granulation led to the definition of different metrics to infer the $\log g$ of stars. The first one is the Flicker (Bastien et al. 2013) computed on a light curve smoothed over 8 hours. But that metric is limited to a small range of red giant stars and can be biased for stars with oscillations close to the 8-hour time scale.

Bugnet et al. (2018) developed another metric computed on the PSD: the FliPer (Flicker in Power), which is the total power in the PSD between a given minimum frequency ($f_{\rm min}$ and the Nyquist frequency to which the photon noise is subtracted. The photon noise for the *Kepler* data is computed following the relation from Jenkins et al. (2010). The FliPer computed for the *Kepler* targets showed a very good correlation with $\log g$.

In order to avoid any issues from low-frequency peaks related to the surface rotation or spurious spikes, the FliPer is computed with several $f_{\rm min}$ values. These values along with the effective temperature and *Kepler* magnitude were then used as input of a Random Forest algorithm (Breiman 2001). After training the algorithm, the test set showed that the RF provided $\log g$ values with uncertainties of 0.046 dex on average, much better that spectroscopic uncertainties and close to the seismic ones.

This metric is very powerful, especially because the modes do not need to be detected. So it can be applied to stars with short light curves and with low SNR. It can also be used to classify stars (Bugnet et al. 2019).

3. Rotation

In this section, we will focus on the rotation of stars from the main-sequence to the red giant branch.

3.1. *Surface rotation*

The need to understand the transport of angular momentum in stars is crucial as it impacts the mixing of elements and hence the ages of stars. We will see how asteroseismology can help understand the evolution of surface rotation with age. In 1972, Skumanich studied two clusters and the Sun and showed that the surface rotation rate decreased as a square root of the age. This was later generalized with more clusters and field stars and led to the definition of rotation-age relations, also known as gyrochronology (Barnes 2007).

With the high-quality of the *Kepler* data, surface rotation periods could be measured through the modulation of the light curves due to the presence of spots and active regions (e.g. Nielsen et al. 2013; Reinhold et al. 2013; McQuillan et al. 2014; García et al. 2014a). In addition, for a subsample of stars, asteroseismic analyses were possible, which provided ages of stars (e.g. Mathur et al. 2012; Metcalfe et al. 2014; Creevey et al. 2017; Silva Aguirre et al. 2017). Seismic ages are obtained by fitting spectroscopic (effective temperature, metallicity) and seismic observables (including global seismic parameters and individual frequencies of the modes) with stellar models. Lebreton and Goupil (2014) showed that by including the individual frequencies of the modes, the inference of the masses, radii, and ages of the stars were improved with a better precision.

By combining surface rotation periods with precise seismic ages, van Saders et al. (2016) compared the evolution of rotation with age of solar-like stars with models including gyrochronology prescriptions. They found that stars older than the Sun were rotating faster than expected given their ages. This discovery suggested that at a given moment (close to the Rossby number of the Sun) during the main-sequence life of a star, the magnetic braking weakens. This means that gyrochronology is only applicable to stars below that Rossby threshold. It was also noticed that the Sun seems to be closed to that transition phase. The exact origin of that phenomenon is not yet understood but hints on change in the differential rotation and magnetic topology are being investigated (e.g. Metcalfe et al. 2020).

3.2. *Latitudinal differential rotation*

Internal rotation of the Sun and stars can be measured through rotational splittings of the acoustic modes. Indeed, rotation lifts the degeneracy and yields several 2l+1 components of modes with degree l above 1. The extracted splittings also depend on the inclination angle of the stars as they impact which components are visible (Ballot et al. 2006).

For the *Kepler* solar-like targets with the highest SNR, several works looked for signature of latitudinal differential rotation (e.g. Lund et al. 2014; Schunker et al. 2016) by using rotational splittings measurement. They mostly concluded that the current data did not allow to have a firm measurement of differential rotation due to the small number of low-degree modes detected.

We highlight her the work by Benomar et al. (2018) who looked at the Clebsch–Gordon a coefficients of the rotation splittings. The third coefficient, a_3 contains information on the latitudinal differential rotation. They measured that a_3 coefficient for 40 solar-like stars observed by *Kepler* and computed the detection probability for stars to have solar or anti-solar rotation. They found 13 candidates with the most significant detection with a solar differential rotation.

3.3. *Internal rotation*

For more evolved stars, the detection of mixed modes that result from the coupling of acoustic- and gravity-mode cavities allowed the study of the core rotation of subgiants and red giants (Beck et al. 2012; Deheuvels et al. 2014). The cores of these evolved stars were found to rotate faster than the envelopes on average. While Mosser et al. (2012b) suggested that the core of red-giant branch (RGB) stars were spinning down towards the red clump (RC) stars, an extended analysis by Gehan et al. (2018) did not confirm such trend. The cores of red-giant stars seem to be constant on the RGB without dependence on mass. The spinning down from RGB to RC can partially explain the expansion of the non-degenerated helium core but not the full slow down.

4. Stellar magnetism

Stellar magnetism can also be studied with asteroseismology for both surface and internal magnetism.

4.1. *Impact of magnetic activity on solar-like oscillations*

For the Sun, it is well known that during its magnetic activity cycle, the acoustic modes are affected in the following way: while magnetic activity increases, the frequencies of the modes increase and their amplitudes decrease (e.g. Woodard and Noyes 1985; Elsworth et al. 1990; Jain et al. 2009). Similar behavior was first observed in another solar-like stars, HD 49933, observed by the CoRoT mission (García et al. 2010). An anti-correlation between the temporal changes of the amplitude and the frequency shifts of the modes was observed. Complementary CaHK observations at the Cerro Tololo observatory done a few years later confirmed that the star was active with an S-index of 0.31. Later with *Kepler*, such phenomenon was detected in more than 50 targets (e.g. Salabert et al. 2016; Kiefer et al. 2017; Santos et al. 2018) where changes were correlated with the photometric activity index, $S_{\rm ph}$ (Mathur et al. 2014). These detections can provide important constraints on the mechanisms involved in the dynamo of these stars.

One example is the *Kepler* target KIC 8006161 for which frequency shifts and changes in the amplitudes of the modes were detected. That star was also followed up decades before the *Kepler* observations with the Keck providing chromospheric index of magnetic activity, in particular a detection of a magnetic activity cycle of 7.4 yrs. From seismology the stellar parameters of KIC 8006161 are very similar to the Sun. Having a stronger chromospheric emission than the Sun and a shorter cycle period could be explained with the fact that KIC 8006161 is a meta-rich star. Karoff et al. (2018) showed how the metallicity impacted the magnetic activity of the star.

Santos et al. (2019) studied more thoroughly the sample of *Kepler* targets with significant frequency shifts as they were also modeled with seismology. As expected, magnetic activity decreases with age (e.g. Wilson 1978). Magnetic activity was also found to be stronger with higher metallicity.

Since a higher level of magnetic activity leads to smaller amplitude of the acoustic modes, this also means that stars with a very high magnetic activity will have much smaller amplitudes of the modes (Chaplin et al. 2011).

The rotation study of ~ 1000 *Kepler* stars observed in short cadence, where mode detection was expected but was not achieved, yielded a sample of 323 stars with reliable rotation periods (Mathur et al. 2019). The results were compared to the ones of ~ 300 stars with solar-like oscillations detected and reliable rotation. Among the stars without detected oscillations, 32% have a photometric magnetic activity index, $S_{\rm ph}$, larger than the maximum level of the Sun, explaining the non detection of the modes. However, there remains a large fraction of the sample with magnetic activity similar to the Sun and for which modes were not observed. The reason behind the non-detection is not clear but hints are possibly: metallicity and impact of the inclination angle of the rotation axis. But larger sample with spectroscopic observations is needed to confirm such trends.

4.2. *Internal magnetic field of evolved stars*

Another sample of stars with unusual mode pattern was highlighted by the *Kepler* mission: stars with abnormal low-amplitude dipole modes (Mosser et al. 2012a; García et al. 2014b) in reg-giant stars. One of the theories developed suggests that there is a strong magnetic field in the radiative zone of those stars, which interacts with the l=1 modes and scatters them into high-degree modes that are trapped in the radiative zone (Fuller et al. 2015). The typical critical magnetic field responsible for such phenomenon was estimated to 10^5 G. The analysis of a larger statistical sample showed that stars with these *depressed* modes had a mass above $1.2\,{\rm M}_\odot$, corresponding to stars that developed a convective core (Stello et al. 2016) agreeing with the aforementioned theory.

The same way rotation lifts the degeneracy of the modes, magnetic field was also shown to have such an impact (Duvall et al. 1984; Goode and Thompson 1992). Recently several teams studied the effect of an axisymmetric field on mixed modes (Mathis et al. 2021; Bugnet et al. 2021; Loi 2021). They showed that this would lead to an asymmetric triplet and suggested that such asymmetry could be searched for in the *Kepler* data. The generalization of the theory to non-axisymmetric magnetic field led to the discovery asymmetries in the splittings of red giants attributed to fields of strength 30 to 100 kG (Li et al. 2022; Deheuvels et al. 2023).

With the coming launch of the ESA PLATO mission in late 2026, even more breakthroughs should be made, helping us in our comprehension of stellar interior dynamics evolution.

5. Acknowledgements

S.M. acknowledges support by the Spanish Ministry of Science and Innovation with the Ramon y Cajal fellowship number RYC-2015-17697, the grant number PID2019-107187GB-I00, the grant no. PID2019-107061GB-C66, and through AEI under the Severo Ochoa Centres of Excellence Programme 2020–2023 (CEX2019-000920-S).

References

Baglin, A., Michel, E., Auvergne, M., & The COROT Team. The seismology programme of the CoRoT space mission. In *Proceedings of SOHO 18/GONG 2006/HELAS I, Beyond the spherical Sun* 2006,, volume 624 of *ESA Special Publication*.

Ballot, J., García, R. A., & Lambert, P. 2006, Rotation speed and stellar axis inclination from p modes: how CoRoT would see other suns. *MNRAS*, 369, 1281–1286.

Barnes, S. A. 2007, Ages for Illustrative Field Stars Using Gyrochronology: Viability, Limitations, and Errors. *ApJ*, 669, 1167–1189.

Bastien, F. A., Stassun, K. G., Basri, G., & Pepper, J. 2013, An observational correlation between stellar brightness variations and surface gravity. *Nature*, 500, 427–430.

Beck, P. G., Montalban, J., Kallinger, T., De Ridder, J., Aerts, C., García, R. A., Hekker, S., Dupret, M.-A., Mosser, B., Eggenberger, P., Stello, D., Elsworth, Y., Frandsen, S., Carrier, F., Hillen, M., Gruberbauer, M., Christensen–Dalsgaard, J., Miglio, A., Valentini, M., Bedding, T. R., Kjeldsen, H., Girouard, F. R., Hall, J. R., & Ibrahim, K. A. 2012, Fast core rotation in red-giant stars as revealed by gravity-dominated mixed modes. *Nature*, 481, 55–57.

Benomar, O., Bazot, M., Nielsen, M. B., Gizon, L., Sekii, T., Takata, M., Hotta, H., Hanasoge, S., Sreenivasan, K. R., & Christensen–Dalsgaard, J. 2018, Asteroseismic detection of latitudinal differential rotation in 13 Sun-like stars. *Science*, 361(6408), 1231–1234.

Borucki, W. J., Koch, D., Basri, G., Batalha, N., Brown, T., Caldwell, D., Caldwell, J., Christensen–Dalsgaard, J., Cochran, W. D., DeVore, E., Dunham, E. W., Dupree, A. K., Gautier, T. N., Geary, J. C., Gilliland, R., Gould, A., Howell, S. B., Jenkins, J. M., Kondo, Y., Latham, D. W., Marcy, G. W., Meibom, S., Kjeldsen, H., Lissauer, J. J., Monet, D. G., Morrison, D., Sasselov, D., Tarter, J., Boss, A., Brownlee, D., Owen, T., Buzasi, D., Charbonneau, D., Doyle, L., Fortney, J., Ford, E. B., Holman, M. J., Seager, S., Steffen, J. H., Welsh, W. F., Rowe, J., Anderson, H., Buchhave, L., Ciardi, D., Walkowicz, L., Sherry, W., Horch, E., Isaacson, H., Everett, M. E., Fischer, D., Torres, G., Johnson, J. A., Endl, M., MacQueen, P., Bryson, S. T., Dotson, J., Haas, M., Kolodziejczak, J., Van Cleve, J., Chandrasekaran, H., Twicken, J. D., Quintana, E. V., Clarke, B. D., Allen, C., Li, J., Wu, H., Tenenbaum, P., Verner, E., Bruhweiler, F., Barnes, J., & Prsa, A. 2010, Kepler Planet-Detection Mission: Introduction and First Results. *Science*, 327, 977–.

Breiman, L. 2001, Random Forests. *Machine Learning*, 45, 5–32.

Bugnet, L., García, R. A., Davies, G. R., Mathur, S., Corsaro, E., Hall, O. J., & Rendle, B. M. 2018, FliPer: A global measure of power density to estimate surface gravities of main-sequence solar-like stars and red giants. *A&A*, 620, A38.

Bugnet, L., García, R. A., Mathur, S., Davies, G. R., Hall, O. J., Lund, M. N., & Rendle, B. M. 2019, FliPer$_{Class}$: In search of solar-like pulsators among TESS targets. *A&A*, 624, A79.

Bugnet, L., Prat, V., Mathis, S., Astoul, A., Augustson, K., García, R. A., Mathur, S., Amard, L., & Neiner, C. 2021, Magnetic signatures on mixed-mode frequencies. I. An axisymmetric fossil field inside the core of red giants. *A&A*, 650, A53.

Campante, T. L., Handberg, R., Mathur, S., Appourchaux, T., Bedding, T. R., Chaplin, W. J., García, R. A., Mosser, B., Benomar, O., Bonanno, A., Corsaro, E., Fletcher, S. T., Gaulme, P., Hekker, S., Karoff, C., Régulo, C., Salabert, D., Verner, G. A., White, T. R., Houdek, G., Brandão, I. M., Creevey, O. L., Doğan, G., Bazot, M., Christensen–Dalsgaard, J., Cunha, M. S., Elsworth, Y., Huber, D., Kjeldsen, H., Lundkvist, M., Molenda-Żakowicz, J., Monteiro, M. J. P. F. G., Stello, D., Clarke, B. D., Girouard, F. R., & Hall, J. R. 2011, Asteroseismology from multi-month Kepler photometry: the evolved Sun-like stars KIC 10273246 and KIC 10920273. *A&A*, 534, A6.

Chaplin, W. J., Basu, S., Huber, D., Serenelli, A., Casagrande, L., Silva Aguirre, V., Ball, W. H., Creevey, O. L., Gizon, L., Handberg, R., Karoff, C., Lutz, R., Marques, J. P., Miglio, A., Stello, D., Suran, M. D., Pricopi, D., Metcalfe, T. S., Monteiro, M. J. P. F. G., Molenda-Żakowicz, J., Appourchaux, T., Christensen–Dalsgaard, J., Elsworth, Y., García, R. A., Houdek, G., Kjeldsen, H., Bonanno, A., Campante, T. L., Corsaro, E., Gaulme,

P., Hekker, S., Mathur, S., Mosser, B., Régulo, C., & Salabert, D. 2014, Asteroseismic Fundamental Properties of Solar-type Stars Observed by the NASA Kepler Mission. *ApJS*, 210, 1.

Chaplin, W. J., Bedding, T. R., Bonanno, A., Broomhall, A.-M., García, R. A., Hekker, S., Huber, D., Verner, G. A., Basu, S., Elsworth, Y., Houdek, G., Mathur, S., Mosser, B., New, R., Stevens, I. R., Appourchaux, T., Karoff, C., Metcalfe, T. S., Molenda-Żakowicz, J., Monteiro, M. J. P. F. G., Thompson, M. J., Christensen–Dalsgaard, J., Gilliland, R. L., Kawaler, S. D., Kjeldsen, H., Ballot, J., Benomar, O., Corsaro, E., Campante, T. L., Gaulme, P., Hale, S. J., Handberg, R., Jarvis, E., Régulo, C., Roxburgh, I. W., Salabert, D., Stello, D., Mullally, F., Li, J., & Wohler, W. 2011, Evidence for the Impact of Stellar Activity on the Detectability of Solar-like Oscillations Observed by Kepler. *ApJL*, 732, L5.

Claverie, A., Isaak, G. R., McLeod, C. P., van der Raay, H. B., & Cortes, T. R. 1979, Solar structure from global studies of the 5-minute oscillation. *Nature*, 282, 591–594.

Creevey, O. L., Metcalfe, T. S., Schultheis, M., Salabert, D., Bazot, M., Thévenin, F., Mathur, S., Xu, H., & García, R. A. 2017, Characterizing solar-type stars from full-length Kepler data sets using the Asteroseismic Modeling Portal. *A&A*, 601, A67.

Davies, G. R., Silva Aguirre, V., Bedding, T. R., Handberg, R., Lund, M. N., Chaplin, W. J., Huber, D., White, T. R., Benomar, O., Hekker, S., Basu, S., Campante, T. L., Christensen–Dalsgaard, J., Elsworth, Y., Karoff, C., Kjeldsen, H., Lundkvist, M. S., Metcalfe, T. S., & Stello, D. 2016, Oscillation frequencies for 35 Kepler solar-type planet-hosting stars using Bayesian techniques and machine learning. *MNRAS*, 456, 2183–2195.

Deheuvels, S., Doğan, G., Goupil, M. J., Appourchaux, T., Benomar, O., Bruntt, H., Campante, T. L., Casagrande, L., Ceillier, T., Davies, G. R., De Cat, P., Fu, J. N., García, R. A., Lobel, A., Mosser, B., Reese, D. R., Regulo, C., Schou, J., Stahn, T., Thygesen, A. O., Yang, X. H., Chaplin, W. J., Christensen–Dalsgaard, J., Eggenberger, P., Gizon, L., Mathis, S., Molenda-Żakowicz, J., & Pinsonneault, M. 2014, Seismic constraints on the radial dependence of the internal rotation profiles of six Kepler subgiants and young red giants. *A&A*, 564, A27.

Deheuvels, S., Li, G., Ballot, J., & Lignières, F. 2023, Strong magnetic fields detected in the cores of 11 red giant stars using gravity-mode period spacings. *A&A*, 670, L16.

Duvall, Jr., T. L., Dziembowski, W. A., Goode, P. R., Gough, D. O., Harvey, J. W., & Leibacher, J. W. 1984, Internal rotation of the sun. *Nature*, 310, 22–25.

Elsworth, Y., Howe, R., Isaak, G. R., McLeod, C. P., & New, R. 1990, Variation of low-order acoustic solar oscillations over the solar cycle. *Nature*, 345, 322–324.

Fuller, J., Cantiello, M., Stello, D., Garcia, R. A., & Bildsten, L. 2015, Asteroseismology can reveal strong internal magnetic fields in red giant stars. *Science*, 350, 423–426.

García, R. A. & Ballot, J. 2019, Asteroseismology of solar-type stars. *Living Reviews in Solar Physics*, 16(1), 4.

García, R. A., Ceillier, T., Salabert, D., Mathur, S., van Saders, J. L., Pinsonneault, M., Ballot, J., Beck, P. G., Bloemen, S., Campante, T. L., Davies, G. R., do Nascimento, Jr., J.-D., Mathis, S., Metcalfe, T. S., Nielsen, M. B., Suárez, J. C., Chaplin, W. J., Jiménez, A., & Karoff, C. 2014,a Rotation and magnetism of Kepler pulsating solar-like stars. Towards asteroseismically calibrated age-rotation relations. *A&A*, 572a, A34.

García, R. A., Mathur, S., Salabert, D., Ballot, J., Régulo, C., Metcalfe, T. S., & Baglin, A. 2010, CoRoT Reveals a Magnetic Activity Cycle in a Sun-Like Star. *Science*, 329, 1032–.

García, R. A., Pérez Hernández, F., Benomar, O., Silva Aguirre, V., Ballot, J., Davies, G. R., Doğan, G., Stello, D., Christensen–Dalsgaard, J., Houdek, G., Lignières, F., Mathur, S., Takata, M., Ceillier, T., Chaplin, W. J., Mathis, S., Mosser, B., Ouazzani, R. M., Pinsonneault, M. H., Reese, D. R., Régulo, C., Salabert, D., Thompson, M. J., van Saders, J. L., Neiner, C., & De Ridder, J. 2014,b Study of KIC 8561221 observed by Kepler: an early red giant showing depressed dipolar modes. *A&A*, 563b, A84.

Gehan, C., Mosser, B., Michel, E., Samadi, R., & Kallinger, T. 2018, Core rotation braking on the red giant branch for various mass ranges. *A&A*, 616, A24.

González-Cuesta, L., Mathur, S., García, R. A., Pérez Hernández, F., Delsanti, V., Breton, S. N., Hedges, C., Jiménez, A., Della Gaspera, A., El–Issami, M., Fox, V., Godoy–Rivera,

D., Pitot, S., & Proust, N. 2023, Multi-campaign asteroseismic analysis of eight solar-like pulsating stars observed by the K2 mission. *A&A*, 674, A106.

Goode, P. R. & Thompson, M. J. 1992, The effect of an inclined magnetic field on solar oscillation frequencies. *ApJ*, 395, 307–315.

Harvey, J. High-Resolution Helioseismology. In Rolfe, E. & Battrick, B., editors, *Future Missions in Solar, Heliospheric & Space Plasma Physics* 1985,, volume 235 of *ESA Special Publication*, 199.

Hatt, E., Nielsen, M. B., Chaplin, W. J., Ball, W. H., Davies, G. R., Bedding, T. R., Buzasi, D. L., Chontos, A., Huber, D., Kayhan, C., Li, Y., White, T. R., Cheng, C., Metcalfe, T. S., & Stello, D. 2023, Catalogue of solar-like oscillators observed by TESS in 120-s and 20-s cadence. *A&A*, 669, A67.

Howell, S. B., Sobeck, C., Haas, M., Still, M., Barclay, T., Mullally, F., Troeltzsch, J., Aigrain, S., Bryson, S. T., Caldwell, D., Chaplin, W. J., Cochran, W. D., Huber, D., Marcy, G. W., Miglio, A., Najita, J. R., Smith, M., Twicken, J. D., & Fortney, J. J. 2014, The K2 Mission: Characterization and Early Results. *PASP*, 126, 398–408.

Huber, D., Chaplin, W. J., Chontos, A., Kjeldsen, H., Christensen–Dalsgaard, J., Bedding, T. R., Ball, W., Brahm, R., Espinoza, N., Henning, T., Jordán, A., Sarkis, P., Knudstrup, E., Albrecht, S., Grundahl, F., Fredslund Andersen, M., Pallé, P. L., Crossfield, I., Fulton, B., Howard, A. W., Isaacson, H. T., Weiss, L. M., Handberg, R., Lund, M. N., Serenelli, A. M., Rørsted Mosumgaard, J., Stokholm, A., Bieryla, A., Buchhave, L. A., Latham, D. W., Quinn, S. N., Gaidos, E., Hirano, T., Ricker, G. R., Vanderspek, R. K., Seager, S., Jenkins, J. M., Winn, J. N., Antia, H. M., Appourchaux, T., Basu, S., Bell, K. J., Benomar, O., Bonanno, A., Buzasi, D. L., Campante, T. L., Çelik Orhan, Z., Corsaro, E., Cunha, M. S., Davies, G. R., Deheuvels, S., Grunblatt, S. K., Hasanzadeh, A., Di Mauro, M. P., García, R. A., Gaulme, P., Girardi, L., Guzik, J. A., Hon, M., Jiang, C., Kallinger, T., Kawaler, S. D., Kuszlewicz, J. S., Lebreton, Y., Li, T., Lucas, M., Lundkvist, M. S., Mann, A. W., Mathis, S., Mathur, S., Mazumdar, A., Metcalfe, T. S., Miglio, A., Monteiro, M. J. P. F. G., Mosser, B., Noll, A., Nsamba, B., Ong, J. M. J., Örtel, S., Pereira, F., Ranadive, P., Régulo, C., Rodrigues, T. S., Roxburgh, I. W., Silva Aguirre, V., Smalley, B., Schofield, M., Sousa, S. G., Stassun, K. G., Stello, D., Tayar, J., White, T. R., Verma, K., Vrard, M., Yıldız, M., Baker, D., Bazot, M., Beichmann, C., Bergmann, C., Bugnet, L., Cale, B., Carlino, R., Cartwright, S. M., Christiansen, J. L., Ciardi, D. R., Creevey, O., Dittmann, J. A., Do Nascimento, Jr., J.-D., Van Eylen, V., Fürész, G., Gagné, J., Gao, P., Gazeas, K., Giddens, F., Hall, O. J., Hekker, S., Ireland, M. J., Latouf, N., LeBrun, D., Levine, A. M., Matzko, W., Natinsky, E., Page, E., Plavchan, P., Mansouri–Samani, M., McCauliff, S., Mullally, S. E., Orenstein, B., Garcia Soto, A., Paegert, M., van Saders, J. L., Schnaible, C., Soderblom, D. R., Szabó, R., Tanner, A., Tinney, C. G., Teske, J., Thomas, A., Trampedach, R., Wright, D., Yuan, T. T., & Zohrabi, F. 2019, A Hot Saturn Orbiting an Oscillating Late Subgiant Discovered by TESS. *AJ*, 157, 245.

Huber, D., Pinsonneault, M., Beck, P., Bedding, T. R., Bland–Hawthorn, J., Breton, S. N., Bugnet, L., Chaplin, W. J., Garcia, R. A., Grunblatt, S. K., Guzik, J. A., Hekker, S., Kawaler, S. D., Mathis, S., Mathur, S., Metcalfe, T., Mosser, B., Ness, M. K., Piro, A. L., Serenelli, A., Sharma, S., Soderblom, D. R., Stassun, K. G., Stello, D., Tayar, J., van Belle, G. T., & Zinn, J. C. 2023, Asteroseismology with the Roman Galactic Bulge Time-Domain Survey. *arXiv e-prints*,, arXiv:2307.03237.

Huber, D., White, T. R., Metcalfe, T. S., Chontos, A., Fausnaugh, M. M., Ho, C. S. K., Van Eylen, V., Ball, W. H., Basu, S., Bedding, T. R., Benomar, O., Bossini, D., Breton, S., Buzasi, D. L., Campante, T. L., Chaplin, W. J., Christensen–Dalsgaard, J., Cunha, M. S., Deal, M., García, R. A., García Muñoz, A., Gehan, C., González-Cuesta, L., Jiang, C., Kayhan, C., Kjeldsen, H., Lundkvist, M. S., Mathis, S., Mathur, S., Monteiro, M. J. P. F. G., Nsamba, B., Ong, J. M. J., Pakštienė, E., Serenelli, A. M., Silva Aguirre, V., Stassun, K. G., Stello, D., Norgaard Stilling, S., Lykke Winther, M., Wu, T., Barclay, T., Daylan, T., Günther, M. N., Hermes, J. J., Jenkins, J. M., Latham, D. W., Levine, A. M., Ricker, G. R., Seager, S., Shporer, A., Twicken, J. D., Vanderspek, R. K., & Winn, J. N. 2022, A 20

Second Cadence View of Solar-type Stars and Their Planets with TESS: Asteroseismology of Solar Analogs and a Recharacterization of π Men c. *AJ*, 163(2), 79.

Jain, K., Tripathy, S. C., & Hill, F. 2009, Solar Activity Phases and Intermediate-Degree Mode Frequencies. *ApJ*, 695(2), 1567–1576.

Jenkins, J. M., Caldwell, D. A., Chandrasekaran, H., Twicken, J. D., Bryson, S. T., Quintana, E. V., Clarke, B. D., Li, J., Allen, C., Tenenbaum, P., Wu, H., Klaus, T. C., Van Cleve, J., Dotson, J. A., Haas, M. R., Gilliland, R. L., Koch, D. G., & Borucki, W. J. 2010, Initial Characteristics of Kepler Long Cadence Data for Detecting Transiting Planets. *ApJL*, 713, L120–L125.

Kallinger, T., De Ridder, J., Hekker, S., Mathur, S., Mosser, B., Gruberbauer, M., García, R. A., Karoff, C., & Ballot, J. 2014, The connection between stellar granulation and oscillation as seen by the Kepler mission. *A&A*, 570, A41.

Karoff, C., Metcalfe, T. S., Santos, Â. R. G., Montet, B. T., Isaacson, H., Witzke, V., Shapiro, A. I., Mathur, S., Davies, G. R., Lund, M. N., Garcia, R. A., Brun, A. S., Salabert, D., Avelino, P. P., van Saders, J., Egeland, R., Cunha, M. S., Campante, T. L., Chaplin, W. J., Krivova, N., Solanki, S. K., Stritzinger, M., & Knudsen, M. F. 2018, The Influence of Metallicity on Stellar Differential Rotation and Magnetic Activity. *ApJ*, 852, 46.

Kiefer, R., Schad, A., Davies, G., & Roth, M. 2017, Stellar magnetic activity and variability of oscillation parameters: An investigation of 24 solar-like stars observed by Kepler. *A&A*, 598, A77.

Lebreton, Y. & Goupil, M. J. 2014, Asteroseismology for "à la carte" stellar age-dating and weighing. Age and mass of the CoRoT exoplanet host HD 52265. *A&A*, 569, A21.

Li, G., Deheuvels, S., Ballot, J., & Lignières, F. 2022, Magnetic fields of 30 to 100 kG in the cores of red giant stars. *Nature*, 610(7930), 43–46.

Loi, S. T. 2021, Topology and obliquity of core magnetic fields in shaping seismic properties of slowly rotating evolved stars. *MNRAS*, 504(3), 3711–3729.

Lund, M. N., Chaplin, W. J., Casagrande, L., Silva Aguirre, V., Basu, S., Bieryla, A., Christensen–Dalsgaard, J., Latham, D. W., White, T. R., Davies, G. R., Huber, D., Buchhave, L. A., & Handberg, R. 2016, Asteroseismic Properties of Solar-type Stars Observed with the NASA K2 Mission: Results from Campaigns 1-3 and Prospects for Future Observations. *PASP*, 128(12), 124204.

Lund, M. N., Miesch, M. S., & Christensen–Dalsgaard, J. 2014, Differential Rotation in Main-sequence Solar-like Stars: Qualitative Inference from Asteroseismic Data. *ApJ*, 790(2), 121.

Lund, M. N., Silva Aguirre, V., Davies, G. R., Chaplin, W. J., Christensen–Dalsgaard, J., Houdek, G., White, T. R., Bedding, T. R., Ball, W. H., Huber, D., Antia, H. M., Lebreton, Y., Latham, D. W., Handberg, R., Verma, K., Basu, S., Casagrande, L., Justesen, A. B., Kjeldsen, H., & Mosumgaard, J. R. 2017, Standing on the Shoulders of Dwarfs: the Kepler Asteroseismic LEGACY Sample. I. Oscillation Mode Parameters. *ApJ*, 835, 172.

Mathis, S., Bugnet, L., Prat, V., Augustson, K., Mathur, S., & Garcia, R. A. 2021, Probing the internal magnetism of stars using asymptotic magneto-asteroseismology. *A&A*, 647, A122.

Mathur, S., García, R. A., Ballot, J., Ceillier, T., Salabert, D., Metcalfe, T. S., Régulo, C., Jiménez, A., & Bloemen, S. 2014, Magnetic activity of F stars observed by Kepler. *A&A*, 562, A124.

Mathur, S., García, R. A., Breton, S., Santos, A. R. G., Mosser, B., Huber, D., Sayeed, M., Bugnet, L., & Chontos, A. 2022, Detections of solar-like oscillations in dwarfs and subgiants with Kepler DR25 short-cadence data. *A&A*, 657, A31.

Mathur, S., García, R. A., Bugnet, L., Santos, Â. R. G., Santiago, N., & Beck, P. G. 2019, Revisiting the impact of stellar magnetic activity on the detection of solar-like oscillations by Kepler. *Frontiers in Astronomy and Space Sciences*, 6, 46.

Mathur, S., Handberg, R., Campante, T. L., García, R. A., Appourchaux, T., Bedding, T. R., Mosser, B., Chaplin, W. J., Ballot, J., Benomar, O., Bonanno, A., Corsaro, E., Gaulme, P., Hekker, S., Régulo, C., Salabert, D., Verner, G., White, T. R., Brandão, I. M., Creevey, O. L., Doğan, G., Elsworth, Y., Huber, D., Hale, S. J., Houdek, G., Karoff, C., Metcalfe,

T. S., Molenda-Żakowicz, J., Monteiro, M. J. P. F. G., Thompson, M. J., Christensen–Dalsgaard, J., Gilliland, R. L., Kawaler, S. D., Kjeldsen, H., Quintana, E. V., Sanderfer, D. T., & Seader, S. E. 2011,a Solar-like Oscillations in KIC 11395018 and KIC 11234888 from 8 Months of Kepler Data. *ApJ*, 733a, 95–+.

Mathur, S., Hekker, S., Trampedach, R., Ballot, J., Kallinger, T., Buzasi, D., García, R. A., Huber, D., Jiménez, A., Mosser, B., Bedding, T. R., Elsworth, Y., Régulo, C., Stello, D., Chaplin, W. J., De Ridder, J., Hale, S. J., Kinemuchi, K., Kjeldsen, H., Mullally, F., & Thompson, S. E. 2011,b Granulation in Red Giants: Observations by the Kepler Mission and Three-dimensional Convection Simulations. *ApJ*, 741b, 119.

Mathur, S., Metcalfe, T. S., Woitaszek, M., Bruntt, H., Verner, G. A., Christensen–Dalsgaard, J., Creevey, O. L., Doğan, G., Basu, S., Karoff, C., Stello, D., Appourchaux, T., Campante, T. L., Chaplin, W. J., García, R. A., Bedding, T. R., Benomar, O., Bonanno, A., Deheuvels, S., Elsworth, Y., Gaulme, P., Guzik, J. A., Handberg, R., Hekker, S., Herzberg, W., Monteiro, M. J. P. F. G., Piau, L., Quirion, P.-O., Régulo, C., Roth, M., Salabert, D., Serenelli, A., Thompson, M. J., Trampedach, R., White, T. R., Ballot, J., Brandão, I. M., Molenda-Żakowicz, J., Kjeldsen, H., Twicken, J. D., Uddin, K., & Wohler, B. 2012, A Uniform Asteroseismic Analysis of 22 Solar-type Stars Observed by Kepler. *ApJ*, 749, 152.

McQuillan, A., Mazeh, T., & Aigrain, S. 2014, Rotation Periods of 34,030 Kepler Main-sequence Stars: The Full Autocorrelation Sample. *ApJS*, 211, 24.

Metcalfe, T. S., Chaplin, W. J., Appourchaux, T., García, R. A., Basu, S., Brandão, I., Creevey, O. L., Deheuvels, S., Doğan, G., Eggenberger, P., Karoff, C., Miglio, A., Stello, D., Yıldız, M., Çelik, Z., Antia, H. M., Benomar, O., Howe, R., Régulo, C., Salabert, D., Stahn, T., Bedding, T. R., Davies, G. R., Elsworth, Y., Gizon, L., Hekker, S., Mathur, S., Mosser, B., Bryson, S. T., Still, M. D., Christensen–Dalsgaard, J., Gilliland, R. L., Kawaler, S. D., Kjeldsen, H., Ibrahim, K. A., Klaus, T. C., & Li, J. 2012, Asteroseismology of the Solar Analogs 16 Cyg A and B from Kepler Observations. *ApJL*, 748, L10.

Metcalfe, T. S., Creevey, O. L., Doğan, G., Mathur, S., Xu, H., Bedding, T. R., Chaplin, W. J., Christensen–Dalsgaard, J., Karoff, C., Trampedach, R., Benomar, O., Brown, B. P., Buzasi, D. L., Campante, T. L., Çelik, Z., Cunha, M. S., Davies, G. R., Deheuvels, S., Derekas, A., Di Mauro, M. P., García, R. A., Guzik, J. A., Howe, R., MacGregor, K. B., Mazumdar, A., Montalbán, J., Monteiro, M. J. P. F. G., Salabert, D., Serenelli, A., Stello, D., Steslicki, M., Suran, M. D., Yıldız, M., Aksoy, C., Elsworth, Y., Gruberbauer, M., Guenther, D. B., Lebreton, Y., Molaverdikhani, K., Pricopi, D., Simoniello, R., & White, T. R. 2014, Properties of 42 Solar-type Kepler Targets from the Asteroseismic Modeling Portal. *ApJS*, 214, 27.

Metcalfe, T. S., van Saders, J. L., Basu, S., Buzasi, D., Chaplin, W. J., Egeland, R., Garcia, R. A., Gaulme, P., Huber, D., Reinhold, T., Schunker, H., Stassun, K. G., Appourchaux, T., Ball, W. H., Bedding, T. R., Deheuvels, S., González-Cuesta, L., Handberg, R., Jiménez, A., Kjeldsen, H., Li, T., Lund, M. N., Mathur, S., Mosser, B., Nielsen, M. B., Noll, A., Çelik Orhan, Z., Örtel, S., Santos, Â. R. G., Yildiz, M., Baliunas, S., & Soon, W. 2020, The Evolution of Rotation and Magnetic Activity in 94 Aqr Aa from Asteroseismology with TESS. *ApJ*, 900(2), 154.

Mosser, B., Elsworth, Y., Hekker, S., Huber, D., Kallinger, T., Mathur, S., Belkacem, K., Goupil, M. J., Samadi, R., Barban, C., Bedding, T. R., Chaplin, W. J., García, R. A., Stello, D., De Ridder, J., Middour, C. K., Morris, R. L., & Quintana, E. V. 2012,a Characterization of the power excess of solar-like oscillations in red giants with Kepler. *A&A*, 537a, A30.

Mosser, B., Goupil, M. J., Belkacem, K., Marques, J. P., Beck, P. G., Bloemen, S., De Ridder, J., Barban, C., Deheuvels, S., Elsworth, Y., Hekker, S., Kallinger, T., Ouazzani, R. M., Pinsonneault, M., Samadi, R., Stello, D., García, R. A., Klaus, T. C., Li, J., Mathur, S., & Morris, R. L. 2012,b Spin down of the core rotation in red giants. *A&A*, 548b, A10.

Nielsen, M. B., Gizon, L., Schunker, H., & Karoff, C. 2013, Rotation periods of 12 000 main-sequence Kepler stars: Dependence on stellar spectral type and comparison with v sin i observations. *A&A*, 557, L10.

Ong, J. M. J., Basu, S., Lund, M. N., Bieryla, A., Viani, L. S., & Latham, D. W. 2021, Mixed Modes and Asteroseismic Surface Effects. II. Subgiant Systematics. *ApJ*, 922(1), 18.

Rauer, H., Catala, C., Aerts, C., Appourchaux, T., Benz, W., Brandeker, A., Christensen-Dalsgaard, J., Deleuil, M., Gizon, L., Goupil, M.-J., Güdel, M., Janot-Pacheco, E., Mas-Hesse, M., Pagano, I., Piotto, G., Pollacco, D., Santos, Ċ., Smith, A., Suárez, J.-C., Szabó, R., Udry, S., Adibekyan, V., Alibert, Y., Almenara, J.-M., Amaro-Seoane, P., Eiff, M. A.-v., Asplund, M., Antonello, E., Barnes, S., Baudin, F., Belkacem, K., Bergemann, M., Bihain, G., Birch, A. C., Bonfils, X., Boisse, I., Bonomo, A. S., Borsa, F., Brandão, I. M., Brocato, E., Brun, S., Burleigh, M., Burston, R., Cabrera, J., Cassisi, S., Chaplin, W., Charpinet, S., Chiappini, C., Church, R. P., Csizmadia, S., Cunha, M., Damasso, M., Davies, M. B., Deeg, H. J., Díaz, R. F., Dreizler, S., Dreyer, C., Eggenberger, P., Ehrenreich, D., Eigmüller, P., Erikson, A., Farmer, R., Feltzing, S., de Oliveira Fialho, F., Figueira, P., Forveille, T., Fridlund, M., García, R. A., Giommi, P., Giuffrida, G., Godolt, M., Gomes da Silva, J., Granzer, T., Grenfell, J. L., Grotsch-Noels, A., Günther, E., Haswell, C. A., Hatzes, A. P., Hébrard, G., Hekker, S., Helled, R., Heng, K., Jenkins, J. M., Johansen, A., Khodachenko, M. L., Kislyakova, K. G., Kley, W., Kolb, U., Krivova, N., Kupka, F., Lammer, H., Lanza, A. F., Lebreton, Y., Magrin, D., Marcos-Arenal, P., Marrese, P. M., Marques, J. P., Martins, J., Mathis, S., Mathur, S., Messina, S., Miglio, A., Montalban, J., Montalto, M., Monteiro, M. J. P. F. G., Moradi, H., Moravveji, E., Mordasini, C., Morel, T., Mortier, A., Nascimbeni, V., Nelson, R. P., Nielsen, M. B., Noack, L., Norton, A. J., Ofir, A., Oshagh, M., Ouazzani, R.-M., Pápics, P., Parro, V. C., Petit, P., Plez, B., Poretti, E., Quirrenbach, A., Ragazzoni, R., Raimondo, G., Rainer, M., Reese, D. R., Redmer, R., Reffert, S., Rojas-Ayala, B., Roxburgh, I. W., Salmon, S., Santerne, A., Schneider, J., Schou, J., Schuh, S., Schunker, H., Silva-Valio, A., Silvotti, R., Skillen, I., Snellen, I., Sohl, F., Sousa, S. G., Sozzetti, A., Stello, D., Strassmeier, K. G., Švanda, M., Szabó, G. M., Tkachenko, A., Valencia, D., Van Grootel, V., Vauclair, S. D., Ventura, P., Wagner, F. W., Walton, N. A., Weingrill, J., Werner, S. C., Wheatley, P. J., & Zwintz, K. 2014, The PLATO 2.0 mission. *Experimental Astronomy*, 38, 249–330.

Reinhold, T., Reiners, A., & Basri, G. 2013, Rotation and differential rotation of active Kepler stars. *A&A*, 560, A4.

Ricker, G. R., Winn, J. N., Vanderspek, R., Latham, D. W., Bakos, G. Á., Bean, J. L., Berta-Thompson, Z. K., Brown, T. M., Buchhave, L., Butler, N. R., Butler, R. P., Chaplin, W. J., Charbonneau, D., Christensen-Dalsgaard, J., Clampin, M., Deming, D., Doty, J., De Lee, N., Dressing, C., Dunham, E. W., Endl, M., Fressin, F., Ge, J., Henning, T., Holman, M. J., Howard, A. W., Ida, S., Jenkins, J. M., Jernigan, G., Johnson, J. A., Kaltenegger, L., Kawai, N., Kjeldsen, H., Laughlin, G., Levine, A. M., Lin, D., Lissauer, J. J., MacQueen, P., Marcy, G., McCullough, P. R., Morton, T. D., Narita, N., Paegert, M., Palle, E., Pepe, F., Pepper, J., Quirrenbach, A., Rinehart, S. A., Sasselov, D., Sato, B., Seager, S., Sozzetti, A., Stassun, K. G., Sullivan, P., Szentgyorgyi, A., Torres, G., Udry, S., & Villasenor, J. 2015, Transiting Exoplanet Survey Satellite (TESS). *Journal of Astronomical Telescopes, Instruments, and Systems*, 1(1), 014003.

Salabert, D., Régulo, C., García, R. A., Beck, P. G., Ballot, J., Creevey, O. L., Pérez Hernández, F., do Nascimento, Jr., J.-D., Corsaro, E., Egeland, R., Mathur, S., Metcalfe, T. S., Bigot, L., Ceillier, T., & Pallé, P. L. 2016, Magnetic variability in the young solar analog KIC 10644253. Observations from the Kepler satellite and the HERMES spectrograph. *A&A*, 589, A118.

Santos, A. R. G., Campante, T. L., Chaplin, W. J., Cunha, M. S., Lund, M. N., Kiefer, R., Salabert, D., García, R. A., Davies, G. R., Elsworth, Y., & Howe, R. 2018, Signatures of Magnetic Activity in the Seismic Data of Solar-type Stars Observed by Kepler. *ApJS*, 237, 17.

Santos, A. R. G., Campante, T. L., Chaplin, W. J., Cunha, M. S., van Saders, J. L., Karoff, C., Metcalfe, T. S., Mathur, S., García, R. A., Lund, M. N., Kiefer, R., Silva Aguirre, V., Davies, G. R., Howe, R., & Elsworth, Y. 2019, Signatures of Magnetic Activity: On the Relation between Stellar Properties and p-mode Frequency Variations. *ApJ*, 883(1), 65.

Schunker, H., Schou, J., Ball, W. H., Nielsen, M. B., & Gizon, L. 2016, Asteroseismic inversions for radial differential rotation of Sun-like stars: ensemble fits. *A&A*, 586, A79.

Silva Aguirre, V., Davies, G. R., Basu, S., Christensen–Dalsgaard, J., Creevey, O., Metcalfe, T. S., Bedding, T. R., Casagrande, L., Handberg, R., Lund, M. N., Nissen, P. E., Chaplin, W. J., Huber, D., Serenelli, A. M., Stello, D., Van Eylen, V., Campante, T. L., Elsworth, Y., Gilliland, R. L., Hekker, S., Karoff, C., Kawaler, S. D., Kjeldsen, H., & Lundkvist, M. S. 2015, Ages and fundamental properties of Kepler exoplanet host stars from asteroseismology. *MNRAS*, 452, 2127–2148.

Silva Aguirre, V., Lund, M. N., Antia, H. M., Ball, W. H., Basu, S., Christensen–Dalsgaard, J., Lebreton, Y., Reese, D. R., Verma, K., Casagrande, L., Justesen, A. B., Mosumgaard, J. R., Chaplin, W. J., Bedding, T. R., Davies, G. R., Handberg, R., Houdek, G., Huber, D., Kjeldsen, H., Latham, D. W., White, T. R., Coelho, H. R., Miglio, A., & Rendle, B. 2017, Standing on the Shoulders of Dwarfs: the Kepler Asteroseismic LEGACY Sample. II.Radii, Masses, and Ages. *ApJ*, 835, 173.

Skumanich, A. 1972, Time Scales for CA II Emission Decay, Rotational Braking, and Lithium Depletion. *ApJ*, 171, 565.

Stello, D., Cantiello, M., Fuller, J., Huber, D., García, R. A., Bedding, T. R., Bildsten, L., & Silva Aguirre, V. 2016, A prevalence of dynamo-generated magnetic fields in the cores of intermediate-mass stars. *Nature*, 529, 364–367.

Trampedach, R. & Stein, R. F. 2011, The Mass Mixing Length in Convective Stellar Envelopes. *ApJ*, 731, 78–+.

van Saders, J. L., Ceillier, T., Metcalfe, T. S., Aguirre, V. S., Pinsonneault, M. H., García, R. A., Mathur, S., & Davies, G. R. 2016, Weakened magnetic braking as the origin of anomalously rapid rotation in old field stars. *Nature*, 529, 181–184.

Wilson, O. C. 1978, Chromospheric variations in main-sequence stars. *ApJ*, 226, 379–396.

Wilson, R. F., Barclay, T., Powell, B. P., Schlieder, J., Hedges, C., Montet, B. T., Quintana, E., Mcdonald, I., Penny, M. T., Espinoza, N., & Kerins, E. 2023, Transiting Exoplanet Yields for the Roman Galactic Bulge Time Domain Survey Predicted from Pixel-level Simulations. *ApJS*, 269(1), 5.

Woodard, M. F. & Noyes, R. W. 1985, Change of solar oscillation eigenfrequencies with the solar cycle. *Nature*, 318, 449–450.

Dynamics of Solar and Stellar Convection Zones and Atmospheres
Proceedings IAU Symposium No. 365, 2024
A. V. Getling & L. L. Kitchatinov, eds.
doi:10.1017/S1743921324000425

Solar Inertial Modes

Laurent Gizon[1,2], Yuto Bekki[1], Aaron C. Birch[1], Robert H. Cameron[1], Damien Fournier[1], Jordan Philidet[1], B. Lekshmi[1] and Zhi–Chao Liang[1]

[1]Max–Planck-Institut für Sonnensystemforschung, Göttingen, Germany

[2]Institut für Astrophysik und Geophysik, Georg–August-Universität Göttingen, Germany
email: gizon@mps.mpg.de

Abstract. The Sun's global inertial modes are very sensitive to the solar differential rotation and to properties of the deep solar convection zone which are currently poorly constrained. These properties include the superadiabatic temperature gradient, the latitudinal entropy gradient, and the turbulent viscosity. The inertial modes also play a key role in controlling the Sun's large-scale structure and dynamics, in particular the solar differential rotation. This paper summarizes recent observations and advances in the (linear and nonlinear) modeling of the solar inertial modes.

Keywords. Solar inertial modes, solar rotation, solar convection

1. Introduction

The modes of stellar oscillation are broadly classified into two categories (e.g., Unno et al. 1989). The *spheroidal modes* of oscillation are associated with both horizontal and radial motions and include the acoustic and the gravity modes. In the presence of rotation, *toroidal modes* of oscillation can also exist. The toroidal modes are associated with horizontal motion and are described by their radial vorticity eigenfunctions. The toroidal modes have frequencies in the inertial frequency range, i.e. of order the rotation frequency of the star. In the special case of uniform rotation, the toroidal modes are retrograde Rossby modes restored by the Coriolis force.

The above classification is a simplification: the actual modes of solar oscillation in the inertial frequency range which have recently been observed are more complicated due to the Sun's differential rotation and, for example, the superadiabatic stratification of its convection zone. In this paper we present an overview of recent work on these hydrodynamic modes. Our aim is to outline progress in the observation, interpretation, and diagnostic potential of these modes. We also present preliminary observations of solar-cycle variations in the mode parameters.

The new field of *inertial-mode helioseismology* has emerged out of three key observational papers: (i) Löptien et al. (2018) , (ii) Gizon et al. (2021) , and (iii) Hanson et al. (2022). The first paper reported the detection of equatorial (sectoral) Rossby modes on the Sun. The second, the detection and identification of additional modes which owe their existence to the differential rotation, among which the $m = 1$ high-latitude mode and other critical latitude modes. The third paper reported modes at mid latitudes with north-south antisymmetric vorticity. The above papers are complemented by a number of useful observational studies by Liang et al. (2019), Proxauf et al. (2020), Mandal et al. (2021), to name a few.

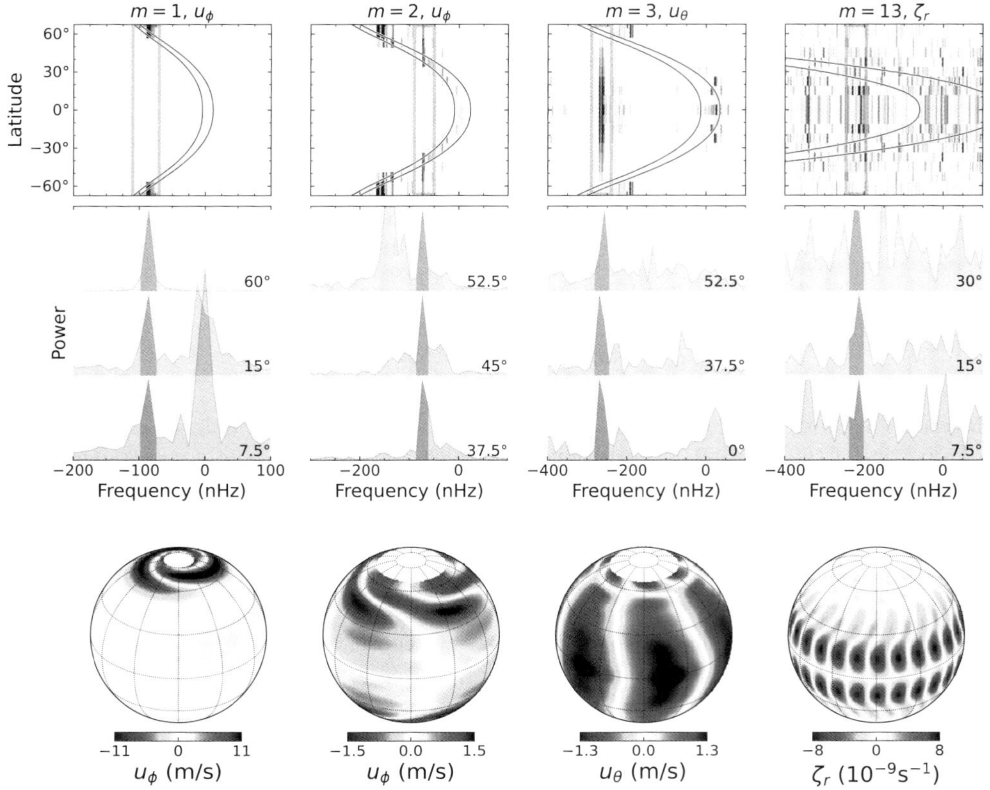

Figure 1. Example inertial modes observed in the power spectra of near-surface horizontal flow components at different latitudes, and the corresponding observed eigenfunctions. These modes are among many other modes observed at any given value of m. The four columns show (from left to right) the $m = 1$ high-latitude mode at -86 nHz (Gizon et al. 2021), the $m = 2$ mid-latitude mode at -73 nHz (Gizon et al. 2021), the $m = 3$ equatorial (sectoral) Rossby mode at -269 nHz (Löptien et al. 2018), and the $m = 13$ high-frequency retrograde mode at -220 nHz (Hanson et al. 2022). The data were analysed in the Carrington frame of reference (a negative frequency indicates retrograde propagation in this frame). For clarity, the power spectra at fixed latitudes (middle row) have a reduced frequency resolution of 12 nHz. The first three modes from the left have u_ϕ antisymmetric across the equator and u_θ is symmetric, while the fourth mode on the right has antisymmetric radial vorticity ζ_r.

Using numerical modeling, the solar inertial modes have been shown to be important not only to probe properties of the deep convection zone that are hardly accessible to the acoustic modes (incl. superadiabaticity and turbulent diffusivities, see Gizon et al. 2021), but also in terms of their dynamical effect on the axisymmetric flows (Bekki et al. 2024). Like acoustic modes, most inertial modes are linearly stable and stochastically excited by turbulent convection (Philidet and Gizon 2023). Some inertial modes, however, are linearly unstable and reach relatively large velocity amplitudes at high latitudes (Fournier et al. 2022; Bekki et al. 2022b).

2. Observations of global modes of oscillation in the inertial frequency range

Figure 1 shows low-frequency power spectra of solar oscillations for four particular values of the azimuthal wavenumber m. The power spectrum is computed for each colatitude

θ according to

$$P(\omega,\theta)=\left|\sum_{\phi,t}u_j(\theta,\phi,t)e^{-\mathrm{i}(m\phi-\omega t)}\right|^2, \tag{1}$$

where u_j is either the colatitudinal component of velocity (u_θ) or the longitudinal component (u_ϕ) inferred from ring-diagram analysis near the solar surface applied to 10 years of HMI observations. The top row shows excess power corresponding to four selected modes, including the $m=1$ high-latitude mode identified by Gizon et al. (2021) and the $m=13$ mode reported by Hanson et al. (2022). As shown in the second row, the excess power associated with a mode is visible at the same frequency at multiple latitudes. We are thus in the presence of global modes of oscillation. The last row of plots displays the eigenfunctions in either u_ϕ, u_θ or the radial vorticity ζ_r extracted from narrow frequency windows centered on the mode frequencies. The typical mode lifetime is half a year (for a 20 nHz linewidth). Most of the modes that have been detected have velocity amplitudes in the range 0.5–2 m/s. The largest amplitude mode is the $m=1$ high-latitude mode, with a velocity reaching ~ 15 m/s on average and up to 20 m/s during quiet Sun periods. The modes are quasi-toroidal, i.e. their motion is dominantly horizontal and the two components of velocity can be described via a streamfunction. The modes reported by Hanson et al. (2022) have low amplitudes of a few 0.1 m/s.

3. 2D Linear analysis: families of toroidal modes

Before looking at the full problem, it is useful to consider the much simpler problem of toroidal modes on a spherical surface. Such modes are described in terms of their stream function

$$\Psi=\psi(\theta)e^{\mathrm{i}(m\phi-\omega t)}. \tag{2}$$

We work in a frame that rotates with the Sun at the rate Ω_0 (either the Carrington rotation rate or the equatorial rotation rate). In this rotating frame, the governing equation follows from the radial component of the curl of the linearized Navier–Stokes equation (Fournier et al. 2022):

$$(\omega-m\Delta\Omega)\,L_m\psi-m(2\Omega_0-\Delta\Omega'')\psi=\mathrm{i}E\Omega_0L_m^2\psi, \tag{3}$$

where the quantity $\Delta\Omega(\theta)=\Omega(\theta)-\Omega_0$ is the latitudinal differential rotation measured in the rotating frame, L_m is the Laplace-Beltrami operator on the unit sphere, $\Delta\Omega''=\frac{1}{\sin\theta}\frac{d}{d\theta}\left(\frac{1}{\sin\theta}\frac{d}{d\theta}\left(\Delta\Omega\sin^2\theta\right)\right)$, and $E=\nu/(r^2\Omega_0)$ is the Ekman number. A nonzero E accounts for the turbulent viscosity ν, which is an essential ingredient of the problem.

3.1. *Uniform rotation*

In the special case of uniform rotation ($\Omega=\Omega_0$, $\Delta\Omega=0$), we obtain the classical Rossby modes with

$$\psi(\theta)=P_l^m(\cos\theta)\quad\text{and}\quad\omega=-\frac{2m\Omega_0}{l(l+1)}-\mathrm{i}l(l+1)E\Omega_0, \tag{4}$$

where the P_l^m are associated Legendre polynomial of degree l and azimuthal order m. The sectoral ($l=m$) modes are observed on the Sun to have frequencies close to the model frequencies $\Re[\omega]=-2\Omega_0/(m+1)$ (Löptien et al. 2018). The condition $E\lesssim 2\times10^{-3}$ follows from the observed mode lifetimes (Philidet and Gizon 2023), implying a turbulent viscosity $\nu\lesssim 2\times10^{13}$ cm^2/s of order the mixing length estimate (Muñoz-Jaramillo et al. 2011). The observed eigenfunctions are close to $P_m^m(\cos\theta)\propto(\sin\theta)^m$, but with notable differences due to the presence of critical latitudes (Gizon et al. 2020), as we now discuss.

3.2. *Importance of differential rotation: critical latitudes*

Differential rotation means the rotation rate of the Sun is a function $\Omega(\theta, r)$ of colatitude θ and radius r. The factor $(\omega - m\Delta\Omega)$ in the first term of Eq. (3) leads to the presence of critical latitudes θ_c where

$$\frac{\omega}{m} = \Delta\Omega(\theta_c, r) \quad \text{in the corotating frame.} \tag{5}$$

In the absence of viscosity, we have a singular eigenvalue problem and the eigenfunctions are not regular at the critical latitudes (see, e.g., Balmforth and Morrison 1995). See also Watson (1981) and Charbonneau et al. (1999) for a discussion of stability when $E = 0$. The inclusion of viscosity in Eq. (3) leads to a regular eigenvalue problem of the Orr–Sommerfeld type (fourth-order eigenvalue problem). The top panels of Fig. 2 show the corresponding eigenvalues for $E = 4 \times 10^{-4}$ and in the case of a smooth differential rotation profile $\Delta\Omega(\theta) = -\Omega_2 \cos^2\theta$ that approximates the Sun's angular velocity at low latitudes. For these parameters, all the modes are linearly stable and belong to three families of solutions in addition to the Rossby modes: the high-latitude modes, the critical-latitude (or center) modes, and the highly-damped modes, which are analogous to the A, P and S families of modes of a plane parabolic shear flow (see Drazin and Reid 2004; Gizon et al. 2020). The lower panels of Fig. 2 show the modes when the observed surface differential rotation of the Sun is used instead of the two-term profile. Because the solar rotation rate drops sharply at high latitudes, an unstable mode appears in each of the $m = 1$ and $m = 2$ spectra. The number of unstable modes as a function of the Ekman number was discussed by Fournier et al. (2022). The $m = 1$ mode R_2 with a streamfunction that resembles $P_3^1(\cos\theta)$ is unstable and has a frequency very close to that of the largest amplitude mode in the observations (see Fig. 1, left column). The high-latitude mode A_2 has a similar frequency but is linearly stable. Neither mode has an eigenfunction that matches the observed spiral pattern — this requires 3D modeling and additional physics, as explained in the next section. The observed $m = 2$ critical-latitude mode is in the range of frequencies corresponding to the P family (also see figure 2 in Gizon et al. 2021). The frequencies and eigenfunctions of the equatorial Rossby mode are satisfactorily reproduced in 2D, including their departure from sectoral spherical harmonics (Gizon et al. 2020; Fournier et al. 2022). The 2D model however does not give the modes reported by Hanson et al. (2022) because these modes are not purely toroidal, see Sect. 7.

4. Linear analysis in three dimensions

4.1. *Sectoral Rossby modes*

For the case of uniform rotation in three dimensions, the only modes that can easily be described analytically are the sectoral Rossby modes, which are purely toroidal. Their streamfunction separates, $\Psi = f(r)(\sin\theta)^m e^{\mathrm{i}m\phi + 2\mathrm{i}\Omega t/(m+1)}$. Combining the radial and latitudinal components of the equation of motion, we obtain the radial dependence $f(r) \propto r^{m+1}$. Proxauf et al. (2020) find that this model is not inconsistent with the radial dependence inferred from ring-diagram helioseismology, although only the top 8 Mm could be studied with confidence. For a more sophisticated discussion of the sectoral Rossby modes, see Provost et al. (1981) and Damiani et al. (2020).

4.2. *Quasi-toroidal modes of the solar convection zone*

The Sun has radial structure and its inertial modes are not separable in terms of radius and latitude (e.g., Baruteau and Rieutord 2013). To study the linear stability of the eigenmodes of oscillation in the convection zone, we consider the linearized hydrodynamic equations (conservation of mass, momentum, and energy). Here we outline the linear

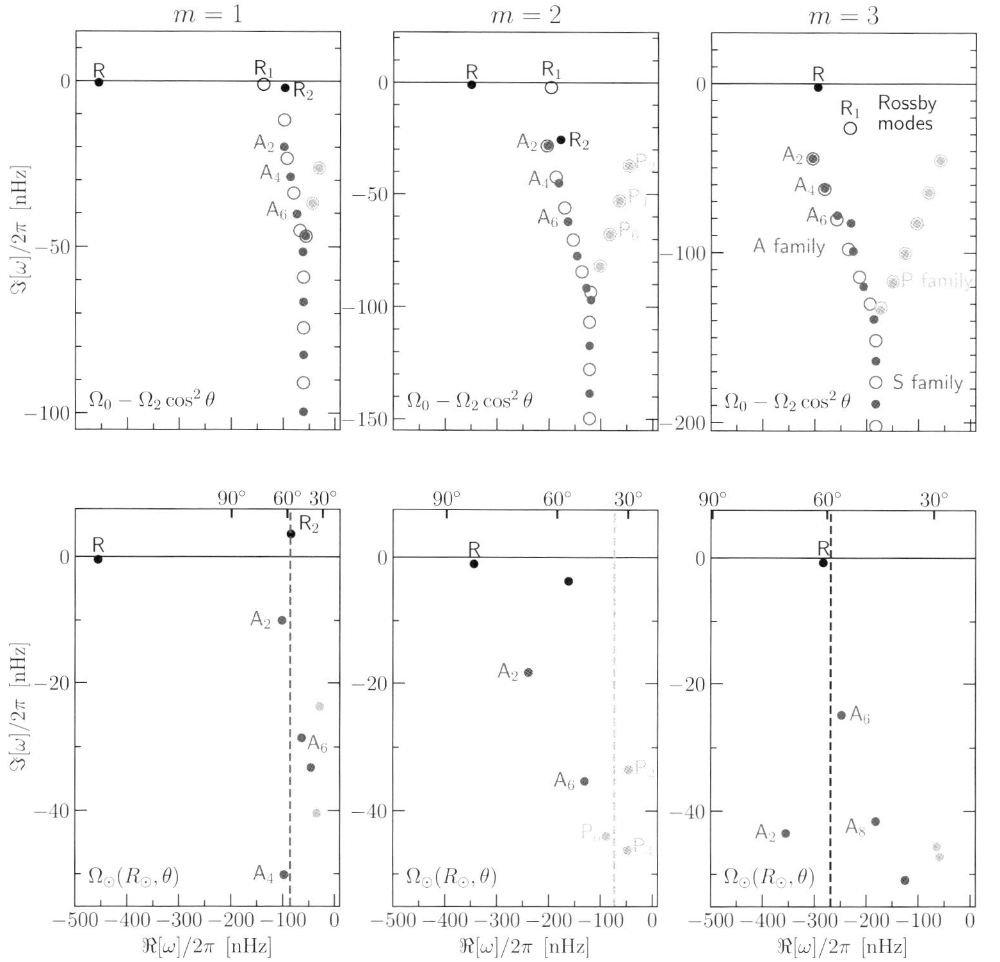

Figure 2. Eigenvalues in the complex plane for small-amplitude toroidal modes on a differentially rotating spherical surface (Fournier et al. 2022). Viscosity is specified via the Ekman number $E = 4 \times 10^{-4}$. The top panels show the real and imaginary parts of the eigenvalues for $m = 1, 2, 3$ and a differential rotation profile $\Delta\Omega = -\Omega_2 \cos^2\theta$ with $\Omega_2/2\pi = -114$ nHz. Using the notations of Drazin and Reid (2004), the various families of hydrodynamic modes are labelled R for the Rossby modes (black), A for the wall modes (in red), S for the highly damped modes (blue), and P for the center modes (orange). The full circles mark modes with radial vorticity that is symmetric across the equator (such as the first three modes shown in Fig. 1), and the open circles are for modes that are antisymmetric. The bottom panels show the eigenvalues of the symmetric modes for the differential rotation profile measured at the Sun's surface (the labels follow the tracks obtained by slowly tracking the modes from the top panel). The vertical dashed lines correspond to the observed frequencies of the first three example modes from Fig. 1.

model presented by Bekki et al. (2022b). Ignoring the meridional flow for simplicity, the governing equations in spherical-polar coordinates (r, θ, ϕ) are

$$\frac{D\rho'}{Dt} = -\nabla \cdot (\rho \boldsymbol{u}'), \tag{6}$$

$$\rho \frac{D\boldsymbol{u}'}{Dt} = -\nabla p' + \rho' \boldsymbol{g} + 2\rho \boldsymbol{u}' \times (\Omega \boldsymbol{e}_z) - \rho r \sin\theta (\boldsymbol{u}' \cdot \nabla\Omega) \boldsymbol{e}_\phi + \nabla \cdot \mathcal{D}, \tag{7}$$

$$\frac{Ds'}{Dt} = c_p \frac{\delta}{H_p} u_r' - \frac{u_\theta'}{r} \frac{\partial s}{\partial \theta} + \frac{1}{\boldsymbol{\rho T}} \nabla \cdot (\rho T \kappa \nabla s'), \tag{8}$$

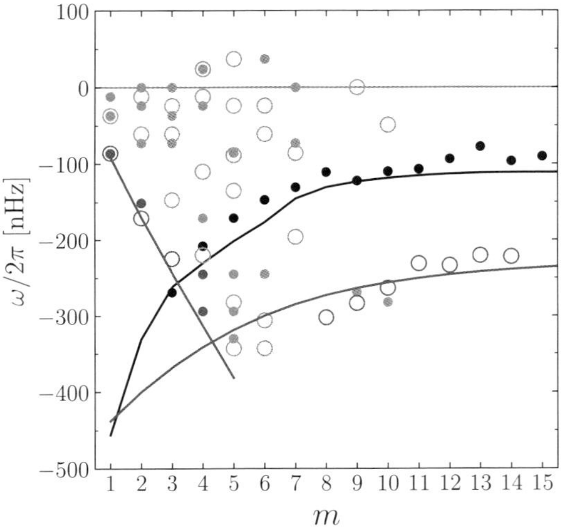

Figure 3. Comparison of the observed mode frequencies in the Carrington frame (circles) and the real part of the mode frequencies from the 3D linear eigenvalue problem (curves). The filled circles mark modes with radial vorticity that is symmetric across the equator while the open circles are for modes that are antisymmetric. Black refers to the equatorial Rossby modes, red to the high-latitude modes, orange to the critical-latitude modes, and blue to the high-frequency retrograde mode. The observations are from Löptien et al. (2018), Gizon et al. (2021), and Hanson et al. (2022). The model frequencies are from Bekki et al. (2022b) and Bekki (2024).

with the linearized equation of state

$$\frac{p'}{p} = \gamma\frac{\rho'}{\rho} + \frac{s'}{c_v}. \tag{9}$$

The perturbations from the background are denoted by primes. A realistic solar background stratification is considered. The total derivative along the solar differential rotation is given by

$$\frac{D}{Dt} = \frac{\partial}{\partial t} + \Delta\Omega\frac{\partial}{\partial\phi}, \tag{10}$$

where $\Delta\Omega = \Omega(r,\theta) - \Omega_0$ is the rotational profile from global helioseismology (Larson and Schou 2018). The superadiabaticity $\delta = -(H_p/c_p)\partial s/\partial r$ describes the small deviations from the adiabatic stratification. The latitudinal entropy gradient is estimated under the assumption that the Sun's differential rotation is in accordance with the thermal wind balance (Rempel 2005; Miesch et al. 2006),

$$\frac{\partial}{\partial\theta}\left(\frac{s}{c_p}\right) = \frac{r^2\sin\theta}{g}\frac{\partial(\Omega^2)}{\partial z}. \tag{11}$$

The viscous stress tensor $\mathcal{D}$ is given by

$$\mathcal{D}_{ij} = \rho\nu\left[\left(\frac{\partial u_i'}{\partial x_j} + \frac{\partial u_j'}{\partial x_i}\right) - \frac{2}{3}(\nabla\cdot\boldsymbol{u}')\delta_{ij}\right]. \tag{12}$$

Here, ν and κ are the (turbulent) viscous and thermal diffusivities, which are assumed to be equal.

The above linearized equations (6)–(8) are combined into an eigenvalue problem assuming that the perturbations are proportional to $\exp\mathrm{i}(m\phi - \omega t)$. The computational domain extends from the base of the convection zone ($r = 0.71R_\odot$) to slightly below the photosphere ($r = 0.985R_\odot$). At both radial boundaries, impenetrable and stress-free boundary conditions are used. At the poles, all the perturbed quantities (ρ', $\boldsymbol{u}'$, and s') are set to be zero for azimuthal orders $m > 1$. For $m = 1$, the special boundary condition $\partial u_\theta / \partial \theta = 0$ is used to allow for the existence of a pole-crossing flow. We numerically solve the eigenvalue equation using a second-order finite difference method.

The modes with eigenfrequencies in the inertial frequency range $|\Re(\omega)| \lesssim 2\Omega_0$ and with low damping rates are plotted in Fig. 3. We see that the main families of inertial modes are identified. Gizon et al. (2021) concluded that it is essential to include the latitudinal entropy gradient $\partial s / \partial \theta$ from Eq. (11) in the model to reproduce the correct spiral patterns of the u_ϕ eigenfunctions of the high-latitude modes. This also requires the conditions $\langle \delta \rangle < 2 \times 10^{-7}$ and $\langle \nu \rangle \leq 10^{12}$ cm^2/s on average in the convection zone. Both these upper limits are an order of magnitude below the mixing length estimates. See Dey et al. (2024) for an analysis of the value of δ in the lower half of the convection zone.

5. Excitation and damping mechanisms

Inertial modes are present in numerical simulations of rotating convection by Bekki et al. (2022a), Matilsky et al. (2022), and Blume et al. (2023). These simulations indicate that the Rossby modes are excited to significant amplitudes in these self-consistent simulations. The high-latitude modes are not present in these simulation because the latitudinal differential rotation is not strong enough. In the following paragraphs we discuss the excitation of the modes under specific setups.

5.1. *Stochastic excitation of equatorial Rossby modes*

According to the studies of Sect. 4, many solar inertial modes are linearly stable, including the equatorial Rossby modes. The observed amplitudes of these modes are the result of a balance between the stochastic excitation of the waves by turbulent convection and their damping by turbulent viscosity. Philidet and Gizon (2023) added a source term on the RHS of Eq. (2) to account for the fluctuations of the divergence of the Reynolds stress tensor of the turbulence. As summarized in this volume (Philidet and Gizon 2024), this excitation mechanism is similar to the one used for p modes (Goldreich and Keeley 1977), except that the vorticity in the turbulence plays a major role here. Treating the turbulent flow as a known input, based on solar surface measurements (Langfellner et al. 2015), this model reproduces the observed amplitudes of the solar equatorial Rossby modes of ~ 1 m/s. It should be noted that there is a qualitative transition in the shape of the power spectrum between $m \lesssim 5$ where the inertial modes are clearly resolved in frequency, and $m \gtrsim 5$ where the modes overlap. This greatly complicates the interpretation of the observation, and implies that a model for the whole shape of the power spectrum is necessary to exploit the full potential of the solar inertial modes.

5.2. *Low-m high-latitude modes are baroclinically unstable*

Using the eigenvalue solver of Sect. 4.2, Bekki et al. (2022b) found that the low-m high-latitude modes are very sensitive to the latitudinal entropy gradient $\partial s / \partial \theta$ in the convection zone (or the corresponding latitudinal temperature gradient $\partial T / \partial \theta$). Although the high-latitude modes are stable in 3D in the absence of $\partial s / \partial \theta$, they become linearly unstable with the presence of $\partial s / \partial \theta$ even when the radial stratification is adiabatic or subadiabatic. Therefore, they are *baroclinically* unstable. This explains why the high-latitude modes have larger velocity amplitudes than the other inertial modes on the Sun (Gizon et al. 2021).

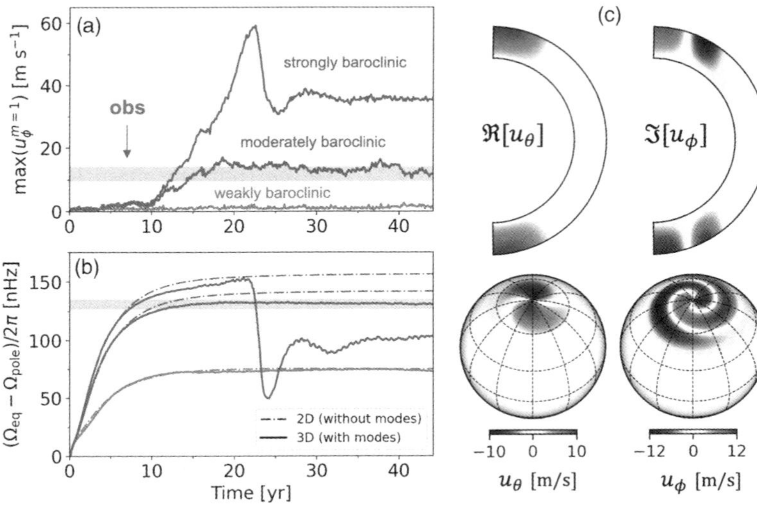

Figure 4. Results from 3D nonlinear simulations of the baroclinically-unstable modes (Bekki et al. 2024). (a) Amplitudes of the $m = 1$ high-latitude mode at the surface as functions of time. Green, red, and blue curves show the results from the simulations with weak, moderate, and strong baroclinic forcing. The horizontal gray shade denote the observed value. (b) Latitudinal differential rotation, $\Omega_{\rm eq} - \Omega_{\rm pole}$, in the middle convection zone as functions of time. The solid and dashed curves denote the results from 3D full-spherical simulations (where the high-latitude modes are present) and from the 2D axisymmetric simulations (where the modes cannot exist), respectively. (c) Extracted eigenfunctions of horizontal velocities (u_θ and u_ϕ) from the 3D nonlinear simulation (moderate case). Top and bottom rows show the meridional cuts and the surface eigenfunctions.

We note that, with a moderate value of turbulent viscosity ($\nu = 10^{12}$ cm^2/s), a very small latitudinal temperature variation of $\Delta T = T_{\rm pole} - T_{\rm eq} \approx 3$ K is enough to make the low-m high-latitude modes unstable. A realistic latitudinal entropy gradient can be estimated from Eq. (11), which implies $\Delta T \approx 7$ K in the middle convection zone. Once this profile of $\partial s/\partial\theta$ is included, the linear model can successfully reproduce the observed eigenfunctions of the high-latitude modes with $1 \leq m < 5$ (Fig. 1, left panels), as well as their observed frequencies (Fig. 3).

6. Baroclinally unstable modes control differential rotation

6.1. *3D nonlinear mean-field simulations*

To study the amplitudes of the high-latitude inertial modes, Bekki et al. (2024) carried out a series of hydrodynamic simulations of the large-scale flows in the Sun using a mean-field approach, where the small-scale convective motions are not explicitly solved for and the turbulent transport effects (incl. the Λ effect) are parameterized. A parameter survey was conducted with varying baroclinic forcings whereby a negative (positive) latitudinal entropy gradient $\partial s/\partial\theta$ is generated in the northern (southern) hemisphere.

6.2. *Dynamical effect of modes on differential rotation*

Bekki et al. (2024) found that the amplitudes of the low-m high-latitude modes are nonlinearly coupled to the latitudinal differential rotation in the convection zone. Figures 4a and b show the temporal evolution of the $m = 1$ mode amplitude and the latitudinal differential rotation for three representative simulation cases where the baroclinic forcing is weak (green), moderate (red), and strong (blue). When the baroclinicity is weak, the high-latitude modes are not excited to large amplitudes. With a moderate amount of baroclinicity, the high-latitude modes become linearly unstable and the $m = 1$ mode velocity amplitude saturates at ≈ 13 m s^{-1}, which is consistent with the solar observation. The extracted velocity eigenfunction of the $m = 1$ mode from this case is shown in Fig. 4c. The observed spiral feature in u_ϕ is nicely reproduced. When the baroclinic forcing is even stronger, the modes are very efficiently baroclinically-excited and their amplitudes are larger than observed on the Sun. We find that the high-amplitude modes play a significant dynamical role in limiting the latitudinal differential rotation via a nonlinear feedback. The reduction of the latitudinal differential rotation is dominantly caused by the equatorward heat transport due to the high-latitude modes, which reduces the baroclinicity and changes the meridional circulation such that the poleward transport of angular momentum is increased. This dynamical effect is not present in axisymmetric models where the non-axisymmetric inertial modes cannot exist (Fig. 4b). We note that the horizontal Reynolds stresses associated with the high-latitude modes imply a transport of angular momentum toward the equator (Hathaway et al. 2013; Mandowara et al. 2024), however this (direct) effect on differential rotation is not as important as the (indirect) effect via heat transport.

6.3. *Latitudinal temperature gradient*

Bekki et al. (2024) find that the high-latitude modes place an upper limit on $\Delta T = T_{\rm pole} - T_{\rm eq}$. Using the 3D nonlinear mean-field simulations mentioned above, a relationship between ΔT and the amplitudes of the $m \leq 3$ high-latitude modes can be established. These simulations show that the value of ΔT cannot exceed 7 K. Interestingly, the observed mode amplitudes imply that this maximum of 7 K is likely reached in the Sun. This conclusion is also consistent with the value estimated from the thermal wind balance approximation, Eq. (11).

7. Modeling high-frequency retrograde modes

The retrograde modes with $l = m + 1$ radial vorticity reported by Hanson et al. (2022) do not follow the dispersion relation of the classical Rossby modes given by Eq. (4), but have about three times the frequencies of the sectoral Rossby modes. Hanson et al. (2022) argued that these modes cannot be identified as magneto-Rossby modes, coupled Rossby-gravity modes, nor thermal Rossby modes. The observations were interpreted by Triana et al. (2022) as a particular class of inertial modes where substantial radial flows are involved in the equatorial regions. In this sense, they may be called *non-toroidal* inertial modes. Although the linear model used by Triana et al. (2022) was highly simplified (e.g., assuming a fluid with constant density), this mode identification was later confirmed by Bhattacharya and Hanasoge (2023) and Bekki (2024) using more realistic linear models that include the solar-like background stratification.

Unlike the modes that are quasi-toroidal, non-toroidal inertial modes manifest a strong sensitivity to the background superadiabaticity $\delta = \nabla - \nabla_{\rm ad}$. Using the model described in Sect. 4.2, Bekki (2024) showed that the mode frequencies are compatible with the observed frequencies only when the bulk convection zone (in the model) is weakly subadiabatic on average. Several branches of enhanced power, with one branch corresponding to the high-frequency retrograde modes, have been reported in non-linear rotating convection simulations by Blume et al. (2023). In these simulations the branches have low quality factors.

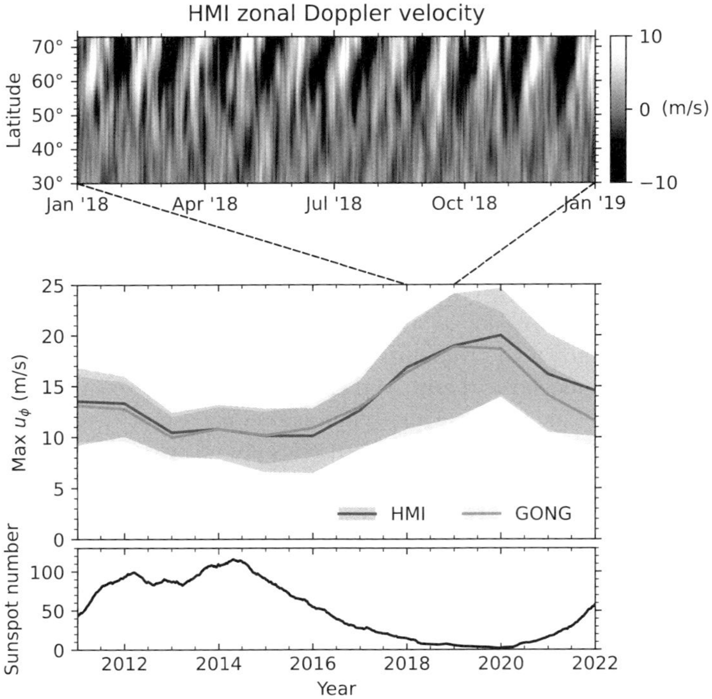

Figure 5. Map of zonal velocity (top) obtained from HMI line-of-sight Doppler signals during solar minimum showing the signature of the $m = 1$ high-latitude mode whose amplitude varies with the solar cycle (bottom). The daily zonal velocity V_{zonal} were antisymmetrized across the equator and combined to form a synoptic map. The $m = 1$ high-latitude mode of oscillation is seen as stripes above 50° latitude with a period of about 34 days as viewed from Earth. The mode amplitude averaged over the latitude range $60° - 75°$ is estimated from V_{zonal} in a 3-yr sliding window. The mode amplitude is scaled by a multiplicative factor of 2.9 so that it is directly comparable to the deprojected longitudinal velocity u_ϕ shown in the bottom left panel of Fig. 1.

8. Solar cycle variations of mode parameters

Because of the close connection between the differential rotation profile and the characteristics of the inertial modes, Goddard et al. (2020) suggested that the frequencies of the equatorial Rossby modes could vary with the sunspot cycle due to the temporal changes in rotation (varying zonal flows or 'torsional oscillations'). The prediction was a positive correlation between the mode frequency shifts and the sunspot number. However, Waidele and Zhao (2023) and Lekshmi et al. (2024) measured an anticorrelation. This strongly suggests that the solar magnetic fields play the dominant role in the modulation of the mode properties. Waidele and Zhao (2023) also noticed that the average power of the equatorial Rossby modes varies in phase with the sunspot cycle. Lekshmi et al. (2024) reach the same conclusion in this volume for the Rossby modes with $6 \leq m \leq 10$.

Here we perform an analysis of the solar cycle dependence of the amplitude of the $m = 1$ high-latitude mode using Dopplergrams as input. Because this mode is associated with horizontal flows in the near polar regions, it is detectable in the direct Doppler data. The top panel of Fig. 5 shows the zonal velocity computed from the HMI Dopplergrams during solar minimum, where the zonal velocity is given by (Ulrich et al. 1988)

$$V_{\text{zonal}}(\theta) = \frac{\sum_\phi \Delta V_{\text{los}} \sin(\phi - \phi_{\text{c}})}{\sum_\phi |\sin(\phi - \phi_{\text{c}})|}. \tag{13}$$

In the above definition, $\Delta V_{\rm los}(\theta, \phi)$ is the Dopplergram corrected for the background signals and $\phi - \phi_{\rm c}$ is the longitude measured from the central meridian. The background signals consist of a fit to the time-averaged limb shift, solar rotation, and meridional flows. In Fig. 5, $V_{\rm zonal}$ displays tilted stripes at high latitudes (these were already visible in the MWO data presented by Ulrich 1993, 2001). The frequency of this pattern is measured to be 338 nHz in the Earth frame, which corresponds exactly to the frequency of the $m = 1$ high-latitudes mode (-86 nHz in the Carrington frame). The middle panel of Fig. 5 shows the velocity amplitude u_ϕ of the mode estimated from $V_{\rm zonal}$ over the latitude range $60° - 75°$ as a function of time. The amplitude in u_ϕ varies between 10 and 20 m/s over this time period. We find that the mode amplitude is clearly anticorrelated with the sunspot number (bottom panel of Fig. 5). These new observations imply that a measurable interaction happens between the $m = 1$ mode and the Sun's global dynamo. Numerical modeling is needed to identify the nature of this interaction and to find out how and where it takes place in the convection zone.

9. Outlook

9.1. *Mode physics*

While helioseismology with acoustic modes is very largely based on the interpretation of the frequencies of the individual modes of oscillation using first-order perturbation theory, the interpretation of the inertial modes cannot be separated from the study of their amplitudes and eigenfunctions — this requires more sophisticated forward modeling. In some regions of frequency space, the spectrum is too dense to be resolved: modes may overlap in frequency space and make the power spectrum difficult to decipher. Excitation and damping mechanisms include the action of turbulent convection, but also the nonlinear interaction of the modes with their environment. Solar inertial modes are not only important as new diagnostics of the solar convection zone, but also as regulating agents of the global dynamics such as the Sun's meridional flow and differential rotation.

9.2. *Opportunities to better understand the solar convection zone*

Figure 6 shows power spectra of the east–west velocity obtained from a recent high-resolution numerical simulation by Hotta et al. (2022) and from two observational studies, one from time-distance helioseismology and another from tracking the granulation pattern in intensity (see Proxauf 2021, and references therein). The order of magnitude discrepancy at low angular degrees (large horizontal scales) between the numerical simulation and the observations is an outstanding problem (Hanasoge et al. 2012; Gizon and Birch 2012).

This problem is connected to the apparent lack of north–south elongated convective modes in the solar surface observations. In numerical models these so-called banana cells are rotationally-constrained modes of rotating convection (e.g., Gilman 1977; Miesch et al. 2008). The banana cells are robust features of the theory; they propagate prograde and can be understood in terms of the compressional β-effect or the conservation law of potential vorticity (e.g., Glatzmaier and Gilman 1981; Verhoeven and Stellmach 2014). They are a class of inertial modes sometimes called *thermal Rossby modes*. Their properties have been studied in the linear regime (Glatzmaier and Gilman 1981; Bekki et al. 2022b; Hindman and Jain 2022; Jain and Hindman 2023) and in nonlinear rotating convection simulations (Bekki et al. 2022a; Blume et al. 2023). Modes of mixed character sharing characteristics of prograde thermal Rossby and retrograde inertial modes have also been discussed by Bekki et al. (2022b,a), Jain and Hindman (2023), and Blume et al. (2023). The impact of missing physics, such as, e.g., the importance of magnetic fields, should be assessed.

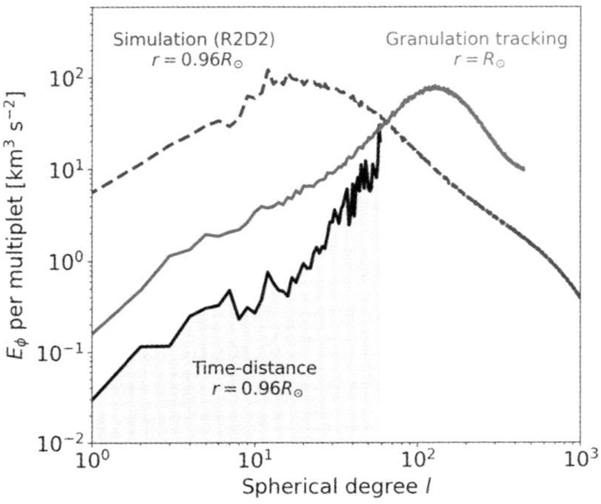

Figure 6. Power spectra of the longitudinal velocity near the solar surface. The black solid curve shows the observational upper limit of $E_\phi(l)$ at $r = 0.96R_\odot$ inferred by time-distance helioseismology (Hanasoge et al. 2012) and revised by Proxauf (2021). The quantity E_ϕ represents the spatial power of the daily average of u_ϕ, summed over all m values at fixed angular degree l. The red solid curve shows E_ϕ at the surface obtained by local correlation tracking of surface granulation (Proxauf 2021). These observational data are available online (Birch 2023). The blue dashed curve shows the spectrum from a numerical simulation of solar rotating magneto-convection (Hotta et al. 2022).

The inertial modes give us new information on the physical conditions deep inside the convection zone. In this summary paper we gave examples of what these modes are sensitive to, including the latitudinal temperature gradient, superadiabaticity, and the strength of the turbulent motions (through constraints on the turbulent viscosity) in the deep convection zone. Looking further ahead, the inertial modes are likely to be affected by – and to affect – the magnetic field of the Sun. Modes of mixed character, gravito-magneto-inertial, also present novel opportunities and challenges to understand the tachocline (see, e.g., Dikpati et al. 2018; Matilsky et al. 2022).

Acknowledgements

The Solar and Stellar Interiors Department of the Max–Planck Institute for Solar System Research acknowledges generous support from the Max–Planck-Gesellschaft zur Förderung der Wissenschaften e.V., the European Research Council (ERC Synergy Grant WHOLE SUN 810218), and the Deutsche Forschungsgemeinschaft (SFB 1456/432680300 Mathematics of Experiment, project C04). Work on mode excitation by turbulence is supported by JP's Max Planck Partner Group at LESIA. The HMI data are courtesy of NASA/SDO and the HMI Science Team. This work utilizes data acquired by GONG instruments operated by NISP/NSO/AURA/NSF with contribution from NOAA. The sunspot numbers are from the World Data Center SILSO, Royal Observatory of Belgium, Brussels.

References

Balmforth, N. J. and Morrison, P. J. (1995). Normal modes and continuous spectra. *Ann. New York Acad. Sci.*, 773:80–94.

Baruteau, C. and Rieutord, M. (2013). Inertial waves in a differentially rotating spherical shell. *J. Fluid Mech.*, 719:47–81.

Bekki, Y. (2024). Numerical study of non-toroidal inertial modes with $l = m + 1$ radial vorticity in the Sun's convection zone. *Astron. Astrophys.*, 682:A39.

Bekki, Y., Cameron, R. H., and Gizon, L. (2022a). Theory of solar oscillations in the inertial frequency range: Amplitudes of equatorial modes from a nonlinear rotating convection simulation. *Astron. Astrophys.*, 666:A135.

Bekki, Y., Cameron, R. H., and Gizon, L. (2022b). Theory of solar oscillations in the inertial frequency range: Linear modes of the convection zone. *Astron. Astrophys.*, 662:A16.

Bekki, Y., Cameron, R. H., and Gizon, L. (2024). The Sun's differential rotation is controlled by high-latitude baroclinically unstable inertial modes. *Science Adv.*, accepted.

Bhattacharya, J. and Hanasoge, S. M. (2023). A Spectral Solver for Solar Inertial Waves. *Astrophys. J. Suppl. Ser.*, 264(1):21.

Birch, A. (2023). Convection spectra from the thesis of B. Proxauf, https://doi.org/10.17617/3.DFU3SQ.

Blume, C. C., Hindman, B. W., and Matilsky, L. I. (2023). Inertial Waves in a Nonlinear Simulation of the Sun's Convection Zone and Radiative Interior, doi:10.48550/arxiv.2312.14270.

Charbonneau, P., Dikpati, M., and Gilman, P. A. (1999). Stability of the Solar Latitudinal Differential Rotation Inferred from Helioseismic Data. *Astrophys. J.*, 526(1):523–537.

Damiani, C., Cameron, R. H., Birch, A. C., and Gizon, L. (2020). Rossby modes in slowly rotating stars: depth dependence in distorted polytropes with uniform rotation. *Astron. Astrophys.*, 637:A65.

Dey, P., Bekki, Y., and Gizon, L. (2024). Probing the superadiabaticity of the solar convection zone with inertial modes. In *this volume*.

Dikpati, M., McIntosh, S. W., Bothun, G., Cally, P. S., Ghosh, S. S., Gilman, P. A., and Umurhan, O. M. (2018). Role of Interaction between Magnetic Rossby Waves and Tachocline Differential Rotation in Producing Solar Seasons. *Astrophys. J.*, 853(2):144.

Drazin, P. G. and Reid, W. H. (2004). *Hydrodynamic Stability*. Cambridge University Press, Cambridge.

Fournier, D., Gizon, L., and Hyest, L. (2022). Viscous inertial modes on a differentially rotating sphere: Comparison with solar observations. *Astron. Astrophys.*, 664:A6.

Gilman, P. A. (1977). Nonlinear Dynamics of Boussinesq Convection in a Deep Rotating Spherical Shell. I. *Geophys. Astrophys. Fluid Dyn.*, 8:93–135.

Gizon, L. and Birch, A. C. (2012). Helioseismology challenges models of solar convection. *Proc. Nat. Acad. Sci.*, 109(30):11896–11897.

Gizon, L., Cameron, R. H., Bekki, Y., Birch, A. C., Bogart, R. S., Brun, A. S., Damiani, C., Fournier, D., Hyest, L., Jain, K., Lekshmi, B., Liang, Z.-C., and Proxauf, B. (2021). Solar inertial modes: Observations, identification, and diagnostic promise. *Astron. Astrophys.*, 652:L6.

Gizon, L., Fournier, D., and Albekioni, M. (2020). Effect of latitudinal differential rotation on solar Rossby waves: Critical layers, eigenfunctions, and momentum fluxes in the equatorial β plane. *Astron. Astrophys.*, 642:A178.

Glatzmaier, G. A. and Gilman, P. A. (1981). Compressible Convection in a Rotating Spherical Shell - Part Three - Analytic Model for Compressible Vorticity Waves. *Astrophys. J. Suppl. Ser.*, 45:381.

Goddard, C. R., Birch, A. C., Fournier, D., and Gizon, L. (2020). Predicting frequency changes of global-scale solar Rossby modes due to solar cycle changes in internal rotation. *Astron. Astrophys.*, 640:L10.

Goldreich, P. and Keeley, D. A. (1977). Solar seismology. II. The stochastic excitation of the solar p-modes by turbulent convection. *Astrophys. J.*, 212:243–251.

Hanasoge, S. M., Duvall, T. L., and Sreenivasan, K. R. (2012). Anomalously weak solar convection. *Proc. Nat. Acad. Sci.*, 109(30):11928–11932.

Hanson, C. S., Hanasoge, S., and Sreenivasan, K. R. (2022). Discovery of high-frequency retrograde vorticity waves in the Sun. *Nature Astron.*, 6:708–714.

Hathaway, D. H., Upton, L., and Colegrove, O. (2013). Giant Convection Cells Found on the Sun. *Science*, 342(6163):1217–1219.

Hindman, B. W. and Jain, R. (2022). Radial Trapping of Thermal Rossby Waves within the Convection Zones of Low-mass Stars. *Astrophys. J.*, 932(1):68.

Hotta, H., Kusano, K., and Shimada, R. (2022). Generation of Solar-like Differential Rotation. *Astrophys. J.*, 933(2):199.

Jain, R. and Hindman, B. W. (2023). Latitudinal Propagation of Thermal Rossby Waves in Stellar Convection Zones. *Astrophys. J.*, 958(1):48.

Langfellner, J., Gizon, L., and Birch, A. C. (2015). Spatially resolved vertical vorticity in solar supergranulation using helioseismology and local correlation tracking. *Astron. Astrophys.*, 581:A67.

Larson, T. P. and Schou, J. (2018). Global–Mode Analysis of Full-Disk Data from the Michelson Doppler Imager and the Helioseismic and Magnetic Imager. *Solar Phys.*, 293(2):29.

Lekshmi, B., Gizon, L., Jain, K., Liang, Z.-C., and Philidet, J. (2024). Temporal Variation of Solar Equatorial Rossby Modes with Azimuthal Orders $6 \leq m \leq 10$. In *this volume*.

Liang, Z.-C., Gizon, L., Birch, A. C., and Duvall, T. L. (2019). Time-distance helioseismology of solar Rossby waves. *Astron. Astrophys.*, 626:A3.

Löptien, B., Gizon, L., Birch, A. C., Schou, J., Proxauf, B., Duvall, T. L., Bogart, R. S., and Christensen, U. R. (2018). Global-scale equatorial Rossby waves as an essential component of solar internal dynamics. *Nature Astron.*, 2:568–573.

Mandal, K., Hanasoge, S. M., and Gizon, L. (2021). Detection of Rossby modes with even azimuthal orders using helioseismic normal-mode coupling. *Astron. Astrophys.*, 652:A96.

Mandowara, Y., Bekki, Y., and Gizon, L. (2024). Reynolds stress of the $m = 1$ high latitude inertial mode on the Sun. In *this volume*.

Matilsky, L. I., Hindman, B. W., Featherstone, N. A., Blume, C. C., and Toomre, J. (2022). Confinement of the Solar Tachocline by Dynamo Action in the Radiative Interior. *Astrophys. J.*, 940(2):L50.

Miesch, M. S., Brun, A. S., DeRosa, M. L., and Toomre, J. (2008). Structure and Evolution of Giant Cells in Global Models of Solar Convection. *Astrophys. J.*, 673(1):557–575.

Miesch, M. S., Brun, A. S., and Toomre, J. (2006). Solar Differential Rotation Influenced by Latitudinal Entropy Variations in the Tachocline. *Astrophys. J.*, 641(1):618–625.

Muñoz-Jaramillo, A., Nandy, D., and Martens, P. C. H. (2011). Magnetic Quenching of Turbulent Diffusivity: Reconciling Mixing-length Theory Estimates with Kinematic Dynamo Models of the Solar Cycle. *Astrophys. J.*, 727(1):L23.

Philidet, J. and Gizon, L. (2023). Interaction of solar inertial modes with turbulent convection. A 2D model for the excitation of linearly stable modes. *Astron. Astrophys.*, 673:A124.

Philidet, J. and Gizon, L. (2024). A 2D model for the excitation of the linearly stable inertial modes of the Sun by turbulent convection. In *this volume*.

Provost, J., Berthomieu, G., and Rocca, A. (1981). Low Frequency Oscillations of a Slowly Rotating Star - Quasi Toroidal Modes. *Astron. Astrophys.*, 94:126.

Proxauf, B. (2021). *Observations of large-scale solar flows*. PhD thesis, Georg–August-Universität Göttingen, Germany.

Proxauf, B., Gizon, L., Löptien, B., Schou, J., Birch, A. C., and Bogart, R. S. (2020). Exploring the latitude and depth dependence of solar Rossby waves using ring-diagram analysis. *Astron. Astrophys.*, 634:A44.

Rempel, M. (2005). Solar Differential Rotation and Meridional Flow: The Role of a Subadiabatic Tachocline for the Taylor–Proudman Balance. *Astrophys. J.*, 622(2):1320–1332.

Triana, S. A., Guerrero, G., Barik, A., and Rekier, J. (2022). Identification of Inertial Modes in the Solar Convection Zone. *Astrophys. J.*, 934(1):L4.

Ulrich, R. K. (1993). The controversial Sun. In Weiss, W. W. and Baglin, A., editors, *IAU Colloq. 137: Inside the Stars*, volume 40 of *Astron. Soc. Pacific Conf. Ser.*, page 25.

Ulrich, R. K. (2001). Very Long Lived Wave Patterns Detected in the Solar Surface Velocity Signal. *Astrophys. J.*, 560:466.

Ulrich, R. K., Boyden, J. E., Webster, L., Snodgrass, H. B., Padilla, S. P., Gilman, P., and Shieber, T. (1988). Solar Rotation Measurements at Mt. Wilson – Part Five. *Solar Phys.*, 117:291.

Unno, W., Osaki, Y., Ando, H., Saio, H., and Shibahashi, H. (1989). *Nonradial oscillations of stars*. Tokyo: University of Tokyo Press, 2nd edition.

Verhoeven, J. and Stellmach, S. (2014). The compressional beta effect: A source of zonal winds in planets? *Icarus*, 237:143–158.

Waidele, M. and Zhao, J. (2023). Observed Power and Frequency Variations of Solar Rossby Waves with Solar Cycles. *Astrophys. J.*, 954:L26.

Watson, M. (1981). Shear instability of differential rotation in stars. *Geophys. Astrophys. Fluid Dyn.*, 16:285.

Contributed papers

Dynamics of Solar and Stellar Convection Zones and Atmospheres
Proceedings IAU Symposium No. 365, 2024
A. V. Getling & L. L. Kitchatinov, eds.
doi:10.1017/S1743921324000309

A 2D model for the excitation of the linearly stable inertial modes of the Sun by turbulent convection

J. Philidet and L. Gizon

Max–Planck Institute for Solar System Research, Göttingen, Germany
email: jordan.philidet@obspm.fr

Abstract. The newly discovered inertial modes in the Sun offer the opportunity to probe the solar convective zone down to the tachocline. While linear analysis predicts the frequencies and eigenfunctions of the modes, it gives no information about their excitation or their amplitudes. We present here a theoretical formalism for the stochastic excitation of the solar inertial modes by turbulent convection. The amplitudes predicted by our model are in complete agreement with observations, thus supporting the assumption that they are stochastically excited. Our work also uncovers a qualitative transition in the shape of the inertial mode spectrum, between $m \lesssim 5$ where the modes are clearly resolved in frequency, and $m \gtrsim 5$ where the modes overlap. This complicates the interpretation of the high-m data, and suggests that a model for the whole shape of the power spectrum is necessary to exploit the full seismic potential of solar inertial modes.

Keywords. waves – turbulence – Sun: oscillations – Sun: interior – Sun: helioseismology

1. Introduction

Internal rotation in the Sun allows for the development of inertial modes, which propagate through the Coriolis force. While theory has predicted their existence for a long time (Papaloizou and Pringle 1978), sectoral ($l = m$) Rossby modes have only recently been observed in solar data (Löptien et al. 2018; Liang et al. 2019). This was followed shortly thereafter by the report of additional families of inertial modes, due to the internal differential rotation (Gizon et al. 2021; Hanson et al. 2022).

These newly observed inertial modes offer the possibility to probe the dynamics and the structure of the solar convective envelope in a way that complements p-mode seismology. For example, they are expected to be much more sensitive on the superadiabatic temperature gradient or the turbulent viscosity throughout the convective zone. To exploit this potential, theoretical modelling of the solar inertial modes is necessary. So far, efforts have focused on linear analysis to model their eigenfrequencies and eigenfunctions (Gizon et al. 2020; Fournier et al. 2022; Bekki et al. 2022; Triana et al. 2022), which allows for the identification of the observed modes. However, this approach gives no information about the amplitude of the modes, that is to say about their excitation mechanism. Most of them are linearly stable, meaning that they are most likely stochastically excited by turbulent convection, like p-modes. Having a model for the excitation of the solar inertial modes would allow us not only to predict which modes are expected to be visible and identifiable, but also to put stronger constraints on the dynamics of the solar convective zone, rather than only its equilibrium structure.

We present here such a model, tailored for purely toroidal inertial modes – which the observed solar inertial modes are –, with the inclusion of latitudinal differential rotation and turbulent viscosity, and placing ourselves in the equatorial β-plane approximation. The model is similar to the commonly accepted picture for p-modes (e.g. Samadi and Goupil 2001), where turbulent emission, meaning the non-linear term in the momentum equation (Lighthill 1967), provides the necessary mechanical work to excite the modes. This presentation highlights the main characteristics of the model, as well as the results that we obtained when applying it to the solar case.

2. Synthetic power spectrum

We focus here on quasi-toroidal vorticity modes, for which the horizontal part of the wave equation can be separated from its radial part. This allows us to adopt a 2D setting, and study the excitation of vorticity waves in a 2D shear flow representing the solar differential rotation. We place ourselves in the equatorial β-plane approximation, that is transforming the latitude λ and longitude ϕ into Cartesian coordinates $x \equiv R\phi$ and $y \equiv R \sin\lambda$, where R is the radius of the spherical shell under consideration. The incompressibility of the flow allows us to describe it by means of a stream function Ψ, related to the velocity $\mathbf{u}$ by

$$\mathbf{u} = \boldsymbol{\nabla} \times (\Psi \mathbf{e_z}) \ , \tag{1}$$

where $\mathbf{e_z}$ is the unit vector normal to the surface. After some algebra, we get a linear wave equation with a source term (Philidet and Gizon 2023)

$$\left(\frac{\partial}{\partial t} + U\frac{\partial}{\partial x}\right)\Delta\Psi_{\mathrm{osc}} + (\beta - U'')\frac{\partial \Psi_{\mathrm{osc}}}{\partial x} - \nu_{\mathrm{turb}}\Delta^2\Psi_{\mathrm{osc}} = \frac{\partial \Psi_{\mathrm{turb}}}{\partial x}\frac{\partial \Delta\Psi_{\mathrm{turb}}}{\partial y} - \frac{\partial \Psi_{\mathrm{turb}}}{\partial y}\frac{\partial \Delta\Psi_{\mathrm{turb}}}{\partial x} \ , \tag{2}$$

where U'' is the second derivative of $U(y)$ (which is the velocity due to differential rotation), Δ is the Laplacian operator, and ν_{turb} the turbulent viscosity. This was written under the assumption that we could separate the field into an oscillation part Ψ_{osc} and a turbulence part Ψ_{turb}. We can then go in Fourier domain in time t and azimuth x, so that the wave equation becomes a 1D ordinary differential equation (ODE).

Because the oscillation and turbulence parts are decoupled, the latter can be treated as a source term, which we will consider as an input to the model, and will be taken from observation of the solar surface. Then Eq. (2) becomes a linear ODE with a known source term, whose solution can therefore be obtained by convolving the source term with the Green function $G(y_1, y_2)$ associated to the linear left-hand side. Eventually, one obtains the following expression for the expected power spectra in terms of latitudinal velocity, azimuthal velocity and vorticity respectively (Philidet and Gizon 2023)

$$\left\langle |\widehat{u}_{x,\mathrm{osc}}(y_{\mathrm{obs}})|^2 \right\rangle = \int_{-R}^{R} \mathrm{d}y_s \ \left|\frac{\partial G}{\partial y_{\mathrm{obs}}}\right|^2 \mathcal{I}(y_s) \ , \tag{3}$$

$$\left\langle |\widehat{u}_{y,\mathrm{osc}}(y_{\mathrm{obs}})|^2 \right\rangle = \int_{-R}^{R} \mathrm{d}y_s \ |k_x G(y_{\mathrm{obs}}, y_s)|^2 \, \mathcal{I}(y_s) \ , \tag{4}$$

$$\left\langle \left|\widehat{\zeta}_{\mathrm{osc}}(y_{\mathrm{obs}})\right|^2 \right\rangle = \int_{-R}^{R} \mathrm{d}y_s \ \left|\left(k_x^2 G - \frac{\partial^2 G}{\partial y_{\mathrm{obs}}^2}\right)\right|^2 \mathcal{I}(y_s) \ , \tag{5}$$

where y_{obs} is the latitudinal coordinate at which the power spectrum is evaluated, and $\langle \, . \, \rangle$ denotes an ensemble average. The function $\mathcal{I}(y_s)$ denotes the source covariance, and

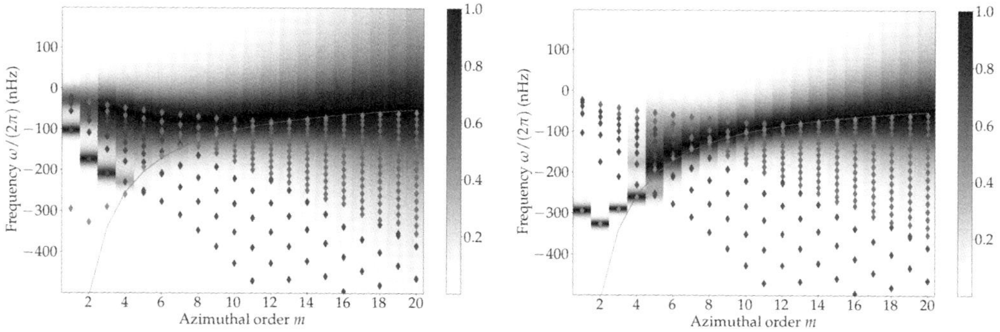

Figure 1. Equatorial power spectrum in the m-ω plane, in terms of azimuthal velocity (**left**) and latitudinal velocity (**right**). Each vertical slice is normalised separately such that the maximum is unity. The diamonds show the real part of the eigenfrequencies of the linear homogeneous problem. The colour code refers to the mode categories: in particular, the red diamonds represent the sectoral Rossby modes. The solid red line shows the theoretical dispersion relation for sectoral Rossby modes.

is defined by

$$\mathcal{I}(y_s) \equiv \int \mathrm{d}Y \; \left\langle \widehat{S}(y_s) \; \widehat{S}^*(y_s + Y) \right\rangle \, , \tag{6}$$

where $\widehat{S}$ is the Fourier transform in t and x of the right-hand side of Eq. (2).

The only assumption required to write Eqs. (3) to (5) is that the correlation scale of the source is much smaller than the typical variation scale of the Green function. Since most of the turbulent energy comes from the granulation scales at $\ell \sim 120$ and the inertial modes under consideration are low degree modes, this assumption is valid. Some additional algebra is required to specify the source covariance; we eventually find (Philidet and Gizon 2023)

$$\mathcal{I} = \frac{1}{4\pi^3} \int \mathrm{d}\omega' \, \mathrm{d}^2\mathbf{k}' \; k_x^3 k_x' k_y'^2 \, |\mathbf{k}' + \mathbf{k}/2|^2 \times \mathcal{E}_\Psi\left(\omega' - \frac{\omega}{2}, \mathbf{k}' - \frac{\mathbf{k}}{2}\right) \mathcal{E}_\Psi^*\left(\omega' + \frac{\omega}{2}, \mathbf{k}' + \frac{\mathbf{k}}{2}\right) \, . \tag{7}$$

The function $\mathcal{E}_\Psi$ represents the turbulent stream function spectrum, and is defined by

$$\mathcal{E}_\Psi(\omega, \mathbf{k}) \equiv \int \mathrm{d}\tau \, \mathrm{d}^2\mathbf{x} \, \langle \Psi_{\mathrm{turb}}(T, \mathbf{X}) \Psi_{\mathrm{turb}}(T + \tau, \mathbf{X} + \mathbf{x}) \rangle \; e^{\mathrm{j}(\omega\tau - \mathbf{k}\cdot\mathbf{x})} \, . \tag{8}$$

The two ingredients needed to apply Eqs. (3) to (5) to the solar case are the turbulent spectrum $\mathcal{E}_\Psi$ and the Green function G. The former is extracted from observations, and the latter is computed numerically using spectral methods (see Philidet and Gizon 2023, for more details). Once the differential rotation profile is set, the turbulent viscosity is the only adjustable parameter.

3. Results

Figure 1 highlights how the shape of the inertial mode power spectrum predicted by our model, and as it would be observed at the equator, depends on azimuthal order m. Naturally, the region of excess power (black regions) align with the linear eigenfrequencies of the left-hand side of the wave equation. A striking feature, however, is that the spectrum seems to undergo a qualitative transition at around $m \sim 5$. Below, the regions of excess power are very thin, with clearly resolved and identifiable modes, especially in

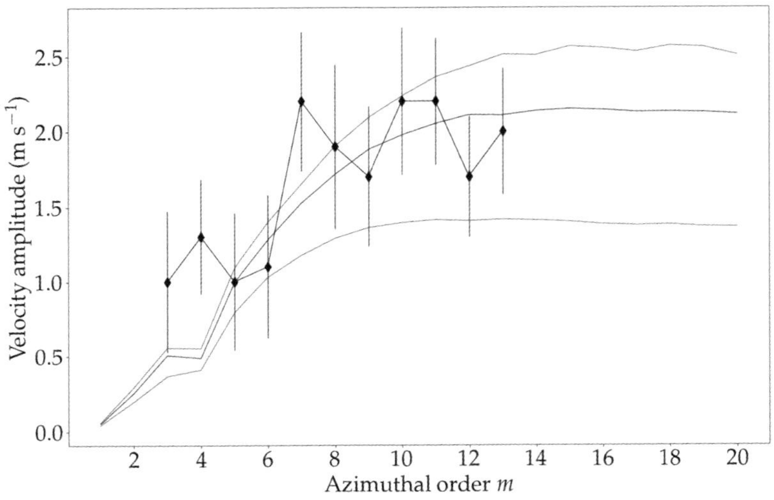

Figure 2. Prediction for the Rossby mode amplitudes (coloured solid lines) compared to observations (black line with error bars), as a function of azimuthal order m. The colour code refers to the value of the turbulent Reynolds number: $\mathrm{Re}_{\mathrm{turb}} = 300$ (red), 700 (blue), and 1000 (green).

the latitudinal velocity spectrum, which is where the sectoral Rossby modes will be most visible. Above $m \sim 5$, the regions of excess power start blending with each other, forming a continuum of overlapping modes where modes cannot be resolved and identification proves more delicate.

From these equatorial power spectra for each m, we can predict the amplitude of the Rossby mode, like so

$$A = \left(\int_{\omega_a}^{\omega_b} \mathrm{d}\omega \; P(\omega) \right)^{1/2} , \tag{9}$$

where the boundaries ω_a and ω_b enclose the whole peak in the power spectrum. We show the result in Fig. 2. The solid lines represent our model predictions, with three different values of the turbulent viscosity ν_{turb}, or alternatively of the turbulent Reynolds number $\mathrm{Re}_{\mathrm{turb}} \equiv \overline{U} R / \nu_{\mathrm{turb}}$ ($\overline{U} = 244$ m/s being the velocity due to differential rotation at the equator, and R the solar radius). Also shown in the plot are the observed solar Rossby modes amplitudes, in black. There is not only qualitative, but also quantitative agreement between the two, with Rossby mode amplitudes starting out at a fraction of m/s for very low order modes, then increasing with m, until they reach a plateau at around $m \sim 10$, with modes reaching an amplitude between 1.5 and 2.5 m/s depending on $\mathrm{Re}_{\mathrm{turb}}$.

4. Discussion and conclusion

The general qualitative and quantitative agreement between the Rossby mode amplitudes predicted by our model and the solar observations is striking, and supports the hypothesis that the linearly-stable solar inertial modes are indeed stochastically excited by turbulent convection. The mechanism is similar to the one traditionally invoked to explain the observed the observed amplitudes of acoustic modes in low-mass stars (Goldreich and Keeley 1977; Samadi and Goupil 2001), where the small-scale turbulent motions in the convective zone exert mechanical work on (and therefore inject energy into) the modes. A key difference is that while p-modes are mainly excited by vertical turbulent motions, inertial modes are much more sensitive to toroidal turbulent motions.

The qualitative transition that we find at $m \sim 5$ between clearly resolved structures for low orders and overlapping modes for higher orders is an indication that the interpretation of the data in the latter case may be much more complex than in the former. In particular, it would suggest that in order to exploit the full probing potential of inertial modes, it is necessary to model not only their individual, discrete eigenfrequencies (as well as their respective eigenfunctions), but also to model the whole shape of the power spectral density, a task for which linear analysis is not sufficient.

References

Y. Bekki, R. H. Cameron, and L. Gizon. Theory of solar oscillations in the inertial frequency range: Amplitudes of equatorial modes from a nonlinear rotating convection simulation. *arXiv e-prints*, art. arXiv:2208.11081, Aug. 2022.

D. Fournier, L. Gizon, and L. Hyest. Viscous inertial modes on a differentially rotating sphere: Comparison with solar observations. *A&A*, 664:A6, Aug. 2022. doi: 10.1051/0004-6361/202243473.

L. Gizon, D. Fournier, and M. Albekioni. Effect of latitudinal differential rotation on solar Rossby waves: Critical layers, eigenfunctions, and momentum fluxes in the equatorial β plane. *A&A*, 642:A178, Oct. 2020. doi: 10.1051/0004-6361/202038525.

L. Gizon, R. H. Cameron, Y. Bekki, A. C. Birch, R. S. Bogart, A. S. Brun, C. Damiani, D. Fournier, L. Hyest, K. Jain, B. Lekshmi, Z.-C. Liang, and B. Proxauf. Solar inertial modes: Observations, identification, and diagnostic promise. *A&A*, 652:L6, Aug. 2021. doi: 10.1051/0004-6361/202141462.

P. Goldreich and D. A. Keeley. Solar seismology. II. The stochastic excitation of the solar p-modes by turbulent convection. *ApJ*, 212:243–251, Feb. 1977. doi: 10.1086/155043.

C. S. Hanson, S. Hanasoge, and K. R. Sreenivasan. Discovery of high-frequency retrograde vorticity waves in the Sun. *Nature Astronomy*, 6:708–714, Mar. 2022. doi: 10.1038/s41550-022-01632-z.

Z.-C. Liang, L. Gizon, A. C. Birch, and T. L. Duvall. Time-distance helioseismology of solar Rossby waves. *A&A*, 626:A3, June 2019. doi: 10.1051/0004-6361/201834849.

M. J. Lighthill. Predictions on the Velocity Field Coming from Acoustic Noise and a Generalized Turbulence in a Layer Overlying a Convectively Unstable Atmospheric Region. In R. N. Thomas, editor, *Aerodynamic Phenomena in Stellar Atmospheres*, volume 28, page 429, Jan. 1967.

B. Löptien, L. Gizon, A. C. Birch, J. Schou, B. Proxauf, T. L. Duvall, R. S. Bogart, and U. R. Christensen. Global-scale equatorial Rossby waves as an essential component of solar internal dynamics. *Nature Astronomy*, 2:568–573, May 2018. doi: 10.1038/s41550-018-0460-x.

J. Papaloizou and J. E. Pringle. Non-radial oscillations of rotating stars and their relevance to the short-period oscillations of cataclysmic variables. *MNRAS*, 182:423–442, Feb. 1978. doi: 10.1093/mnras/182.3.423.

J. Philidet and L. Gizon. Interaction of solar inertial modes with turbulent convection. A 2D model for the excitation of linearly stable modes. *A&A*, 673:A124, May 2023. doi: 10.1051/0004-6361/202245666.

R. Samadi and M. J. Goupil. Excitation of stellar p-modes by turbulent convection. I. Theoretical formulation. *A&A*, 370:136–146, Apr. 2001. doi: 10.1051/0004-6361:20010212.

S. A. Triana, G. Guerrero, A. Barik, and J. Rekier. Identification of Inertial Modes in the Solar Convection Zone. *ApJ*, 934(1):L4, July 2022. doi: 10.3847/2041-8213/ac7dac.

Dynamics of Solar and Stellar Convection Zones and Atmospheres
Proceedings IAU Symposium No. 365, 2024
A. V. Getling & L. L. Kitchatinov, eds.
doi:10.1017/S1743921324000450

Temporal Variation of Solar Equatorial Rossby Modes with Azimuthal Orders $6 \leq m \leq 10$

B. Lekshmi[1], Laurent Gizon[1,2], Kiran Jain[3], Zhi–Chao Liang[1] and Jordan Philidet[1]

[1]Max–Planck-Institut für Sonnensystemforschung, 37077 Göttingen, Germany
email: lekshmib@mps.mpg.de

[2]Institut für Astrophysik, Georg–August-Universität Göttingen, 37077 Göttingen, Germany

[3]National Solar Observatory, Boulder, CO 80303, USA

Abstract. We use nearly two decades of helioseismic data obtained from the GONG (2002–2020) and HMI (2010–2020) ring-diagram pipelines to examine the temporal variations of the properties of individual equatorial Rossby modes with azimuthal orders in the range $6 \leq m \leq 10$. We find that the mode parameters obtained from GONG and HMI are consistent during the data overlapping period of 2010–2020. The power and the frequency of each mode exhibit significant temporal variations over the full observing period. Using the GONG data during solar cycles 23 and 24, we find that the mode power averaged over $6 \leq m \leq 10$ shows a positive correlation with the sunspot number (0.42), while the averaged frequency shift is anti-correlated (-0.91). The anti-correlation between the average mode power and frequency shift is -0.44.

Keywords. helioseismology, solar inertial modes, sunspot cycle

---

Introduction

The solar equatorial Rossby modes were first observed using helioseismic observations from the Helioseismic and Magnetic Imager (HMI) onboard the Solar Dynamics Observatory (Löptien 2018; Liang et al. 2019; Gizon et al. 2021) and the Global Oscillation Network Group (GONG) (Hanson et al. 2020). In the Carrington frame, these modes closely follow the standard dispersion relation for classical sectoral modes, $\nu \approx -2\nu_0/(m+1)$ where ν_0 is the equatorial rotational frequency, m is the azimuthal order, and the negative sign indicates retrograde propagation.

Waidele and Zhao (2023), using velocities obtained from time-distance and ring-diagram analysis of HMI data, measured solar-cycle variations in the mode amplitudes. They found that the average mode amplitude over $3 \leq m \leq 16$ positively correlates with the sunspot number. They also reported that the mode frequencies vary over the solar cycle, with more negative frequencies during solar maximum.

The present study focuses on the temporal variation of the power and the frequency of individual equatorial Rossby modes with $m = 6, 7, \cdots, 10$. In addition, we consider both the HMI and the GONG datasets.

Data and Analysis

We use the horizontal flows in the co-latitudinal direction (u_θ) from the ring diagram (RD) pipelines of GONG (Corbard et al. 2003) and HMI (Bogart et al. 2011a), covering the periods January 2002 to June 2020 and January 2011 to June 2020, respectively. The data extend over $\pm 52.5°$ in latitude and central meridian distance for GONG and $\pm 67.5°$ for HMI. We remove the zero and annual frequencies in the flows following the method of Proxauf et al. (2020). An additional one-day periodicity aliased as 8.33 days in the GONG measurements is also removed. We obtain the horizontal flows at a depth of ~ 2 Mm in the Carrington frame. We denote the component of velocity in the colatitudinal direction symmetrized across the equator by $u_\theta^+(\theta, \phi, t)$, where θ is the colatitude, ϕ the Carrington longitude, and t time.

The GONG and HMI time series are divided into overlapping segments of duration $T = 4$ years, shifted by multiples of six months. For GONG, the central times of the segments are given by $t_n = 1.1.2004 + n \times 6$ months with $n = 0, 1, \cdots, 31$. For HMI, the central times are $t_n = 1.1.2013 + n \times 6$ months with $n = 0, 1, \cdots, 12$.

The corresponding power spectra are given by,

$$P_m(\nu, t_n) = \frac{1}{N_\theta} \sum_{|\theta - 90°| \leq 30°} \left| \sum_{\phi, t'} \mathrm{rect}\left(\frac{t' - t}{T}\right) u_\theta^+(\theta, \phi, t') e^{\mathrm{i}(m\phi - 2\pi\nu t')} \right|^2, \qquad (1)$$

where 'rect' is a symmetric rectangular window function of width 1, $N_\theta = 3$ is the number of independent latitude bins over which the power is averaged.

We also compute a reference power spectrum for each m obtained by averaging over the data segments. The peak of power associated with the Rossby mode in each average power spectrum is fitted with a Lorentzian profile,

$$L_m(\nu) = \overline{A}_m \left[1 + \left(\frac{\nu - \overline{\nu}_m}{\overline{\Gamma}_m/2}\right)^2\right]^{-1} + \overline{B}_m, \qquad (2)$$

where $\overline{A}_m$, $\overline{\nu}_m$, $\overline{\Gamma}_m$ and $\overline{B}_m$ are the mode amplitude, frequency, full width at half maximum, and background power respectively. The probability density function of the power spectrum at fixed frequency is approximated by a χ^2 with $2 \times N \times N_\theta$ degrees of freedom, where N is the number of independent time segments and the factor N_θ is for the number of independent latitude bins.

The GONG and HMI power spectra, together with the fit to the GONG power spectra, are shown in Fig. 1. The fitting frequency range is chosen for each mode separately (see vertical dashed lines in Fig. 1). We find that the mode parameters extracted from the GONG and HMI power spectra are consistent.

To study the temporal variation of the mode parameters, we compute the mode power (E_m) and the mode frequency (ν_m) of the modes in each time segment as follows:

$$E_m(t_n) = \sum_{\nu \in W_m} [P_m(\nu, t_n) - B_m(t_n)] \, \Delta\nu, \qquad (3)$$

$$\nu_m(t_n) = \frac{\sum_{\nu \in W_m} \nu [P_m(\nu, t_n) - B_m(t_n)]}{\sum_{\nu \in W_m} [P_m(\nu, t_n) - B_m(t_n)]}. \qquad (4)$$

In the above equations, the sums are computed over a frequency interval $W_m = [\overline{\nu}_m - 3\overline{\Gamma}_m/2, \overline{\nu}_m + 3\overline{\Gamma}_m/2]$, which is centered at the average mode frequency $\overline{\nu}_m$. The frequency resolution is denoted by $\Delta\nu \approx 8$ nHz and the background power for the n-th time segment is denoted by $B_m(t_n)$. The background power depends on the phase of the solar cycle and is obtained by computing the median power outside the peak of power. The time stamps for which the mode power is not significant are discarded. A sample power spectrum of the $m = 8$ mode in the time window centered at 2010.5 is shown in Fig. 2.

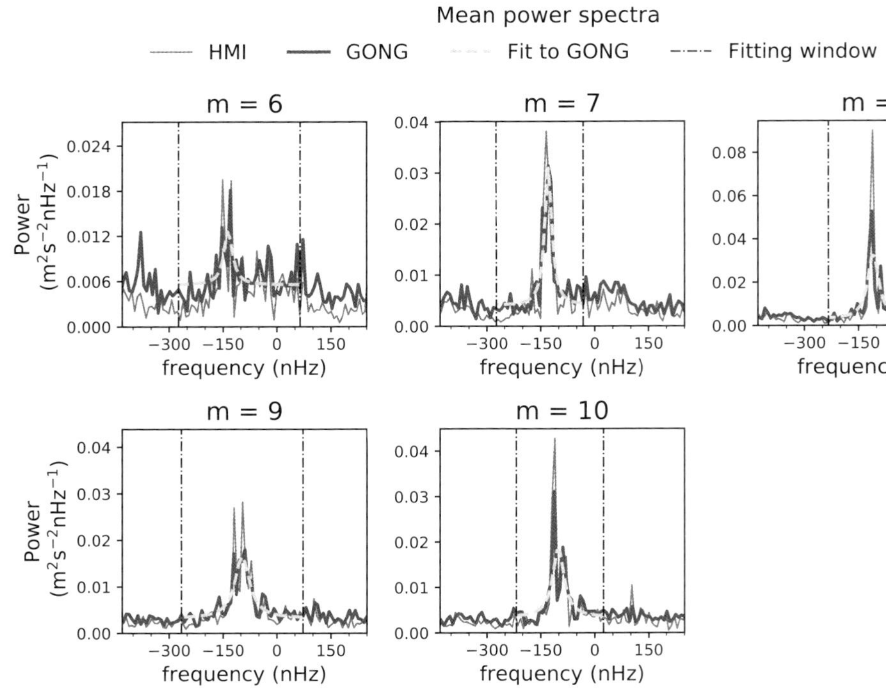

Figure 1. Power spectra of the equatorial Rossby modes for each m. The power spectra are averaged over all segments of duration $T = 4$ years (frequency resolution is approximately 8 nHz). The dark blue curves are for GONG, the magenta curves for HMI. The Lorentzian fits to the GONG power spectra are overplotted in cyan. The fitting window is bounded by the two vertical dashed lines.

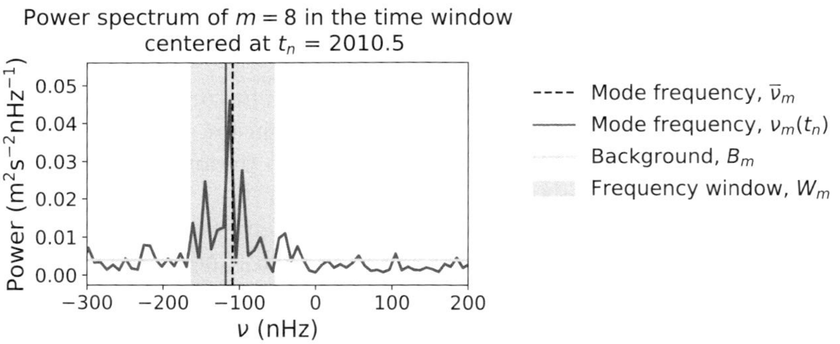

Figure 2. The GONG power spectrum of the $m = 8$ mode in the four year time window centered at $t_n = 2010.5$. The average mode frequency $\overline{\nu}_m$ for the entire data set is given by the vertical black line. The mode frequency $\nu_m(t_n)$ obtained from Eq. (3) for this particular time stamp is marked by the vertical red line. The cyan horizontal line shows the estimated background $B_m(t_n)$.

The E_m and ν_m of different modes as functions of time are plotted in Fig. 3. The shaded regions represent the 68% confidence interval on each parameter, estimated from Monte Carlo simulations.

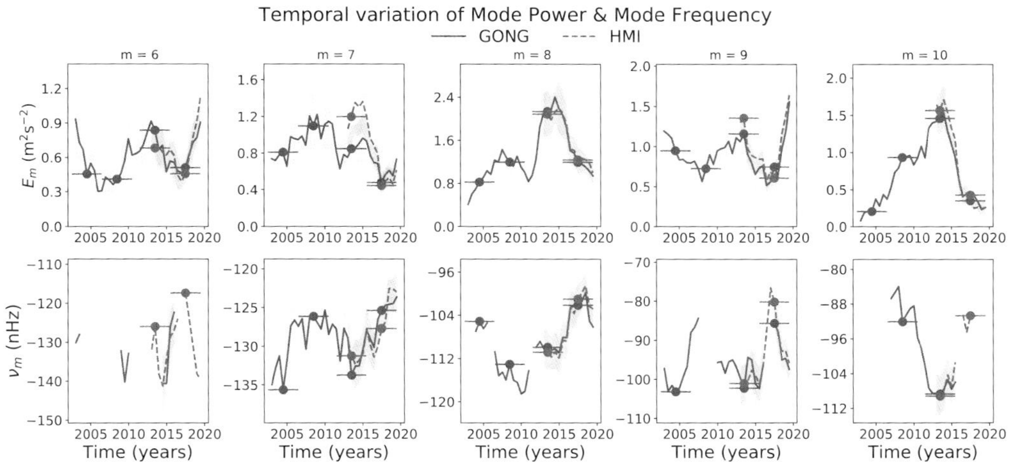

Figure 3. The temporal variation of mode power E_m (top row) and mode frequency ν_m (bottom row) of the equatorial Rossby modes with m from 6 to 10 from GONG and HMI data. The shaded regions represent the 68% confidence bounds estimated from Monte Carlo simulations.

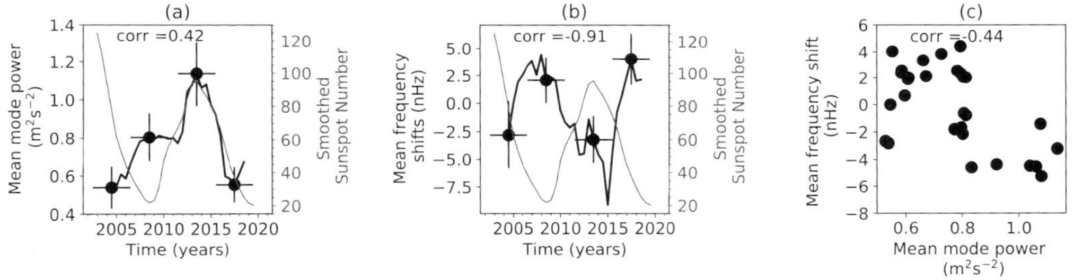

Figure 4. (a) Mode power $\langle E_m(t_n)\rangle$ and (b) mode frequency shift $\langle \nu_m(t_n) - \overline{\nu}_m\rangle$ averaged over m from 6 to 10 as functions of time. The Pearson correlation coefficients between these and the smoothed sunspot number (thin curve) are given on the plots. Only the non-overlapping time segments (points with error bars) were used in the computation of the correlation coefficients. (c) Scatter plot of mode frequency shifts versus mode power. The Pearson correlation coefficient is -0.44.

Discussion

To study the relationship between the mode parameters and the solar cycle, we obtain the sunspot numbers (SSN) from the SILSO World Data Center (2020). In order to reduce the noise level at any given time stamp, we average the mode parameters over m (6 to 10). These m-averaged parameters are plotted in Fig. 4. The Pearson correlation coefficients between the parameters and the SSN are computed using only non-overlapping (statistically independent) time segments. We find a positive correlation coefficient (0.42) between the mode power and SSN, while the frequency shifts show a very strong anti-correlation (-0.91) with the SSN. An anti-correlation (-0.44) is observed between the mode power and the frequency shift.

Our observations align broadly with the HMI findings from Waidele and Zhao (2023). Using the GONG data, we confirm that the anticorrelation between the frequency shifts and SSN is very strong through both cycles 23 and 24. However, we find that the correlation obtained between the mode power and the SSN is not as strong during cycle 23 as during cycle 24. We note that the observed frequency shifts have a sign opposite to that given by a model (Goddard et al. 2020) based on the perturbation of the modes by the torsional oscillations (time-varying rotation). Thus we infer that magnetic effects must play an important role in the solar-cycle variations of the equatorial Rossby mode parameters.

Acknowledgements

We acknowledge the support from ERC Synergy Grant WHOLE SUN 810218 and Deutsche Forschungsgemeinschaft (DFG, German Research Foundation) through SFB 1456/432680300 Mathematics of Experiment, project C04. Data were acquired by GONG instruments operated by NISP/NSO/AURA/NSF with contribution from NOAA. The HMI data are courtesy of NASA/SDO and the HMI Science Team.

References

Löptien, Björn & Gizon, Laurent & Birch, Aaron C. &Schou, Jesper & Proxauf, Bastian and Duvall, Thomas L. & Bogart, Richard S. & Christensen, Ulrich R.,2018, *Nature Astronomy*, 2, 568–573

Liang, Z.-C., Gizon, L., Birch, A. C., & Duvall, T. L. 2019, Time-distance helioseismology of solar Rossby waves *Astron. Astrophys.*, 626, A3.

Gizon, L., Cameron, R. H., Bekki, Y., Birch, A. C., Bogart, R. S., Brun, A. S., Damiani, C., Fournier, D., Hyest, L., Jain, K., Lekshmi, B., Liang, Z.-C., & Proxauf, B. 2021, Solar inertial modes: Observations, identification, and diagnostic promise. *Astron. Astrophys.*, 652, L6.

Hanson, C. S., Gizon, L., & Liang, Z.-C. 2020, Solar Rossby waves observed in GONG++ ring-diagram flow maps. *Astron. Astrophys.*, 635, A109.

Waidele, M. & Zhao, J. 2023, Observed power and frequency variations of solar rossby waves with solar cycles, *Astrophys. J.*, 954, L26.

Corbard, T., Toner, C., Hill, F., Hanna, K. D., Haber, D. A., Hindman, B. W., & Bogart, R. S. Ring-diagram analysis with GONG++. In Sawaya–Lacoste, H., editor, *GONG+ 2002. Local and Global Helioseismology: the Present and Future* 2003, Vol. 517, *ESA Special Publication*, pp. 255–258.

Bogart, R. S., Baldner, C., Basu, S., Haber, D. A., & Rabello–Soares, M. C. 2011, HMI ring diagram analysis I. The processing pipeline. *J. Phys. Conf. Series*, 271, 012008.

Proxauf, B., Gizon, L., Löptien, B., Schou, J., Birch, A. C., & Bogart, R. S. 2020, Exploring the latitude and depth dependence of solar Rossby waves using ring-diagram analysis, *Astron. Astrophys.*, 634, A44.

SILSO, World Data Center - Sunspot Number and Long-term Solar Observations, Royal Observatory of Belgium, on-line Sunspot Number catalogue: http://www.sidc.be/SILSO/, 2002–2020

Goddard, C. R. and Birch, A. C. and Fournier, D. and Gizon, L. 2020, Predicting frequency changes of global-scale solar Rossby modes due to solar cycle changes in internal rotation, *Astron. Astrophys.*, 640, L10.

Dynamics of Solar and Stellar Convection Zones and Atmospheres
Proceedings IAU Symposium No. 365, 2024
A. V. Getling & L. L. Kitchatinov, eds.
doi:10.1017/S1743921324000401

Observational Study of Reynolds Stresses Associated with Solar Inertial Modes

Yash Mandowara[1], Yuto Bekki[1], Richard S. Bogart[2] and Laurent Gizon[1,3]

[1]Max–Planck-Institut für Sonnensystemforschung, 37077 Göttingen, Germany
email: yash.uni13n@gmail.com

[2]Stanford University, Stanford, CA 94305-4085, USA

[3]Institut für Astrophysik, Georg–August-Universität Göttingen, 37077 Göttingen, Germany

Abstract. We study the $m=1$ high-latitude inertial mode and its contribution to the latitudinal transport of the Sun's angular momentum. Ring-diagram helioseismology applied to 5° tiles is used to obtain the horizontal flows near the surface of the Sun. Using 10 years of data from SDO/HMI, we report on the horizontal eigenfunction and Reynolds stress $Q_{\theta\phi} = \langle u'_\theta u'_\phi \rangle$ for the $m=1$ high-latitude inertial mode (frequency -86.3 nHz, critical latitudes $\pm 58^\circ$). We find that $Q_{\theta\phi}$ takes significant values above the critical latitude and is positive (negative) in the northern (southern) hemisphere, implying equatorward transport of angular momentum. The $Q_{\theta\phi}$ peaks above latitude 70° with a value of 38 m^2/s^2.

Keywords. Solar inertial modes, solar oscillations, helioseismology, solar rotation

1. Introduction

Inertial modes are low-frequency modes of global oscillations whose restoring force is the Coriolis force. Although inertial modes were predicted to exist in the Sun since 1970s, they were only recently detected on the Sun. Using 10 years of data from SDO/HMI, Löptien et al. (2018) unambiguously detected the equatorial Rossby modes, a particular class of quasi-toroidal inertial modes. Gizon et al. (2021) further detected and identified a large number of additional modes at middle and high latitudes in the inertial frequency range. The modes were identified with the help of a linear eigenmode solver (Bekki et al. 2022a). More recently Hanson et al. (2022) also reported another family of inertial modes on the Sun.

All of these observed modes propagate in the retrograde direction (opposite to rotation) when seen from the Carrington frame of reference. For these modes, the Sun's latitudinal differential rotation plays a critical role as it introduces critical latitudes where the phase speed of the mode is equal to the local differential rotation speed (Gizon et al. 2020). Theoretical studies suggest that the inertial modes have substantial Reynolds stresses near the critical latitudes (Gizon et al. 2020; Bekki et al. 2022a,b). Theory also shows that the $m=1$ mode is linearly unstable to the sharp latitudinal gradient of rotation at high latitudes (Fournier et al. 2022) and to the associated latitudinal entropy gradient (Bekki et al. 2022a) — this explains why it has a large velocity amplitude (Gizon et al. 2021).

In this study, we report the observational measurement of the horizontal Reynolds stresses associated with $m=1$ high-latitude mode, which has the largest velocity amplitude among all the observed solar inertial modes. This will help us to assess the contribution from the observed inertial modes to the latitudinal transport of the angular momentum of the Sun.

2. Methods

2.1. *Data*

In this study, we use ring-diagram local helioseismology (Hill 1988) to obtain the near-surface flow maps of the Sun. The Dopplergrams are obtained from the Helioseismic and Magnetic Imager (HMI) onboard the Solar Dynamics Observatory (SDO). The ring-diagram method involves computing 3D (k_x, k_y, ω) local power spectra from small patches (tiles) on the solar surface. Here, we use the highest-resolution ring-diagram data with tile size of 5°, which covers the latitudinal range from 85° N to 85° S (Bogart et al. 2011a,b, 2015). The flows near the surface cause Doppler shifts in the f- and p-mode frequencies in the local power spectrum. Fitting for these shifts and averaging the fitted parameters from lower to upper turning points gives the underlying horizontal flows.

Due to lack of sufficient usable data at higher latitudes, we use a cropped version of this data cube, focusing on latitudes 75° N to 75° S. We use a 10-year-long data set from May 2010 to Feb 2021 with a temporal cadence of 9.08 hr (1/72 of the Carrington rotation period). This results in a data cube with $N_t \times N_\theta \times N_\phi = 10418 \times 61 \times 144$ grid points. The flow maps consist of the longitudinal velocity u_x and the latitudinal velocity u_y, both as functions of time t, colatitude θ, and longitude ϕ.

After removing outliers, we remove the systematic effects due to the B_0 angle by modeling it as a harmonic signal with a one-year period. This gives us flow maps free of the most significant systematic errors.

2.2. *Mode analysis*

In contrast to the analysis of Gizon et al. (2021), we do not perform north-south symmetrization of flow maps for mode extraction. The HMI ring-diagram products (u_x, u_y) can be easily converted to standard (u_θ, u_ϕ) in spherical coordinate,

$$u_\theta = -u_y, \qquad u_\phi = u_x. \tag{1}$$

From the processed data above, we generate a power spectrum by performing the Fourier transform in time and longitude:

$$j \in \{\theta, \phi\}, \qquad \hat{u}_j(\theta, \nu) = \frac{1}{f(\theta)} \sum_t \sum_\phi u_j(\theta, \phi, t) e^{\mathrm{i}(2\pi\nu t - m\phi)}, \tag{2}$$

where ν and $m = 1$ are the frequency and the azimuthal order, respectively, and $f(\theta)$ is a correction for missing data,

$$f(\theta) = \frac{\text{No. of measurements available at fixed } \theta}{\text{total number of points at fixed } \theta}. \tag{3}$$

2.3. *Extraction of eigenfunctions*

To filter out a mode identified by its azimuthal order m and peak frequency ν_0, we apply a rectangular filter 'rect' is unity in the range $[\nu_0 - \gamma, \nu_0 + \gamma]$ and zero elsewhere, and γ is the FWHM of the line profile of the mode in the power spectrum. This mode has a peak frequency and line-width of -86.3 nHz and 7.8 nHz, respectively, and can be seen clearly in the power spectrum of longitudinal velocity u_ϕ (Gizon et al. 2021, see Table A.1 therein). To obtain the filtered data in the time domain we apply an inverse

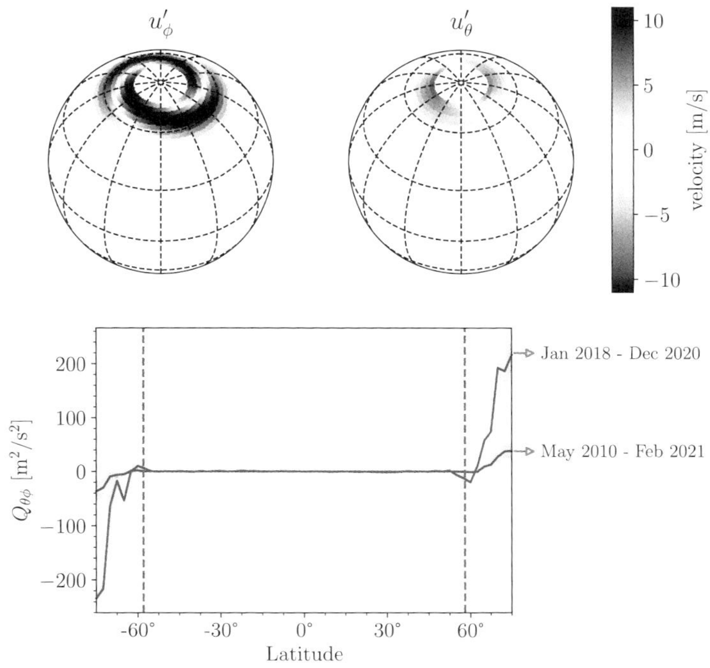

Figure 1. Observed velocity eigenfunctions and horizontal Reynolds stress of the $m = 1$ high-latitude mode (frequency: -86.3 nHz, critical latitude: 58°). The eigenfunctions are obtained from 10 years of SDO/HMI 5° ring-diagram data. Bottom panel: Reynolds stress $Q_{\theta\phi}$ during the entire observation period 2010–2021 (blue curve) and during the quiet sun period 2018–2020 (red curve). The green dashed lines mark the critical latitudes for this mode.

Fourier transform in time,

$$U^j_{qp} = \frac{1}{N_t} \sum_\nu \hat{u}_j(\theta_p, \nu) \text{ rect}\left(\frac{\nu - \nu_0}{\gamma}\right) e^{-2\mathrm{i}\pi\nu t_q}, \quad p = 1, \cdots, N_\theta, \quad q = 1, \cdots, N_t. \tag{4}$$

Next we compute the SVD of the matrix $\left[U^\theta \; U^\phi\right] \in \mathbb{C}^{N_t \times 2N_\theta}$ and retain only one term in the sum, $s_1 w[v_\theta^H \; v_\phi^H]$, where s_1 is the largest singular value and the vectors $w \in \mathbb{C}^{N_t}$ and $v_j \in \mathbb{C}^{N_\theta}$ are the corresponding singular vectors. The superscript H denotes the conjugate transpose. The vectors v_θ and v_ϕ encapsulate the θ-dependence of the mode eigenfunctions. Finally, we normalize the eigenfunctions by the factor K, such that $u'_j(\theta, \phi)$ is the effective root mean square representation of the velocity profile of the the mode.

$$u'_j(\theta, \phi) := \Re\left\{K s_1 v_j^*(\theta) e^{\mathrm{i}m\phi}\right\}, \quad K = 2N_\phi^{-1} N_t^{-1/2}. \tag{5}$$

To obtain the latitudinal profile of the horizontal Reynolds stress we compute

$$Q_{\theta\phi}(\theta) = \langle u'_\theta(\theta, \phi) u'_\phi(\theta, \phi)\rangle = K^2 s_1^2 \; \Re\{v_\theta^*(\theta) v_\phi(\theta)\}/2 \tag{6}$$

where the angular brackets denote a longitudinal average.

3. Results

Figure 1 shows the obtained velocity eigenfunctions of the $m = 1$ high-latitude mode, which has a large longitudinal velocity of ≈ 13 m/s above 60° and shows a spiral pattern around the poles. The bottom panel of figure 1 shows the computed horizontal Reynolds stress $Q_{\theta\phi}$ of the $m = 1$ high-latitude mode, which is positive (negative) in the northern (southern) hemisphere, indicating equatorward transport of angular momentum. Hathaway and Upton (2021, their figure 7) reported a similar latitudinal profile.

For this mode, the Reynolds stress, $Q_{\theta\phi}$ has a peak value of about 38 m^2/s^2, which it attains above latitude 70°. To understand the cause of these Reynolds stresses, we use the solar differential rotation data (Larson and Schou 2018) to compute the critical latitudes $\lambda_{\rm crit}$. The locations of the critical latitudes ($\lambda_{\rm crit} = \pm 58°$ for the $m = 1$ high-latitude mode are shown by green dotted lines in Fig. 1. We note that most of the non-zero $Q_{\theta\phi}$ is found above the critical latitude $\lambda_{\rm crit}$.

We also compute the Reynolds stress focusing only on the quiet Sun period (Jan 2017 - Dec 2020) during which this mode has a much higher amplitudes. The angular momentum transported by the $m = 1$ high-latitude mode is much more significant during the quiet Sun period, see the red curve in Fig. 1. The importance of the contribution of the $m = 1$ high-latitude mode to the overall angular momentum balance is discussed in an upcoming paper by Bekki et al. (2024).

Acknowledgements

We thank Z-C. Liang and D. Fournier for helpful discussions. The HMI data are courtesy of NASA/SDO and the HMI Science team. Y.M. acknowledges support from the International Max–Planck Research School (IMPRS) for Solar System Science at the University of Göttingen. We acknowledge support from ERC Synergy Grant WHOLESUN 81218 and Deutsche Forschungsgemeinschaft (DFG, German Research Foundation) through SFB 1456/432680300 Mathematics of Experiment, project C04.

References

Basu, S., Antia, H. M., & Tripathy, S. C. 1999, Ring Diagram Analysis of Near-Surface Flows in the Sun. *ApJ*, 512, 458–470.

Bekki, Y., Cameron, R. H., & Gizon, L. 2022a, Theory of solar oscillations in the inertial frequency range: Linear modes of the convection zone. *A&A*, 662, A16.

Bekki, Y., Cameron, R. H., & Gizon, L. 2022b, Theory of solar oscillations in the inertial frequency range: Amplitudes of equatorial modes from a nonlinear rotating convection simulation. *A&A*, 666, A135.

Bekki, Y., Cameron, R. H., & Gizon, L. 2024, The Sun's differential rotation is controlled by baroclinically unstable high-latitude inertial modes. *Science Adv.*, under review.

Bogart, R. S., Baldner, C., Basu, S., Haber, D. A., & Rabello–Soares, M. C. HMI ring diagram analysis I. The processing pipeline. In *GONG-SoHO 24: A New Era of Seismology of the Sun and Solar-Like Stars* 2011, Vol. 271 of *J. Phys. Conf. Series*, 012008.

Bogart, R. S., Baldner, C., Basu, S., Haber, D. A., & Rabello–Soares, M. C. HMI ring diagram analysis II. Data products. In *GONG-SoHO 24: A New Era of Seismology of the Sun and Solar-Like Stars* 2011, Vol. 271, *J. Phys. Conf. Series*, 012009.

Bogart, R. S., Baldner, C. S., & Basu, S. 2015, Evolution of Near-surface Flows Inferred from High-resolution Ring-diagram Analysis. *ApJ*, 807, 125.

Fournier, D., Gizon, L., & Hyest, L. 2022, Viscous inertial modes on a differentially rotating sphere: Comparison with solar observations. *A&A*, 664, A6.

Gizon, L., Cameron, R. H., Bekki, Y., Birch, A. C., Bogart, R. S., Brun, A. S., Damiani, C., Fournier, D., Hyest, L., Jain, K., Lekshmi, B., Liang, Z.-C., & Proxauf, B. 2021, Solar inertial modes: Observations, identification, and diagnostic promise. *A&A*, 652, L6.

Gizon, L., Fournier, D., & Albekioni, M. 2020, Effect of latitudinal differential rotation on solar Rossby waves: Critical layers, eigenfunctions, and momentum fluxes in the equatorial β plane. *A&A*, 642, A178.

Hanson, C. S., Hanasoge, S., & Sreenivasan, K. R. 2022, Discovery of high-frequency retrograde vorticity waves in the Sun. *Nature Astron.*, 6, 708–714.

Hathaway, D. H. & Upton, L. A. 2021, Hydrodynamic Properties of the Sun's Giant Cellular Flows. *ApJ*, 908, 160.

Hill, F. 1988, Rings and Trumpets—Three-dimensional Power Spectra of Solar Oscillations. *ApJ*, 333, 996.

Larson, T. P. & Schou, J. 2018, Global-Mode Analysis of Full-Disk Data from the Michelson Doppler Imager and the Helioseismic and Magnetic Imager. *Solar Phys.*, 293, 29.

Löptien, B., Gizon, L., Birch, A. C., Schou, J., Proxauf, B., Duvall, T. L., Bogart, R. S., & Christensen, U. R. 2018, Global-scale equatorial Rossby waves as an essential component of solar internal dynamics. *Nature Astron.*, 2, 568–573.

Dynamics of Solar and Stellar Convection Zones and Atmospheres
Proceedings IAU Symposium No. 365, 2024
A. V. Getling & L. L. Kitchatinov, eds.
doi:10.1017/S1743921324000413

Probing the superadiabaticity of the solar convection zone with inertial modes

Prithwitosh Dey[1], **Yuto Bekki**[1], **and Laurent Gizon**[1,2]

[1]Max–Planck-Institut für Sonnensystemforschung, 37077 Göttingen, Germany
email: dey@mps.mpg.de

[2]Institut für Astrophysik, Georg–August-Universität Göttingen, 37077 Göttingen, Germany

Abstract. Our understanding of solar convection is incomplete. A crucial gap is the unknown superadiabaticity in the solar convection zone, $\delta = \nabla - \nabla_{\rm ad}$. Global modes of oscillations in the inertial frequency range are sensitive to δ and serve as a novel tool to explore solar convection. Here, we address the forward problem where the superadiabaticity $\delta(r)$ varies with radius. We solve the 2.5D eigenvalue problem, considering the linearized equations for momentum, mass and energy conservation with respect to a realistic solar model. We find that the frequency and eigenfunction of the $m = 1$ high-latitude mode are influenced by δ in the lower convection zone. Our prescribed setup suggests that the superadiabaticity in the lower half of the convection zone is below 2.4×10^{-7} to reach a qualitative agreement with the observed eigenfunction.

Keywords. Sun: interior, Sun: oscillations, Sun: helioseismology, Sun: convection

1. Introduction

In the outer 30% of the Sun's interior, the thermal energy is transported by convection. Since convection can efficiently mix the entropy, the mean stratification in the convection zone (CZ) is believed to be very close to adiabatic. A tiny deviation from the adiabatic stratification is measured by the superadiabaticity

$$\delta = \nabla - \nabla_{\rm ad} = -\frac{H_{\rm p}}{c_{\rm p}}\frac{ds_0}{dr},$$

where $\nabla = d\ln T/d\ln p$ is the double-logarithmic temperature gradient, $H_{\rm p}$ is the pressure scale height, $c_{\rm p}$ is the specific heat at constant pressure, and s_0 denotes the horizontal mean of the background entropy. The stratification is convectively unstable when $\delta > 0$ (superadiabatic) and is convectively stable when $\delta < 0$ (subadiabatic). According to the local mixing-length model (Böhm-Vitense 1958), δ is estimated to be of the order of 10^{-6}. Acoustic modes, which are the subject of conventional p-mode helioseismology are largely insensitive to the value of δ.

Evidence indicates that the solar convection cannot be described by local mixing-length models. Time-distance helioseismology has provided an observational upper limit on the subsurface convective velocity at large scales which are much smaller than the typical convective speed estimated by the mixing-length model (Hanasoge et al. 2012). Furthermore, recent numerical simulations (which are consistent with the local mixing-length model) have difficulty in reproducing the observed differential rotation of the Sun (e.g., Hotta et al. 2023). These problems are referred to as the convective conundrum (O'Mara et al. 2016). It has been implied that the solar convective energy transport is a highly non-local process (Brandenburg 2016) and the mean stratification in the CZ could be much less superadiabatic than typically assumed (Cossette and Rast 2016).

Recent numerical simulations also show that a weakly subadiabatic layer can exist near the bottom of the CZ as a result of non-local convective heat transport (Hotta et al. 2022; Käpylä 2023).

Recently, various kinds of inertial modes have been detected and identified on the Sun. They are global-scale low-frequency modes of oscillation restored by Coriolis force. The observed solar inertial modes include the equatorial Rossby modes (Löptien et al. 2018; Liang et al. 2019; Proxauf et al. 2020; Mandal and Hanasoge 2020), high-latitude and mid-latitude modes (Gizon et al. 2021) and others (Hanson et al. 2022). Theoretical models of the solar inertial modes predict that some modes are highly sensitive to the superadiabaticity δ in the CZ (Bekki et al. 2022; Bekki 2024). It was demonstrated through numerical studies by Gizon et al. (2021) that the mode structure can be reproduced with $\delta \leq 2\times10^{-7}$. However, these studies have used the simplifying assumption of a spatially constant δ. Here, we go a step further and aim to constrain the superadiabaticity in the CZ as a function of radius $\delta(r)$ by solving the forward problem of the solar inertial modes.

2. The forward problem

The model equations consist of the linearized equations of motion, continuity, and energy conservation, in a frame rotating at the Carrington angular velocity Ω_0. The magnetic field is not included. We consider a realistic solar background stratification (Christensen–Dalsgaard et al. 1996), the differential rotation measured by global helioseismology (Larson and Schou 2018), and the latitudinal entropy gradient required to sustain the observed differential rotation via thermal wind balance. For simplicity, uniform viscous and thermal diffusivities of 5×10^{11} cm^2 s^{-1} are chosen.

The function $\delta(r)$ enters the linearised energy equation:

$$\frac{Ds_1}{Dt} = c_{\mathrm{p}}\delta\frac{v_r}{H_{\mathrm{p}}} - \frac{v_\theta}{r}\frac{\partial s_0}{\partial\theta} + \frac{1}{\rho_0 T_0}\nabla\cdot(\kappa\rho_0 T_0\nabla s_1), \tag{1}$$

where the variables have their usual meaning, the subscript 0 is for the background values, and the subscript 1 is for the wave perturbations (Gizon et al. 2021, their appendix B.1). The quantities v_r and v_θ are the radial and colatitudinal components of wave velocity. Furthermore, we parameterize the function $\delta(r)$ as follows:

$$\delta(r) = \delta_{\mathrm{lower}} + \left(\frac{\delta_{\mathrm{upper}} - \delta_{\mathrm{lower}}}{2}\right)\left[1 + \tanh\left(\frac{r - r_{\mathrm{mid}}}{d}\right)\right], \tag{2}$$

where δ_{upper} and δ_{lower} denote the superadiabaticity in the upper and lower CZ, respectively. The transition from δ_{lower} to δ_{upper} occurs at $r = r_{\mathrm{mid}}$ with a transition width d. In this study, we fix $d = 0.03R_\odot$ and δ_{lower}, δ_{upper}, r_{mid} are parameters to be varied.

We use the linear eigenvalue solver of the differentially-rotating CZ from Bekki et al. (2022). It is a 2.5D solver, which computes the eigenmodes of oscillation in the inertial frequency range at fixed longitudinal wavenumber m, over a staggered grid in radius and colatitude. The radial domain is $0.710 \leq r/R_\odot \leq 0.985$.

3. Outlook

Our preliminary results concern the high-latitude mode with $m = 1$ and the equatorial Rossby mode with $m = 3$. The real part of the eigenfrequency, ω^{sim}, and the horizontal velocity eigenfunction at the surface, $\mathbf{v}^{\mathrm{sim}}(\theta,\phi)$, are computed for various combinations of $(\delta_{\mathrm{lower}}, \delta_{\mathrm{upper}}, r_{\mathrm{mid}})$. The function $\delta(r)$ can then be constrained using the observations ω^{obs} and $\mathbf{v}^{\mathrm{obs}}(\theta,\phi)$ reported by Gizon et al. (2021, their table A1).

We find that the surface eigenfunction of the $m = 1$ high-latitude mode is strikingly sensitive to the value of the superadiabaticity in the lower half of the CZ, δ_{lower}. Figure 1

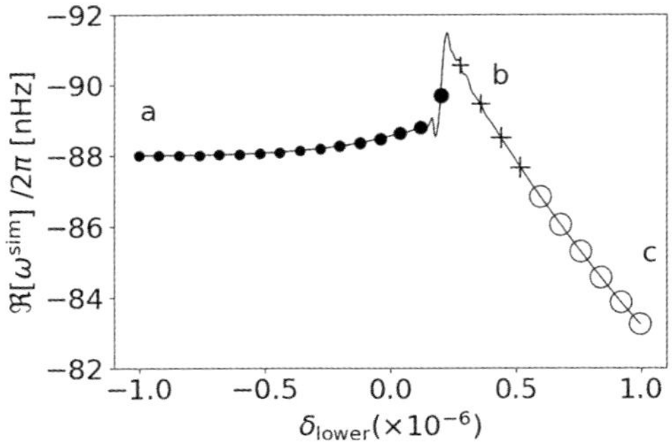

Figure 1. Variation of the frequency of the $m = 1$ high-latitude mode as a function of $\delta_{\rm lower}$. Filled circles mark the modes whose eigenfunctions have the same direction of spiralling as the observations (cf. Fig. 2a), plus symbols mark the modes with no distinct spiralling (cf. Fig. 2b), and open circles show modes with reverse spiralling (cf. Fig. 2c). The size of the marker is proportional to the (positive) growth rate of the modes. The value of $\delta_{\rm lower}$ covers the range $\pm 10^{-6}$, while the parameters $\delta_{\rm upper} = 0$ and $r_{\rm mid} = 0.85 R_{\odot}$ are fixed.

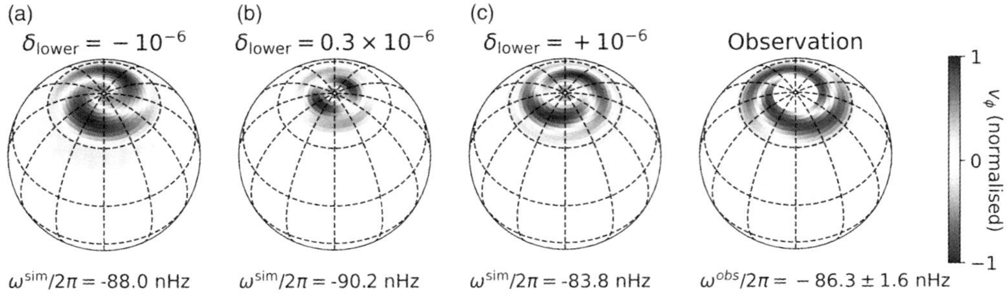

Figure 2. Longitudinal component of the velocity eigenfunction of the $m = 1$ high-latitude mode, indicated by different symbols in Fig. 1. Left to right: mode with the correct spiralling, marked with filled circles; mode with indistinct spiralling, marked with plus symbols; mode with reverse spiralling, marked with open circles; observation. The observation shown uses data from the HMI ring diagram pipeline with 5° tiles (Gizon et al. 2021).

illustrates how the frequency and the inclination of the spiral pattern of the v_ϕ eigenfunction depends on $\delta_{\rm lower}$, for fixed $\delta_{\rm upper} = 0$ and $r_{\rm mid} = 0.85 R_{\odot}$. We find that the direction of spiralling matches the observation for $\delta_{\rm lower} < 2.4 \times 10^{-7}$, and reverses sign for $\delta_{\rm lower} > 5.5 \times 10^{-7}$. For $\delta_{\rm lower}$ between these values, the direction of spiral is unclear. Sample eigenfunctions for all three cases are shown in Fig. 2. For this particular set-up, we can therefore conclude that in order to recreate the observed latitudinal profile of the mode, the superadiabaticity in the lower half of the convection zone must be less than 2.4×10^{-7}.

Unlike the $m = 1$ high-latitude mode, we find that the $m = 3$ equatorial Rossby mode with zero radial node is rather insensitive to changes in $\delta_{\rm lower}$, as shown in Fig. 3, likely due to their toroidal nature. In future work, we will carry out a comprehensive study of the sensitivity of selected inertial modes to more general radial profiles of $\delta(r)$ and of the turbulent viscosity profile $\nu_{\rm turb}(r)$.

Acknowledgements

We acknowledge support from ERC Synergy Grant WHOLESUN 810218 and Deutsche Forschungsgemeinschaft (DFG, German Research Foundation) through SFB 1456/432680300 Mathematics of Experiment, project C04. P.D. acknowledges support

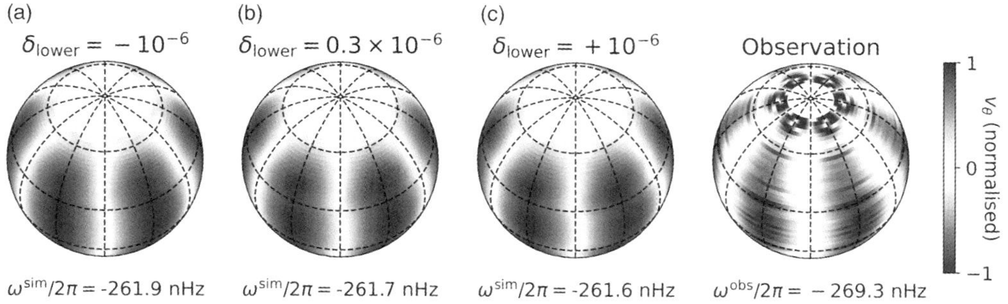

Figure 3. Colatitudinal component of the velocity eigenfunction of the $m = 3$ Rossby mode.

from the International Max–Planck Research School (IMPRS) for Solar System Science at the University of Göttingen. The HMI data are courtesy of NASA/SDO and the HMI Science Team.

References

Bekki, Y. 2024, Numerical study of non-toroidal inertial modes with $l = m + 1$ radial vorticity in the Sun's convection zone. *Astron. Astrophys.*, 682, A39.

Bekki, Y., Cameron, R. H., & Gizon, L. 2022, Theory of solar oscillations in the inertial frequency range: Linear modes of the convection zone. *Astron. Astrophys.*, 662, A16.

Bogart, Richard S. and Baldner, Charles S. and Basu, Sarbani 2015, Evolution of Near-surface Flows Inferred from High-resolution Ring-diagram Analysis. *Astrophys. J.*, 807, 125.

Böhm-Vitense, E. 1958, Über die Wasserstoffkonvektionszone in Sternen verschiedener Effektivtemperaturen und Leuchtkräfte. *Zeitschrift Astrophys.*, 46, 108.

Brandenburg, A. 2016, Stellar Mixing Length Theory with Entropy Rain. *Astrophys. J.*, 832, 6.

Christensen–Dalsgaard, J., Dappen, W., Ajukov, S. V., Anderson, E. R., Antia, H. M., Basu, S., Baturin, V. A., Berthomieu, G., Chaboyer, B., Chitre, S. M., Cox, A. N., Demarque, P., Donatowicz, J., Dziembowski, W. A., Gabriel, M., Gough, D. O., Guenther, D. B., Guzik, J. A., Harvey, J. W., Hill, F., Houdek, G., Iglesias, C. A., Kosovichev, A. G., Leibacher, J. W., Morel, P., Proffitt, C. R., Provost, J., Reiter, J., Rhodes, Jr., E. J., Rogers, F. J., Roxburgh, I. W., Thompson, M. J., & Ulrich, R. K. 1996, The Current State of Solar Modeling. *Science*, 272, 1286–1292.

Cossette, J.-F. & Rast, M. P. 2016, Supergranulation as the Largest Buoyantly Driven Convective Scale of the Sun. *Astrophys. J.*, 829, L17.

Gizon, L., Cameron, R. H., Bekki, Y., Birch, A. C., Bogart, R. S., Sacha Brun, A., Damiani, C., Fournier, D., Hyest, L., Jain, K., Lekshmi, B., Liang, Z.-C., & Proxauf, B. 2021, Solar inertial modes: Observations, identification, and diagnostic promise. *Astron. Astrophys.*, 652, L6.

Hanasoge, S. M., Duvall, T. L., & Sreenivasan, K. R. 2012, Anomalously weak solar convection. *PNAS*, 109, 11928–11932.

Hanson, C. S., Hanasoge, S., & Sreenivasan, K. R. 2022, Discovery of high-frequency retrograde vorticity waves in the Sun. *Nat. Astron.*, 6, 708–714.

Hotta, H., Bekki, Y., Gizon, L., Noraz, Q., & Rast, M. 2023, Dynamics of Large-Scale Solar Flows. *Space Sci. Rev.*, 219, 77.

Hotta, H., Kusano, K., & Shimada, R. 2022, Generation of Solar-like Differential Rotation. *Astrophys. J.*, 933, 199.

Käpylä, P. J. 2023, Convective scale and subadiabatic layers in simulations of rotating compressible convection. Preprint, https://arxiv.org/abs/2310.12855.

Larson, T. P. & Schou, J. 2018, Global-Mode Analysis of Full-Disk Data from the Michelson Doppler Imager and the Helioseismic and Magnetic Imager. *Sol. Phys.*, 293, 29.

Liang, Z.-C., Gizon, L., Birch, A. C., & Duvall, T. L. Jr. 2019, Time-distance helioseismology of solar Rossby waves. *Astron. Astrophys.*, 626, A3.

Löptien, B., Gizon, L., Birch, A. C., Schou, J., Proxauf, B., Duvall, T. L., Bogart, R. S., & Christensen, U. R. 2018, Global-scale equatorial Rossby waves as an essential component of solar internal dynamics. *Nat. Astron.*, 2, 568–573.

Mandal, K. & Hanasoge, S. 2020, Properties of Solar Rossby Waves from Normal Mode Coupling and Characterizing Its Systematics. *Astrophys. J.*, 891, 125.

O'Mara, B., Miesch, M. S., Featherstone, N. A., & Augustson, K. C. 2016, Velocity amplitudes in global convection simulations: The role of the Prandtl number and near-surface driving. *Adv. Space Res.*, 58, 1475–1489.

Proxauf, B., Gizon, L., Löptien, B., Schou, J., Birch, A. C., & Bogart, R. S. 2020, Exploring the latitude and depth dependence of solar Rossby waves using ring-diagram analysis. *Astron. Astrophys.*, 634, A44.

Dynamics of Solar and Stellar Convection Zones and Atmospheres
Proceedings IAU Symposium No. 365, 2024
A. V. Getling & L. L. Kitchatinov, eds.
doi:10.1017/S1743921323004994

Effects of small-scale magnetic fields in the photosphere on surface effects for solar-like stars

Guifang Lin[1,2,5], Yan Li[1,2,3,4,5], Jie Su[1,2,3,5], Tao Wu[1,2,3,5] and Yuetong Wang[1,2,5]

[1]Yunnan Observatories, Chinese Academy of Sciences, Kunming 650216, P. R. China

[2]Key Laboratory for the Structure and Evolution of Celestial Objects, Chinese Academy of Sciences, P. R. China

[3]Center for Astronomical Mega-Science, Chinese Academy of Sciences, Beijing, 100012, P. R. China

[4]University of Chinese Academy of Sciences, Beijing 100049, P. R. China

[5]International Centre of Supernovae, Yunnan Key Laboratory, Kunming 650216, P. R. China

Abstract. Magnetic fields are important physics in stellar evolutionary theory, which seriously affects the stellar structure and evolutionary statues. The small-scale magnetic fields in the photosphere are ubiquitous, and float on the stellar surface, which usually couple with the acoustic waves, affecting the propagation of the acoustic waves. Considering the effect of the magnetic fields in the stellar photosphere on the oscillation frequencies, we calculate the asteroseismology for solar-like star KIC 11295426 and KIC 10963065. We obtain the stellar fundamental parameters, especially the strength of small-scale magnetic fields in the stellar photosphere. We find that the small-scale magnetic fields in the stellar photosphere may obviously improve the agreement between the observations and the theoretical models for two stars. The magnetic strength for KIC 11295426 and KIC 10963065 from asteroseismology are in agreement with the stellar period-activity relation.

Keywords. Asteroseismology, Solar-like stars:oscillations, stars:atmosphere, magnetic fields

1. Introduction

Over the past decades, many space telescope (i.e. MOST, CoRoT, Kepler, K2 and TESS) have observed more and more uninterrupted high-precision photometric data of solar-like stars and red giant stars (e.g. Koch et al. 2010; Howell et al. 2014). This will facilitate better constraining stellar parameters and deeper exploring the stellar inner structure (e.g. Soriano & Vauclair 2008; Beck et al. 2012; Bellinger et al. 2019).

Due to the complex surface structure of the sun, there was a systematical offset between the observed oscillation frequencies and the theoretical ones (Christensen–Dalsgaard et al. 1988; Dziembowski et al. 1988), which was independent on the spherical degree ℓ and increased with the frequencies. This is known as the near-surface effect of solar p-mode oscillations. Until now, we have not fully understood the physics of the near-surface effect.

In order to eliminate the systematical offset, many researchers have used the different numeric fitting formulae to correct the frequencies. They used the form of single power law or the form of ν^3/I and ν^{-1}/I to correct the frequencies (Kjeldsen et al. 2008; Ball & Gizon 2014). On the side of stellar structure, the averaged structure of 3D hydrodynamical simulation are used to replace the stellar surface structure (Ball et al. 2016; Houdek et al. 2017; Schou & Birch 2020). They considered the physical effects of non-adiabaticity and turbulent pressure, and got the frequency deviations with $\leq 3\mu$Hz for the sun. Considering the magnetic pressure in the quiet solar photosphere, Li et al. (2021) found that the solar p-mode oscillations were reflected at the magnetic-arch splicing layer. They got the smallest systematical deviations to date between the observed and theoretical frequencies.

The small-scale magnetic field in the photosphere is very important. It couples with the acoustic waves, and may affect the propagation of the acoustic waves (Rosenthal et al. 2002; Bogdan et al. 2003). In our work, based on Li et al. (2021), we mainly focus on the effects of the small-scale magnetic field in the photosphere on the p-mode oscillations frequencies for the solar-like stars. Considering the magnetic fields in the photosphere, we construct the stellar models for the solar-like star KIC 11295426 and KIC 10963065.

2. Stellar model with small-scale magnetic fields

We use MESA evolutionary code, version r-10398 to construct the stellar structure (e.g. Paxton et al. 2018). We adopt the GS98 (Grevesse & Sauval 1998) chemical composition and the OPAL opacity tables from Iglesias & Rogers (1996). In the stellar evolutionary models, the convection is treated as the standard mixing length theory (Böhm-Vitense 1958), and the convective overshooting is not considered. We also include the element diffusion. We use the "adipls" package (Christensen–Dalsgaard 2008) in MESA to solve the adiabatic oscillation equations and get the theoretical p-mode oscillations frequencies.

Based on Li et al. (2021), we consider the effects of magnetic fields in the Eddington gray-atmosphere model. We use the unified $T-\tau$ relation (Li et al. 2021). In the Hopf-like function $q(\tau)$, we add a term to account for the effects of magnetic fields:

$$q(\tau) = q_{\rm ori}(\tau) + a\exp(-b\sqrt{\tau}) \tag{1}$$

In the Eddington gray-atmosphere model, $q_{\rm ori}$ is 2/3. Parameter a represents the magnetic strength and b determines the location of the magnetic-arch splicing layer. For the reflected standing waves, we adopt the surface mechanical boundary condition as $P^{'} = 0$. For more details, please refer to Li et al. (2021).

We apply this method to two solar-like stars: KIC 11295426 and KIC 10963065. These two stars are host stars with a few transiting planets (Gilliland et al. 2013; Marcy et al. 2014). Using Kepler short cadence data, we compare the theoretical p-mode frequencies with observed 38 mode frequencies for KIC 11295426 and 44 ones for KIC 10963065 identified from Davies et al. (2016). We use the value of χ^2 which is defined as the average of squared differences between the observed and theoretical frequencies weighted by the observational errors to evaluate the matching goodness. The χ^2-minimization model is used as the best-fit model.

3. Asteroseismology analysis

In our models, in order to reduce the amount of calculations, we use a method of loop calculation to get a self-consistent magnetic parameters a and b. We use the internal structure of the models without magnetic fields to constrain the initial fundamental parameters by comparing the ratios of small- to large- separations between the observed and theoretical ones. Using the initial fundamental parameters, we can get the values of a and b. Using the determined values of a and b, we calculate the stellar structure and evolution of the models with magnetic fields and get the χ^2-minimization model. Through the iterative calculations, when the initial input fundamental parameters are the same with the ones of χ^2-minimization model, the values of a and b are adopted as

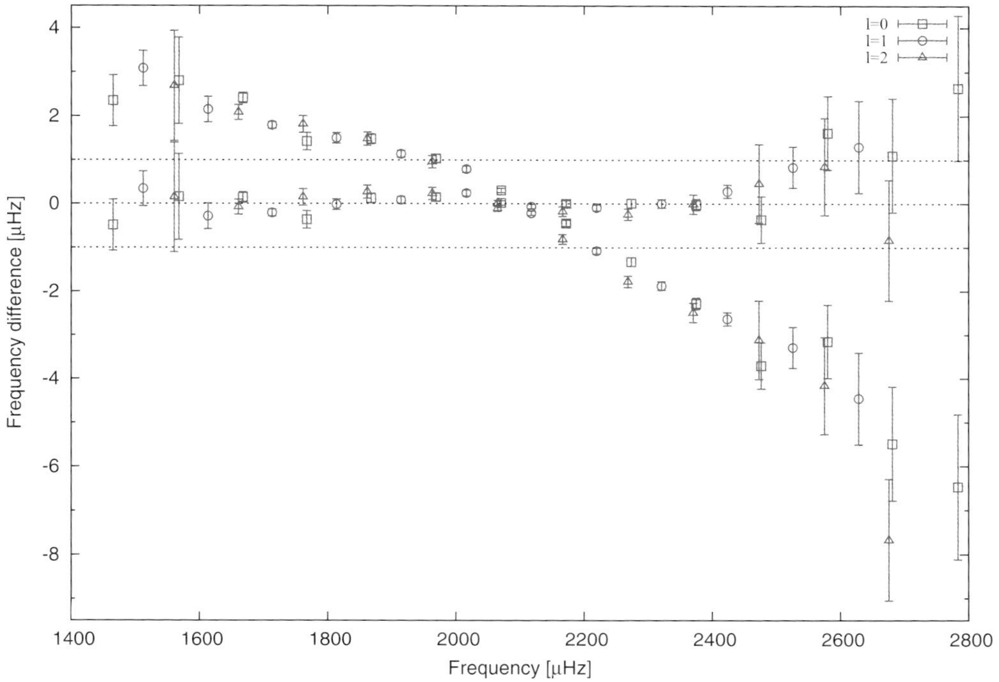

Figure 1. The frequency differences between observations and best-fit model with (open red points) and without magnetic fields (open blue points) for KIC 11295426. The horizontal dotted lines represent $y = 0, \pm 1\mu$Hz.

the optimal ones. For KIC 11295426, the optimal magnetic parameters a and b are 280 and 70, respectively, and the ones for KIC 10963065 are 480 and 54.

For KIC 11295426, the parameters of the best-fit model with magnetic fields and minimum χ^2 are $M = 1.09M_{\odot}$, $Z_{\text{init}} = 0.0255$, $\alpha_{\text{MLT}} = 1.78$. The corresponding differences between the observed frequencies and the theoretical ones are shown in Fig. 1. The fundamental parameters of the best-fit model with magnetic fields for KIC 10963065 are $M = 1.095M_{\odot}$, $Z_{\text{init}} = 0.015$, $\alpha_{\text{MLT}} = 1.72$. The corresponding frequencies differences are shown in Fig. 2. From our results, we can find that except for a few higher frequencies, the most frequency differences of the best-fit models with magnetic fields for two solar-like stars are less than 1μHz. In other words, the most frequencies of the best-fit models with magnetic fields are all in agreement with the observed ones. The frequencies for the models without magnetic fields deviate systematically and significantly from the observation. This means that the small-scale magnetic fields in the photosphere can obviously reduce the systematical deviations between the observed frequencies and the theoretical ones.

In addition, according to the values of optimal magnetic parameters a and b for the two solar-like stars, using the relationship that gas pressure equals to magnetic pressure (Rosenthal et al. 2002; Cally 2007), we independently derive the strength of small-scale magnetic fields in the photosphere for KIC 11295426 and KIC 10963065. The magnetic strength for KIC 11295426 is about 89 G and the one for KIC 10963065 is about 127 G. Through the spectroscopy observations, KIC 11295426 is a relatively slower rotating star (Gilliland et al. 2013) than the sun, while KIC 10963065 is a faster rotating star with rotation period 12.5 days (Suto et al. 2019). Although the magnetic strength for KIC 11295426 is very close to the one of the sun (Li et al. 2021), the magnetic strength of the two solar-like stars are in agreement with the stellar period-activity relation.

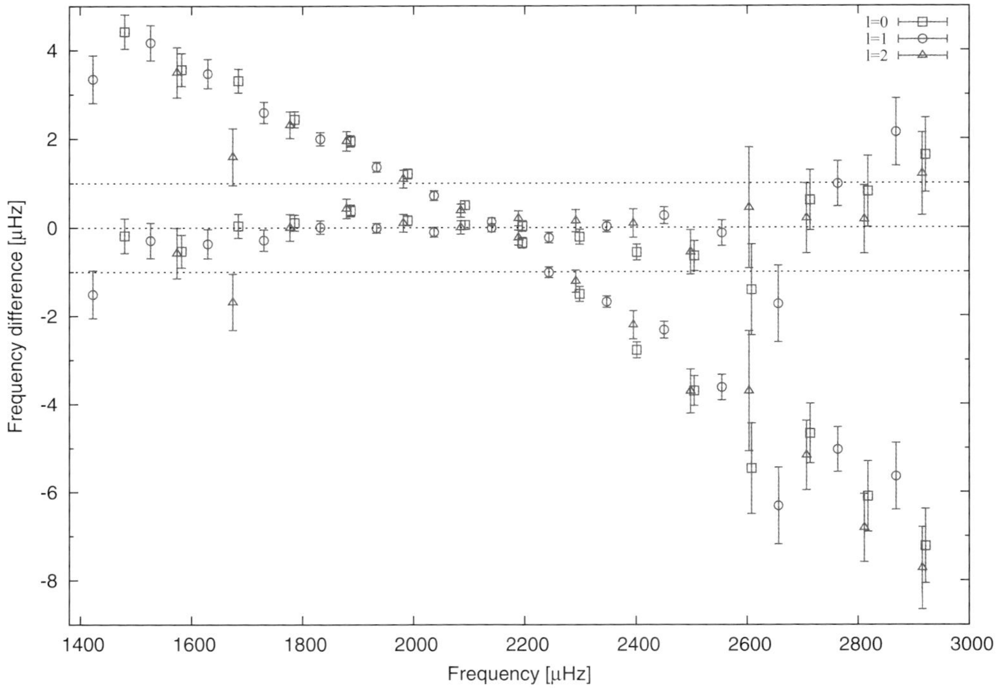

Figure 2. The frequency differences between observations and best-fit model with (open red points) and without magnetic fields (open blue points) for KIC 10963065.

4. Summary and discussion

Due to the convective shell for the solar-like stars, the stellar surface structure is very complex. This results in a systematic deviation between the observation frequencies and the theoretical ones. By considering the effects of small-scale magnetic fields in the photosphere on stellar oscillation frequencies, we calculate the asteroseismology for KIC 11295426 and KIC 10963065.

We consider magnetic pressure with two parameters a and b in the stellar Eddington gray-atmosphere model, and construct the asteroseismic models with magnetic fields. We derive the stellar fundamental parameters for the two solar-like stars KIC 11295426 and KIC 10963065. The results are consistent with other results from spectroscopic observations and the other theoretical models (e.g. Gilliland et al. 2013; Marcy et al. 2014; Silva Aguirre et al. 2015; Kayhan et al. 2019).

We find that magnetic fields in the photosphere obviously reduces the frequency deviations between observations and the best-fit model with minimum χ^2. The small-scale magnetic fields in the photosphere only change the stellar surface structure. The improvement on the systematical offset between observation and the best-fit models is mainly due to the effect of magnetic field on the stellar surface physics. In addition, using the asteroseismology, we independently derive the magnetic strength for the two solar-like stars. Compared with the one of the sun, the magnetic strength for KIC 11295426 and KIC 10963065 are consistent with the stellar period-activity relation.

Acknowledgement This research is supported in part by the National Key R&D Program of China (Grant No. 2021YFA1600400/2021YFA1600402), by the B-type Strategic Priority Program No. XDB41000000 funded by the Chinese Academy of Sciences (CAS), by the National Natural Science Foundation of China (NSFC) under grants 12133011, 12288102 and 11833006, and supported by International Centre of Supernovae, Yunnan Key Laboratory (No. 202302AN360001).

References

Ball, W. H., Beeck, B., Cameron, R. H., & Gizon, L. 2016, *A&A*, 592, 159
Ball, W. H., & Gizon, L. 2014, *A&A*, 568, A123
Beck, P. G., Montalban, J., Kallinger, T., *et al.* 2012, *Nature*, 481, 55
Bellinger, E. P., Basu, S., Hekker, S., & Christensen–Dalsgaard, J. 2019, *ApJ*, 885, 143
Bogdan, T. J., Carlsson, M., Hansteen, V. H., *et al.* 2003, *ApJ*, 599, 626
Böhm-Vitense, E. 1958, *Z. Astrophys.*, 46, 108
Cally, P. S. 2007, *Astron. Nachr.*, 328, 286
Christensen–Dalsgaard, J. 2008, *Ap&SS*, 316, 113
Christensen–Dalsgaard, J., Däppen, W., & Lebreton, Y. 1988, *Nature*, 336, 634
Davies, G. R., Silva Aguirre, V., Bedding, T.R., *et al.* 2016, *MNRAS*, 456, 2183
Dziembowski, W. A., Paterno, L., & Ventura, R. 1988, *A&A*, 200, 213
Gilliland, R. L., Marcy, G. W., Rowe, J. F., *et al.* 2013, *ApJ*, 766, 40
Grevesse, N., & Sauval, A. J. 1998, *Space Sci. Rev.*, 85, 161
Houdek, G., Trampedach, R., Aarslev, M. J., & Christensen–Dalsgaard, J. 2017, *MNRAS*, 464, L124
Howell, S. B., Sobeck, C., Haas, M., *et al.* 2014, *PASP*, 126, 398
Iglesias, C. A., & Rogers, F. J. 1996, *ApJ*, 464, 943
Kayhan, C., Yıldız, M., & Celik Orhan, Z. 2019, *MNRAS*, 490, 1509
Kjeldsen, H., Bedding, T. R., & Christensen–Dalsgaard, J. 2008, *ApJL*, 683, L175
Koch, D. G., Borucki, W. J., Basri, G., *et al.* 2010, *ApJL*, 713, L79
Li, Y., Zhang, Q. S., Wu, T., *et al.* 2021, *ApJ*, 916, 107
Marcy, G.W., Isaacson, H., Howard, A.W., *et al.* 2014, *ApJS*, 210, 20
Paxton, B., Schwab, J., Bauer, E. B., *et al.* 2018, *ApJS*, 234, 34
Rosenthal, C. S., Bogdan, T. J., Carlsson, M., *et al.* 2002, *ApJ*, 564, 508
Schou, J., & Birch, A. C. 2020, *A&A*, 638, 51
Silva Aguirre, V., Davies, G. R., Basu, S., *et al.* 2015, *MNRAS*, 452, 2127
Soriano, M., & Vauclair, S. 2008, *A&A*, 488, 975
Suto, Y., Kamiaka, S., & Benomar, O. 2019, *AJ*, 157, 172

Dynamics of Solar and Stellar Convection Zones and Atmospheres
Proceedings IAU Symposium No. 365, 2024
A. V. Getling & L. L. Kitchatinov, eds.
doi:10.1017/S1743921323004209

The effect of Coulomb interactions on acoustic oscillations in the outer layers of low-mass stars

Ana Brito[1,2] **and Ilídio Lopes Fuentes**[2]

[1]Instituto Superior de Gestão
Rua Prof. Reinaldo dos Santos 46, 1500-522, Lisbon, Portugal
email: anabrito@isg.pt

[2]Centro de Astrofísica e Gravitação - CENTRA, Departamento de Física, Instituto Superior Técnico, Universidade de Lisboa
Av. Rovisco Pais 1, 1049-001, Lisbon, Portugal
email: ilidio.lopes@tecnico.ulisboa.pt

Abstract. As is well known, low-mass stars constitute the most abundant class of stars in our galaxy. In stars less massive than the Sun, the density within stellar interiors increases as the stellar mass decreases. Therefore, for low-mass stars, the significance of electrostatic effects in stellar interiors cannot be neglected, as these interactions can alter the properties of matter.

In our study, we focus on exploring the outer layers of stars less massive than the Sun. We have computed a range of stellar models, ranging from 0.4 to 0.9 solar masses, to investigate the effects of two physical processes on the acoustic oscillations in the envelopes of these stars: partial ionization of chemical elements and electrostatic interactions between particles in the outer layers. In addition to partial ionization, we demonstrate that Coulomb effects also influence the acoustic oscillation spectrum. Our investigation reveals the following findings:

1. Coulomb effects can indeed influence the acoustic oscillations in low-mass stars.
2. The model with a mass of $0.6\,M_\odot$ serves as a transition point. For models less massive than $0.6\,M_\odot$, their acoustic spectrum is more affected by electrostatic interactions, whereas models more massive than $0.6\,M_\odot$ have their acoustic spectrum more impacted by partial ionization processes.

Our work unveils the promising possibilities that future discoveries related to the detection of solar-like oscillations in stars less massive than the Sun could offer in terms of understanding the connections between the internal structure of low-mass stars and their observable characteristics.

Keywords. Asteroseismology, Low-mass stars, Stellar interiors, Coulomb effects, Partial ionization

1. Introduction

It is known from observations that stellar formation mechanisms favor the creation of low-mass stars, which constitute the most abundant class of stars in our Milky Way Bochanski et al. (2010). Specifically, M dwarfs make up approximately 70% of the solar neighborhood. However, this class of stars is the least understood due to their faintness and complex atmospheres, making it challenging to obtain reliable observational data. These stars possess convective envelopes (or are fully convective) and therefore have the potential to exhibit solar-like oscillations. Nevertheless, only a limited number of stars less massive than the Sun have had solar-like oscillations detected, primarily due to the small amplitudes of these oscillations (Borucki et al. 2010 - *Kepler*). Nonetheless, efforts to detect solar-like oscillations persist with upcoming missions such as PLATO (e.g., Rauer et al. 2014), ground-based initiatives like the Stellar Observations Network Group (Grundahl et al. 2017), and high-precision telescopes such as the Keck Planet Finder (Gibson et al. 2016).

It is well-known that as stellar masses decrease, the densities within their interiors increase. With increasing density in a stellar interior, we cannot disregard the electrostatic interactions between the particles comprising the stellar plasma. This is because electrostatic interactions between particles alter the properties of matter. For example, electrostatic interactions lead to a reduction in pressure and ionization potential, resulting in a physical process known as pressure ionization in the deeper layers of stars. Additionally, the electron shielding effect decreases the repulsion between nuclei, facilitating nuclear reactions. While electron conduction and crystallization are also possible phenomena, they are not applicable to the types of stars studied in this research.

In this study, we investigate the impact of electrostatic effects on the acoustic oscillation spectrum of six theoretical stellar models with masses ranging from $0.9\,M_\odot$ to $0.4\,M_\odot$. It's important to note that all these models have a radiative interior and a convective envelope. These models were computed using the stellar evolution code MESA (Paxton et al. 2011, 2013, 2015, 2018; Jermyn et al. 2019), and for calculating the adiabatic oscillation mode frequencies, we employed the stellar pulsation code GYRE (e.g., Townsend & Teitler 2013). Further details regarding the input physics of these models can be found in the source paper Brito & Lopes (2021).

In the next section, section 2, we describe the outer convective layers of the models using thermodynamic quantities. Then, in section 3, we analyse from a seismic point of view the outer layers of the six theoretical models. Lastly, our conclusions (section 4) will be presented, along with a preview of our forthcoming work, which also emphasizes the significance of electrostatic interactions, although in the internal radiative zones of low-mass stars.

2. The adiabatically stratified convective zones

In this section, we will explore the outer convective envelopes of the six theoretical models. To achieve this, we will examine the profiles of four thermodynamic quantities: the first adiabatic exponent, the acoustic potential, the sound speed gradient, and the plasma coupling parameter. It is well known that the convective envelopes of low-mass stars exhibit a quasi-adiabatic stratification. The first adiabatic exponent, denoted as $\Gamma_1 = (\partial \ln P/\partial \ln \rho)_{ad}$, and represented in Figure 1, is determined by the equation of state and defines the structure of the adiabatic convective zones. As shown in Figure 1, the behavior of Γ_1 around its monoatomic ideal gas value is primarily influenced by two physical processes: partial ionisation of atomic species and electrostatic interactions between particles. An increase in the degree of ionization results in a decrease in Γ_1 (this is particularly evident in the ionization zones of light elements). Instead, electrostatic interactions between particles lead to an increase in the values of Γ_1.The first adiabatic exponent also plays a special role in asteroseismology. This is because, for acoustic oscillations, the adiabatic sound speed is determined by the equation $c^2 = \Gamma_1 P/\rho$.

In addition to Γ_1, another important quantity to investigate is the acoustic potential. By using the acoustic depth ($\tau = \int_r^R dr/c$) as an independent variable, which represents the wave's travel time from the star's surface to a specific radius depth r, we can describe the propagation of adiabatic acoustic oscillations in the outer layers of low-mass stars using a Schrödinger-type equation: $d^2\psi/d\tau^2 + (\omega^2 - U^2)\psi = 0$. Here, ω represents the angular frequency of the acoustic mode, and U^2 denotes the acoustic potential. This quantity, the acoustic potential, is intricately linked to the thermodynamics of stellar

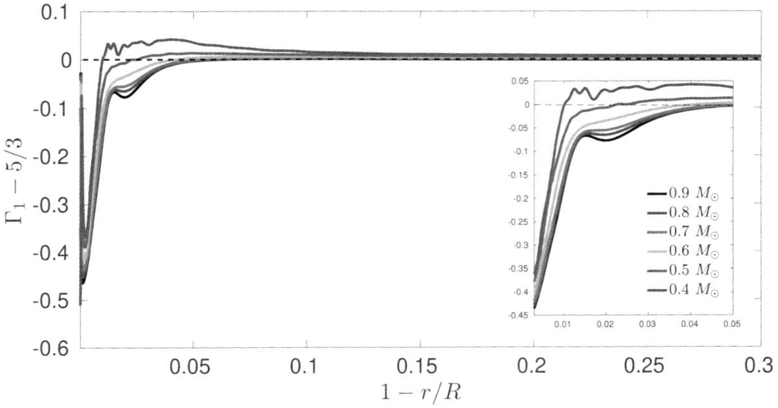

Figure 1. The first adiabatic exponent, Γ_1 is plotted for the outer layers of the six theoretical models.

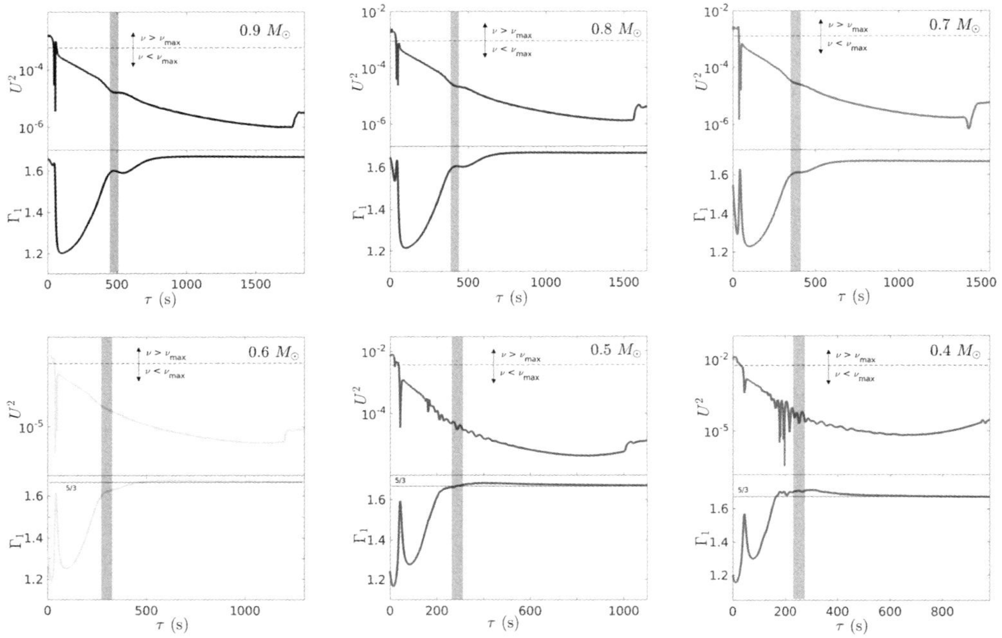

Figure 2. The acoustic potentials are plotted as functions of the acoustic depth in the upper layers of the theoretical models (upper panels). For comparison, Γ_1 is also represented as a function of the acoustic depth (lower panels).

interiors and is influenced by physical processes such as partial ionization of chemical elements and Coulomb interactions between particles. There is another compelling reason why the acoustic potential holds great significance, which will become evident in the upcoming Section 3. Figure 2 illustrates the acoustic potentials plotted as functions of acoustic depth for all six models. In the case of more massive models, the dip corresponding to helium ionization is clearly visible. However, for the 0.6 $M_\odot$ model, this dip nearly vanishes due to the increasing importance of electrostatic effects. Finally, in the less massive models, a distinct noisy pattern is observable, reflecting the impact of electrostatic interactions between charged particles.

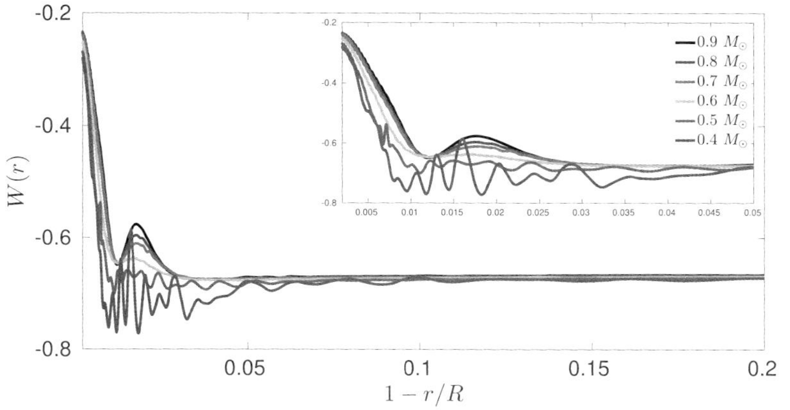

Figure 3. The sound-speed gradient, W is plotted for the outer layers of the six theoretical models.

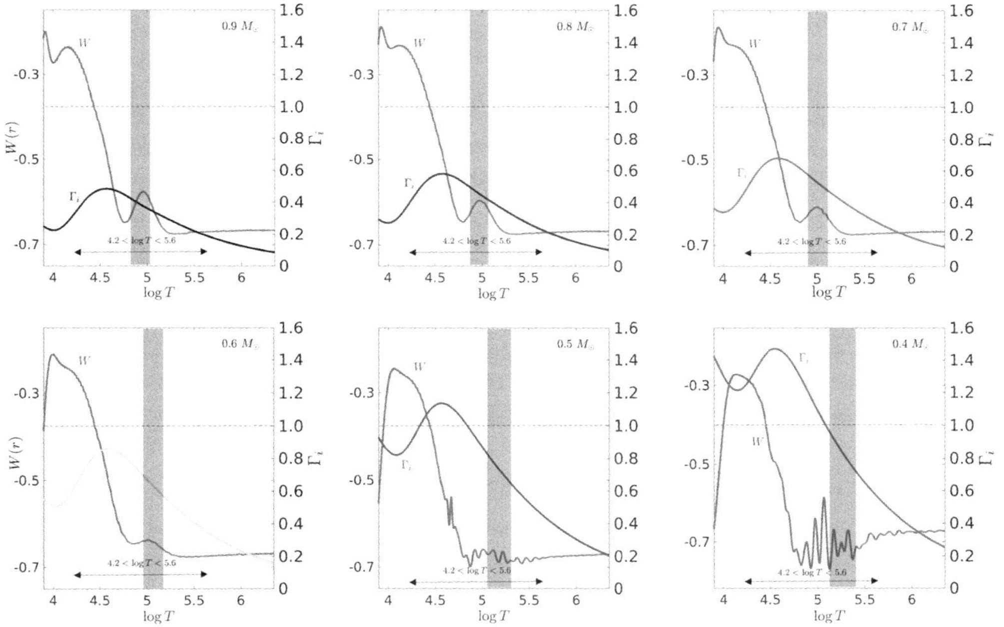

Figure 4. The ionic plasma coupling parameter, Γ_i, is plotted as a function of temperature for all six theoretical models. The temperature range $4.2 \lesssim \log T \lesssim 5.6$ is especially important for electrostatic interactions because it contains the intervals where the two helium ionisations occur. Additionally, we have overlaid the sound-speed gradient profiles in grey for comparison.

Another diagnostic quantity that is sensitive to Γ_1 is the sound speed gradient represented in figure 3 for all the six models. Here as well, we observe that the perturbations in the acoustic spectrum of more massive stars are more affected by partial ionization processes, resulting in a moderate variation in the sound speed gradient. In contrast, for less massive models, the acoustic spectrum is more influenced by electrostatic interactions, leading to a pronounced oscillatory component in the profile of the sound speed gradient.

To complete the study of the convective outer layers of the models, we also computed the plasma coupling parameter. The plasma parameter represents the ratio of the average potential energy to the thermal kinetic energy of the gas. Figure 4 illustrates the plasma

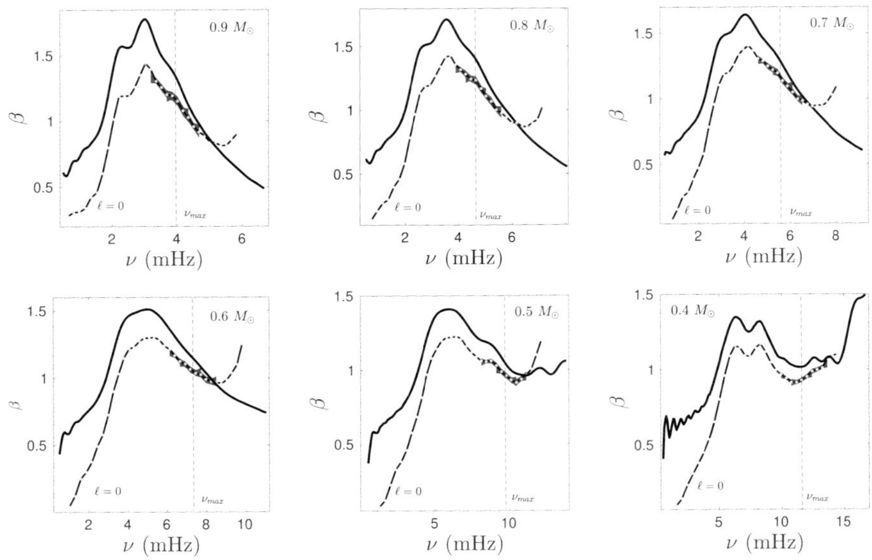

Figure 5. The seismic diagnostic β is numerically computed from the structural model parameters (indicated by the solid, thicker black line). The thinner, solid black line represents the diagnostic β computed from a theoretical table of frequencies ($\ell = 0$), with the mode frequencies highlighted in cyan. In this figure, two shaded regions are shown: one in blue and the other in red. These regions result from tests that validate the stability of the algorithm.

interaction parameter plotted within the temperature range where the two helium ionisations occur, which is also the temperature range where Coulomb effects are known to be more significant. The value of the plasma parameter increases as stellar mass decreases. Moreover, for the less massive stellar models, the plasma parameter value is above one or close to one in this temperature range. This suggests that the strong oscillatory signal in the sound speed gradient can indeed be attributed to Coulomb effects.

3. Seismic analysis of the outer layers

Helio- and asteroseismology have demonstrated that abrupt variations in the sound speed produce a quasi-periodic oscillatory component in the frequencies of stellar acoustic oscillations (e.g., Gough & Toomre 1991). This type of signal contains information that helps in our understanding of the physical processes taking place within stellar interiors. In this section, we will analyse the outer layers of the six theoretical models from a seismic perspective. To accomplish this, we will employ a seismic diagnostic denoted as β, which is particularly suited for studying the outer convective layers of low-mass main-sequence stars (e.g., Brodskii & Vorontsov 1987; Vorontsov & Zharkov 1989; Brito & Lopes 2019, 2018). In the previous section, when we introduced the concept of the acoustic potential, we mentioned that the acoustic potential is of interest for two primary reasons. The second reason will now become clear, and it is as follows: we can establish a connection between the shape of the acoustic potential and the shape of the seismic diagnostic β. Figure 5 illustrates the seismic diagnostic $\beta(\nu)$, where ν represents the cyclic frequency, numerically computed from the structural model parameters (indicated by the thicker black solid line) for all the models. The quasi-periodic signal in $\beta(\nu)$ observed in the more massive models (0.9, 0.8, and 0.7 $M_\odot$) is associated with the dip in the acoustic potential found in the region of partial ionization of light elements, specifically the second helium ionization. The 0.6 $M_\odot$ model serves as a transitional model in the sense that the quasi-periodic oscillations linked to helium ionization completely disappear. For models less massive than $0.6\,M_\odot$, the shape of the seismic diagnostic $\beta(\nu)$ is entirely different from that of $\beta(\nu)$ for the more massive models, reflecting the significance of electrostatic interactions between particles in the acoustic spectrum. Figure 5 also presents the seismic parameter $\beta(\nu)$ obtained from theoretical tables of mode frequencies for each model. These frequencies were computed using the stellar pulsation code GYRE (e.g., Townsend & Teitler 2013).

The frequencies near ν_{max} are of particular interest because it is in the vicinity of this value that the detection of observational modes of oscillation is more likely. If we consider a local linear approximation in the vicinity of ν_{max}, we observe that as the significance of Coulomb effects increases, the slope of the line changes, allowing us to differentiate between the cases. It is evident that the diagnostic potential of β significantly improves with the number of modes used in its calculation. However, even with only five modes per degree, the behavior of β for a $0.9\,M_\odot$ star can be distinguished from that of a $0.4\,M_\odot$ star (see figure 5).

4. Conclusions

In this study, we investigate the oscillatory patterns present in the oscillation mode frequencies of six theoretical low-mass main-sequence stellar models ($0.4\,M_\odot < M < 0.9\,M_\odot$), resulting from two distinct physical processes: the partial ionization of light elements and the electrostatic interactions between the particles constituting the stellar plasma.

The $0.6\,M_\odot$ stellar model serves as a reference model, with models more massive than $0.6\,M_\odot$ having their acoustic spectrum primarily influenced by partial ionization processes, while models less massive than $0.6\,M_\odot$ have their acoustic spectrum predominantly affected by Coulomb effects. Therefore, this work unveils the potential for future discoveries related to the detection of solar-like oscillations in stars less massive than the Sun to provide insights into the study of the relationships between the internal thermodynamics of low-mass stars and their observable characteristics.

Finally, we would like to emphasize that the interiors of low-mass stars constitute what is commonly referred to as a hot and dense plasma. By definition, a plasma is associated with two key concepts: the concept of quasi-neutrality and the concept of collective behavior. Quasi-neutrality implies that the plasma maintains an overall neutral charge, but it is not so neutral that all the significant electromagnetic forces disappear (e.g., Chen 2016). This means that stellar interiors can exhibit small local imbalances of electric charge. Consequently, the study of the electrostatic properties of stellar interiors holds great potential for enhancing our understanding of the observable patterns of rotation and magnetic activity in low-mass main-sequence stars.

References

Bochanski, J., Hawley, S., Covey, K., *et al.* 2010, *AJ*, 30, 490
Borucki, W., Koch, D., Basri, G., *et al.* 2010, *Science* 377, 977
Brito A. & Lopes I. 2018, *ApJ* 853, 183
Brito A. & Lopes I. 2019, *MNRAS* 488, 1558–1571
Brito A. & Lopes I. 2021, *MNRAS* 507, 5747–5757
Brodskii M. & Vorontsov S. 1987, *SvAL* 13, 179
Chen, F. 2016, *Introduction to Plasma Physics and Controlled Fusion*
Gibson, S., Howard, A., Marcy, G., *et al.* 2016, *Society of Photo-Optical Instrumentation Engineers (SPIE) Conference Series* 9908, 990870
Gough D. & Toomre J. 1991, *ARAA* 29, 627–684
Grundahl, F., Fredslund Andersen, M., Christensen–Dalsgaard, J., *et al.* 2017, *ApJ* 836, 142
Jermyn, A., Bauer, E., Schwab, J., *et al.* 2019, *ApJS* 265, 15

Paxton, B., Bildsten, L., Dotter, A., *et al.* 2011, *ApJS* 192, 3
Paxton, B., Cantiello, M., Arras, P., *et al.* 2013, *ApJS* 208, 4
Paxton, B., Marchant, P., Schwab, J., *et al.* 2015, *ApJS* 220, 15
Paxton, B., Schwab, J., Bauer, E., *et al.* 2018, *ApJS* 234, 34
Paxton, B., Smolec, R., Schwab, J., *et al.* 2019, *ApJS* 243, 10
Rauer, H., Catala, C., Aerts, C., *et al.* 2014, *Experimental Astronomy* 38, 249–330
Townsend, R. & Teitler, S. 2013, *MNRAS* 435, 3406–3418
Vorontsov, S. & Zharkov, V. 1989, *APSPR* 13, 179

Section 5. Local Processes of Magnetic-Flux Emergence. Sunspot and Starspot Formation

Invited talks

Dynamics of Solar and Stellar Convection Zones and Atmospheres
Proceedings IAU Symposium No. 365, 2024
A. V. Getling & L. L. Kitchatinov, eds.
doi:10.1017/S174392132400036X

Signature of local (turbulent) dynamo on middle and small scales

V. I. Abramenko

Crimean Astrophysical Observatory of Russian Academy of Sciences, Nauchny, Russia
email: vabramenko@gmail.com

Abstract. The solar dynamo is a physical process of magnetic field generation due to conversion of kinetic energy of plasma flows into magnetic energy. However, in the mean-field dynamo theory, one needs to segregate scales and consider separately large-scale dynamo and small-scale dynamo. The large-scale dynamo produces the large-scale mean field and unavoidable fluctuations of the mean field. Both are cycle-dependent. The small-scale dynamo is supposed to produce only the small-scale field, and this field is cycle-independent. There is no sharp boundary between the intervals of the large-scale and small-scale dynamos. An unavoidable presence of a smooth transition implies that there is a region where the properties of the large-scale global dynamo and fluctuations inherent to small-scale dynamo co-exist on some intermediate scales. Recent achievements in observations of the small-scale dynamo operation on the smallest observable scales and on the intermediate scales of typical active regions are discussed in the review.

Keywords. Sun:magnetic fields, Sun:dynamo, turbulence

1. Introduction

The whole variety of observed solar magnetic fields is produced by the solar dynamo, which manifests itself differently on different scales. Although we know that there are no two absolutely similar solar cycles, yet, self-organization of the large-scale global dynamo remains impressive. In opposite, we observe a drastically different picture of chaos on smallest scales (the magnetic carpet), and something between chaos and order in the intermediate scales (typical active regions). Strictly speaking, the dynamo, as a physical process of magnetic field generation, has a unique underlying basis: conversion of kinetic energy of plasma flows into magnetic energy. In this sense, dynamo is a uniform mechanism. However, when we need to deal with a theoretical concept, say, the mean-field dynamo theory, we need to segregate scales and consider separately large-scale dynamo and small-scale dynamo. The large-scale dynamo produces the large-scale mean field and unavoidable fluctuations of the mean field. Both are cycle-dependent. The small-scale dynamo is supposed to produce only the small-scale field, and this field is cycle-independent. Of cause, there is no sharp boundary between the intervals of the large-scale and small-scale dynamos. An unavoidable presence of a smooth transition implies that there is a region where the properties of the large-scale global dynamo and fluctuations inherent to small-scale dynamo co-exist on some intermediate scales.

The helicity concept provides a specific view on the origin of the large-scale and small-scale dynamos. Depending on the presence or absence of the net flow helicity, two types of dynamo can be suggested, Meneguzzi *et al.* (1981). With the net helicity, magnetic energy grows on scales larger than the energy-containing scale of the fluid motions providing the large-scale dynamo (LSD), or mean-field dynamo. The production of large-scale magnetic energy is ensured by the alpha-effect (see, e.g., Brandenburg (2003), Charbonneau (2020)).

Without the net helicity, the dynamo action is harder to achieve and the magnetic energy grows at scales smaller than the forcing scale. This process defines the small-scale dynamo (SSD), or the fluctuation dynamo Batchelor (1950), Kazantsev (1968), Petrovay and Szakaly (1993), Brandenburg and Subramanian (2005).

2. Small-Scale Dynamo (SSD)

Near the solar surface, the convective time scale is much shorter than the rotation period, therefore, the effects of rotation can be neglected, and a flow with no net helicity results. Any surface dynamo (local dynamo) will thus be the SSD.

Numerical simulations provide evidence of the SSD driven by turbulence at the solar surface (likely deeper as well). Since the pioneering publications by Petrovay and Szakaly (1993), and by Cattaneo (1999) a variety of approaches were suggested, for example, Vögler (and Schüssler2007), Pietarila Graham, Cameron, and Schüssler (2010), Rempel and Cheung (2014), Kitiashvili *et al.* (2015). Appearance of the magnetic filigree out of the seed field due to the turbulent convection is well proved from Figures 1 in works by Rempel and Cheung (2014) and Kitiashvili *et al.* (2015). The results of numerical simulations of SSD became more close to reality (as compared to the pioneering studies), embarrassing broader ranges as over the solar surface, so in depth and in height. The existence of the SSD is thought to be proved in numerical simulations.

Along with theoretical and numerical efforts, the observational studies of the small-scale magnetic features was continued. Thus, in the paper by Jin *et al.* (2011) on the basis of the MDI full disk data it was shown that the total (over the disk) unsigned magnetic flux from small-scale magnetic elements does not depend on the solar cycle. The authors applied the magnetic flux density of 15 Mx cm^{-2} as a threshold to outline active regions and then to segregate the quiet-sun flux from the active regions flux, see Figure 2 in Jin *et al.* (2011). This experiment clearly separates the total solar magnetic flux into: i) the cycle-dependent component, associated with active regions and caused presumably by the large-scale dynamo, and ii) the cycle independent component produced presumably by the small-scale local dynamo. The weak but still noticeable enhancement of the quiet-sun flux during the solar maximum might indicate a presence of the mean-field fluctuations in the small-scale flux.

Another similar experiment was undertaken by Obridko, Livshits, and Sokoloff (2017) on the basis of the same MDI data but another flux segregation criteria. The authors used three thresholds: 33, 100, and 3000 Mx cm^{-2}. The total unsigned flux from all pixels below a threshold was calculated (see Figure 3 in Obridko, Livshits, and Sokoloff (2017)). The lowest threshold data show no dependence on the cycle, the one-hundred threshold data show a very weak enhancement during the cycle maximum, and the 3000-threshold data actually correspond to the total flux from the entire disk and follow the cycle. So, the stronger the field, the more strong is the connection with the global dynamo. Thus, the authors concluded that the weak field of magnitude less than 100 Mx cm^{-2} apparently arise as a result of a small-scale sub-surface process, while the stronger fields are connected with the cycle and are generated by the classical mean-field solar dynamo.

3. Magnetic Power Spectra as a Tool to Diagnose SSD

However, the cycle-independence *per se* does not allow us to decide whether the observed effect is due to real field generation (SSD), or it is purely a result of turbulence, of direct turbulent cascade, as it is argued by Stenflo and Kosovichev (2012).

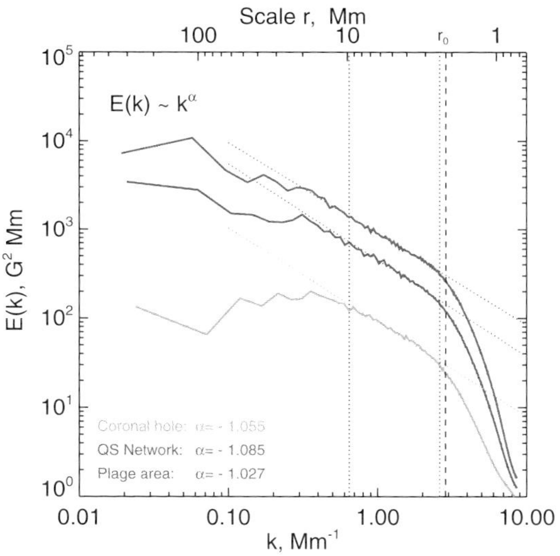

Figure 1. Magnetic power spectra calculated from three areas on the solar disk outside active regions. The full disk HMI/LOS magnetograms (hmi.M_720s series) taken on 10 February 2015 (Schou *et al.* (2012)) were used. The linear inertial range of (2.4-10) Mm is marked by the vertical dotted segments, and the scale r_0 corresponding to the triple resolution of HMI, is indicated with the vertical dashed line. The spectral indexes, α, derived from the best linear fit to the data points inside the inertial range, are listed in the lower left corner of the graph.

An analysis of the magnetic energy spectrum can help in this question. According to the classical turbulence theory by Kolmogorov (1941) (K41 theory), the energy input occurs at large scales and energy cascades down toward smallest scales so that the index of the power spectrum is -5/3. The cascade works as shredding and fragmentation of eddies. In this case, no additional energy input in intermediate scales is happen. The total amount of energy on small scales is assumed to originate from the direct turbulent cascade. However, in the case of additional energy input on some interval of scales due to some process, for example, the turbulent dynamo action, this energy becomes transferred up and down along the spectrum, the energy peak becomes smoothed and, as a result, the spectrum becomes more shallow than -5/3. So, the shape and slope of the magnetic power spectrum can tell us whether we deal with the turbulent cascade along, or some other processes are involved into the formation of the magnetic elements.

The magnetic power spectra determined from the SDO/HMI/LOS magnetograms for a coronal hole, a quiet sun, and a plage areas (Figure 1) exhibit the same spectral index of -1 on a broad range of spatial scales from 1020 Mm down to 2.4 Mm Abramenko and Yurchyshyn (2020). This implies that the same mechanism(s) of magnetic field generation operate everywhere in the undisturbed photosphere, and this mechanism is not the direct turbulent cascade only. The most plausible additional mechanism is the local turbulent dynamo. The similar inference on the basis of Hinode data was reported by other authors, for example, Ishikawa and Tsuneta (2010), Jin and Wang (2015).

To estimate qualitatively the productivity of the local small-scale dynamo, one needs to compare the magnetic spectra from the data obtained simultaneously by different instruments from the same area on the solar disk. On June 19, 2017 the Goode Solar Telescope at Big Bear Solar observatory, using the instrument Near InfraRed Imaging Spectrapolarimeter (NIRIS, Cao *et al.* (2012)), recorded a quiet-sun area on the disk center. A set of LOS magnetograms with the spatial resolution of 0.08 arcsec per pixel was acquired and used to calculate the average magnetic power spectrum (red line in

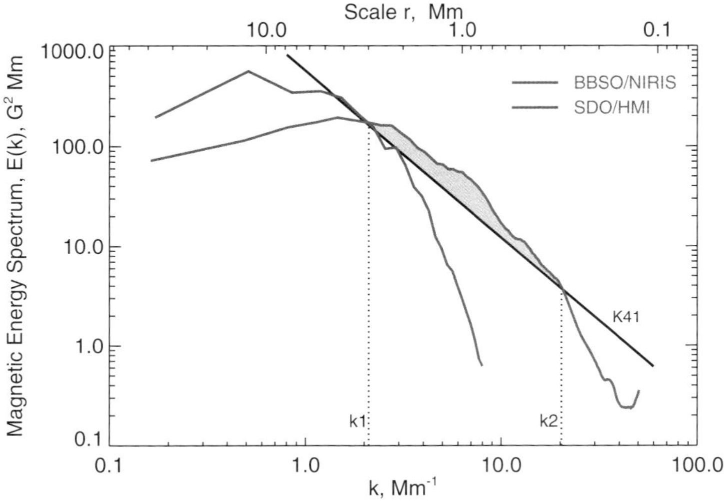

Figure 2. HMI (blue) and NIRIS (red) magnetic power spectra calculated over the same area of the solar photosphere. The black solid line shows the Kolmogorov-type K41 spectrum with the slope of -5/3. The shaded area shows the excess of magnetic energy above the K41 line within the $k_1 - k_2$ interval, which contains about 35% of the NIRIS energy in this interval.

Figure 2). Simultaneous maps of the LOS-field measured by HMI in the same area were used to derive the counterpart HMI-spectrum for comparison (blue line in Figure 2). If we consider an idea that following the K41 theory energy cascades from large scales (in this case, from scales of 3–5 Mm) down to small spatial scales, then the NIRIS spectrum should follow the K41 line (black solid line in Figure 2). However, within the interval between k_1 and k_2 the observed spectrum is well above the K41 line, which implies that besides the energy cascading to smaller scales along the inertial interval, an additional energy may be injected in this spectral range. Our estimates (Abramenko and Yurchyshyn (2020)) showed that the energy excess constitutes about 35% of the total NIRIS energy concentrated within the interval of (3.5 - 0.3) Mm. We emphasis that this is only a lower estimate of the energy excess because the high-frequency part of the spectrum is definitely lowered on the plot due to instrumental issues.

Comparison of magnetic spectra in quiet sun derived from different instruments (Abramenko, Yurchyshyn, and Goode (2012)) demonstrated that the spectrum tends to extent to the small-scale end as the resolution of an instrument improves (Figure 3). The spectrum from Nakagawa and Priest (1973) (black line), obtained from low-resolution Kit Peak data, shows only the energy on super-granula scales; the MDI spectrum (green line) shows the narrow inertial range with a slope close to -1 on scales around 10 Mm, the HMI spectrum follows with the same slope down to 4 Mm, and the Hinode spectra follow with the same slope down to 1 Mm. The spectrum, which is more shallow than the Kolmogorov -5/3 spectrum (the dashed K41 line), seems to be met on all observable scales.

Thus, in the entire undisturbed photosphere, the small-scale dynamo produces a substantial part of the total magnetic energy. The rest is presumably supplied by turbulent cascading of large-scale fields generated by the global dynamo.

4. Fluctuation Dynamo on Intermediate Scales

As mentioned in Introduction and argued by Stenflo and Kosovichev (2012), there exists no sharp transition between the global dynamo and SSD. Therefore, there should

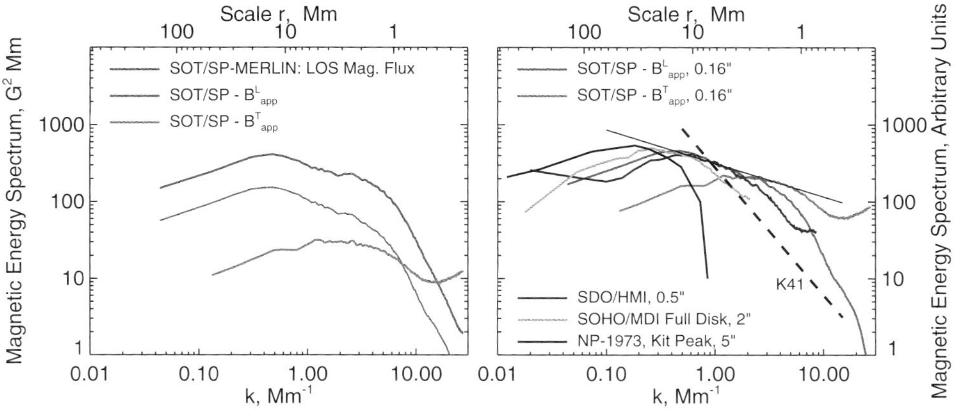

Figure 3. Magnetic energy spectra from quiet sun magnetograms obtained with different instruments. For better comparison, in the right panel the spectra are shifted along the vertical axis. Black line: Nakagawa and Priest (1973) spectrum. As the resolution improves, the cut-off of the spectra shifts toward the higher wave-numbers. The thin straight line shows a tentative behavior of the spectra as the resolution improves. The Kolmogorov spectrum with the slope of -5/3 is shown with the dashed line K41.

exist certain processes that introduce chaos on intermediate scales, for example, on the typical scales of active regions.

In general, there exists a possibility for the fluctuation dynamo to operate on typical scales of active regions (e.g., Brandenburg and Subramanian (2005), Sokoloff, Khlystova, and Abramenko (2015)). Continuous turbulence in the convection zone might affect the coherent toroidal field while it rises to the surface. Namely, the turbulence can distort the toroidal flux tubes and, even more, it can generate the additional flux. To explore the contribution of these processes, one has to segregate fluxes from active regions produced by the coherent global toroidal field (regular ARs), on one hand, and fluxes from active regions that were distorted by sub-photospheric turbulence (irregular, or complex ARs). Motivated by this reasoning, a new classification of active regions was suggested in Abramenko (2021). In brief, the underlying arguments to elaborate the magneto-morphological classification (MMC) were as follows.

A majority of active regions of all sizes are bipolar active regions and they follow the empirical laws, compatible with the mean-field dynamo theory, namely, the Hale polarity law, the Joy's law, and the rule of the leading sunspot prevalence (e.g., Babcock (1961), van Driel–Gesztelyi, Green (2015)). In the MMC, these active regions are classified as A-class, or regular active regions. Two sub-classes A1 and A2 were introduced to segregate in the A2-class those active regions, which contain a small (as compered to the leading sunspot) δ-structure. Scanty unipolar active regions were segregated to the separate class U. All the rest belong to the class of irregular active regions, B-class. Active regions of this class were also separated between three sub-classes: B1, B2, and B3. The sub-class B1 contains the wrong bipolar active regions, those violating at least one of the aforementioned empirical laws (emergence of a single toroidal flux tube rotated and/or inclined owing to the turbulent dynamo action); B2 - multipolar active regions consisting of several co-aligned bipoles (as a result of fragmentation and distortion of a single flux tube), or tight strong δ-structure (as a result of strong twist of a single flux tube by the turbulent dynamo action); B3 - multipolar ARs with chaotically distributed spots of both polarities (emergence of several intertwined by turbulent dynamo flux tubes). Examples of active regions of the A1-class (NOAA 12674) and of the B3-class (NOAA 12673) are shown in Figure 4.

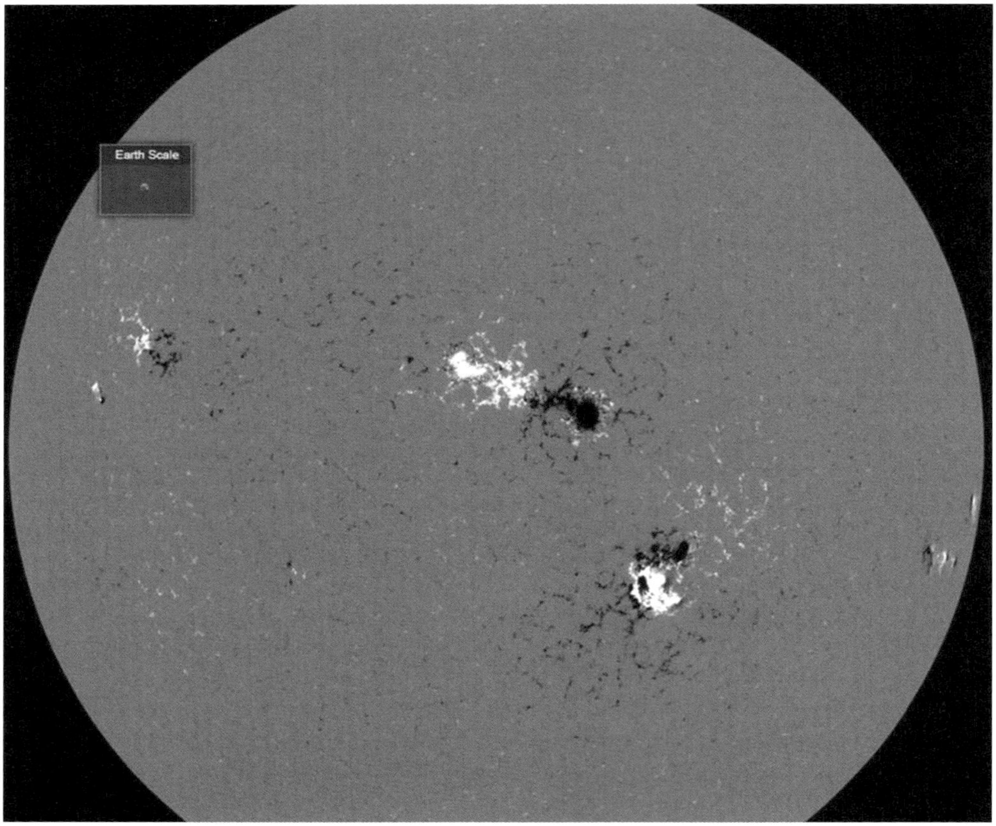

Figure 4. Examples of active regions of different magneto-morphological classes. In the Northern hemisphere the NOAA AR 12674 is observed: a bipolar magnetic structure complying with the Hale polarity lay, having the tilt angle in accordance with the Joy's law, and having the dominating leading sunspot; this active region belongs to the A1 class. In the Southern hemisphere the NOAA AR 12673 is observed: a multipolar structure with chaotically distributed sunspots; this active region was classified as B3-class active region.

In Abramenko *et. al* (2023) a set of independent snapshots of the solar disk magnetograms was compiled an utilized to explore how the summed (over the disk) magnetic flux from active regions of a given MMC class varies during the Solar Cycles 23 and 24. This experiment (see Figure 5) allowed the authors to reveal how the turbulent component of the dynamo manifests itself on the intermediate scales. It was found that the overall shape of the cycle is better correlated with the flux from all regular active regions, whereas the fine structure of the solar maximum (in particular, the origin of the secondary peaks in 2002 and 2014) is determined by the irregular active regions. Note also that the extreme flaring is caused by the active regions of B2 and B3 classes, Abramenko (2021).

During the phase of solar activity maximum, about a half of the total flux comes from irregular ARs and about the same amount comes from regular ARs, see Figure 5). So, about a half of the active regions flux is *distorted* by turbulence, however, not necessary *generated* by turbulence, because there exists (Abramenko *et. al* (2023)) a high correlation of the both fluxes with the cycle (the correlation coefficient CC between the A1+A2 (B1+B2+B3) and the total flux consists of 0.98 (0.94)) and between each other (CC between the A1+A2 and B1+B2+B3 fluxes consists of 0.88). This implies that the regular and irregular fluxes have a common origin: the coherent toroidal magnetic field generated by the global dynamo.

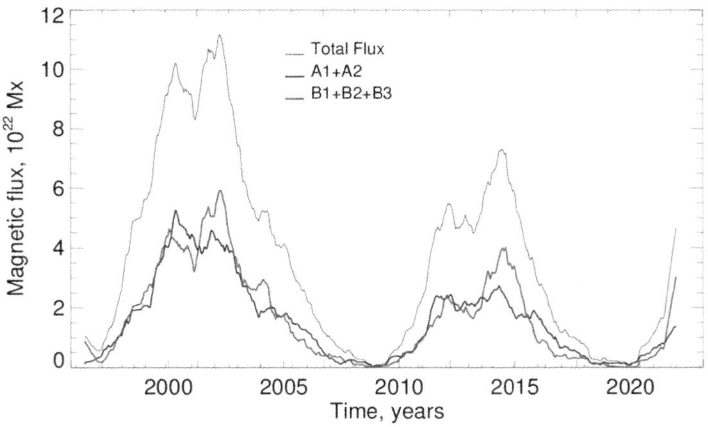

Figure 5. Time variations of the summed over the disk fluxes of active regions belonging to different MMC classes. Blue - fluxes of regular active regions (summed flux from A1 and A2 classes); red - fluxes of all irregular active regions (summed flux from B1, B2, and B3 classes). The thin black line shows the total over the disk flux from all active regions, including unipolar sunspots.

The deepest solar minima occur simultaneously for all classes. Moreover, during the minima, the irregular ARs are much more scanty than regular ones (see Suleymanova and Abramenko, this issue). So, the active regions generation during the solar minima is dominated by the global dynamo action with minor influence of the turbulence in the convection zone.

Therefore, a reservoir for all active regions' magnetic flux is the cycle-dependent global dynamo, whereas the turbulent convection hardly generates much flux, distorting a half of this flux during the active phase of a solar cycle.

5. Concluding Remarks

In summary, a manifestation of the local (small-scale, turbulent, or fluctuation) dynamo from the observational standpoint can be formulated as follows.

On the intra-network and network scales, the small-scale (local) dynamo is at work and generates more than 35% of the total magnetic energy, augmenting the direct turbulent cascade.

On intermediate scales of typical active regions, the solar dynamo operates as a unique process and generates ARs of all classes, whereas the turbulent component of dynamo (local dynamo) distorts a half of the emerging toroidal flux, determines the fine structure of the time profile of a cycle during a cycle maximum, causes the extreme flaring.

However, it would be very useful to keep in mind that from the physical standpoint, the segregation of the process of the magnetic flux generation into global (large-scale) dynamo and local (small-scale, turbulent, fluctuation) dynamo is very relative.

References

Abramenko, V., Yurchyshyn, V., and Goode, P.R. 2012, *4th Hinode Science Meeting: Unsolved Problems and Recent Insights*, 455, 17.

Abramenko, V.I. 2021, *MNRAS*, 507, 3698.

Abramenko, V.I. and Yurchyshyn, V.B.: 2020, *MNRAS*, 497, 5405.

Abramenko, V.I., Suleymanova, R.A., Zhukova, A.V. 2023, *MNRAS*,518,, 4746.

Babcock, H.W. 1961, *A&A*, 133, 572.

Batchelor, G.K. 1950, RSPSA, 201,405.

Brandenburg, A. 2003, *Turbulence and Magnetic Fields in Astrophysics*, Edited by E. Falgarone, and T. Passot., *LNP*, 614, 402.

Brandenburg, A. and Subramanian, K. 2005, *Phys. Rep.*, 417, 1.

Cao, W., Goode, P.R., Ahn, K., Gorceix, N., Schmidt, W., and Lin, H. 2012, *Second ATST-EAST Meeting: Magnetic Fields from the Photosphere to the Corona*, 463, 291.

Cattaneo, F. 1999, *ApJ*, 515, L39.

Charbonneau, P. 2020, *LRSP*, 17, 4.

Ishikawa, R. and Tsuneta, S. 2010, *ApJ*, 718, L171.

Jin, C. and Wang, J.: 2015, *ApJ*, 806, 174.

Jin, C.L., Wang, J.X., Song, Q., and Zhao, H. 2011, *ApJ*, 731, 37.

Kazantsev, A.P. 1968, *JETP*, 26, 1031.

Kitiashvili, I.N., Kosovichev, A.G., Mansour, N.N., and Wray, A.A. 2015, *ApJ*, 809, 84.

Kolmogorov, A. 1941, *Akad Nauk SSSR Doklady*, 30, 301.

Meneguzzi, M., Frisch, U., and Pouquet, A. 1981, *Phys. Rev.*, 47, L1060.

Nakagawa, Y. and Priest, E.R. 1973, *ApJ*, 179, 949.

Obridko, V.N., Livshits, I.M., and Sokoloff, D.D. 2017, *MNRAS*, 472, 2575.

Petrovay, K. and Szakaly, G. 1993, *A&A*, 274, 543.

Pietarila Graham, J., Cameron, R., and Schüssler, M. 2010, *ApJ*, 714, 1606.

Rempel, M. 2014, *ApJ*, 789, 132.

Schou, J., Borrero, J.M., Norton, A.A., Tomczyk, S., Elmore, D., and Card, G.L. 2012, *SoPh*, 275, 327.

Sokoloff, D., Khlystova, A., and Abramenko, V. 2015, *MNRAS*, 451, 1522.

Stenflo, J.O. 2012, *A&A*, 541, A17.

van Driel–Gesztelyi, L., Green, L.M. 2015, *LRSP*,12, 1.

Vögler, A., Schüssler, M. 2007, *A&A*, 465, L43.

Contributed papers

Dynamics of Solar and Stellar Convection Zones and Atmospheres
Proceedings IAU Symposium No. 365, 2024
A. V. Getling & L. L. Kitchatinov, eds.
doi:10.1017/S1743921323005355

Magnetic field and radial velocity fluxes at the initial stages of the evolution of solar active regions based on measurements at the photospheric level

A. M. Sadykov and S. A. Krasotkin

Department of Physics, Moscow State University, Moscow, Russian Federation
email: sadykov.am19@physics.msu.ru

Abstract. In this article, the physical processes occurring in the convective layer and the photosphere of the Sun and their connection to the formation of active regions (ARs) and the development of the corresponding magnetic field are explored. Specifically, we test the magnetic flux emergence hypothesis and based on the line-of-sight magnetic field and Doppler shift data obtained from the Global Oscillation Network Group (GONG) observations. The study encompasses the analysis of 24 ARs observed during the period from 2011 to 2022. We find a strong correlation between the magnetic flux and the imbalance of radial velocity fluxes. The results indicate that the magnetic flux emergence hypothesis cannot fully explain the evolution of ARs during their early stages of development.

Keywords. Solar magnetic flux emergence, Solar active region velocity fields, Solar active region magnetic fields

1. Introduction

As of today, there are no universally accepted views on the physical processes occurring in the convective layer and the photosphere of the Sun related to the emergence of active regions (ARs) and corresponding magnetic field development. One of the most common physical theories regarding the evolution of ARs is the hypothesis of magnetic flux emergence, or alternatively, the emerging magnetic flux tube hypothesis (Parker 1955). This concept of emerging magnetic flux tubes is the default assumption in a significant number of theoretical works. Consequently, there is an evident need to verify the accuracy of this concept based on solar magnetic field and Doppler observations.

According to the emerging magnetic flux tube model, the ARs appear in a certain area of the surface because the magnetic flux tube loop rises from the convective layer to the photosphere due to magnetic buoyancy. According to calculations by Caligari et al. (1995), the radial velocity of the rising part of the magnetic loop increases as it moves closer to the surface. At the level of the photosphere, the radial velocity of the magnetic tube is expected to exceed $\approx$500 m/s. Such a velocity is greater than the radial velocity of the plasma motion at the center of a granule, which is approximately $\approx$300 m/s (Zirin 1966). Consequently, if the emerging magnetic flux tube hypothesis is correct, a local increase in the radial velocity flux of matter should be observed on the photosphere, accompanied by an enhancement of the magnetic field.

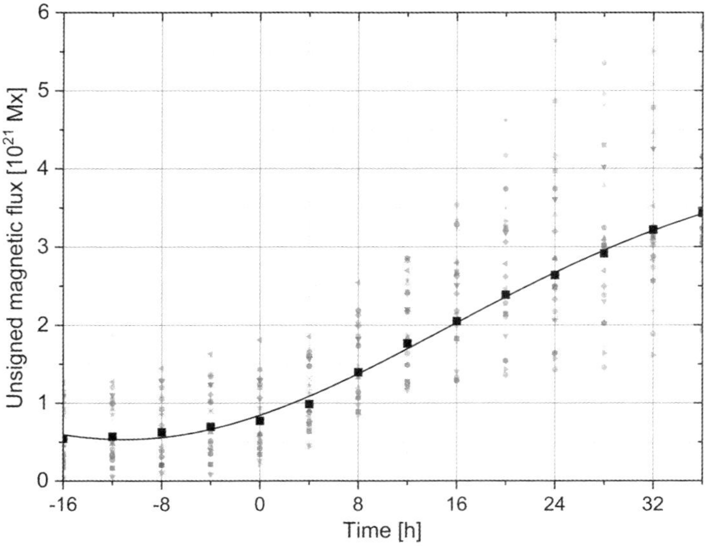

Figure 1. Unsigned magnetic flux in a unified time reference frame. The moment t=0 hours corresponds to the moment of sharp increase of the magnetic flux.

2. Observations and data processing

To conduct the research, we selected 24 ARs observed during the period from 2011 to 2022. The properties of their appearance on the Sun and subsequent development were studied through the analysis of a series of line-of-sight magnetograms and Dopplergrams by the network of ground-based Global Oscillation Network Group (GONG) telescopes, with a temporal resolution of 4 hours over a period of 5–6 days. Each of the selected ARs initially resided on the eastern hemisphere of the Sun and later evolved into a distinct structure consisting of two or more spots.

The time series data for all of the examined ARs were synchronized to a common time reference textbfframe using an epoch overlay method, ensuring that the onset of the sharp growth of the magnetic flux corresponds to t=0 h of the AR evolution time. The moment t=0 h was determined as the earliest time moment there the following condition is satisfied:

$$J_{mag}(t_0 + 8\,h.) - J_{mag}(t_0) \geq 0.5 * 10^{21}\,Mx, \tag{1}$$

where J_{mag} is the unsigned magnetic field flux over the selected frame of the Solar disk $(4' \times 4')$, which is calculated by multiplying the magnetic field strength value in each pixel by the area of the photospheric corresponding to this pixel. Thanks to the alignment in a unified time reference system, it became possible to perform averaging for each moment of the considered magnetic field characteristics. The evolution of the unsigned magnetic flux averaged over the studied regions is presented in Figure 1.

For demonstrating the relationship between the fluxes of the upward and downward radial velocity of matter, we consider a quantity known as the imbalance of radial velocity fluxes. This quantity is defined as the difference between the fluxes of upward and downward radial velocity, normalized to their sum:

$$D = \frac{|J_+| - |J_-|}{|J_+| + |J_-|}, \tag{2}$$

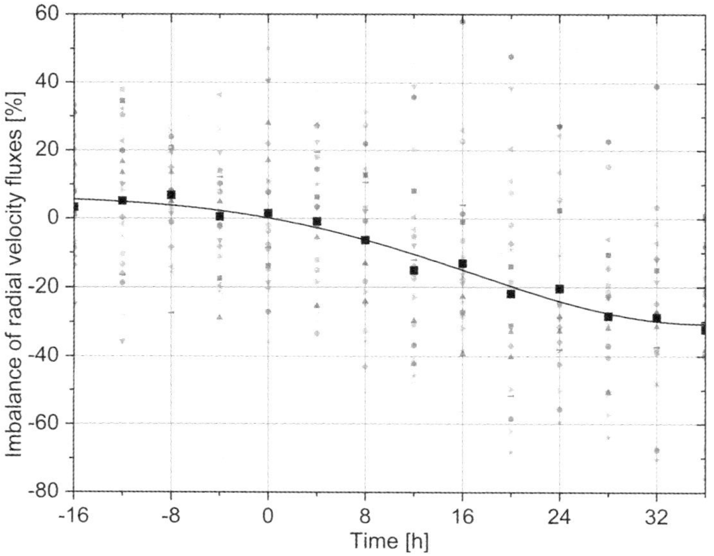

Figure 2. Imbalance of radial velocity fluxes as a function of time in a unified time reference frame.

where J_+ and J_- represent the fluxes of the upward and downward radial velocity of matter integrated over the selected area, respectively. The fluxes of radial velocity are calculated in a similar manner to the magnetic field flux. As evident from the definition, the imbalance of radial velocity fluxes is dimensionless (does not require adherence to a specific system of units) and ranges from -1 to 1 or from -100% to 100% (percentage values are used in this work).

3. Results

As can be seen from Figure 1, the magnetic field flux basically monotonically increases with time. Moreover, it vividly illustrates the method of superimposing epochs of AR data sampling at the moment of a sharp increase in the magnetic flux (Eq. 1), which corresponds to the time label at $t_0 = 0$ h in a unified time reference frame.

The graph of the radial velocity flux imbalance over time (Fig. 2) shows a strong descending trend over time. Before the moment of the sharp increase in the magnetic field flux corresponding to $t_0 = 0$ h the imbalance of fluxes was around zero, indicating an equilibrium between the upflow and downflow of material in the system. Then, after $t_0 = 0$ hours, a monotonic shift of the flux imbalance towards negative values is observed. This means that in the studied ARs, the downward motions of the material start to dominate over the upward motions.

By using the binary relationship graphs, one can visually observe how the velocity flux imbalance changes with different values of the magnetic field flux. For the radial velocity flux imbalance (Fig. 3), there is a notably strong correlation with the total magnetic field flux. For averaged values, the correlation coefficient is r$= -0,99$. This implies that as the magnetic field flux increases, the material's downward velocity flux becomes increasingly dominant. At low values of the magnetic field flux, the velocity flux imbalance remains close to zero, indicating no tendency towards upward or downward material movement.

4. Conclusion

Based on the obtained results, it can be concluded that, on average, in the early stages of active region (AR) evolution, there is a change in the radial velocity flux of material.

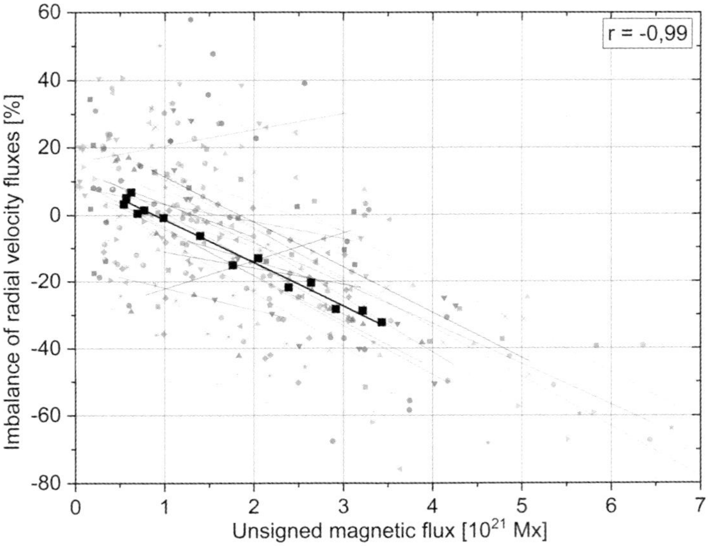

Figure 3. Imbalance of radial velocity fluxes as a function of the magnetic flux.

A sharp transition is observed from an equilibrium between the upflow and downflow radial velocity to the dominance of the downflow radial velocity as the magnetic field flux increases. The hypothesis of magnetic flux emergence implies an increase in the magnetic field flux synchronized with the increase in the upflow radial velocity. In terms of the velocity flux imbalance, this would lead to a shift towards positive values. However, in the examined ARs, no corresponding synchronous changes potentially corresponding to the emerging magnetic flux model were observed. Therefore, the magnetic flux tube emergence model should be used with caution when describing the evolution of AR, at least in the initial phases of their development, indicating a local rather than a global nature of the processes occurring during the AR formation phase.

References

Caligari, P., Moreno–Insertis, F., & Schussler, M. 1995, *ApJ*, 441, 886.
Parker, E. N. 1955, *ApJ*, 121, 491.
Zirin, H. 1966,. *The Solar Atmosphere*. Blaisdell Publishing Company.

Dynamics of Solar and Stellar Convection Zones and Atmospheres
Proceedings IAU Symposium No. 365, 2024
A. V. Getling & L. L. Kitchatinov, eds.
doi:10.1017/S1743921324000991

Subsurface Flows Associated with Formation and Flaring Activity of Solar Active Regions

Alexander G. Kosovichev[1,2] **and Viacheslav M. Sadykov**[3]

[1]New Jersey Institute of Technology, Newark, NJ 07102, U.S.A.
email: alexander.g.kosovichev@njit.edu

[2]NASA Ames Research Center, Moffett Field, CA 95035, U.S.A.

[3] Physics & Astronomy Department, Georgia State University, Atlanta, GA 30303, U.S.A.

Abstract. We investigate the evolution of subsurface flows during the emergence and the active phase of sunspot regions using the time–distance helioseismology analysis of the full-disk Dopplergrams from the Helioseismic and Magnetic Imager (HMI) onboard the Solar Dynamics Observatory (SDO). We present an analysis of emerging active regions of various types, including delta-type active regions and regions with the reverse polarity order ('anti-Hale active regions'). The results reveal strong vortical and shearing flows during the emergence of magnetic flux, as well as the process of formation of large-scale converging flow patterns around developing active regions, predominantly in the top 6 Mm deep layers of the convection zone. Our analysis revealed a significant correlation between the flow divergence and helicity in the active regions with their flaring activity, indicating that measuring characteristics of subsurface flows can contribute to flare forecasting.

Keywords. Sun: activity, Sun: helioseismology, Sun: magnetic fields, Sun: interior, MHD

1. Introduction

Most manifestations of solar activity are associated with magnetic sunspot regions emerging from the Sun's interior. The bipolar structure of active regions indicates that they originate from the toroidal magnetic flux generated by solar dynamo, but the physical processes of the flux emergence and formation of active regions and sunspots are not understood. In addition, the origin of complex magnetic configurations, such as delta-type sunspots with strong magnetic field gradients across the polarity inversion line, leading to the impulsive release of magnetic energy in the form of solar flares and coronal mass ejection, remains unclear.

Observations of photospheric magnetic fields show that the formation of active regions starts from the emergence of fragmented magnetic patches that are subsequently collected in compact, stable magnetic structures, observed as pores and sunspots. However, it remains unclear whether this process reflects 'monolithic' magnetic structures formed inside the Sun prior to the flux emergence or magnetic self-organization controlled by the near-surface plasma flows associated with emerging magnetic structures.

The process of magnetic self-organization on the Sun was demonstrated in 3D radiative MHD simulations by Kitiashvili et al. (2010). In these simulations, the initial uniformly distributed 100 G field collapsed into a pore-like structure with the magnetic field strength reaching 1.5 kG on the surface and ~ 6 kG below. This structure is maintained by self-forming converging downdrafts confirming the idea initially suggested by Parker (1955). A similar spontaneous formation of a pore-like structure was described by

Stein and Nordlund (2012) who simulated the emergence of initially a uniform horizontal magnetic field and confirmed that the self-organization is an intrinsic feature of magnetic fields emerging on the solar surface.

However, attempts to model the spontaneous formation of stable sunspot-like structures have not been successful. Rempel and Cheung (2014) modeled the flux emergence driven by advecting a semi-torus-like structure of the magnetic field at the bottom boundary. This model showed the formation of a bipolar magnetic structure at the surface. However, it quickly decayed due to the turbulent diffusion. Hotta and Iijima (2020) simulated the formation of sunspot-like structures from an initially horizontal twisted magnetic flux tube located at a depth of about 35 Mm, a part of which was advected by convective upflows. It was demonstrated that the converging downdrafts are a key mechanism of the structure formation. Also, it was found that the temperature distribution and flow patterns beneath the structures resemble the helioseismology results (Kosovichev et al. 2000; Zhao et al. 2001; Zhao and Kosovichev 2003a). However, like in the previous simulations of Rempel and Cheung (2014), the decay of the sunspot-like structures started immediately after their formation when the downflows in these structures turned into upflows. To overcome this difficulty, Brandenburg et al. (2016) investigated the formation of magnetic structures in a forced stratified turbulent layer and demonstrated that the effective magnetic pressure may become negative and result in magnetic field concentrations. However, these concentrations were found to be too deep and too narrow to explain the sunspots (Perri and Brandenburg 2018).

It has been known for a long time that complex active regions, in which strong magnetic fields of the opposite polarity form polarity inversion lines with high magnetic field gradients, are a primary source of large solar flares and eruptions (Severnyj 1958; Severny 1964). Magnetic structures of this type (called delta-type sunspots) can be formed when magnetic fields with opposite polarities emerge close to each other or pushed toward each other after the emergence (Zirin and Liggett 1987). Such magnetic configurations have been reproduced in the simulations of Kaneko et al. (2022) who employed the flux-tube model of Hotta and Iijima (2020) for various distributions of upwelling convective flows. However, contrary to observations, in these simulations, the delta-type sunspot-like structures were forming more frequently than the simple beta-type sunspots.

In general, the MHD simulations demonstrated that subsurface flows associated with the emergence and evolution of the magnetic field play a critical role in the formation, stability, and activity of sunspot regions. The observational information about the subsurface flows is provided by local helioseismology techniques: ring-diagram analysis (Gough and Toomre 1983; Komm et al. 2005), time-distance helioseismology (Duvall et al. 1993; Kosovichev 1996), and acoustic holography (Braun and Lindsey 2000). The subsurface flow maps obtained by the ring-diagram technique detected large-scale converging flows around developed active regions in the top 7 Mm-deep layer and diverging flows at a depth of 14 Mm (Haber et al. 2004). These flows, which may significantly affect the meridional magnetic flux transport, were confirmed by the time-distance helioseismology (Zhao and Kosovichev 2004).

The primary goals of our study are to improve the characterization and physical understanding of the emergence and evolution of active regions and explore the potential of utilizing physical descriptors of subsurface flows for predicting periods of flaring activity. The processes by which the magnetic energy is stored, enters, and leaves active solar regions, are critically linked to the flow patterns therein. Large-scale organized flows are developed spontaneously in subsurface layers due to the emergence of magnetic flux and its interaction with the existing magnetic field of active regions. This process forms stressed magnetic configurations that trigger solar eruptions.

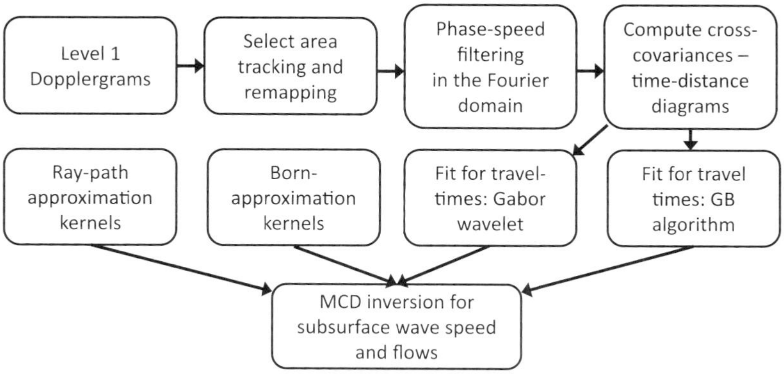

Figure 1. A schematic representation of the time-distance helioseismology pipeline (Zhao et al. 2012).

In Section 1, we briefly review the time-distance helioseismology pipeline methodology implemented in the Solar Dynamics Observatory's (SDO) Joint Science Operation Center (JSOC), and the procedure for tracking the evolution of subsurface flows in emerging active regions. In Section 2, we present and discuss the evolution of subsurface flows for several active regions, as well as the relationship of various flow characteristics to the flaring activity. We present initial results on the correlation of flow characteristics with the flare productivity of active regions and Section 3. In Section 4, we discuss the outcome of this study and future perspectives.

2. Measurements of Subsurface Flows by Time-Distance Helioseismology

The computational pipeline (Fig. 1) for studying the subsurface dynamics of active regions takes the Carrington coordinates of active regions at the central meridian from the Solar Region Summary (SRS) database, compiled by the National Oceanic and Atmospheric Administration's (NOAA) Space Weather Prediction Center (SWPC), and uses these coordinates as the central points of 30-degree areas tracked for 10 days during their passage on the solar disk. This setup allows us to follow the evolution of active region areas even before the magnetic flux emergence (i.e., retroactively) and after the decay. The 3D subsurface flow maps are calculated from the tracked Dopplergrams that are remapped onto the heliographic coordinates using Postel's projection (transverse cylindrical projection that preserves the distance along great circles). Each tracked, 8-hour-long datacube consists of 640 Dopplergrams of pixels with a spatial resolution of 0.06 degree/pixel and a 45-second time cadence. The tracked datacubes are processed through the Time-Distance Helioseismology Pipeline (Zhao et al. 2012; Couvidat et al. 2012). The output represents acoustic travel-time maps calculated with 0.12-deg sampling for the whole tracked area (512×512 pixels).

The travel times are calculated for eleven annuli located at different distances from central points corresponding to 2x2-binned original Dopplergram pixels. The signals of acoustic waves traveling between the central points and the surrounding annuli are calculated from the HMI Doppler velocity measurements as the corresponding cross-covariance functions. The cross-covariances are computed in the Fourier space, and phase-space filters are applied to isolate the signals corresponding to each of the travel distances.

$$\psi(\tau, \Delta) = \int_0^T f(t, \boldsymbol{r}) f^*(t + \tau, \boldsymbol{r} + \Delta) dt$$

The travel times are calculated by two different methods: 1) the Gabor wavelet fitting (Kosovichev and Duvall, 1997)

$$G\left(A, \omega_0, \delta\omega, \tau_{\mathrm{p}}, \tau_{\mathrm{g}}; t\right) = A\cos\left\{\omega_0\left(t-\tau_{\mathrm{p}}\right)\right\}\exp\left(-\frac{\delta\omega^2}{4}\left(t-\tau_{\mathrm{g}}\right)^2\right)$$

and 2) a cross-correlation with a reference (Gizon & Birch 2002).

The travel times are used to infer the 3D maps of subsurface flows by solving an inverse problem. It is formulated as a set of linear integral equations whose kernels are calculated by using either the ray-path theory or the Born approximation:

$$\delta\tau\left(x_1, x_2\right) = \int_V \mathrm{d}\boldsymbol{r}\boldsymbol{K}\left(\boldsymbol{r}; \boldsymbol{x}_1, \boldsymbol{x}_2\right)\cdot\boldsymbol{v}(\boldsymbol{r})$$

The system of integral equations can be discretized and formulated as the set of linear equations for grid points (i, j, k) for each surface point (λ, μ) and travel distance, ν:

$$\delta\tau_l^{\lambda\mu\nu} = \sum_{ijk,l} A_{ijk,l}^{\lambda\mu\nu}\tilde{v}_l^{ijk}.$$

Here, $\tilde{v}_l^{ijk} = \left[v_x^{ijk}, v_y^{ijk}, v_z^{ijk}\right]/c^k$ is the vector of velocity at the horizontal grid points i, j, and vertical grid points, k, relative to the sound speed at these points, calculated from the standard solar model. Regularized solutions are determined by the Multi-Channel Deconvolution (MCD) method (Couvidat et al. 2005), and the regularization parameters are chosen to suppress noise and represent a smooth solution.

We used the SDO/HMI time-distance helioseismology pipeline (Fig. 1) to infer 3D subsurface flow maps during the emergence and evolution of Active Regions. The travel times are used for the reconstruction of subsurface flows in 8 subsurface layers in the depth ranges: 0–1, 1–3, 3–5, 5–7, 7–10, 10–13, 13–17, and 17–21 Mm, and with the horizontal spatial sampling of 0.12 degrees (1.5 Mm). The horizontal and vertical resolution is determined by the averaging kernels calculated in the inversion procedures. The characteristic width of the averaging kernels is roughly proportional to the wavelength of acoustic waves and increases within this range of depth from $\sim$2 to 10 Mm (Couvidat et al. 2005; Parchevsky et al. 2014). Therefore, the flow maps obtained from the helioseismic inversions represent smoothed versions of the actual subsurface flows. Combinations of two travel-time measurement techniques and two types of sensitivity kernels provided four flow maps for each time interval. The flow patterns in these maps are very similar. Some differences in the flow amplitude (mainly in the top layers) can be attributed to differences in the sensitivity and averaging kernels. In this paper, we present the flow maps obtained by using the Gabor-wavelet technique for travel-time measurements and the Born-approximation kernels.

The primary source of noise is due to the random excitation of solar oscillations, the so-called "realization noise." In the flow maps obtained for a particular depth, it causes random variations with the characteristic scale corresponding to the wavelength of acoustic waves with the turning point located at this depth. These variations are significantly weaker than the signal from convective flows for the depths up to 8 Mm but become greater than the convective velocities in the deeper layers. However, the flows associated with active regions are significantly stronger than the quiet-Sun convection in these layers.

The helioseismic inversions for all active regions presented in this paper show large-scale divergent flows at depths greater than 6 Mm, similar to the flows around isolated sunspots (Kosovichev et al. 2000; Zhao and Kosovichev 2003a; Kosovichev et al. 2011). Therefore, we focus on the dynamics of the shallower 4 Mm deep layers which seems to play a key role in the formation and dynamics of sunspots and active regions. We primarily discuss the horizontal velocity maps because the inversions for the vertical velocity may have systematic errors due to a potential cross-talk with the horizontal flow components, particularly in the regions with regular horizontal velocity patterns and small vertical velocities, e.g. in the near-surface layers of supergranulation cells. However, the testing with artificial data showed that the inferences of stronger vertical velocities in the deeper layers are not affected by the cross-talk (Zhao and Kosovichev 2003b).

3. Subsurface Dynamics During Emergence and Evolution of Active Regions

In this Section, we present four examples of emerging active regions of various types:

• AR 11158, a delta-type region formed by the interaction of two emerging bipolar magnetic regions,

• AR 12673, a delta-type region formed by the collision of a large sunspot with emerging bipolar magnetic flux,

• AR 12282, an anti-Hale active region formed in the vicinity of a large sunspot,

• AR 13006: a small delta-type anti-Hale active region that produced an X-class flare,

• AR 13179, a beta-type active region formed by the continuous emergence of small-scale bipolar magnetic elements.

While these examples do not cover all active region types, they provide interesting examples of the formation of flare-productive delta-type regions, relatively simple regular beta-type regions, and the mysterious 'anti-Hale' regions with the polarity order violating Hale's polarity law (Hale et al. 1919).

In Figures 2–4, we present the flow maps for each of these active regions for two moments of time, during the emergence and the developed state, for three top layers at the depths, 0–1 Mm, 1–3 Mm, and 3–5 Mm. We use the corresponding central locations 0.5, 2, and 4 Mm for the identification of these layers. The corresponding surface line-of-sight magnetograms are shown in the color images. The magnetic field of the positive polarity is shown in red, and the negative polarity is shown in blue. The titles at the top of each panel in these figures show the midtime of the flow maps, obtained from the 8-hour Dopplergram series, the depth d, the latitude ϕ, longitude relative to the central meridian λ, and the Carrington longitude λ_c of the center of the images at $x = y = 0$. For the time-distance analysis, the areas on the solar disk are remapped onto the heliographic coordinates, using Postel's projection centered at the center of these images. For convenience, the heliographic coordinates are converted in the linear distances along the surface, and shown in megameters along the x and y axes. The horizontal velocity vectors are shown by arrows, the arrow scale is shown at the bottom of each panel (kept the same for all active regions). The label Gabor–Born indicates that the travel times were determined by the technique of fitting the Gabor wavelets to the oscillation cross-covariance function (Zhao et al. 2012), and the Born-approximation kernels were used for the travel-time inversion (Birch and Kosovichev 2000; Birch et al. 2004; Birch and Gizon 2007).

3.1. *AR 11158: Delta-type Region Formed by Interaction of Two Emerging Bipolar Magnetic Regions*

Active region 11158 initially emerged on 2011.02.09 as a simple bipolar magnetic region (BMR). About 20 hours later, a second bipolar structure emerged about 50 Mm from the first BMR. The leading positive polarity was rapidly moving apparently by local subsurface flows, and on 2011.02.12, it reached the trailing negative polarity of the first BMR (Fig. 2, left column). During the subsequent magnetic flux emergence in both

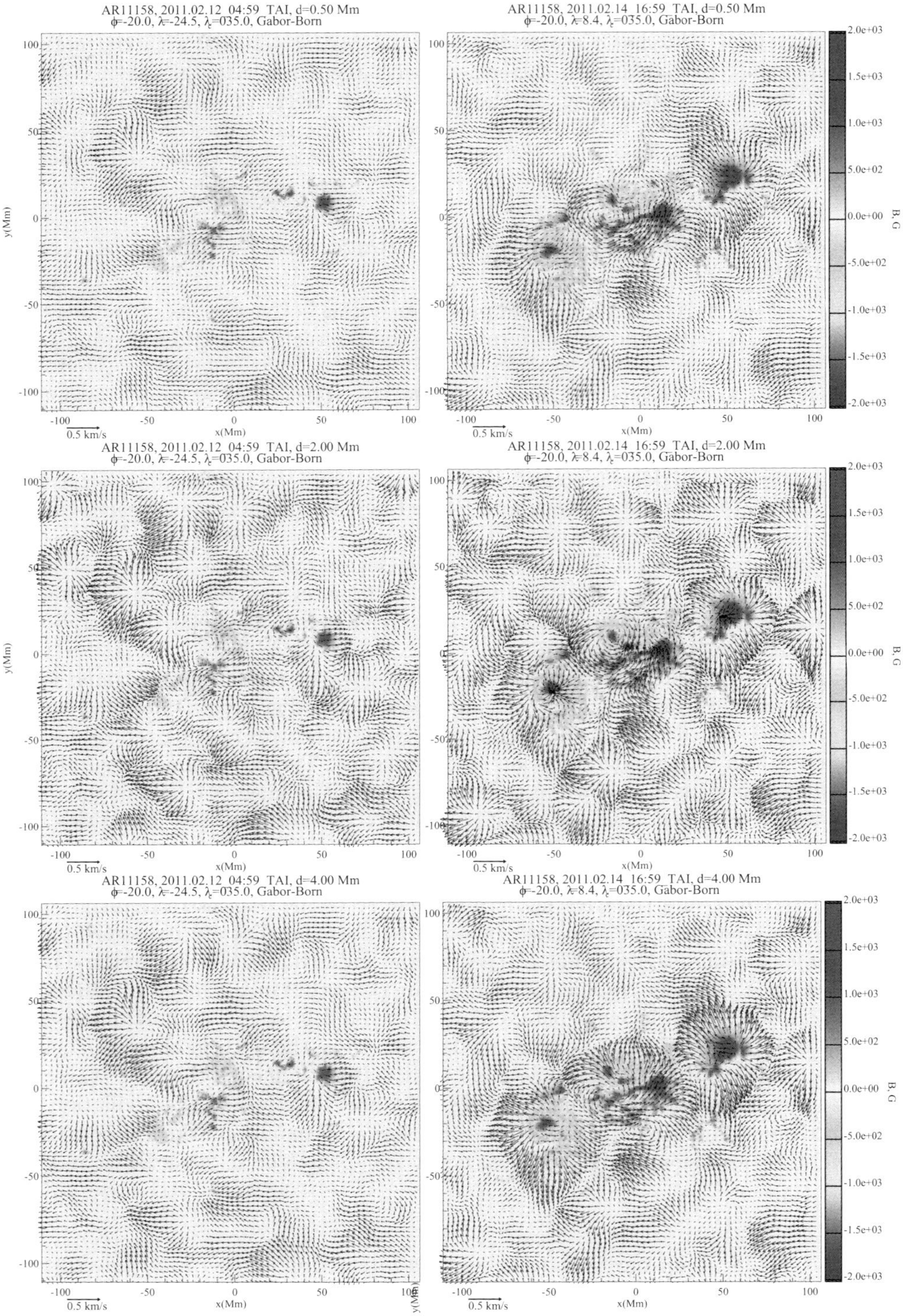

Figure 2. Flow maps of AR 11158 during the emergence (left column) and before the X2.2 flare (right) at three depth levels centered at 0.5, 2, and 4 Mm below the solar surface (from top to bottom). The background images show the corresponding surface magnetograms.

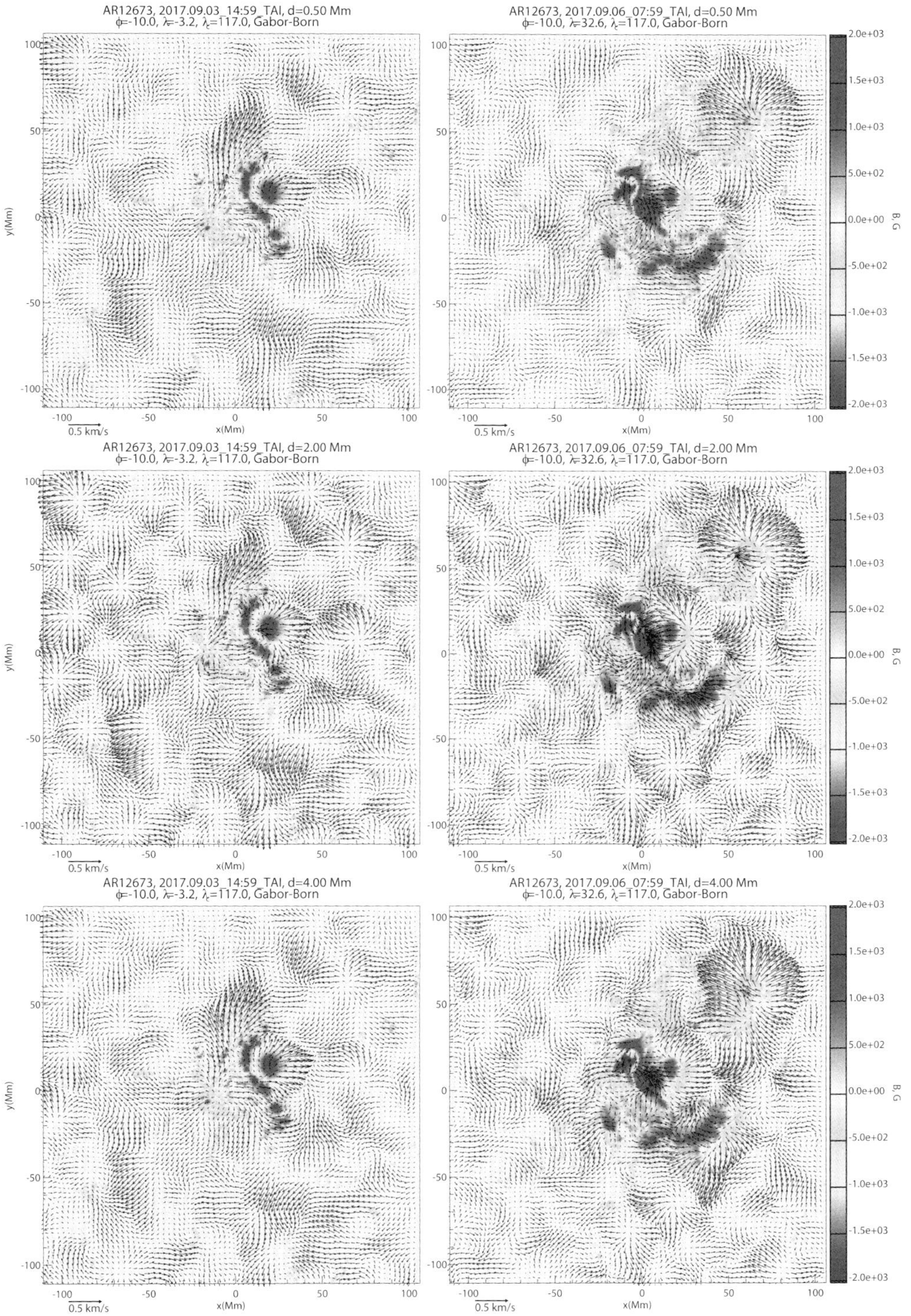

Figure 3. Flow maps of AR 12673 during the emergence of bipolar magnetic flux in the form of fragmented ribbons near the existing positive polarity sunspot (left column) and before the X9.3 flare (right) at three depth levels centered at 0.5, 2, and 4 Mm (from top to bottom). The background images show the corresponding surface magnetograms.

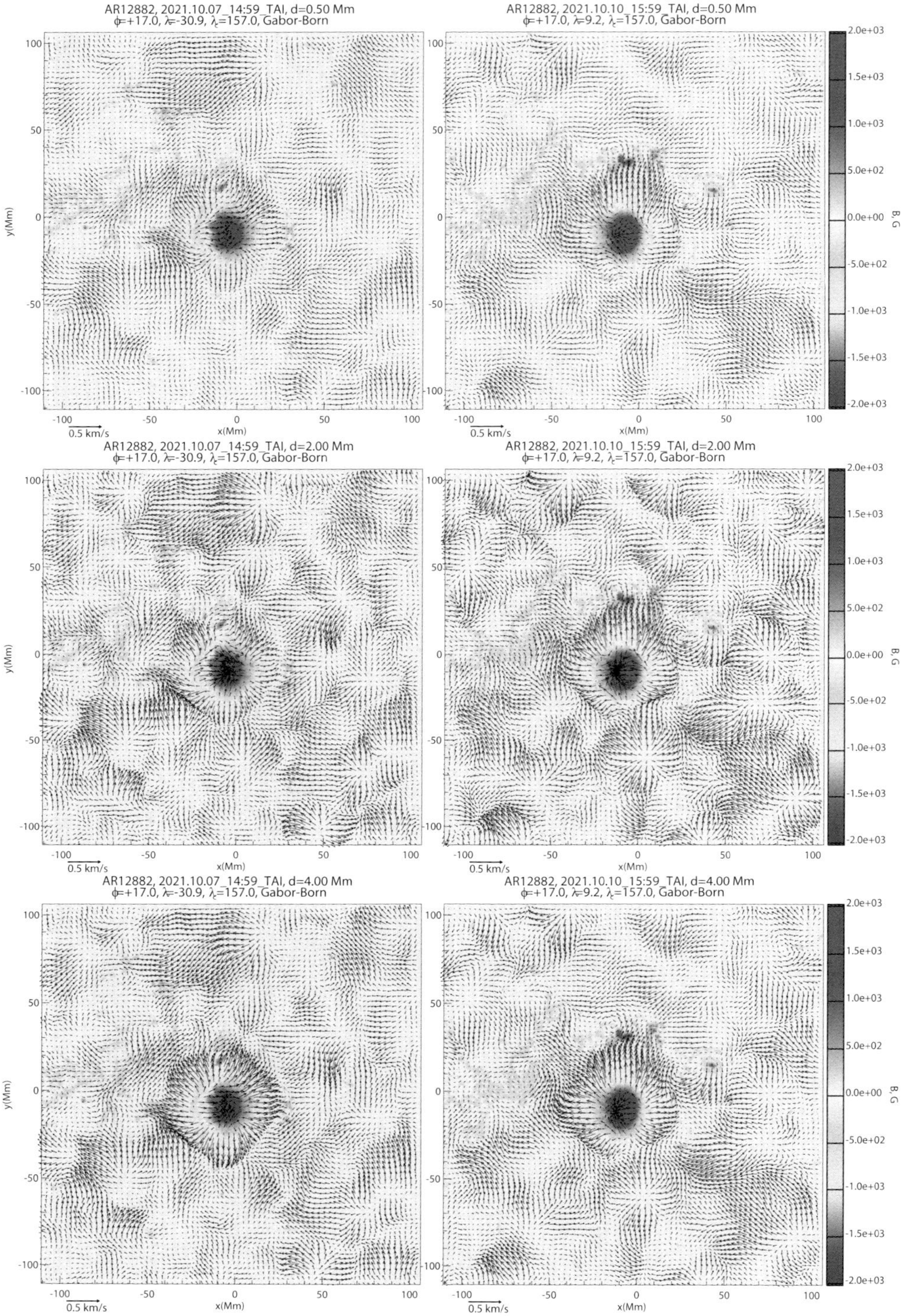

Figure 4. Subsurface flow maps of AR 12882 during the initial emergence of a small bipolar magnetic region just above of a large positive polarity sunspot (left), and at the developed stage of an anti-Hale region (right). The background images show the corresponding surface magnetograms.

BMRs, the positive polarity of the second BMR overtook the trailing polarity region of the first BMR. The interaction of these magnetic structures created a delta-type configuration with a strong magnetic field gradient along the polarity inversion line, which resulted in X2.2 flare on 2011.02.15. The right panel in Fig. 2 shows the subsurface flows and the surface magnetic field on 2011.02.14, about 9 hours before the flare. The flows around the mixed polarity structure in the middle of this image reveal a large-scale vortex in the top 2 Mm layer. This vortex pushed the magnetic field of the positive polarity towards the negative polarity spot, enhancing the magnetic field gradient across the polarity inversion line. The flows converging towards the polarity inversion line are also visible in the 4 Mm layer. This indicates that the subsurface flows may play a significant role in the formation and flaring activity of the delta-type active regions.

The flows outside the isolated sunspots are predominately diverging, while they are converging in these layers beneath the sunspots, corresponding to the previous results of Kosovichev et al. (2000); Zhao et al. (2001).

The relationship between the surface and subsurface flows in this active regions was studied by Liu et al. (2013) who compared flows obtained by applying the Differential Affine Velocity Estimator for Vector Magnetograms (DAVE4VM) technique with the near-surface velocity from the time-distance helioseismology measurements in the 0.5 Mm deep layer. They found general similarities between surface and subsurface flows but also significant differences in some areas. These differences may be due to rapid variations of plasma flows with depth or due to systematic differences in the measurement techniques. In particular, the DAVE4VM technique determines the flow velocity from the vector magnetic field variations by assuming that the magnetic field is frozen into plasma (Schuck 2008), which is not necessarily the case because of the turbulent diffusion. On the other hand, the spatial and temporal resolutions of the time-distance technique are significantly lower than those of the DAVE4VM method.

3.2. *AR 12673: Delta-type Region Formed by Collision with Emerging Emerging Bipolar Region*

Active region 12673 presents another interesting example of the formation and flaring activity of a delta-type region with a strong magnetic gradient across the polarity inversion line (Fig. 3). The delta-type configuration of this active region was formed by the interaction of new emerging bipolar flux with the existing stable sunspot. Initially, a small bipolar region emerged on 2017.09.02 south to the sunspot, but then the flux emergence quickly expanded to the north forming ring-like structures around the sunspots by 2017.09.03 (left panel in Fig. 3). The northern part of the emerging negative polarity flux was moving fast to the east and formed a stable sunspot (in the upper right corner in the right panel of Fig. 3). The portion of the emerging flux, which was located south of the original sunspot, created an extended sunspot system of positive polarity. The portions of the emerging flux, which were in the latitudinal range as the original sunspot, collided with this sunspot and created a delta-type configuration with a longitudinally extended polarity inversion line. This polarity inversion line was a source of several strong flares, including the X9.3 flare on 2017.09.06, which produced a series of helioseismic events - sunquakes (Sharykin and Kosovichev 2018).

The right panel in Fig. 3 shows the subphotospheric flows just before the big flare. The flow structure in this region is very complicated. We observe significant variations not only in the flow speed but also in the flow direction with the depth. The most notable feature is the converging flows towards the polarity inversion line, which are the strongest in the 2 and 4 Mm deep layers, and the large-scale vortex flow around the negative polarity in the top two layers near the center of the presented flow maps. This vortex flow converges to the polarity inversion line from the south. Curiously, the east-ward flow towards the polarity inversion line stops at about 2 Mm from the polarity inversion line. At this point, it is unclear whether it is a physical effect or a result of systematic errors (e.g. caused by the suppression of acoustic waves in strong field regions). Shearing flows in opposite directions are developed along the polarity inversion line in the top subsurface layer. However, these are not as strong as observed by feature tracking techniques on the surface.

The surface flows in this region were previously studied by Getling (2019); Liu et al. (2023); Verma (2018); Wang et al. (2018) using various techniques of local correlation tracking and the DAVE4VM technique. These studies also revealed strong shearing and vortex flows and high vertical velocities in the vicinity of the polarity inversion line. The subphotospheric flows are generally consistent with these conclusions, but there are some differences, which are likely to be attributed to differences in the spatial and temporal resolutions and to variations of the flow patterns with depth. In particular, the helioseismic maps reveal flows crossing the polarity inversion line, which the surface measurements cannot detect by the nature of these techniques.

3.3. *AR 12882: delta-type Anti-Hale Active Region*

Most active regions follow Hale's polarity law (Hale et al. 1919), according to which the magnetic polarity of the leading sunspots is the opposite in the northern and southern and changes the polarity every 11 years. However, in about 3% of the observed active regions, the leading sunspots have a polarity that is opposite to the majority of active regions, thus, violating the Hale law. Figure 4 shows subsurface flows in AR 12882, in which an anti-Hale active region was formed. The formation of the anti-Hale magnetic structure started from the emergence of a small bipolar magnetic region in the vicinity of a large sunspot, oriented in the South-North direction (Fig. 4, left panels). Then, in the next three days, the bipolar region continued to grow, and the process of the magnetic flux emergence was similar to the emergence of a normal 'Hale' active region, accompanied by the accumulation of small-scale magnetic elements and the polarity separation. However, the negative (blue) polarity quickly moved forward, apparently driven by local west-directed flows, and formed a small negative polarity sunspot (Fig. 4, right panels). At this time, the flows around these sunspots were diverging. Curiously, the interaction of the outflows from this sunspot and the following big positive polarity sunspot generated a vortex between them. The negative polarity spot continued moving away from the positive polarity sunspot and decayed in about 4 days. The fast polarity separation may indicate that the emerging 'anti-Hale' sunspot belonged to a toroidal flux system with the field direction opposite to the primary toroidal flux. Such scenarios of the formation of the anti-Hale bipolar magnetic regions were previously suggested by Stenflo and Kosovichev (2012).

3.4. *AR 13006: delta-type Anti-Hale Active Region With X-class Flare*

AR 13006 is another interesting example of the delta-type active region that produced the X1.5 flare with a strong sunquake on 2022.05.10. This active region was formed near the boundary of the polarity reversal in a very large elongated diffuse magnetic field structure. This active region appeared in this form on the East limb; thus, its initial emergence was not observed. In this region, a small but stable positive polarity sunspot was surrounded by the magnetic field of the opposite polarity (Fig. 5, left panel). Interestingly, the supergranulation flows look substantially stronger south of the elongated magnetic structure than north of it, which is apparently caused by large-scale flows directed toward this structure. This flow is particularly strong in the 2 and 4 Mm

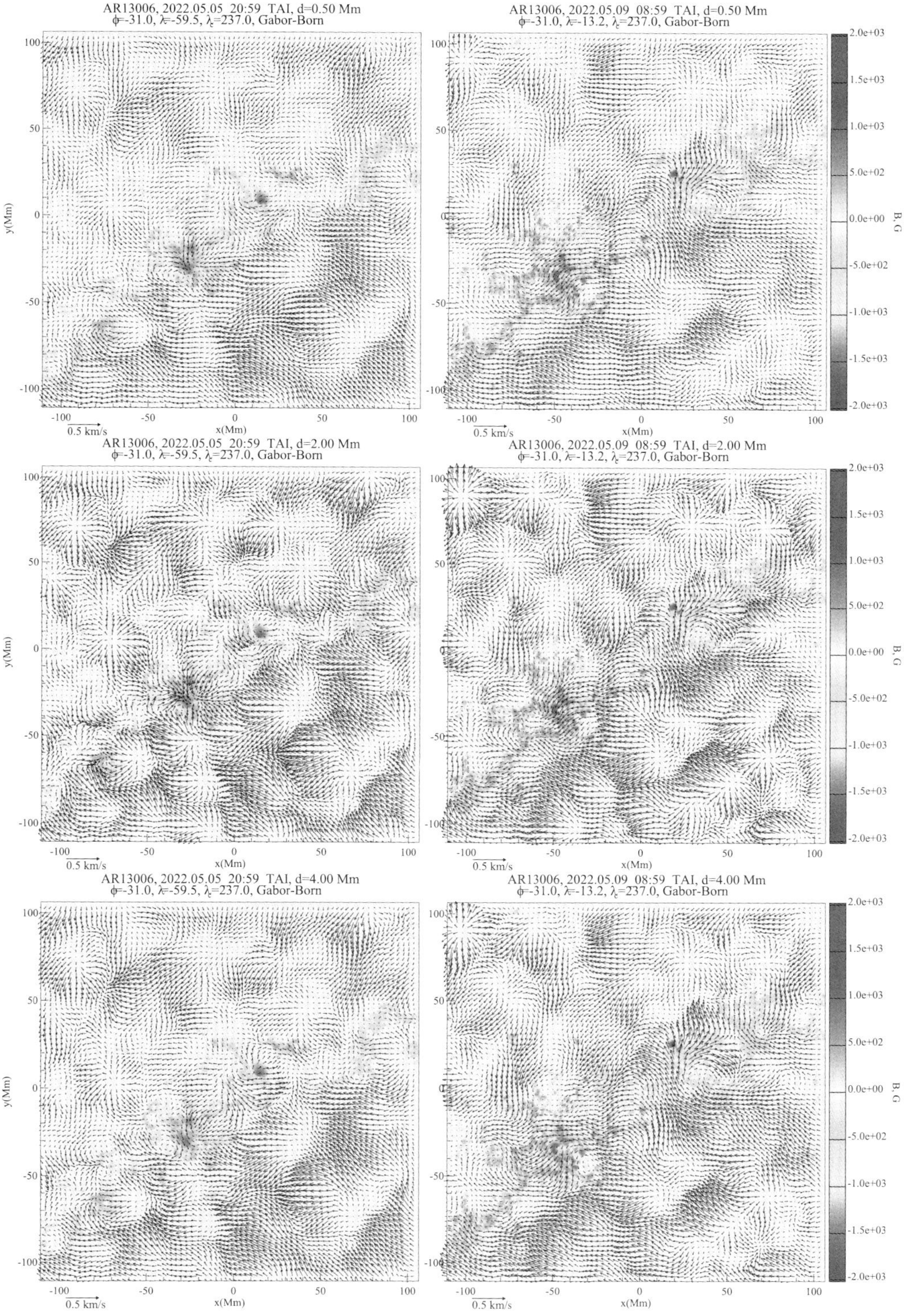

Figure 5. Flow maps in the area of a small delta-type anti-Hale active regions AR 13006, which formed at a boundary of a large-scale diffused bipolar magnetic region (forming an inclined belt of fragmented magnetic flux). The emergence of additional negative polarity flux (blue) around the round positive polarity sunspot (red) caused an X1.2 flare on May 10, 2022. The background images show the corresponding surface magnetograms.

deep layers. In the deeper layers, the supergranulation flows diminish but the large-scale flow directed north-east remains. Probably, it reflects a deep circulation of a giant-cell scale, correlated with large-scale magnetic structures in the convection zone.

As the active region evolved, the negative polarity continued to emerge on the south-east side of the positive polarity sunspot. Such emergence was accompanied by local northwest-directed flows, which pushed this flux closer to the sunspot. One of these flux-emerging events created a ring-like structure around the sunspot, forming a polarity inversion line with a strong magnetic field gradient, which was the source of the X1.5 flare and a strong sunquake (Kosovichev et al. 2023). After the flare, the positive polarity sunspot quickly disappeared.

The right panel in Figure 5 shows the flows and photospheric magnetic field on 2022.05.09, a day before the flare. At that time, the negative polarity of bipolar magnetic flux elements, which were emerging in the area of the polarity inversion of the large-scale magnetic structures, were moved towards the positive polarity spot. It seems that the expansion of continuously emerging small-scale bipolar magnetic elements and their merging into larger-scale magnetic structures is a common mechanism of the active region formation and that the emerging magnetic flux has a fragmented small-scale turbulent structure, which is organized in large-scale sunspot regions in the relatively shallow near-surface layers. Flare-productive delta-type configurations can be formed during this process when the new flux emerges in the vicinity of previously emerged magnetic structures. The subphotospheric flow maps indicate that the plasma flows resulting from the interaction of emerging magnetic flux with the turbulent convection in the top 4-6 Mm deep layer, play a key role in the magnetic self-organization of magnetic field on the solar surface.

3.5. *AR 13179: Formation of a β-type Active Region*

Active region 13179 provides an interesting example of the magnetic self-organization process. It started with the emergence of a compact circular-shaped bipolar region at a boundary between supergranules on 2022.12.29 (Fig. 6, left panel). It quickly expanded, apparently reflecting a large underlying magnetic structure. This expansion was accompanied by the flows diverging from the central area of the active region. The subsequent emergence occurred in the region separating the positive and negative polarities in the form of continuous emergence of small-scale bipolar elements. The positive and negative elements were streaming towards the forming sunspots of the corresponding polarity. The rate of magnetic flux emergence in the central area of the active region was about 2×10^{16} Mx s^{-1}. Thus, the total magnetic flux that emerged in the central area during the three days of the active region formation was about 5×10^{21} Mx, which is about half of the total active region magnetic flux.

At the beginning of the active region emergence, the subsurface flows corresponded to an expansion from the central area towards the growing magnetic polarities (Fig. 6, left panel). When the active region developed, the inflows into the sunspot areas were observed, which presumably increased the accumulation of the magnetic flux. In addition, strong outflows extended outside the sunspot penumbras were developed predominantly in the longitudinal directions (Fig. 6, right panel). Usually, the flow patterns in the sunspot areas, constituting compact converging flows beneath the sunspots in shallow ~ 4 Mm deep layers, and large-scale outflows in the deeper layers, are more symmetrical. In this case, the inversion results show a stream of flows through the trailing sunspot. It was more compact and better developed than the leading fragmented sunspot, which is unusual.

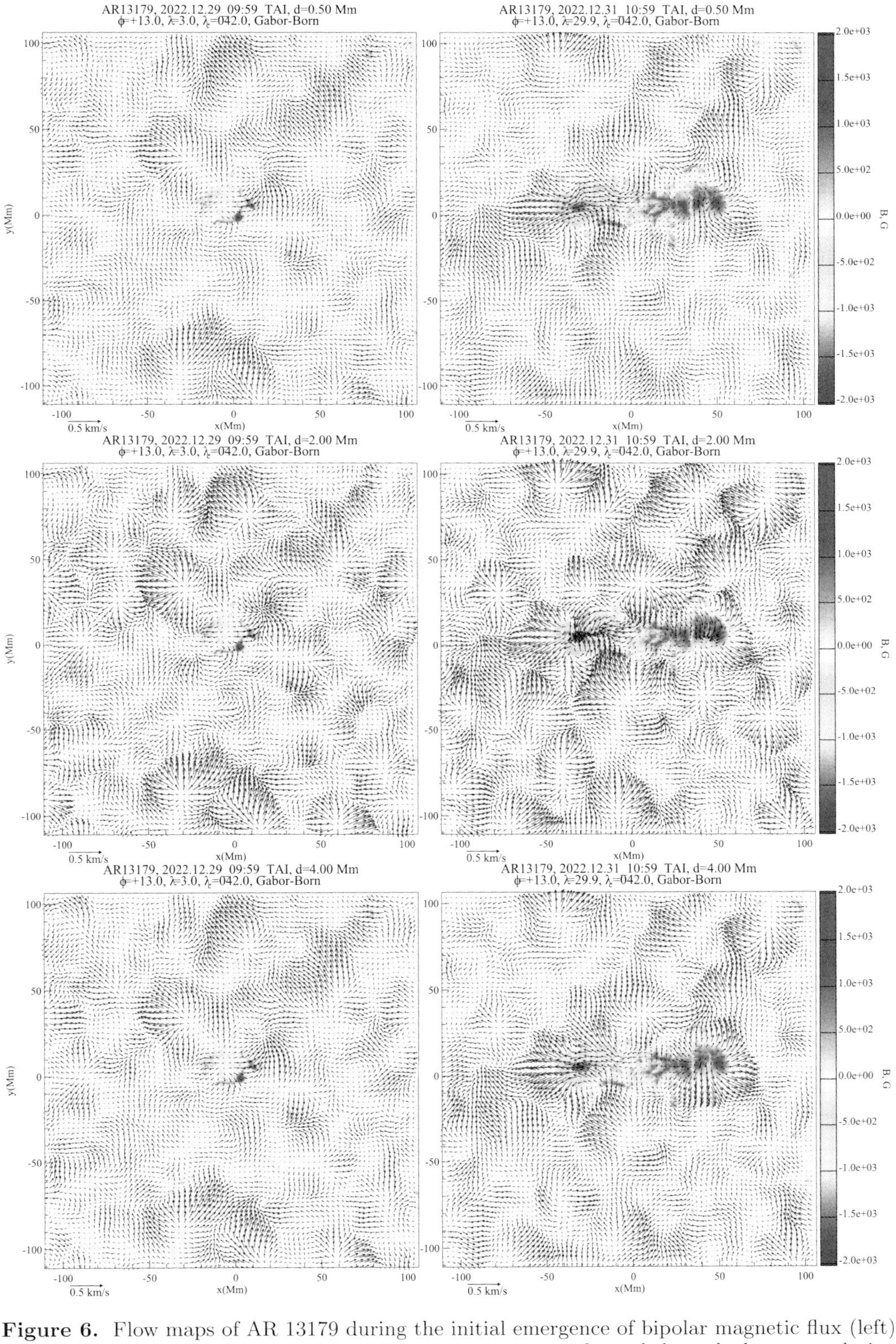

Figure 6. Flow maps of AR 13179 during the initial emergence of bipolar magnetic flux (left) and the developed active region phase. The active region was formed through the accumulation of small-scale bipolar magnetic elements continuously emerging in the middle area of this region. The background images show the corresponding surface magnetograms.

4. Flow Characteristics and Flaring Activity

The subsurface flow maps revealed shearing and converging flows around the polarity inversion lines, which are the places where energy is released in solar flares. For example, the flow maps of AR 12673 reveal shearing flows beneath the active region (Fig. 3, right panels) and, in particular, in the area of the polarity inversion line, which was the source of several flares, including the X9.3 flare of Sept. 6, 2017.

Therefore, it is important to investigate the links between the flow characteristics and the flaring activity. For this analysis, we derived the physical descriptors to characterize the subsurface flow maps: a layer-average horizontal divergence calculated from the horizontal velocity components, a layer-average vertical component of the flow vorticity, and a proxy for the kinetic helicity defined as a product of the horizontal divergence and the vertical component of vorticity.

We used the horizontal components of the velocity for the calculations of the flow divergence, vorticity, and helicity because the vertical component of the subsurface flows is determined with significantly larger uncertainties. Nevertheless, we use the averaged vertical velocity as a separate descriptor. The three subsurface layers closest to the surface (0–1 Mm, 1–3 Mm, and 3–5 Mm deep) are most relevant to the short-term active region evolution.

The comparison of the flow characteristics in the 1-3 Mm deep layer (mean divergence, helicity proxy, and vertical velocity component) with the mean magnetic flux, represented in the plots by mean unsigned magnetic field (Fig. 7, left panels a) and b) shows the significant increase in the flow convergence (corresponding to the negative divergence) and an increase in the kinetic helicity. In addition, we observe the development of downflows beneath the active regions, illustrated for the 3–5 Mm deep layer in Fig. 7, left panel c).

The comparison of the flow characteristics with the soft X-ray flux (1-hour averages from the GOES satellite data) in Figure 7 (right panels) reveals a correlation of the large flares with enhancements of flow divergence. There is evidence of upflows in the vicinity of the polarity inversion line (Fig. 7, right panel d) prior to the big flares, X2.2 and X9.3 on Sept.6, and X1.3 on Sept.7, 2017.

5. Correlation of Flow Characteristics with Flare Productivity

To investigate the potential of using the flow characteristics for the flare prediction, we performed a correlation analysis for eight active regions. We define the flare productivity of the active region as $P = N_C + 10 \times N_M + 100\times\ N_X$ where N_C, N_M, and N_X are the total number of C, M, and X-class flares that were observed by the GOES satellite in the active regions within 24 hours from the considered moments. The flare productivity is used for correlation analysis with the active region magnetic and flow descriptors shown in Fig. 8.

We analyzed correlations of the flow descriptors, the horizontal divergence, the vertical vorticity, and the helicity proxy, averaged over the whole 30×30-degree regions tracked for 10-day during their passage on the solar disk, with the flare productivity of the parental active region within the next 24-hour window. In addition to classically-used Pearson's correlation coefficients, which checks for linear dependence between the set of pairs of parameters, we analyze non-parametric Kendall's tau correlation coefficient defined as:

$$\tau = \frac{2}{n(n-1)} \sum_{i<j} \text{sgn}\,(x_i - x_j)\,\text{sgn}\,(y_i - y_j)$$

where the x_i and y_i are the values of the considered pair of parameters; sgn is a sign operator; n is a number of elements in each data set. Kendall's tau ranges between -1 and 1, and its value is expected to be 0 for independent data sets.

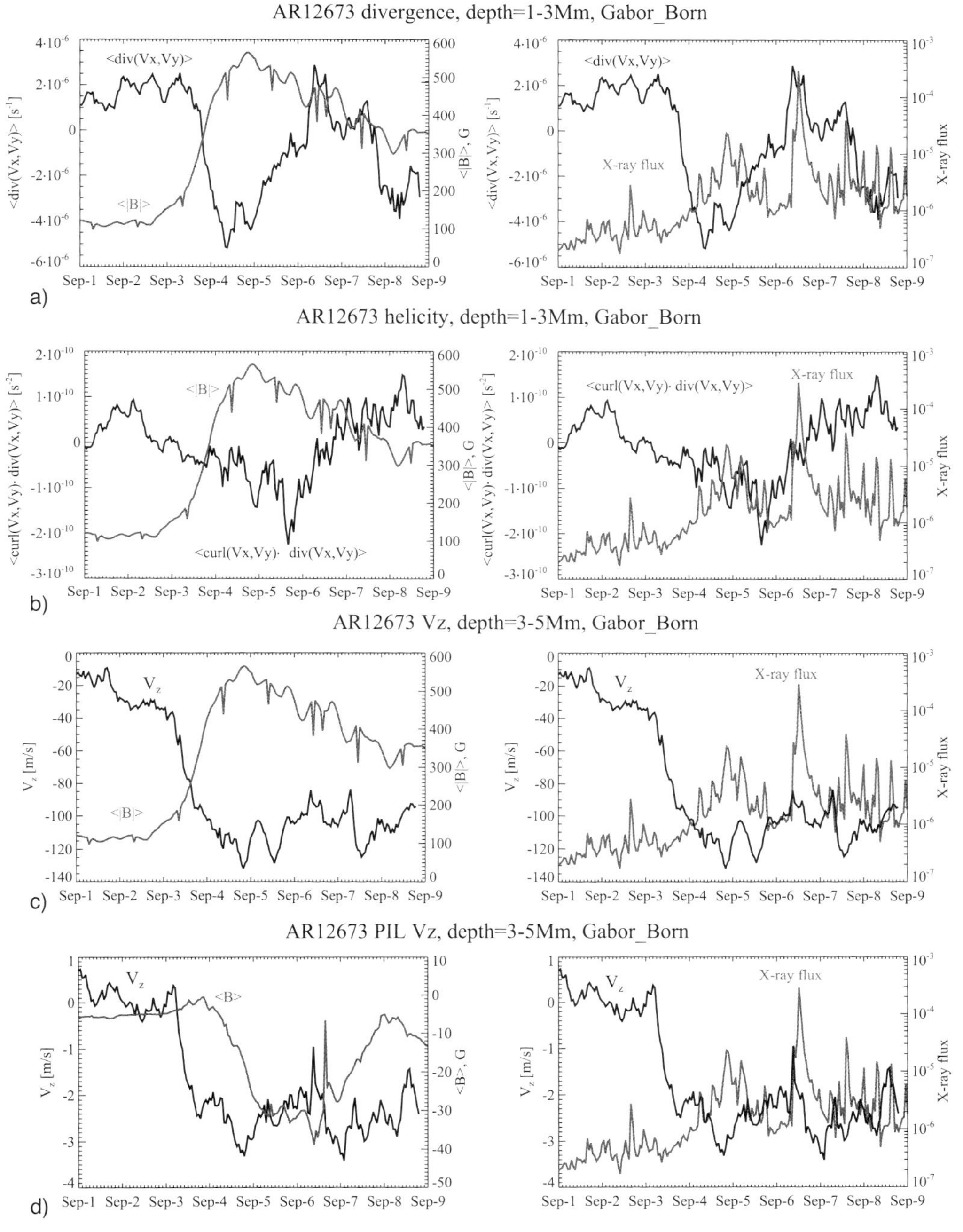

Figure 7. Comparison of the mean magnetic field (left panels) and the flaring activity (right panels) of AR 12673 with subsurface flow characteristics: a) the mean horizontal divergence, $\langle \mathrm{div}(V_x, V_y)\rangle$, averaged in the $1-3$ Mm deep layer; b) the mean kinetic helicity proxy, $\langle \mathrm{div}(V_x, V_y)\mathrm{curl}(V_x, V_y)\rangle$, in the same layer; c) the mean vertical velocity, v_z, in the $3-5$ Mm deep layer; and d) the vertical velocity averaged over the area around the polarity inversion line (near the center of the right panels in Fig. 2).

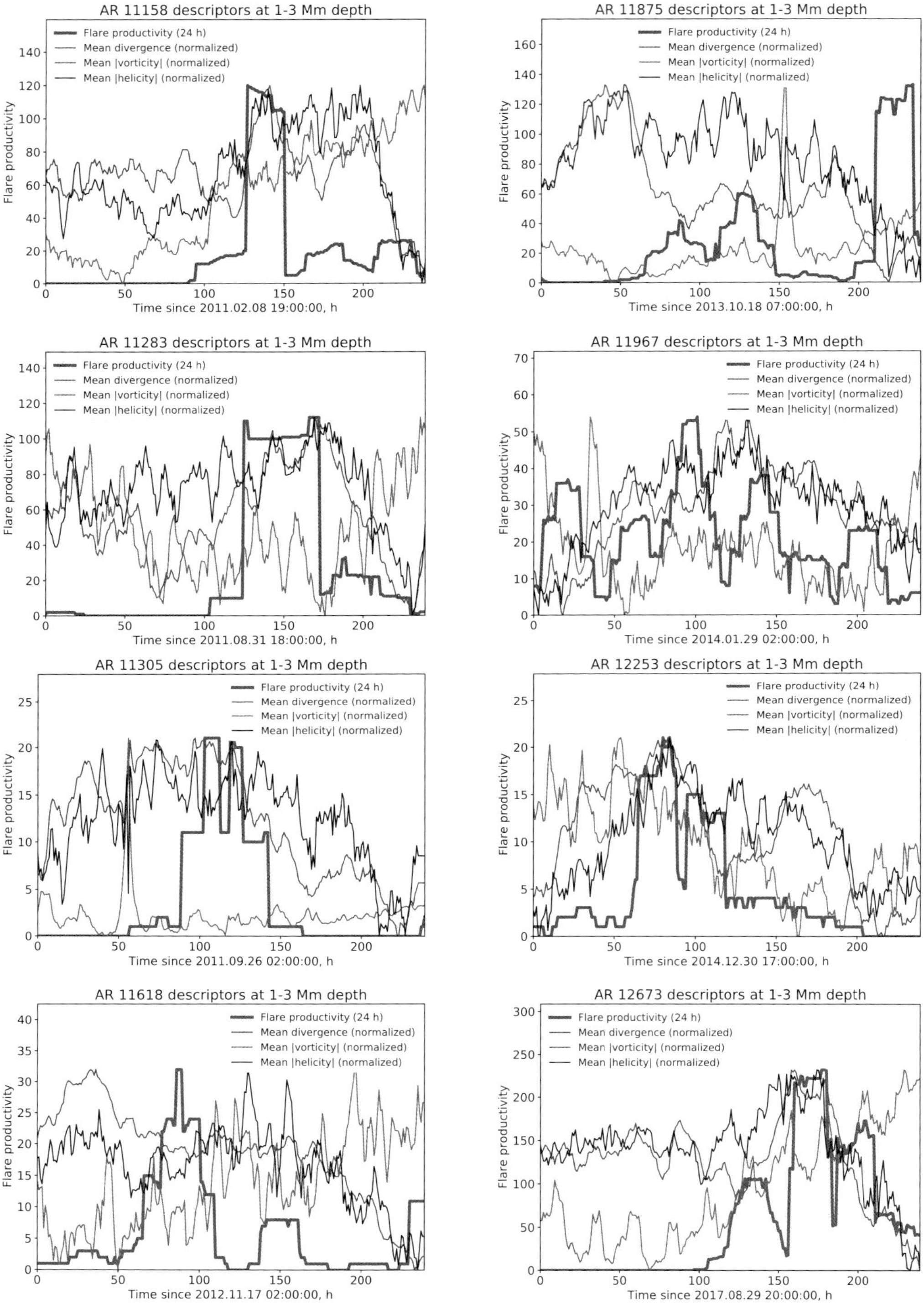

Figure 8. Comparison of the flare productivity during the consecutive 24-hour periods (red lines) with the subsurface flow characteristics at the depth $1-3$ Mm: the mean horizontal divergence (blue), the mean vertical vorticity (green), and the mean helicity proxy (black), for the 8 active regions, listed in Table 1.

Table 1. Strongest Kendall's tau rank correlation coefficients and corresponding time lags (varying from 0 hr to 24 hr) found for subsurface flow map descriptors at 1-3 Mm depth and flare productivity. The correlations are obtained for 8 ARs tracked for 10 days with the 1-hour cadence. The entries in italics indicate cases where correlation coefficients were not statistically significant (the corresponding p-value is larger than 0.05).

Active Region	**Total divergence at 1-3 Mm**	**Total vorticity at 1-3 Mm**	**Total kinetic helicity at 1-3 Mm**
11158	0.14 / 0 h	0.53 / 7 h	0.47 / 0 h
11283	0.61 / 0 h	-0.27 / 18 h	0.40 / 15 h
11305	0.65 / 20 h	-0.25 / 23 h	0.46 / 2 h
11618	0.30 / 24 h	-0.43 / 7 h	*0.08 / 0 h*
11875	-0.38 / 0 h	*0.10 / 7 h*	-0.16 / 22 h
11967	0.15 / 6 h	0.20 / 0 h	0.48 / 11 h
12253	0.41 / 24 h	0.54 / 24 h	0.65 / 2 h
12673	0.14 / 19 h	0.65 / 0 h	0.22 / 20 h

Table 1 shows the strongest Kendall's tau rank correlation coefficients and corresponding time lags (varying from 0 hr to 24 hr) found for subsurface flow map descriptors at 1-3 Mm depth and flare productivity. These results show that, in some cases, a significant correlation can be found 24 hours prior to the flares. Thus, the subsurface flow characteristics even averaged over a large area around the active regions, may provide valuable information for predicting the flaring activity of active regions. These results warrant a more complete statistical analysis of the relationship between the subsurface flows and the flaring activity of active regions.

6. Discussion

Detailed maps of subsurface flows inferred by time-distance helioseismology provide unique information about the subsurface dynamics during the emergence and formation of active regions and the periods of their flaring activity. The presented examples of emerging active regions show that the subsurface large-scale flows play a critical role in the active region formation and evolution. Each flow map was obtained by analyzing an 8-hour series of solar oscillations observed in the SDO/HMI data. To track the active region evolution, the flow maps were obtained with a one-hour cadence and ten subsurface layers in the top 19 Mm of the solar convection zone. The most significant variations of subsurface flows were observed in the top three layers in the depth range of 0–6 Mm, illustrated for five active regions in Figures 2–4. In these layers, we observed the formation of vortexes and shearing flows that contributed to the formation of the delta-type magnetic configurations, which were sources of powerful solar flares. The typical flow pattern during the initial emergence of bipolar magnetic regions consists of the flows diverging from the central area between the magnetic polarities observed in the line-of-sight magnetograms. A significant portion of the total magnetic flux can emerge in the central area between the forming sunspots, e.g. in the case of AR 13189, about half of the total flux emerged in such area in the form of small-scale bipolar elements, which moved to the large-scale magnetic structures (sunspots) of the corresponding polarities, contributing to their formation. This process continued for about 3 days. It means that the emerging magnetic flux represents a large-scale, highly fragmented structure and that the accumulation of magnetic field in sunspots is a result of magnetic self-organization in the near-surface of the Sun. The stability of sunspots is likely to be supported by the compact converging flows observed beneath the sunspots in the top 6 Mm deep layers. Outside the sunspots, we observe large-scale outflows extended to about 20 Mm outside the sunspot penumbrae. The speed of these flows is about 100–200 $\mathrm{m\,s^{-1}}$. In general, such flow pattern in the developed sunspots corresponds to the previous results of Kosovichev et al. (2000) and Zhao et al. (2001). A similar flow structure was also recently found in the numerical simulations of emerging magnetic flux and sunspot formation by Hotta and Iijima (2020).

Prior to and at the beginning of the flux emergence, the flow averaged over the whole tracked region is diverging. However, this flow becomes converging in the subsurface layers once a significant amount of the magnetic flux emerges. However, the large-scale flows are predominately divergent in the deeper layers. Such inflows and outflows around active regions were found in the earlier ring-diagram and time-distance helioseismology results (Haber et al. 2004; Zhao and Kosovichev 2004).

It appears that, in addition to the subsurface flows caused by the interaction of the magnetic field with turbulent convection, the solar differential rotation may play a significant role. In particular, the newly emerging flux typically rotates faster than the already-formed sunspots. This leads to the interaction between the bipolar active regions and causes the formation of the flare-producing delta-type active regions. For example, examples, the delta-type AR 11158 was formed when the fast-moving leading positive polarity magnetic patch of the bipolar region, which emerged behind another bipolar region, collided with its trailing negative polarity spot, creating the flare-producing delta-type configuration. Active region 12673 presents an even more dramatic example of the interaction of an emerging bipolar magnetic region in the vicinity of a large sunspot. A possible explanation of such interactions is that the emerging flux is anchored deeper in the near-surface shear layer than the developed sunspots and, thus, rotates faster due to the increasing rotation rate with the depth (Getling 2019). However, the subsurface flow maps showed that the fast eastward motion of the emerging magnetic flux was caused by the local flows associated with this flux. The physical mechanism driving these flows is not yet understood.

The subsurface maps showed substantial shearing and converging flows in the areas beneath the polarity inversion lines of the delta-type sunspots. Similar shearing flows were observed on the surface by using various local correlation tracking and flow reconstruction techniques. Our results showed that, in addition to the shearing flows along the polarity inversion line, there were flows crossing this line. The subsurface flows might contribute to stressing the magnetic field in the polarity inversion line, intensifying the electric currents, and triggering the release of magnetic energy. In particular, a significant correlation was found between the flow divergence and helicity in the active regions with their flaring activity, although an extensive statistical analysis has not been performed.

To evaluate the potential of the subsurface flow maps for flare prediction, we analyzed correlations of the flow descriptors (the total divergence, vorticity, and helicity averaged over the 30×30-degree patches) with the flare productivity within the next 24-hour window for a sample of eight active regions. The strongest Kendall's tau rank correlation coefficients and corresponding time lags (varying from 0 hr to 24 hr) were found for subsurface flow map descriptors at 1-3 Mm depth and flare productivity. These results indicate that measuring characteristics of subsurface flows can contribute to flare forecasting. However, more extensive statistical analysis and optimization of the flow characteristic calculations (e.g. defining spatial and temporal windows) are required.

In summary, the results show that the subsurface flow dynamics play a significant role in the formation, evolution, and flaring activity of active regions. The flow characteristics have the potential to improve the prediction of periods of flaring activity in active regions.

Acknowledgments
The work was partially supported by NASA grants: NNX14AB70G, 80NSSC19K0268, 80NSSC19K0630, 80NSSC20K0602, 80NSSC20K1870, 80NSSC22M0162, 80NSSC23K0097, and NSF grants 1936361 and 1835958.

Supplementary material

To view supplementary material for this article, please visit http://dx.doi.org/10.1017/S1743921324000991

References

Birch, A. C. & Gizon, L. 2007, Linear sensitivity of helioseismic travel times to local flows. *Astronomische Nachrichten*, 328(3), 228.

Birch, A. C. & Kosovichev, A. G. 2000, Travel time sensitivity kernels. *Solar Physics*, 192, 193–201.

Birch, A. C., Kosovichev, A. G., & Duvall, T. L., J. 2004, Sensitivity of acoustic wave travel times to sound-speed perturbations in the solar interior. *The Astrophysical Journal*, 608(1), 580–600.

Brandenburg, A., Rogachevskii, I., & Kleeorin, N. 2016, Magnetic concentrations in stratified turbulence: the negative effective magnetic pressure instability. *New Journal of Physics*, 18(12), 125011.

Braun, D. C. & Lindsey, C. 2000, Helioseismic holography of active-region subphotospheres - (invited review). *Solar Physics*, 192, 285–305.

Couvidat, S., Gizon, L., Birch, A. C., Larsen, R. M., & Kosovichev, A. G. 2005, Time-distance helioseismology: Inversion of noisy correlated data. *The Astrophysical Journal Supplement Series*, 158(2), 217–229.

Couvidat, S., Zhao, J., Birch, A. C., Kosovichev, A. G., Duvall, T. L., Parchevsky, K., & Scherrer, P. H. 2012, Implementation and comparison of acoustic travel-time measurement procedures for the Solar Dynamics Observatory/Helioseismic and Magnetic Imager time - distance helioseismology pipeline. *Solar Physics*, 275(1-2), 357–374.

Duvall, T. L., J., Jefferies, S. M., Harvey, J. W., & Pomerantz, M. A. 1993, Time-distance helioseismology. *Nature*, 362(6419), 430–432.

Getling, A. V. 2019, Peculiarities of the dynamics of solar noaa active region 12673. *The Astrophysical Journal*, 878(2), 127.

Gough, D. O. & Toomre, J. 1983, On the detection of subphotospheric convective velocities and temperature fluctuations. *Solar Physics*, 82(1-2), 401–410.

Haber, D. A., Hindman, B. W., Toomre, J., & Thompson, M. J. 2004, Organized subsurface flows near active regions. *Solar Physics*, 220(2), 371–380.

Hale, G. E., Ellerman, F., Nicholson, S. B., & Joy, A. H. 1919, The magnetic polarity of sun-spots. *The Astrophysical Journal*, 49, 153.

Hotta, H. & Iijima, H. 2020, On rising magnetic flux tube and formation of sunspots in a deep domain. *Monthly Notices of the Royal Astronomical Society*, 494(2), 2523–2537.

Kaneko, T., Hotta, H., Toriumi, S., & Kusano, K. 2022, Impact of subsurface convective flows on the formation of sunspot magnetic field and energy build-up. *Monthly Notices of the Royal Astronomical Society*, 517(2), 2775–2786.

Kitiashvili, I. N., Kosovichev, A. G., Wray, A. A., & Mansour, N. N. 2010, Mechanism of spontaneous formation of stable magnetic structures on the sun. *The Astrophysical Journal*, 719(1), 307–312.

Komm, R., Howe, R., González Hernández, I., Hill, F., & Toner, C. Large-scale flows in sub-surface layers. In Sankarasubramanian, K., Penn, M., & Pevtsov, A., editors, *Large-scale Structures and their Role in Solar Activity* 2005,, volume 346 of *Astronomical Society of the Pacific Conference Series*, 33.

Kosovichev, A. G. 1996, Tomographic imaging of the sun's interior. *The Astrophysical Journal*, 461, L55.

Kosovichev, A. G., Basu, S., Bogart, R., Duvall, T. L., J., Gonzalez–Hernandez, I., Haber, D., Hartlep, T., Howe, R., Komm, R., Kholikov, S., Parchevsky, K. V., Tripathy, S., & Zhao, J. Local helioseismology of sunspot regions: Comparison of ring-diagram and time-distance results. In *GONG-SoHO 24: A New Era of Seismology of the Sun and Solar-Like Stars* 2011,, volume 271 of *Journal of Physics Conference Series*, 012005.

Kosovichev, A. G., Duvall, T. L. _Jr., J., & Scherrer, P. H. 2000, Time-distance inversion methods and results - (invited review). *Solar Physics*, 192, 159–176.

Kosovichev, A. G., Sadykov, V. M., & Stefan, J. T. 2023, Spectro-polarimetric Properties of Sunquake Sources in X1.5 Flare and Evidence for Electron and Proton Beam Impacts. *The Astrophysical Journal*, 958(2), 160.

Liu, J., Sun, X., Schuck, P. W., Jaeggli, S. A., Welsch, B. T., & Noda, C. Q. 2023, Large photospheric doppler shift in solar active region 12673. i. Field-aligned flows. *The Astrophysical Journal*, 955(1), 40.

Liu, Y., Zhao, J., & Schuck, P. W. 2013, Horizontal flows in the photosphere and subphotosphere of two active regions. *Solar Physics*, 287(1-2), 279–291.

Parchevsky, K. V., Zhao, J., Hartlep, T., & Kosovichev, A. G. 2014, Verification of the helioseismology travel-time measurement technique and the inversion procedure for sound speed using artificial data. *The Astrophysical Journal*, 785(1), 40.

Parker, E. N. 1955, The formation of sunspots from the solar toroidal field. *The Astrophysical Journal*, 121, 491.

Perri, B. & Brandenburg, A. 2018, Spontaneous flux concentrations from the negative effective magnetic pressure instability beneath a radiative stellar surface. *Astronomy and Astrophysics*, 609, A99.

Rempel, M. & Cheung, M. C. M. 2014, Numerical simulations of active region scale flux emergence: From spot formation to decay. *The Astrophysical Journal*, 785(2), 90.

Schuck, P. W. 2008, Tracking vector magnetograms with the magnetic induction equation. *The Astrophysical Journal*, 683(2), 1134–1152.

Severny, A. B. 1964, Solar flares. *Annual Review of Astronomy and Astrophysics*, 2, 363.

Severnyj, A. B. 1958, *Izvestiya Ordena Trudovogo Krasnogo Znameni Krymskoj Astrofizicheskoj Observatorii*, 20, 22.

Sharykin, I. N. & Kosovichev, A. G. 2018, Onset of photospheric impacts and helioseismic waves in x9.3 solar flare of 2017 September 6. *The Astrophysical Journal*, 864(1), 86.

Stein, R. F. & Nordlund, Å. Flux emergence and pore formation: What ATST can see. In Rimmele, T. R., Tritschler, A., Wöger, F., Collados Vera, M., Socas–Navarro, H., Schlichenmaier, R., Carlsson, M., Berger, T., Cadavid, A., Gilbert, P. R., Goode, P. R., & Knölker, M., editors, *Second ATST-EAST Meeting: Magnetic Fields from the Photosphere to the Corona.* 2012,, volume 463 of *Astronomical Society of the Pacific Conference Series*, 83.

Stenflo, J. O. & Kosovichev, A. G. 2012, Bipolar magnetic regions on the sun: Global analysis of the SoHO/MDI data set. *The Astrophysical Journal*, 745(2), 129.

Verma, M. 2018, The origin of two x-class flares in active region noaa 12673. shear flows and head-on collision of new and preexisting flux. *Astronomy and Astrophysics*, 612, A101.

Wang, R., Liu, Y. D., Hoeksema, J. T., Zimovets, I. V., & Liu, Y. 2018, Roles of photospheric motions and flux emergence in the major solar eruption on 2017 September 6. *The Astrophysical Journal*, 869(2), 90.

Zhao, J., Couvidat, S., Bogart, R. S., Parchevsky, K. V., Birch, A. C., Duvall, T. L., Beck, J. G., Kosovichev, A. G., & Scherrer, P. H. 2012, Time-distance helioseismology data-analysis pipeline for Helioseismic and Magnetic Imager onboard Solar Dynamics Observatory (SDO/HMI) and its initial results. *Solar Physics*, 275(1-2), 375–390.

Zhao, J. & Kosovichev, A. G. 2003,a Helioseismic observation of the structure and dynamics of a rotating sunspot beneath the solar surface. *The Astrophysical Journal*, 591a(1), 446–453.

Zhao, J. & Kosovichev, A. G. On the inference of supergranular flows by time-distance helioseismology. In Sawaya–Lacoste, H., editor, *GONG+ 2002. Local and Global Helioseismology: the Present and Future* 2003,b, volume 517 of *ESA Special Publication*, pp. 417–420.

Zhao, J. & Kosovichev, A. G. 2004, Torsional oscillation, meridional flows, and vorticity inferred in the upper convection zone of the sun by time-distance helioseismology. *The Astrophysical Journal*, 603(2), 776–784.

Zhao, J., Kosovichev, A. G., & Duvall, Thomas L., J. 2001, Investigation of mass flows beneath a sunspot by time-distance helioseismology. *The Astrophysical Journal*, 557(1), 384–388.

Zirin, H. & Liggett, M. A. 1987, Delta spots and great flares. *Solar Physics*, 113(1-2), 267–283.

Dynamics of Solar and Stellar Convection Zones and Atmospheres
Proceedings IAU Symposium No. 365, 2024
A. V. Getling & L. L. Kitchatinov, eds.
doi:10.1017/S1743921323004040

Subsurface Flows in Active Regions with Peculiar Magnetic Configurations

Kiran Jain and Sushanta C. Tripathy

National Solar Observatory, 3665 Discovery Dr., Boulder, CO 80303, USA
email: kjain@nso.edu

Abstract. We present analysis of the evolution of subsurface flows in and around active regions with peculiar magnetic configurations and compare their characteristics with the normal active regions. We also study the zonal and meridional components of subsurface flows separately in different polarity regions separately to better understand their role in flux migration. We use the techniques of local correlation tracking and ring diagrams for computing surface and subsurface flows, respectively. Our study manifests an evidence that the meridional component of the flows near anti-Hale active regions is predominantly equatorward which disagrees with the poleward flow pattern seen in pro-Hale active regions. We also find clockwise or anti-clockwise flows surrounding the anti-Joy active regions depending on their locations in the Southern or Northern hemispheres, respectively.

Keywords. Solar physics, Helioseismology, Solar interior, Solar oscillations, Solar activity, Solar convection zone, Solar active region velocity fields, Solar rotation, Solar meridional circulation

1. Introduction

Plasma motion in the convection zone plays a crucial role in the evolution of solar magnetic activity. In particular, the meridional component of the horizontal flow is responsible for redistributing angular momentum and transporting magnetic flux to the poles. It is an important ingredient in the flux-transport dynamo models for determining the duration of the solar cycle, the polar field strength, and the onset time of next solar cycle. Therefore, precise flow measurements are crucial for the better understanding of solar dynamo (Dikpati et al. 2010). Moreover, last two solar minima with extended and exceptionally low magnetic activity have raised new challenges to our understanding of interior dynamics and the magnetic field generation (Jain et al. 2022). Studies suggest that the flows beneath active regions are typically larger than the quiet regions (e.g., Haber et al. 2004; Jain et al. 2015) and in some cases the morphology of the active regions alter flow directions (Komm et al. 2011; González Hernández et al. 2013; Jain et al. 2012, 2017).

The emergence of active regions in each solar cycle is primarily governed by three laws; (i) Spörer's law postulates the latitude of emerging sunspots that decreases with time as the solar cycle goes forward from mid-latitudes toward the equator (Carrington 1858), (ii) Hale's law governs the direction of the polarities in the regions of strong magnetic fields of the Sun (Hale & Nicholson 1925). At any given time, the ordering of positive or negative polarities within the active regions with respect to the direction of rotation is same in one hemisphere, but it reverses in the other hemisphere. With the onset of a new solar cycle, the polarities flip in both hemispheres, and (iii) Joy's law defines the latitudinal dependence of the tilt of the bipolar active regions where leading polarity stays closer to the Equator (Hale et al. 1919). While most active regions follow these rules, there are some exceptions and the decay of such regions is believed to have adverse

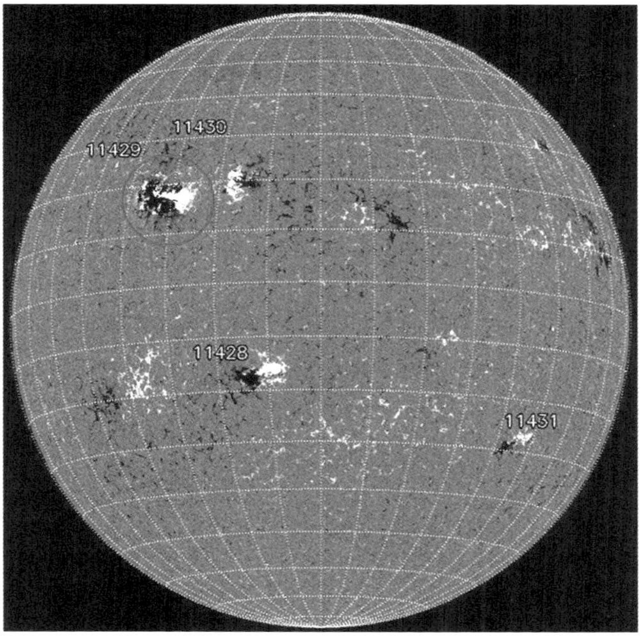

Figure 1. HMI magnetogram on 2012-03-06. The anti-Hale NOAA active region 11429 is marked by a circle. Note that this region did violate the Hale-polarity law while other regions did not. Image credit: Solarmonitor.org.

effect on the magnetic flux transportation to the poles and ultimately on the strength of the polar fields. In this paper, we study flows in and around these peculiar active regions.

2. Analysis and Results

High-resolution Dopplergrams from Helioseimic and Magnetic Imager (HMI) onboard *Solar Dynamics Observatoty (SDO)* at a cadence of 45s are utilized to measure subsurface flows. The Dopplergrams are constructed using observations taken in the 6173.3 Å absorption line in the photosphere with pixel size of 0.5 arcsec. Since HMI Dopplergrams have some known biases, we first correct them for the systematics, e.g., the observer motion and the east-west trends in Doppler velocities. We studied subsurface flows in about 180 active regions consisting of normal and peculiar configurations, here we present the detailed analyses for two active regions with unusual configurations.

2.1. *Subsurface flows around anti-Hale NOAA active region 11429*

NOAA 11429 was significantly large and magnetically strong active region with sunspot area greater than 1000μ Hemi and violated the Hale polarity law during its entire front-disk passage. It appeared on the East limb in Northern hemisphere on 2012-03-03 and crossed the West limb on 2012-03-15. In addition, it had a complex magnetic configuration, $\beta\gamma\delta$, and produced several energetic events, including X5.4-class flare. We studied flows in NOAA 11429 and compared with those in two more regions present at the same time but with Hale-law configuration, i.e., ARs 11430 in the close proximity of NOAA 11429 in the same hemisphere and 11428 in the opposite hemisphere (see Figure 1). This allows us to investigate the similarities as well as differences in flow patterns of active regions possessing different characteristics.

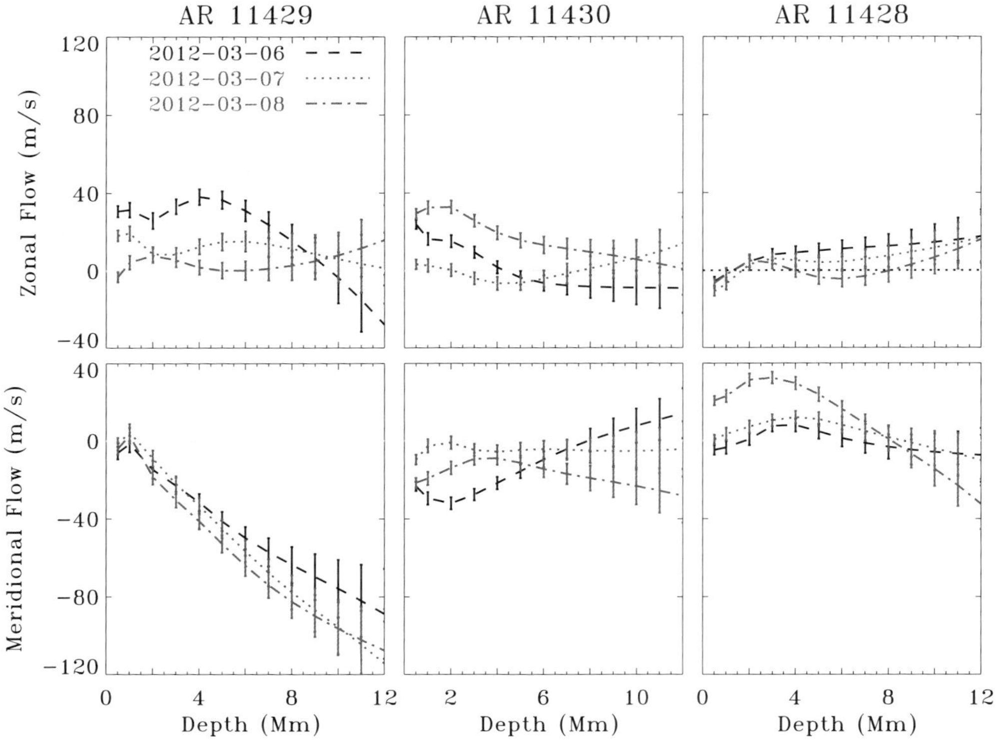

Figure 2. Depth variation of zonal (top row) and meridional (bottom row) flows in NOAA active regions; 11429 (left), 11430 (middle) and 11428 (right) for three consecutive days.

For studying the subsurface flows, we use the local helioseismology technique of ring diagrams (RD; Hill 1988). In this method, high-degree waves propagating in localized regions over the solar surface are used to obtain an average velocity vector for that region. Here, all three regions are tracked for 1396 minutes (i.e., 1862 HMI Dopplergrams) at the Synodgrass tracking rate and remapped, where each active region is centered on a 13° × 13° tile. A 3-D FFT is applied in both spatial and temporal directions to obtain the power spectrum which is fitted to a Lorentzian profile to obtain various mode parameters, including fitted velocities. These velocities are inverted using the regularized least-square method to obtain the depth dependence in both zonal and meridional components. In order to remove large scale flows, such as meridional circulation and rotation residuals, and the contributions from projection effects as discussed by Jain et al. (2013), we subtract ensemble flow averages of five quiet areas exactly at the same heliographic location in a nearby Carrington rotation for each of the individual analyzed region (Tripathy et al. 2009).

We display, in Figure 2, the zonal and meridional components of the flow vector for three consecutive days from March 6 – 8. Since AR 11429 is magnetically stronger than other two active regions, the magnitudes of these components are also higher. The most significant difference can be seen in the meridional components where these are predominantly negative (equatorward) in AR 11429 and do not change its direction with time. This component dominates the north-south direction of the horizontal flows. It should be noted that AR 11429 was in the Northern hemisphere and in ideal scenario, it should have positive meridional flow vector.

We further investigate the flow patterns in three major sub-regions within AR 11429, i.e., leading-polarity, mixed-polarity, and trailing-polarity regions. Here we used tiles of

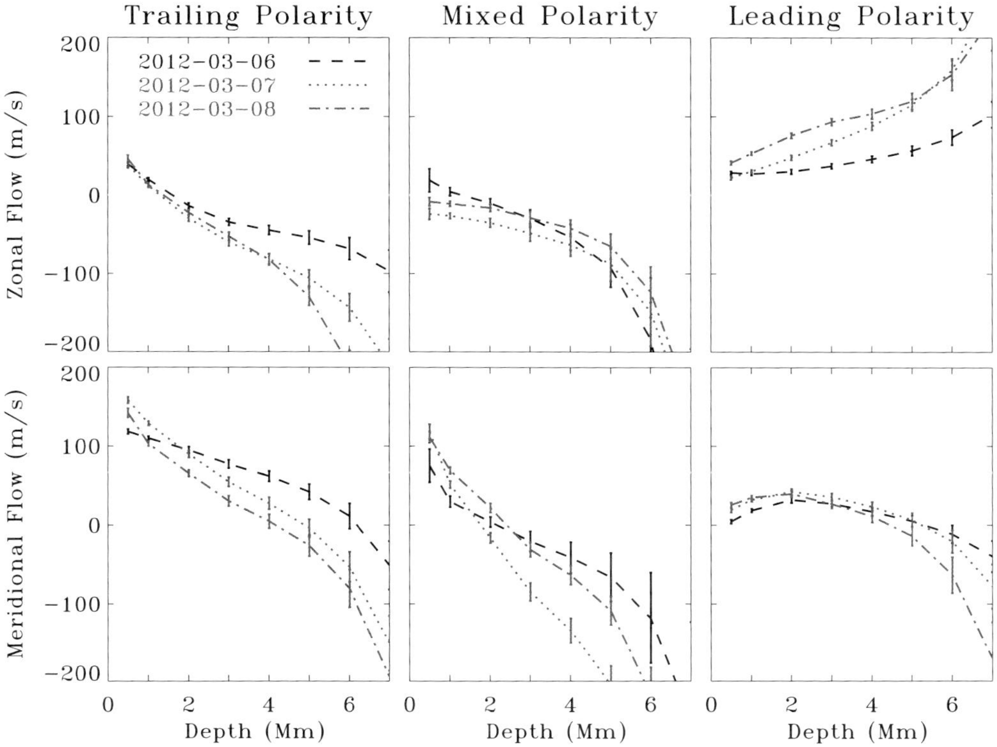

Figure 3. Depth variation of zonal (top row) and meridional (bottom row) flows for three consecutive days for trailing-polarity (left), mixed-polarity (middle) and the leading-polarity (right) regions of AR 11429.

$8^o \times 8^o$ to minimize the effect of the nearby regions. Figure 3 displays the depth variations in zonal and meridional components for all regions. There are several interesting features to note in these plots: (i) there is significant variation in the velocity components in sub-regions as compared to flows shown in Figure 2 where the active region is considered as a whole, (ii) the zonal flows in the leading polarity region are prograde while these become retrograde with depth in other two regions, (iii) the meridional component of the leading polarity region is equatorward at all depth while it is poleward near the surface but becomes equatorward with increasing depth. The horizontal flows are typically poleward in both hemispheres, however this characteristic changes in the case of anti-Hale active regions. Our study shows that the anti-Hale active region violates the normal flow pattern: the meridional flows in and around anti-Hale active region become predominantly equatorward that dominates the overall direction of the horizontal flows. These equatorward flows may have severe consequences on the transport of magnetic flux to the pole (Jiang et al. 2015). It is argued that if several big regions of similar characteristics emerge during the solar cycle, these may affect the strength of magnetic flux at poles and finally the next cycle.

We also explore photospheric flows calculated using the Local Correlation Technique (LCT; Ravindra et al. 2008). In this work, the object image is separated by 4.5 min interval from the reference image. The corresponding zonal and meridional velocities are the averages over the entire period, i.e., 1396 minutes. We find that zonal flows in and around anti-Hale active region 11429 are higher than those in two active regions. However, there are significant differences between the meridional flow patterns in the Northern and

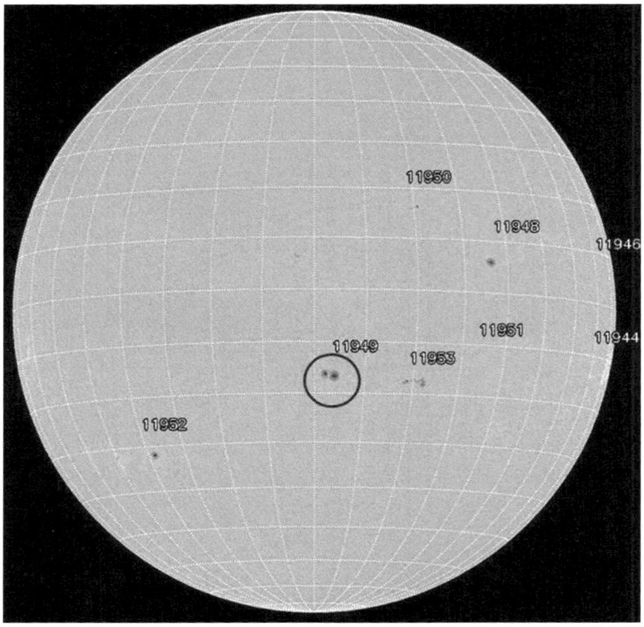

Figure 4. HMI continuum image on 2014-01-14 showing the anti-Joy configuration of the NOAA active region 11949 (marked by a circle). It can be easily seen that the leading polarity lies closer to the pole than the trailing-polarity region. Image credit: Solarmonitor.org.

Southern hemispheres. Flows around two nearby active regions in the Northern hemisphere, i.e., anti-Hale NOAA 11429 and pro-Hale NOAA 11430, are strongly equatorward which is opposite to the normal flow pattern (poleward). Contrary to this, we do not find any unusual flow patterns in AR 11428. These flow measurements are consistent with those obtained with the local helioseismic technique of ring diagrams, as discussed in the previous subsection. Although the spatial scales used in both LCT and RD methods are different, we find a strong positive correlation between the flows obtained from two methods near the surface which is in agreement with previous studies (e.g., Jain et al. 2016).

2.2. *Subsurface flows around anti-Joy NOAA active region 11949*

We selected a relatively isolated active region, i.e., NOAA 11949, that emerged in the Southern hemisphere around mid-January 2014. Its leading polarity is farther away from the equator than the trailing one. This configuration persisted for the entire disk passage from 2014-01-08 to 2014-01-19 (see Figure 4). We studied the subsurface flows for five days, i.e., from 2014-01-12 to 2014-01-16, as the active region crosses the central meridian.

To study the flows around this active region, we divided the surrounding area into 25 patches and followed the analysis as described for the anti-Hale active region. We applied ring-diagram analysis to 24 patches of $8^\circ \times 8^\circ$ surrounding the active region and obtained the horizontal subsurface flows for all five consecutive days. Figure 5 displays temporal averages of the flows surrounding NOAA 11949 for two depths, 2.0 ± 1.24 Mm and 7.0 ± 2.83 Mm, below the solar surface. Since the subsurface properties are known to change in the presence of strong magnetic fields, here we concentrate on the flows around anti-Joy active region and their influence on the nearby areas. We obtain a curly clockwise pattern in these flows that we believe to be resulted from the anti-Joy configuration of NOAA

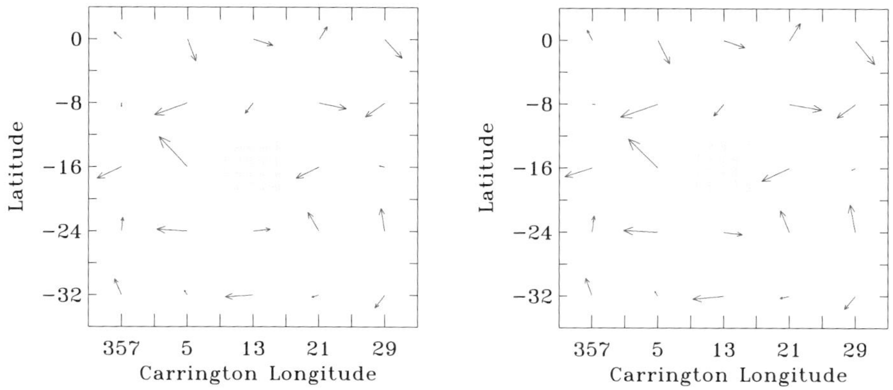

Figure 5. Temporal average of the subsurface horizontal flows surrounding NOAA active region 11949 at 2.0 ± 1.24 Mm (left) and 7.0 ± 2.83 Mm (right) below the surface. The location of the region is shown by the shaded square.

11949. Similar results were reported by González Hernández et al. (2013) in a case study of anti-Joy NOAA active region 11073 but counter-clockwise flows in the surrounding areas. We believe that the opposite flow directions in these two active regions are due to their locations in opposite hemispheres, which may be responsible for driving the leading polarity away from the solar equator. Although these two case studies exhibit similar results, a statistical study comprising a large number of anti-Joy active regions in both hemispheres will provide a clear picture of the influence of anti-Joy regions on the subsurface dynamics or vice versa.

3. Summary

In general, the horizontal velocity in active regions are larger than the quiet regions. Zonal flows in anti-Hale active region do not show any significant difference from those of pro-Hale regions. However, we find that the leading polarity region within an anti-Hale active region rotates faster than the trailing polarity region. This is opposite to the pro-Hale active regions. Our study provides evidence for meridional flows around anti-Hale active regions to be predominantly equatorward (11 out of 14 anti-Hale active regions). However, a thorough study with a larger database is required to understand the zonal and meridional flows in anti-Hale active regions. These equatorward flows may have severe consequences on the transport of magnetic flux to the pole that may affect the strength of polar flux and finally the next solar cycle. We also find curly flows in the surrounding areas of anti-Joy active region, which might be responsible for driving the leading polarity away from the solar equator. We suggest that the morphology of active regions plays an important role in the subsurface flow characteristics.

The data used here are courtesy of NASA/SDO and the HMI Science Team. This work was supported by NASA grants 80NSSC19K0261, 80NSSC20K0194, 80NSSC21K0735, and 80NSSC23K0404 to the National Solar Observatory and the NASA Cooperative Agreement 80NSSC22M0162 to Stanford University for the COFFIES Drive Science Center.

References

Carrington, R. C. 1858, *MNRAS*, 19, 1
Dikpati, M., Gilman, P. A., de Toma, G., & Ulrich, R. K. 2010, *GRL*, 37, L14107
González Hernández, I., Komm, R., van Driel–Gesztelyi, L., et al. 2013, *JPCS*, 440, 012050
Haber, D. A., Hindman, B. W., Toomre, J., & Thompson, M. J. 2004, *Sol Phys*, 220, 371

Hale, G. E., Ellerman, F., Nicholson, S. B., & Joy, A. H. 1919, *ApJ*, 49, 153
Hale, G. E., & Nicholson, S. B. 1925, *ApJ*, 62, 270
Hill, F. 1988, *ApJ*, 333, 996
Jain, K., Jain, N., Tripathy, S. C., & Dikpati, M. 2022, *ApJL*, 924, L20
Jain, K., Komm, R. W., González Hernández, I., Tripathy, S. C., & Hill, F. 2012, *Sol Phys*, 279, 349
Jain, K., Tripathy, S. C., Basu, S., et al. 2013, in Fifty Years of Seismology of the Sun and Stars, ed. K. Jain, S. C. Tripathy, F. Hill, J. W. Leibacher, & A. A. Pevtsov, Vol. CS-478 (San Francisco: Astron. Soc. Pacific), 193
Jain, K., Tripathy, S. C., & Hill, F. 2015, *ApJ*, 808, 60
Jain, K., Tripathy, S. C., & Hill, F. 2017, *ApJ*, 849, 94
Jain, K., Tripathy, S. C., Ravindra, B., Komm, R., & Hill, F. 2016, *ApJ*, 816, 5
Jiang, J., Cameron, R. H., & Schüssler, M. 2015, *ApJL*, 808, L28
Komm, R., Howe, R., Hill, F., & Jain, K. 2011, in IAU Symposium 273 - Physics of Sun and Star Spots, ed. D. P. Choudhary & K. G. Strassmeier, Vol. 273, (Cambridge: University Press), 148
Ravindra, B., Longcope, D. W., & Abbett, W. P. 2008, *ApJ*, 677, 751
Tripathy, S. C., Antia, H. M., Jain, K., & Hill, F. 2009, in Solar-Stellar Dynamos as Revealed by Helio- and Asteroseismology: GONG 2008/SOHO 21, ed. M. Dikpati, T. Arentoft, I. González Hernández, C. Lindsey, & F. Hill, Vol. CS-416 (San Francisco: Astron. Soc. Pacific), 139

Dynamics of Solar and Stellar Convection Zones and Atmospheres
Proceedings IAU Symposium No. 365, 2024
A. V. Getling & L. L. Kitchatinov, eds.
doi:10.1017/S1743921324000188

Complexity of emerging magnetic flux during lifetime of solar ephemeral regions

Hanlin Yang

Key Laboratory of Solar Activity and Space Weather, National Astronomical Observatories, Chinese Academy of Science, Beijing 100101, People's Republic of China
University of Chinese Academy of Sciences, 100049, Beijing, People's Republic of China
email: hlyang@bao.ac.cn

Abstract. As a relatively active region, ephemeral region (ER) exhibits highly complex pattern of magnetic flux emergence. We aim to study detailed secondary flux emergences (SFEs) which we define as bipoles that their locations close to ERs and finally coalesce with ERs after a period. We study the SFEs during the whole process from emergence to decay of 5 ERs observed by the Helioseismic and Magnetic Imager (HMI) aboard Solar Dynamics Observatory (SDO). We find that the maximum unsigned magnetic flux for each of the ERs is around 10^{20} Mx. All ERs have tens of SFEs with an average emerging magnetic flux of approximately 5×10^{18} Mx. The frequency of normalized magnetic flux for all the SFEs follows a power law distribution with an index of -2.08. The majority of SFEs occur between the positive and negative polarities of ER, and their growth time is concentrated within one hour. The magnetic axis of SFEs also exhibits a random characteristic. We suggest that the relationship between SFEs and ERs can be understood by regarding the photospheric magnetic field observations as cross-sections of an emerging magnetic structure. Tracking the ERs' evolution, we propose that the flux emergences are partially emerged Ω-loops, and that the SFEs in ERs may be sequent emergences from the bundle of flux tube of ERs.

Keywords. Sun: Solar Magnetic Field, Sun: Magnetic Bipolarity, Sun: Magnetic Flux, Sun: Solar Activity Effects, Sun: Atmospheric heating, Sun: Magnetic Signatures, Sun: Network Analysis

1. Introduction

Ephemeral regions (ERs) are small, magnetic bipolar configurations on the surface of the Sun. During the emergence phase, their polarities tend to separate from each other, while in the later stages, they either approach and cancel, or separate and diffuse. The average total flux of a single ER is approximately 10^{20} Mx, and they are not biased towards or away from active regions, making them observable at any time during the solar cycle at nearly any position (Harvey and Martin (1973)). This smallest ER appears to consist of a single bipole, although it has been reported that larger-scale ones are composed of a series of bipoles that emerge next to the original ones simultaneously and persistently (Martin (1990)).

Inspired by Martin's concept (Martin (1990)), we have analyzed five typical ephemeral regions observed by HMI aboard SDO and discovered that the bipolar configurations of ephemeral regions do not evolve regularly by growing individually from small to large and eventually disappearing. Each region consists of tens of bipolar emergences, making the phenomenon more complex than previously thought. We define the regions that occupy the vast majority of magnetic flux and exhibit slow and visually discernible topological changes as main polarities, and define these bipolar emergences as secondary flux

Table 1. Mean value for 5 epemeral regions

ER #	Φ^{ER}_{peak} (10^{20}Mx)	Duration (h)	Number of SFEs	Φ^{SFE}_{i} (10^{18}Mx)	$\sum \Phi^{SFE}_{i}$ (10^{20}Mx)	$\Phi^{SFE}_{i}/\Phi^{ER}_{peak}$	T_{gro} (h)	d_i (10^3km)	D (10^3km)
ER1	1.4	69.0	39	0.6-14.6	1.2	0.5%-10%	0.025-5.3	1.1-17.6	18
ER2	1.6	46.0	20	1.0-15.7	1.2	0.6%-10%	0.075-2.1	2.5-15.7	16
ER3	2.1	70.0	41	1.0-40.0	2.2	0.5%-20%	0.1-4.2	0.6-19.5	16
ER4	0.6	30.0	21	0.5-13.0	0.8	0.8%-20%	0.05-1.5	0.5-14.3	12
ER5	2.1	29.0	51	1.3-26.0	2.2	0.6%-10%	0.025-7.6	1.3-28.8	20

Notes:
Several average features of the studied SFEs. Specifically, Φ^{ER}_{peak} denotes the peak magnetic flux of the corresponding ER, while $\sum \Phi^{SFE}_{i}$ indicates the total magnetic flux Φ^{SFE}_{i} for each of these ERs. Additionally, we report the growth time T_{gro} and the distance d_i between each SFE and its corresponding ER, as well as the range of magnetic flux Φ^{SFE}_{i} for the SFEs. Finally, we also provide the maximum separation distance D of the corresponding ER bipole

emergences (SFEs), which are small bipoles that arise close to the main polarities and coalesce with them after a period of evolution and eventually become indistinguishable. In this article, we attempt to address several questions concerning ERs: the number of SFEs, their contribution to the overall ER magnetic flux, their association with the main polarities, the mechanisms of their generation, and their roles in triggering small-scale solar activities. We propose a potential sub-photospheric magnetic field configuration of Ω-shaped loops by analyzing magnetograms over time.

2. Observation and Data Analysis

SDO/HMI measures line-of-sight magnetic fields at the solar photosphere using the 6173 $\mathring{A}$ Fe 1 absorption line. The pixel size is 0.5 $''$. The noise level is estimated to be around 7 Gauss per pixel, which is determined by fitting the magnetic flux density to a Gaussian distribution (Hagenaar (2001); Liu et al. (2004); Jin et al. (2011, 2020)). We select five cases of ERs for nearly four days' observation from November 20, 2010, 13:00:00 UT, to November 24, 2010, 07:00:00 UT. They are far away from existing large-scale magnetic structures and active regions. Furthermore, they are located around the disk center and exhibit a complete evolution process. To counteract the effect of differential rotation, all magnetograms are derotated to fixed positions. To improve the signal-to-noise ratio and accurately capture SFEs, we use a sequence of continuous magnetograms with a cadence of 45 seconds and apply a temporal smoothing technique by averaging images of the three preceding frames, the current frame, and the subsequent three frames, as well as a spatial smoothing technique through overlay averaging with a 2-pixel radius.

3. Results

Flux distribution: Table 1 presents the basic properties of SFEs for five ERs. First, the unsigned magnetic flux of the five ERs is approximately 10^{20} Mx; Secondly, the values of Φ^{SFE}_{i} (magnetic flux of each SFE) range from 10^{17} to 10^{19} Mx. Furthermore, for each ER, the larger the peak magnetic flux (Φ^{ER}_{peak}) of the ER, the stronger the upper limit of individual magnetic flux for SFE (Φ^{SFE}_{i}). We find that the peak magnetic flux of an ER is not positively correlated with the number of SFEs. For example, ER1 has a smaller Φ^{ER}_{peak} compared to ER2, yet it exhibits a higher number of SFEs. This could be attributed to the stronger magnetic field strength of ER2, which counteracts the extent of disruption caused by turbulent convection (Cheung and Isobe (2014)).

Another intriguing observation is that according to Table 1, we observe that $\sum \Phi^{SFE}_{i}$ is comparable to Φ^{ER}_{peak}, and in some cases like ER3, ER4, and ER5, $\sum \Phi^{SFE}_{i}$ exceeds Φ^{ER}_{peak}. This implies that seemingly inconspicuous SFEs contribute to the majority of the magnetic flux in ER. In other words, ER is essentially composed of these smaller magnetic flux

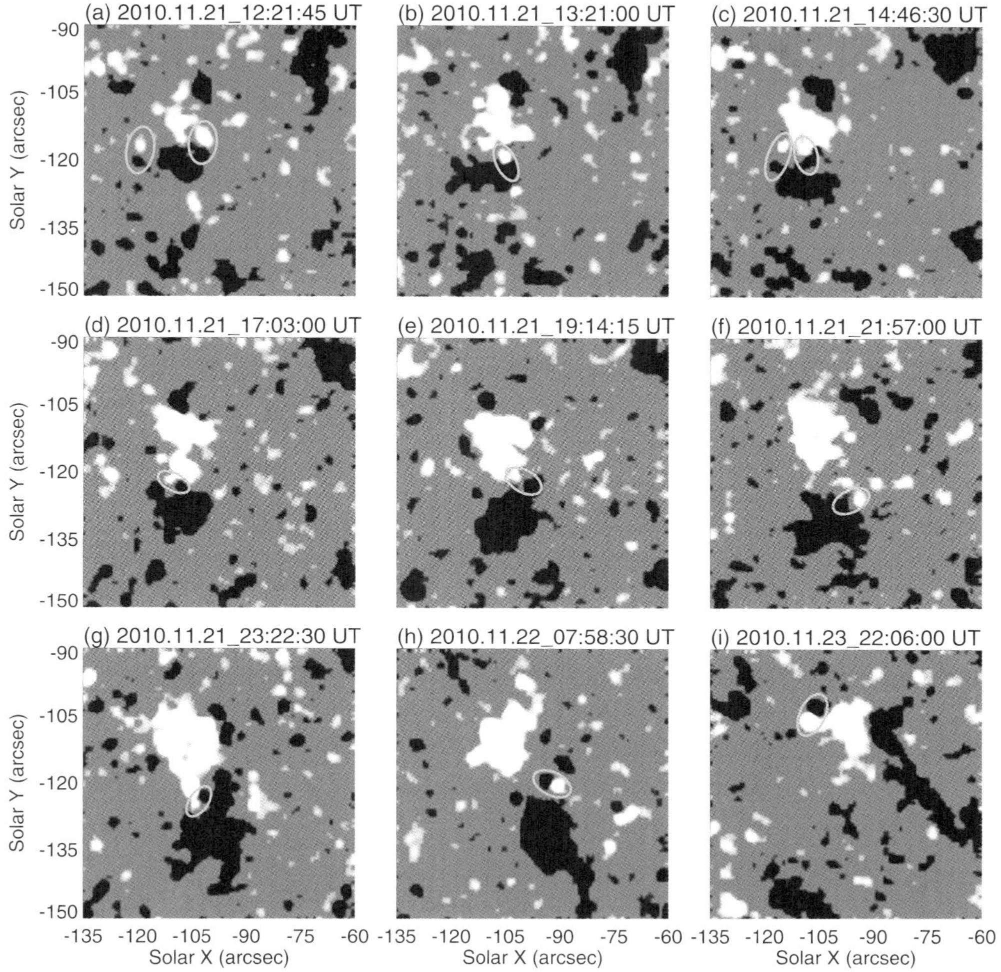

Figure 1. Nine representative instances of secondary flux emergence, all of which are extracted from ER1 due to its location in a relatively quiescent environment and displaying typical ER characteristics (Harvey and Martin 1973). Magnetograms reveal that SFE may arise on all sides surrounding the main bipole. As shown in panels (a)–(h), most of the SFEs eventually merge with the main polarities. However, for a small fraction of SFEs, such as the example in panel (i), magnetic cancellation is the ultimate outcome. To enhance visual contrast, we set pixel values below the noise level ($\pm$ 7 Gauss) to zero. The magnetograms saturate at $\pm$21 Gauss.

tubes converging. The situation where $\sum \Phi_{\mathrm{i}}^{\mathrm{SFE}}$ surpasses $\Phi_{\mathrm{peak}}^{\mathrm{ER}}$ can be attributed to the following three factors. First, SFEs do not all reach their peak strength simultaneously. When we determine the moment of peak strength in ER, some SFEs might not have emerged or reached their maximum strength. Consequently, summing each $\Phi_{\mathrm{i}}^{\mathrm{SFE}}$ results in a magnetic flux larger than the overall flux measured at a specific time. The second point involves that, with the onset of flux cancellation, the magnetic flux in the main polarities diminishes and subsequently undergoes fragmentation (Furusawa and Sakai (1999)), giving rise to SFEs in its proximity. Consequently, flux cancellation also leads to a decrease in $\Phi_{\mathrm{peak}}^{\mathrm{ER}}$ and contributes to $\sum \Phi_{\mathrm{i}}^{\mathrm{SFE}}$. Thirdly, during the emergence of magnetic flux tubes, deformations can occur. If split flux tubes emerge on the photosphere in the form of twisted structures, it essentially leads to repeated statistical counting. This is also responsible for the higher value of $\sum \Phi_{\mathrm{i}}^{\mathrm{SFE}}$.

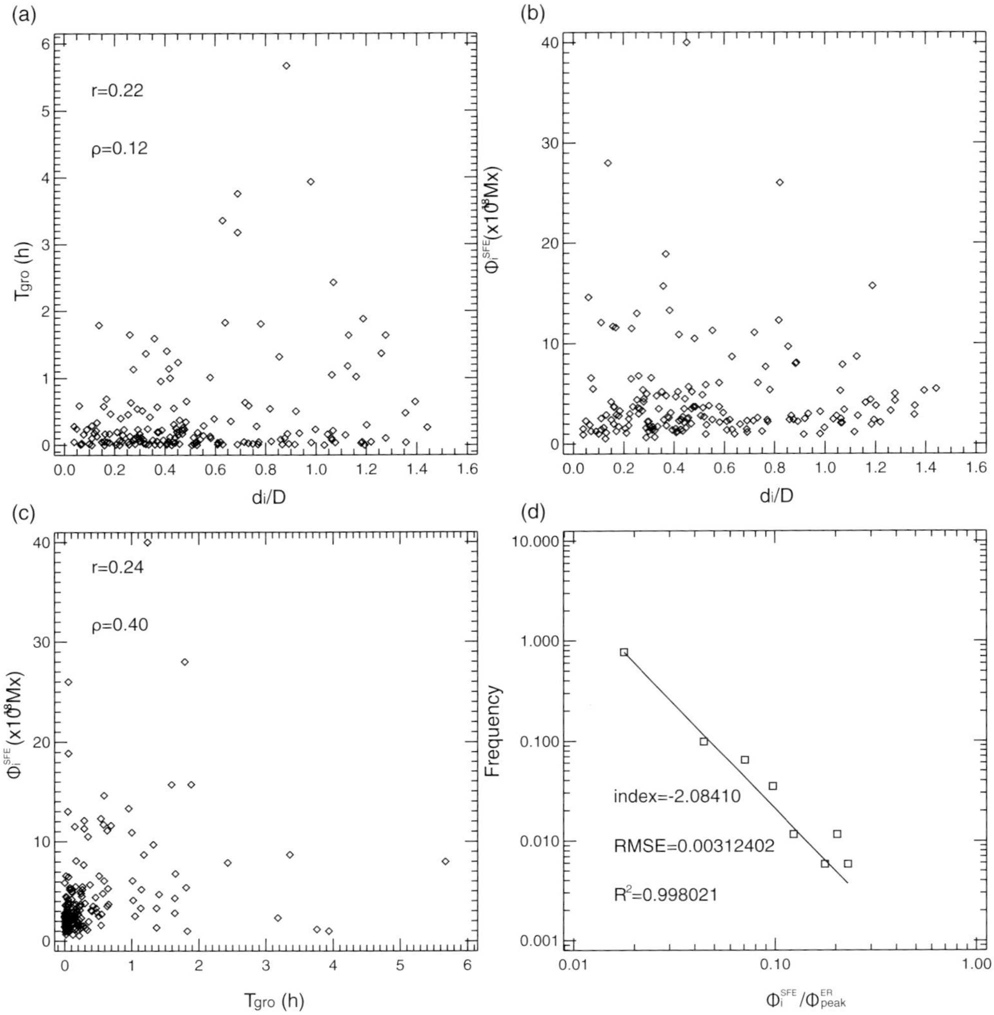

Figure 2. Scattered distributions of T_{gro}-d_i/D, Φ_i^{SFE}-d_i/D, and Φ_i^{SFE}-T_{gro} in panel a, b, c and logarithmic coordinate frequency distribution of $\Phi_i^{SFE}/\Phi_{peak}^{ER}$ in panel d where we employ a bin size of 0.03

If we normalize each Φ_i^{SFE} by the Φ_{peak}^{ER} of its corresponding ER and construct their frequency distribution, we observe that this distribution approximately conforms to a power law. Therefore, we perform a power-law fitting of the data points. After logarithmic transformation of the x and y coordinates, we plot the data points along with the fitted curve in panel d of Figure 2. We employ nine bins, with one bin having a frequency of 0. Due to the occurrence of mathematical errors caused by logarithmic transformation, only eight points are indicated on the graph. Our calculations yield an index of approximately -2.08. To gauge the goodness of fit, we compute the coefficient of determination (R^2) and the root mean square error (RMSE). The resulting values are 0.998 and 0.003, respectively, indicating that the distribution of the relative strength of SFE, i.e., $\Phi_i^{SFE}/\Phi_{peak}^{ER}$, is likely to conform to a power-law distribution with an index of -2.08.

Growth time: Table 1 shows SFEs' T_{gro} (growth time) of 5 ERs. Most SFEs grow to their maximum or merge into the main polarities and lose their identities within an hour or even several minutes. However, there are also a few SFEs that maintain their independence even after several mergers and separations from the main polarities. After a long period, they may take several hours to reach their maximum. From panels a and c in Figure 2, it can be observed that there are instances of T_{gro} extending over several hours. Some undergo multiple mergers at the same location, while others remain stationary for prolonged periods with negligible changes. Additionally, certain SFEs maintain a low magnetic level for a perid before undergoing a sudden enhancement, rapidly merging into main polarities. These occurrences are situated on the periphery of panel a and c. This could be due to their relatively larger distance from the main polarities, allowing us to observe their independent evolutions for a longer period, thereby enabling us to derive their peak strengths from a broader time range. Furthermore, we calculate the Pearson correlation coefficient r (0.22 in panel a, 0.24 in panel c) and the Spearman correlation coefficient ρ (0.12 in panel a, 0.40 in panel c) for $d_i/D - T_{gro}$ and $T_{gro} - \Phi_i^{SFE}$, revealing that in the case of our 172 SFE samples, there is no apparent linear correlation between these parameters. The stochastic behavior indicates that, in the majority of common scenarios, a normal distribution and correlation can be identified between T_{gro} and Φ_i^{SFE} (if eliminating biases introduced by the definition of T_{gro}). Although panel c appears to exhibit no discernible correlation, this could be attributed to the impact of the enhanced scale generated by large values of T_{gro}. Moreover, prolonged growth does not necessarily result in stronger magnetic flux. In specific cases, we can observe that certain SFEs can reach flux exceeding 10^{19} Mx within minutes, while others may grow for a longer period without significantly stronger magnetic flux than most SFEs. This could be attributed to the inherent strength of the magnetic flux tube branch, determining the strength of SFE. Some branches exhibit fast growth, while others grow more slowly. However, the growth time does not fundamentally dictate the amount of magnetic flux involved in SFEs.

Orientation: Figure 3 depicts all SFEs marked with arrows in the same plot. The color bar ranging from purple to red represents the chronological sequence of SFEs. The direction of the arrow points from the positive to the negative polarities. When the magnetic flux of the ER reaches its initial state, peak state, and dissipation state, the arrows connecting the magnetic centroids are marked more prominently in the figure with thicker lines. It can be observed that in the early stage of evolution, SFE tends to appear more often in the middle of the two main polarities. Whereas in the later phase, SFE tends to appear more diffusely around ERs. Another characteristic is that SFEs with larger angles between their magnetic axes and ER's tend to appear around the periphery of main polarities. Conversely, SFEs between the main polarities exhibit seldom large angles. The reason behind this phenomenon might be similar to the aforementioned point. For smaller branches, they are susceptible to distortion or deformation due to drag from turbulent convection during the emergence, resulting in structures such as kinks and twists (this will be discussed in detail in the next section). When they emerge with an inclination, their magnetic axes exhibit larger angles with the magnetic axis of the ER. In contrast, stronger SFEs are better equipped to resist turbulent convection and maintain their shape, allowing them to emerge closer to main polarities, that is, in the middle region between main polarities. After emergence, SFEs naturally follow the rotation of the ERs, but their initial deviations do not align with with the rotational trend of the ER.

4. Discussion

This study presents a detailed investigation of the secondary flux emergence (SFE) process in ephemeral regions (ERs), based on observations of the line-of-sight (LOS) magnetic field obtained from HMI/SDO. Using a visual inspection method, 39, 23, 43, 21, and 48 SFEs are identified in five selected ERs, which is much higher in quantity than previously studies. The magnetic flux of the five ERs is 1.4×10^{20} Mx, 1.6×10^{20} Mx, 2.1×10^{20} Mx, 0.6×10^{20} Mx, and $\times10^{20}$ Mx, respectively. The magnetic flux of SFEs

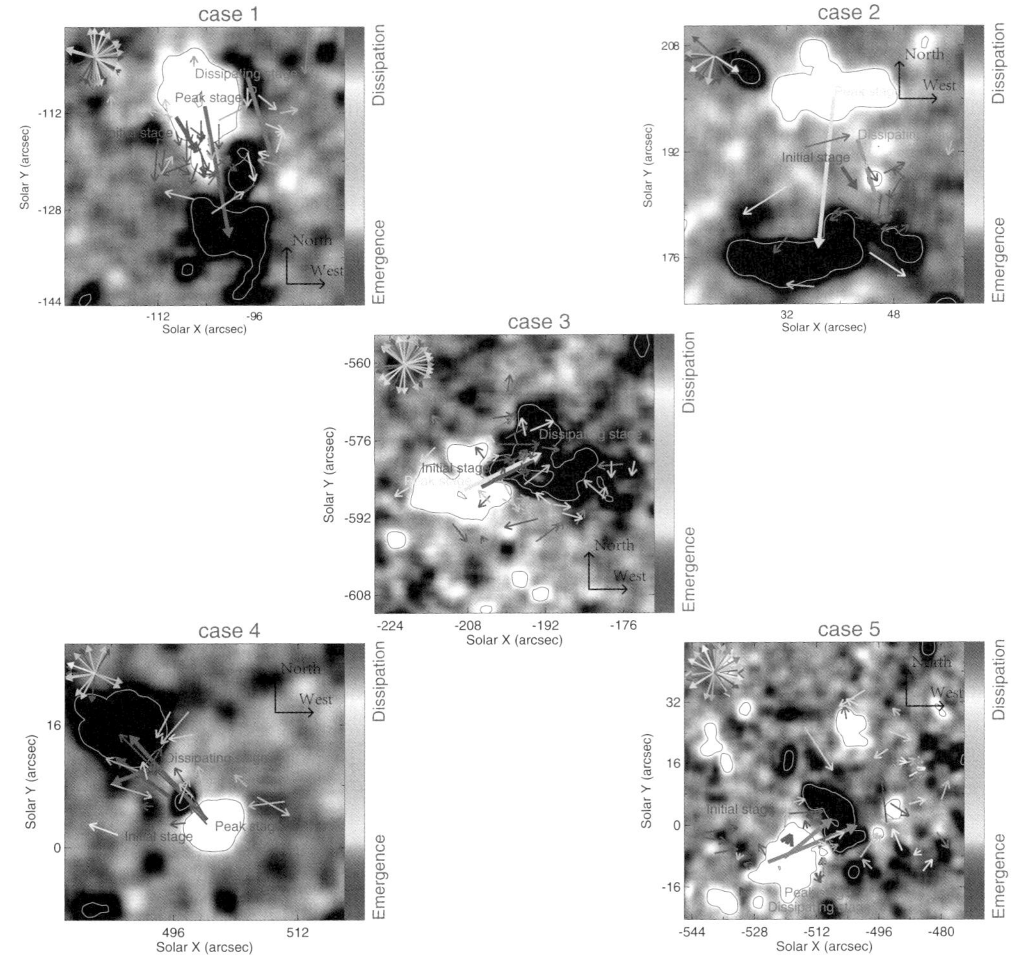

Figure 3. All SFEs with main polarities of ERs marked in magnetograms using arrows. Five magnetograms are selected at the time of maximum magnetic flux for their respective ERs and contour lines with a value of 21 Gauss are used. The arrows start and end at the positive and negative polarities of the SFEs, respectively, and the color indicates the time of SFEs' birth according to the color bar. The main polarities positions during the initial, peak, and dissipation phases are also marked with thick arrows.The arrows in the top left corner of each image represent the direction distribution of SFEs within concentric circles.

relative to their corresponding ER flux follows a power law distribution with an index of -2.08. Previous studies on flux emergence at different scales (from intranetwork to active regions) suggests a power law with an index of -1.85 (Parnell et al. 2009). We attribute this scale-free phenomenon to the influence of near-surface turbulent convection, causing the magnetic flux tubes to systematically break down into a power-law distribution, spanning various scales of magnetic flux emergence. Some studies propose that this might originate from a scale-free dynamo, yet further theoretical and observational validation is required to confirm this hypothesis (Thornton and Parnell (2011)).

Our observations suggest that the ERs are manifestations of the emergence of partially emerged magnetic flux loops from the photosphere, while the SFEs are smaller loops from the corresponding ER structures, as shown in Figure 4. Based on the tracking of the SFEs in the magnetogram, it can be inferred that magnetic flux tubes may exhibit not

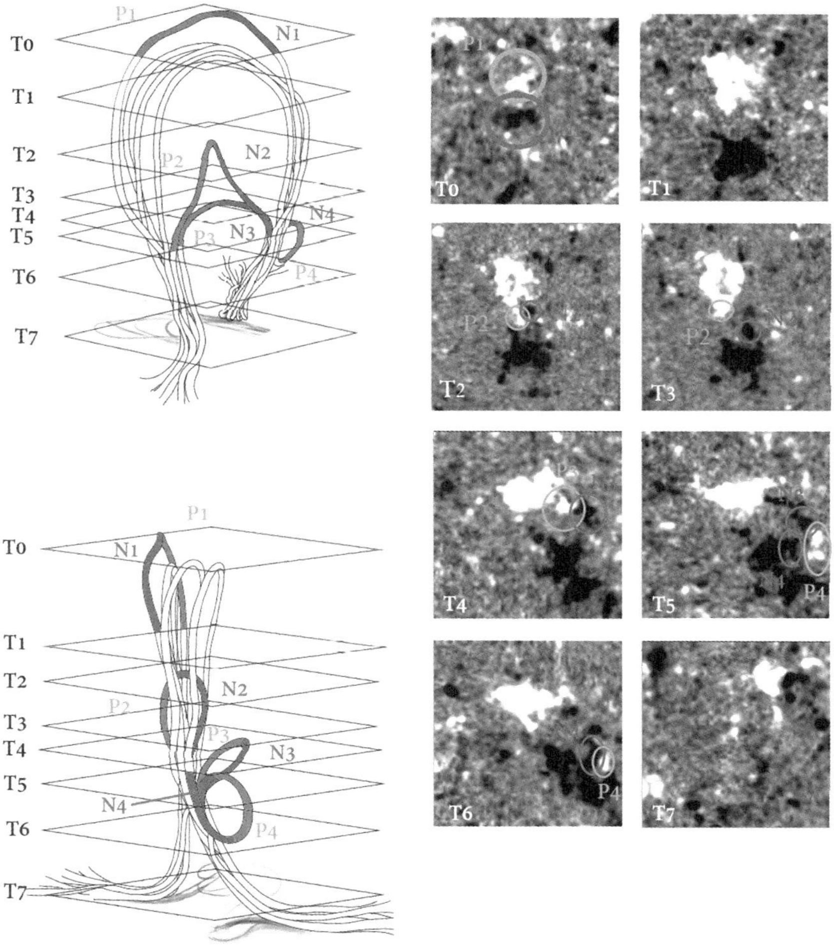

Figure 4. Front and side views of magnetic field configurations of the emerging magnetic flux derived from a series of consecutive magnetograms from ER1. The magnetograms spanning from 21-Nov 11:58:30 UT to 21-Nov 21:27:00 UT for T0 to T1, 21-Nov 22:42:45 UT to 22-Nov 01:08:15 UT for T2 to T3, 22-Nov 10:56:15 UT to 22-Nov 14:25:30 UT for T4 to T6 and 24-Nov 00:04:30 UT for T7 are displayed to construct a schematic for subsurface structures. The left part showcases the positive polarities P1, P2, P3, and P4 in green, while the negative polarities are denoted as N1, N2, N3, and N4 in red. The magnetograms on the right part simultaneously label the corresponding positive and negative polarities of each SFE using green and red circles, respectively.

only regular arch structures, but also various types of twisted and distorted shapes (such as P2, N2, P4, N4). As illustrated in Figure 4, when the magnetic flux tube is below the photosphere, it is tightly confined by the high-density plasma (Longcope and Welsch (2000)). As it rises, the gas pressure decreases, causing some magnetic flux tube branches to detach from the main tube and emerge independently. Since they separate from the same hosting magnetic tube, as the emergence continues, the branches eventually rejoin the loop, demonstrating the connection between the SFE and ER. The frequency distribution of the normalized flux of SFEs reflects the nature that the magnetic flux tube scatters under the influence of turbulent convection. The growth time and distance from the ER represent the emergence rate and the actual coverage range of the ER, respectively. The rotation observed in ER reflects a slight twist of the magnetic flux tube.

References

Cheung, M. C. M. & Isobe, H. 2014, *Living Reviews in Solar Physics*, 11, 3.
Furusawa, K. & Sakai, J.-I. 1999.
Hagenaar, H. J. 2001, *Astrophys. J.*, 555, 448–461.
Harvey, K. L. & Martin, S. F. 1973, *Solar Physics*, 32, 389–402.
Jin, C., Zhou, G., Zhang, Y., & Wang, J. 2020, *Astrophys. J. Lett.*, 889, L26.
Jin, C. L., Wang, J. X., Song, Q., & Zhao, H. 2011, *Astrophys. J.*, 731, 37.
Liu, Y., Zhao, X., & Hoeksema, J. T. 2004, *Solar Phys.*, 219, 39–53.
Longcope, D. W. & Welsch, B. T. 2000, *Astrophys. J.*, 545, 1089–1100.
Martin, S. F. 1990, *Memorie della Societa Astronomica Italiana*, 61, 293–315.
Parnell, C. E., DeForest, C. E., Hagenaar, H. J., Johnston, B. A., Lamb, D. A., & Welsch, B. T. 2009, *Astrophys. J.*, 698, 75–82.
Schrijver, C. J., Title, A. M., Harvey, K. L., Sheeley, N. R., Wang, Y. M., van den Oord, G. H. J., Shine, R. A., Tarbell, T. D., & Hurlburt, N. E. 1998, *Nature*, 394, 152–154.
Thornton, L. M. & Parnell, C. E. 2011, *Solar Phys.*, 269, 13–40.

Dynamics of Solar and Stellar Convection Zones and Atmospheres
Proceedings IAU Symposium No. 365, 2024
A. V. Getling & L. L. Kitchatinov, eds.
doi:10.1017/S1743921324000322

Predicting the Emergence of Solar Active Regions Using Machine Learning

Spiridon Kasapis[1], Irina N. Kitiashvili[1], Alexander G. Kosovichev[1,2], John T. Stefan[2] and Bhairavi Apte[2]

[1]NASA Advanced Supercomputing Division, NASA Ames Research Center, N258, Moffett Field, CA 94035, United States

[2]Department of Physics, New Jersey Institute of Technology, University Heights, Newark, NJ 07102, United States
email: spiridon.kasapis@nasa.gov

Abstract. To create early warning capabilities for upcoming Space Weather disturbances, we have selected a dataset of 61 emerging active regions, which allows us to identify characteristic features in the evolution of acoustic power density to predict continuum intensity emergence. For our study, we have utilized Doppler shift and continuum intensity observations from the Helioseismic and Magnetic Imager (HMI) onboard the Solar Dynamics Observatory (SDO). The local tracking of 30.66×30.66-degree patches in the vicinity of active regions allowed us to trace the evolution of active regions starting from the pre-emergence state. We have developed a machine learning model to capture the acoustic power flux density variations associated with upcoming magnetic flux emergence. The trained Long Short-Term Memory (LSTM) model is able to predict 5 hours ahead whether, in a given area of the solar surface, continuum intensity values will decrease. The performed study allows us to investigate the potential of the machine learning approach to predict the emergence of active regions using acoustic power maps as input.

Keywords. methods: statistical, data analysis, Sun: activity, sunspots

1. Introduction

Building early forecasts of the manifestation of solar activity is crucial to mitigate the impact of Space weather disturbances. Since most research efforts focus on predicting eruptive activities or their consequences, we consider investigating the potential of a machine learning approach to predict the appearance of its source. Capturing signatures of the subsurface activities will allow researchers to anticipate surface phenomena like magnetic flux emergence followed by the formation of eruptive active regions. The initial attempt to observe an emerging active region in the solar interior was made by Kosovichev et al. (2001), who used time–distance helioseismology to image sound-speed variations prior to the appearance of the active region on the surface. The results showed the emerging magnetic flux is fragmented and travels very quickly through the top 18 Mm layer of the convection zone, with a speed of about 1 km/s. The feasibility of using variations in acoustic power as a precursor to detect signatures of emerging magnetic flux was initially explored by Hartlep et al. (2011). Using numerical simulations of acoustic wave propagation within the solar interior, applied to AR10488, they demonstrated that subsurface structures could influence the acoustic power at the overlying photosphere. These findings suggested that changes in acoustic power could be indicative of active regions before they become visible on the surface.

Subsequent studies further explored the potential of helioseismology in detecting pre-emergence signatures of solar active regions. For instance, Leka et al. (2012) analyzed over 100 instances of emerging and non-emerging solar regions, using data from the Global Oscillations Network Group (Harvey et al. 1996, GONG) and SOHO/MDI (Scherrer et al. 1995). Birch et al. (2012) used helioseismic holography on these active region samples and identified statistically significant differences in subsurface flows and wave speeds preceding active region formation. Their findings effectively ruled out extensive flows above 15 m/s in the top 20 Mm of the photosphere prior to emergence, thereby setting constraints on theoretical models of active region development. Barnes et al. (2014) performed a statistical analysis of these ARs' subsurface properties, highlighting the surface magnetic field as a key differentiator between the two populations of regions on the Sun (pre-emergence and without emergence). The research further revealed helioseismic precursors to active region emergence, detectable at least a day prior. These works showed that the subsurface dynamics play an important role in solar active region formation and also acknowledged the limitations posed by the small sample sizes in definitively determining these subsurface properties.

The possibility of using synoptic imaging of subsurface magnetic activity in order to predict large sunspot regions before they become visible was demonstrated by Ilonidis et al. (2011), who detected strong acoustic travel-time anomalies of an order of 12 to 16 seconds as deep as 65 Mm. These anomalies, originating from deeper layers, rise to shallower regions at velocities up to 1 km/s, suggesting their association with acoustic power variations rather than just subsurface flows or wave-speed perturbations (Ilonidis et al. 2013). Similarly, Stefan and Kosovichev (2023) investigated the deviations in the mean phase travel time of acoustic waves before the emergence of 46 large active regions, showing the relationship between subsurface acoustic signals and surface magnetic flux. Despite some limitations in the strength of correlations and the variability in lag times, their findings underscore a discernible connection between the mean phase travel times and the emergence of active regions.

In a similar vein, Attié et al. (2018) observed disruption of the moat flow several hours before the onset of strong flux emergence near active region AR12673. This disruption occurred at the same location where magnetic flux would later emerge. Horizontal divergent flows at the solar surface have also been identified as potential precursors to flux emergence. Using SDO/HMI data, Rees–Crockford et al. (2022) showed that about 37.5% of active regions display divergent flows divergent flows 58 minutes before their emergence. Converging flows with a speed of 20–30 m/s one day before the emergence was also observed by Gottschling et al. (2021).

Strengthening of the solar f-mode prior to active region emergence was observed by Waidele et al. (2023), who utilized the Fourier-Hankel analysis to examine six active regions. This noticeable enhancement in the f-mode power, one to three days before the emergence of these regions, corroborates earlier findings by Singh et al. (2016). However, this enhancement varied significantly across different regions. Importantly, the analysis by Waidele et al. (2023) did not find any directional dependence (in terms of inward and outward moving waves) associated with this f-mode power strengthening. This non-directional nature of the phenomenon and the lack of a strong correlation with surface magnetic flux evolution highlights the complexity of subsurface processes preceding active region emergence. Further analysis by Korpi–Lagg et al. (2022), who applied a new calibration technique, did not reveal a significant enhancement of the f-mode prior to AR emergence.

In this paper, we explore the possibility of detecting emerging active regions before they become visible on the solar surface using a machine learning method trained on observations from the HMI instrument onboard SDO.

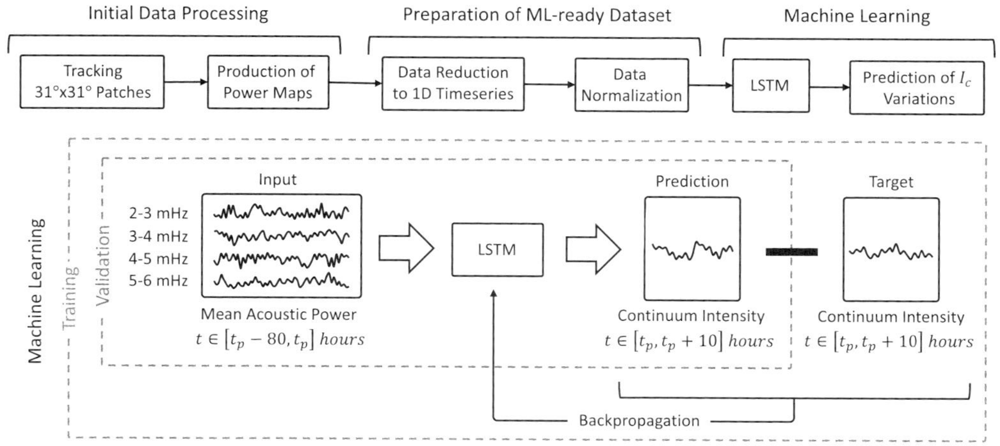

Figure 1. Scheme of the workflow (top panel) and the machine learning operations (bottom panel) to predict the emergence of active regions. Denoted at t_p is the time of prediction in each time step of the LSTM rolling window approach.

2. Data Preparation

To investigate capabilities to predict the appearance of active regions on the solar surface, we analyze the evolution of 61 active regions (ARs) using data from the Helioseismic and Magnetic Imager (Scherrer et al. 2012, HMI) onboard the Solar Dynamics Observatory (SDO). Each selected active region emerged on the solar surface within 30 degrees longitude from the central meridian between 1 March 2010 and 1 June 2023, lived for longer than 4 days, and reached a total area of 200 millionth of the solar hemisphere. The longitude range was set to minimize significant distortion due to the projection and the center-to-limb effects. Our analysis includes three main steps as presented in Figure 1: a) initial data processing, b) preparation of the ML-ready input data and continuum intensity time series to verify results, and c) development and training of an ML-driven model to predict the appearance of active regions using the Long Short-Term Memory (LSTM) method.

To perform data analysis, we tracked the corresponding 30.66×30.66 degrees area with local rotation speed (Figure 1). These patches represent a 512×512-pixel square centered around the target AR. The temporal tracking of ARs is executed through a division into overlapping 8-hour time series. Each of them is comprised of 640 frames, with cadence of 45 seconds. This approach is applied to all HMI data products, including Dopplergrams, line-of-sight magnetograms, and continuum intensity (I_c) maps. The Dopplergrams time series have been used to generate acoustic power maps for four frequency ranges: 2–3, 3–4, 4–5, and 5–6 mHz for the entire 10-day period of the passage of the ARs areas before and after their emergence.

The resulting power maps, magnetograms, and intensity maps were split into a grid of 9 by 9 tiles (Figure 2). The grid size (N) is adjustable and can be used for further exploration to increase the granularity of data, offer improved prediction accuracy, and provide details of the emergence scenario. This splitting process enabled tracing the evolution of the local area and, by calculating the mean value of each tile, it served as a data reduction step (third step, Figure 1), which created structured timelines conducive to machine learning analysis. Because the tracking area was selected with the AR at the center (determined from the NOAA AR database), the grid size was chosen for

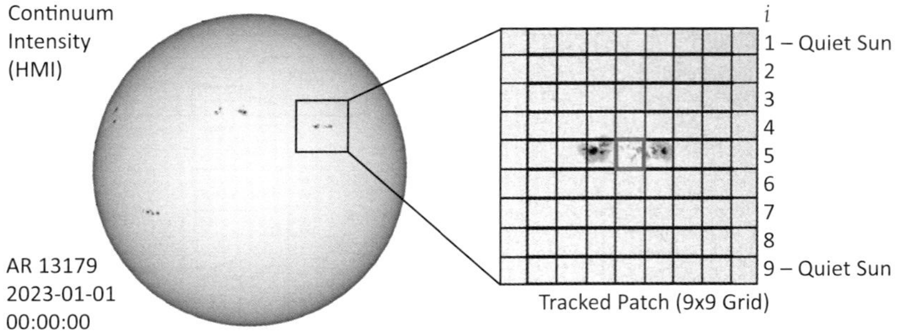

Figure 2. The tracked 30.66 × 30.66 degree patch for AR13179 (left) within the HMI continuum intensity full disc and the 9 by 9 grid it was split in (right). The colored patch corresponds to the variations of the acoustic power and continuum intensity shown in the same color in Figure 3.

convenience of analysis so that the central tile typically corresponds to the center of the developed AR. Thus, dividing a large area into small tiles allows us to investigate the possibility of predicting the time and location of an active region formation.

In the current 9 by 9 configuration, each tile represents a 57 by 57 pixel area, corresponding to 3.36 degrees of heliographic longitude and latitude. After taking the mean over each tile for acoustic power and continuum intensity maps, we obtained an ensemble of time series (or timelines) that describes the evolution of the area within the tile. Similarly, we characterize the local evolution for unsigned magnetic flux. We excluded 20 active regions out of 61 from ML analysis because of data gaps and quality issues that require additional processing and investigation. The resulting timelines in Figure 3, provide the structured time series data necessary for ML-driven forecasts.

To remove the geometric effect, which manifests as systematic variations in the resulting timelines (Figure 3, top panel), we normalize the data using the most distant tiles at the same longitude (top and bottom tiles marked as "Quite Sun" in Figure 2). Because in our dataset these tiles typically represent the quiet-Sun regions, we use them as a baseline to normalize the tiles in the intermediate latitudes relative to zero. These 'reference' tiles are excluded from the ML analysis, bringing the number of tiles (and therefore timelines) for each active region from 81 down to 63. Normalization of the timelines has been performed along the longitudinal direction (Figure 2), using the following expression

$$P_i = P_i - \left(\frac{i}{N} \cdot P_1 + (1 - \frac{i}{N}) \cdot P_N \right), \tag{1}$$

where P_i is a row number for $i = 1, 2, ..., N$. In our case, $N = 9$.

The subtraction of the temporal variations due to passing the tracking area across the solar disk and the normalization based on the quiet-Sun region values allows us to amplify the variations related to the magnetic activity (Figure 3, middle and bottom panels). Because the reference tiles, which are assumed to be 'Quiet Sun' regions, may exhibit some activity that can contribute to these deviations. Significant variations of the mean power may indicate changes in the properties of the acoustic waves and turbulence due to the coming large-scale magnetic flux from solar interiors. Before they are used to train our ML model, the acoustic power maps and intensity time series are further normalized to a range from 0 to 1 to adapt to our LSTM model. In the following section, we discuss the LSTM machine learning method to investigate acoustic power suppression as a precursor of emerging magnetic flux and the formation of an active region.

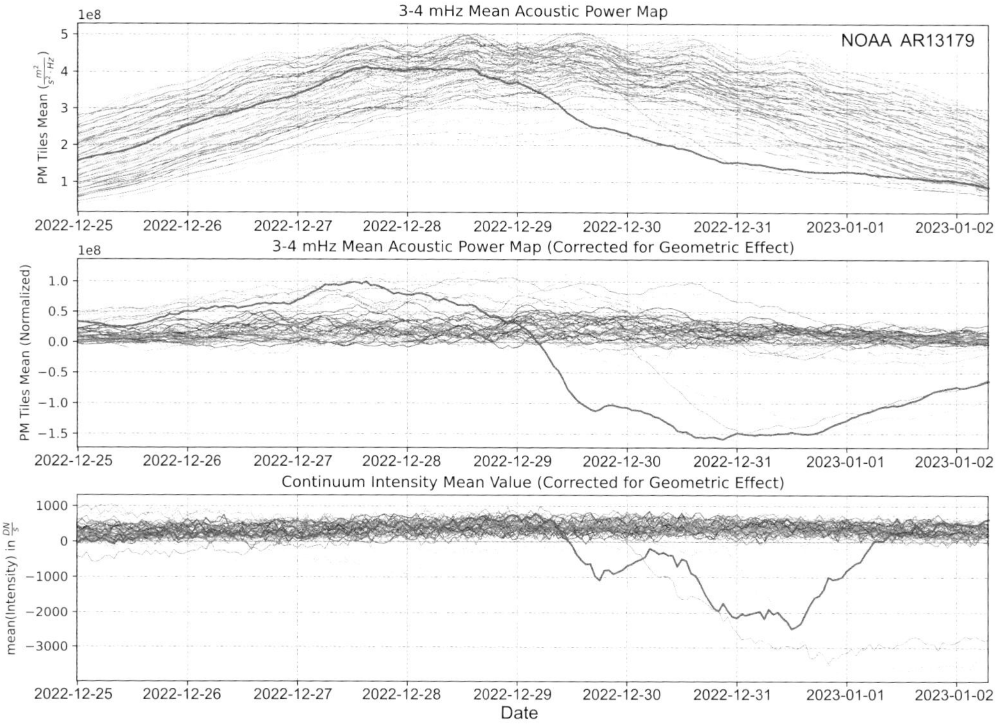

Figure 3. Time evolution of the mean acoustic power for 63 tiles obtained from the 30.66×30.66-degrees area for the 3–4 mHz frequency range (top panel). The middle and bottom panels show the evolution of the normalized acoustic power and the continuum intensity corrected for the geometric effect. The blue curve shows the emergence and evolution of AR12085 in the same color-coded middle tile in Figure 2.

3. Long Short-Term Memory Modeling

Because of the complex links between turbulent motions on the solar photosphere, small-scale dynamo, coming flows, and new magnetic fields from the solar interior, capturing structural and dynamical changes due to the emergence of an active region is not a trivial problem. To address this, we use machine learning to advance the cross-analysis of the local time evolution of the acoustic power for four frequency ranges (2–3, 3–4, 4–5, and 5–6 mHz) and predict a decrease in the mean continuum intensity associated with the appearance of an active region on the solar surface (Figure 1, bottom panel). An example of the data set for the 3–4 mHz frequency range is shown in Figure 3, middle panel.

The strategy to utilize the variations of the oscillatory properties of the solar convection to reconstruct and predict the continuum intensity is based on a general approach to detect active regions at the photosphere when the AR appearance always accompanies the intensity decrease. Because including information about a local evolution of magnetic flux will enhance the predictive capabilities, we will use magnetograms to improve the predictive capabilities of the current ML model. However, using the unsigned magnetic flux as a criterion to define the emergence of an active region makes the results more dependent on a selection of tile size and more sensitive to the enhancement of the magnetic energy on smaller scales, which are out of the scope of this study.

In this work, we employ Long Short-Term Memory (Hochreiter and Schmidhuber 1997) networks, a type of recurrent neural network (RNN) that is particularly adept at handling sequential data (Hua et al. 2019; Tealab 2018; Sak et al. 2014) and aims to resolve the vanishing gradient problem (Basodi et al. 2020). LSTM networks are designed to overcome the limitations of traditional RNNs, particularly in learning long-term dependencies. This is achieved through their unique architecture, which includes memory cells and multiple gates (input, output, and forget gates). These components allow LSTM to regulate the flow of information, effectively remembering important inputs over long sequences while discarding irrelevant data. This capability makes LSTM beneficial for modeling and predicting the local dynamics at the solar surface and, in particular, forecasting the formation and evolution of ARs based on the intricate patterns observed in our acoustic power density and intensity timelines.

The LSTM method employs a sliding window technique to analyze local temporal variations from the 1D time series of the acoustic power to estimate current and future continuum intensity variations. To perform analysis, the 80-hour time window consequently shifted by 1 hour for different prediction times t_p as seen in Figure 1. During the processing of each time window, the LSTM model learned how to predict the mean intensity variations on short-time scales (next 5 hours) using the recent history of acoustic power evolution across different frequencies. During validation (Figure 1), the same process is employed, with the only difference being that this time, the resulting prediction outcome is not compared to the true intensity values, and therefore, this information is not backpropagated to the LSTM and its weights (trainable parameters) are not altered. This analysis is applied to a time series of four frequency ranges and spans across 63 tiles, regardless of their activity level for 40 ARs.

In the LSTM model, we use a hidden layer size of 64, which refers to the number of hidden units within each layer. Every hidden unit (or cell) can be considered as a memory cell that processes input data, retains important information, and passes this information to the next time step or the next layer. Thus, it influences the network's capacity to learn complex patterns. In our configuration, the LSTM comprises three layers, keeping a balance between model complexity and computational efficiency. Our model's learning rate is set at 0.001, while the number of epochs (number of times the training process is repeated) is set to 500, allowing for a slow but steady convergence to the optimal trainable parameter space using the gradient descent approach, with Mean Squared Error (MSE) loss function guiding the model's accuracy improvement. The MSE is a widely used cost function that calculates the average squared difference between the predicted and actual values, providing a measure of the model's prediction accuracy. This relatively low number of epochs also determines the duration of the model training, allowing the LSTM to iteratively refine its predictions in less than 12 hours when run on an NVIDIA A100 GPU node.

4. Prediction of the Continuum Intensity Variations

In this work, we assume that the emergence of the AR starts when the following criteria for the mean normalized intensity are satisfied: 1) intensity decreases below the threshold of 0.8; 2) the time derivative is -0.2 or less; and 3) the previous criteria are satisfied for at least 12 hours. We illustrate the LSTM machine learning model performance for two examples to predict the evolution of the continuum intensity for a tile which manifested the formation of AR13179, a so-called 'active' tile (red rectangle in Figure 4) and quiet-Sun region tile referred to as 'quiet' (green). According to NOAA, the emergence of AR13179 was observed on 29 December 2022, 04:00:00 TAI. To predict intensity for a particular time, the intensity timeline of at least 85 hours is required (80 hours is the defined LSTM input plus 5 hours not seen by the LSTM for the time window). Therefore, in our study, we begin predicting intensity starting at 10:00:00 on 28 December 2022.

Prediction of the continuum intensity for 5 hours ahead (solid curve, Figure 4, bottom plot) that does not exhibit magnetic activity (marked as a green tile) shows a good

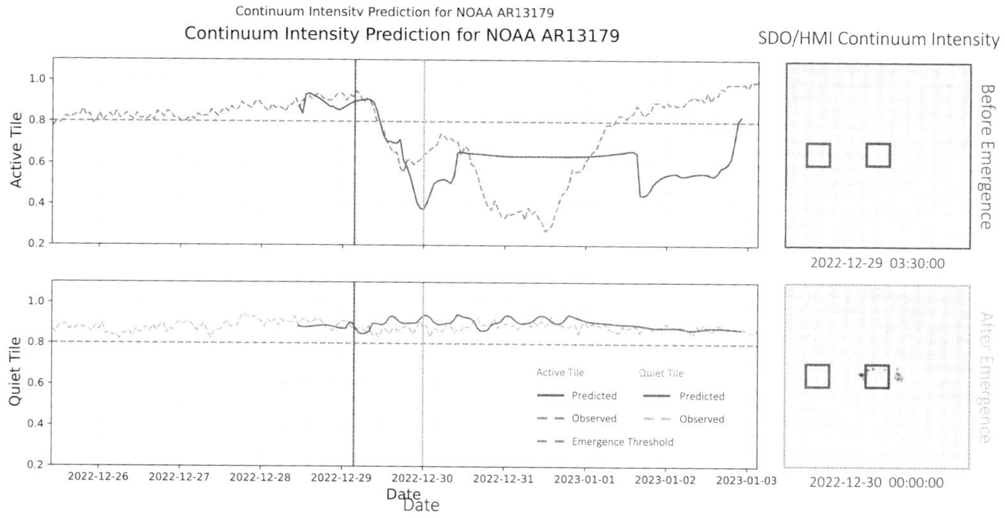

Figure 4. Continuum intensity prediction for AR13179 by the trained LSTM model. The left panels show the predicted (solid curves) and actual (dashed) continuum intensity variations in tiles marked on the right continuum intensity images. The vertical lines indicate the moment half an hour before emergence (blue) and a moment after emergence (orange). The right panels show the whole 30.66 by 30.66 degrees continuum intensity areas corresponding to these moments before (top) and after emergence (bottom image). In the snapshots on the right, enclosed in a red square is a tile that exhibits the formation of an active region (active tile, upper plot) and corresponds to the blue middle tile in Figure 2. Enclosed in the green square is a quiet-Sun region (quiet tile, bottom plot). The grey horizontal dashed line indicates the threshold that intensity has to exceed in order to have emergence in this study.

agreement with the observations (dashed curve). For the active tile, the LSTM model (Figure 4, upper plot) can reproduce well a general trend of intensity decrease, especially at the beginning of the emergence. It is important to note that newly formed and evolving magnetic structures often do not stay in the same place and can migrate from one tile to another. Also, during the emergence and formation of an active region, the local area experiences structural and dynamic changes in flows and magnetic fields, expands to a larger area, and occupies several tiles. The LSTM model is not trained on topological changes of magnetic fields and associated with it the oscillatory properties of the convection. Therefore, discrepancies between the predicted and observed intensity during AR emergence may increase with time (Figure 4, upper plot). Nevertheless, the success of capturing variations of the continuum intensity during the beginning of the flux emergence creates confidence in the ability to apply the LSTM model to predict the appearance of active regions on the solar surface before they become visible.

5. Discussion and Conclusions

To investigate the applicability of the Long Short-Memory (LSTM) algorithm to predict the formation of an active region before it becomes visible on the solar surface, we analyzed 40 active regions observed by SDO/HMI. The formation of active regions can easily be observed in the continuum intensity. Previously founded correlations between changes in the acoustic power and upcoming magnetic activity on the solar surface suggest that this coupling can serve as a reliable predictor. However, the non-linear nature of these correlations requires advanced statistical analysis. Therefore, we developed an LSTM model to predict intensity variations based on local changes in the mean acoustic power. The training was performed for 40 active regions, where time series of the mean continuum intensity and acoustic power for four frequency ranges were utilized. We analyze the time series over 63 tiles for each active region, covering quiet-Sun regions and upcoming emergence areas. The LSTM model was constructed with 64 hidden layers and used a learning rate of 0.001 for 500 epochs.

Our validation results demonstrate the capability of an ML approach and, in particular, the LSTM method to predict emergence at least 5 hours in advance for the active areas of AR13179, which is not used during training and, therefore, we use it for validation. Predicting the continuum intensity variations before and at the beginning of active region formation using changes in the acoustic power as input shows remarkable results even if information about magnetic fields is not available to the model. The large discrepancies between the prediction and observations during the late stage of active region formation suggest that the training strategy requires additional investigation. Thus, the model's capability to accurately forecast significant changes in solar activity underscores its potential as a powerful tool for space weather forecasting. Future efforts will be concentrated on increasing the LSTM model predictive capabilities by including training on magnetic field observations, improving data calibration, and optimizing the model parameters.

Acknowledgements

We want to thank the NAS Visualization Team (Nina McCurdy, Timothy Sandstrom, Christopher Henze) for their help with this project's visualizations. This work is supported by the NASA AI/ML HECC Expansion Program, NASA Heliophysics Supporting Research Program, and the NASA grants 80NSSC19K0630, 80NSSC19K0268, 80NSSC20K1870, and 80NSSC22M0162.

References

Attié, R., Kirk, M. S., Thompson, B. J., Muglach, K., & Norton, A. A. 2018, Precursors of magnetic flux emergence in the moat flows of active region ar12673. *Space Weather*, 16(8), 1143–1155.

Barnes, G., Birch, A., Leka, K., & Braun, D. 2014, Helioseismology of pre-emerging active regions. iii. statistical analysis. *The Astrophysical Journal*, 786(1), 19.

Basodi, S., Ji, C., Zhang, H., & Pan, Y. 2020, Gradient amplification: An efficient way to train deep neural networks. *Big Data Mining and Analytics*, 3(3), 196–207.

Birch, A., Braun, D., Leka, K., Barnes, G., & Javornik, B. 2012, Helioseismology of pre-emerging active regions. ii. average emergence properties. *The Astrophysical Journal*, 762(2), 131.

Gottschling, N., Schunker, H., Birch, A., Löptien, B., & Gizon, L. 2021, Evolution of solar surface inflows around emerging active regions. *Astronomy & Astrophysics*, 652, A148.

Hartlep, T., Kosovichev, A. G., Zhao, J., & Mansour, N. N. 2011, Signatures of emerging subsurface structures in acoustic power maps of the sun. *Solar Physics*, 268, 321–327.

Harvey, J., Hill, F., Hubbard, R., Kennedy, J., Leibacher, J., Pintar, J., Gilman, P., Noyes, R., Title, A., Toomre, J., *et al.* 1996, The global oscillation network group (gong) project. *Science*, 272(5266), 1284–1286.

Hochreiter, S. & Schmidhuber, J. 1997, Long short-term memory. *Neural computation*, 9(8), 1735–1780.

Hua, Y., Zhao, Z., Li, R., Chen, X., Liu, Z., & Zhang, H. 2019, Deep learning with long short-term memory for time series prediction. *IEEE Communications Magazine*, 57(6), 114–119.

Ilonidis, S., Zhao, J., & Hartlep, T. 2013, Helioseismic investigation of emerging magnetic flux in the solar convection zone. *The Astrophysical Journal*, 777(2), 138.

Ilonidis, S., Zhao, J., & Kosovichev, A. 2011, Detection of emerging sunspot regions in the solar interior. *Science*, 333(6045), 993–996.

Korpi–Lagg, M. J., Korpi–Lagg, A., Olspert, N., & Truong, H.-L. 2022, Solar-cycle variation of quiet-sun magnetism and surface gravity oscillation mode. *Astronomy & Astrophysics*, 665, A141.

Kosovichev, A., Duvall, T., & Scherrer, P. 2001, Time-distance inversion methods and results (invited review). *Helioseismic Diagnostics of Solar Convection and Activity*,, 159–176.

Leka, K., Barnes, G., Birch, A., Gonzalez–Hernandez, I., Dunn, T., Javornik, B., & Braun, D. 2012, Helioseismology of pre-emerging active regions. i. overview, data, and target selection criteria. *The Astrophysical Journal*, 762(2), 130.

Rees–Crockford, T., Nelson, C., & Mathioudakis, M. 2022, Preemergence signatures of horizontal divergent flows in solar active regions. *The Astrophysical Journal*, 940(2), 109.

Sak, H., Senior, A. W., & Beaufays, F. 2014, Long short-term memory recurrent neural network architectures for large scale acoustic modeling.

Scherrer, P. H., Bogart, R. S., Bush, R., Hoeksema, J.-a., Kosovichev, A., Schou, J., Rosenberg, W., Springer, L., Tarbell, T., Title, A., *et al.* 1995, The solar oscillations investigation-michelson doppler imager. *The soho mission*,, 129–188.

Scherrer, P. H., Schou, J., Bush, R., Kosovichev, A., Bogart, R., Hoeksema, J., Liu, Y., Duvall, T., Zhao, J., Title, A., *et al.* 2012, The helioseismic and magnetic imager (hmi) investigation for the solar dynamics observatory (sdo). *Solar Physics*, 275, 207–227.

Singh, N. K., Raichur, H., & Brandenburg, A. 2016, High-wavenumber solar f-mode strengthening prior to active region formation. *The Astrophysical Journal*, 832(2), 120.

Stefan, J. T. & Kosovichev, A. G. 2023, Exploring the connection between helioseismic travel time anomalies and the emergence of large active regions during solar cycle 24. *The Astrophysical Journal*, 948(1), 1.

Tealab, A. 2018, Time series forecasting using artificial neural networks methodologies: A systematic review. *Future Computing and Informatics Journal*, 3(2), 334–340.

Waidele, M., Roth, M., Singh, N., & Käpylä, P. 2023, On strengthening of the solar f-mode prior to active region emergence using the fourier-hankel analysis. *Solar Physics*, 298(2), 30.

Dynamics of Solar and Stellar Convection Zones and Atmospheres
Proceedings IAU Symposium No. 365, 2024
A. V. Getling & L. L. Kitchatinov, eds.
doi:10.1017/S1743921323005008

Evolution of a nanoflare-scale magnetic reconnection event in the quiet Sun

Rebecca A. Robinson

Rosseland Centre for Solar Physics, University of Oslo, PO Box 1029, Blindern 0315, Oslo, Norway
email: becca.anne.robinson@gmail.com

Abstract. Low-lying loops in the quiet Sun are a reliable source of energy for atmospheric heating, but the mechanisms by which they evolve are somewhat enigmatic. To address the origins of atmospheric heating events in the quiet Sun, we utilize our stratified, convection-driven, 3D MHD simulation Bifrost to explore the evolution and eventual major reconnection between several magnetic features; one of which is a magnetic flux rope. We zoom in on the buildup of the magnetic flux rope, which self-orders in the corona via an inverse cascade of helicity. We also discover that the flux rope attempts to relax to a linear force-free field according to Taylor's theory, but cannot do so completely. Finally, we demonstrate that the eventual nanoflare-scale reconnection event could potentially be observed in the 171 Å channels of SDO/AIA and the future MUSE mission. We also determine that the spectral resolution of MUSE is sensitive enough to capture the kinematics of the bi-directional plasma jets emanating from the reconnection region.

Keywords. Sun: magnetic fields, Sun: activity, Sun: atmosphere, Sun: corona, Sun: flares

1. Introduction

Nanoflares are small, bright events on the Sun that are associated with localized atmospheric heating in the solar corona. Their potentially ubiquitous distribution makes them good candidates for sustained coronal heating, but their size makes it challenging to observe and understand them. With that, a combination of theory, observations, and simulations can help inform and expand our understanding of how these events occur. We present a simulation of a nanoflare-scale event that is powered by the large-angle reconnection of a magnetic flux rope and magnetic arcade with an overlying, nearly antiparallel horizontal field in the corona.

2. Methods

2.1. *The Bifrost simulation*

The simulation presented here is a quiet Sun simulation produced with the Bifrost code (Gudiksen et al. 2011), a massively parallel MHD code that simulates stellar atmospheres from the convection zone to the corona. The specifics of the simulation are described in detail in Robinson et al. (2022), Robinson et al. (2023), and Robinson and Carlsson (2023), but I will briefly summarize the important characteristics here.

Our simulation domain is a small cube of solar atmosphere with a 12 Mm horizontal extent in each horizontal direction, and a vertical extent from 2.5 Mm below the average solar surface ($z = 0$) to 8 Mm above it. The horizontal resolution is a uniform 23 km, whereas the vertical resolution is variable and sharpest between $z = 2$ and $z = 4$ Mm. The convective boundary allows inflows as long as the effective temperature is consistent with the temperature of photospheric plasma, and the coronal boundary is open. The horizontal boundaries are periodic, meaning that features can exit and reenter the box as physically required. This allows magnetic field lines (for example) to cross the horizontal boundaries but still find magnetic roots in the photosphere, which encourages the buildup of magnetic twist.

The initial magnetic field is a free parameter in Bifrost, and the main phase of this run began with an initially balanced vertical field extrapolated to a potential field. Then, convective drivers process this initially balanced field to evolve it further. Interactions between the magnetic field lines are mitigated by our hyper-diffusion parameters, which prevent diffusive regions from breaching the numerical resolution (see Færder et al. 2023). With these parameters, the energy from magnetic reconnection is still conserved but deposited in a more diffusive manner; this effect is similar to dramatically increasing the resolution of the simulation in diffusive regions.

Bifrost comes with an optional module called `corks`, described in detail in Robinson et al. (2022), Robinson et al. (2023), and Robinson and Carlsson (2023). This module introduces non-interacting test particles into the simulation, which are advected with the ideal flow of the fluid. We also assume that the magnetic field lines follow ideal transport outside of diffusive regions, which means that individual corks can trace individual magnetic field lines. This method of using corks to follow particular magnetic features was used throughout the presented study.

2.2. *Calculating synthetic observables*

To determine whether the nanoflare-scale event could be observed by modern and future instruments, we calculated the synthetic intensities of two iron ions that form under optically thin conditions in the solar corona. The CHIANTI database (Del Zanna et al. 2021) was used to calculate synthetic spectra of Fe IX at 171 Å and Fe XII at 195 Å, integrating both along the z axis as well as the y axis. This provided a top-down and a side view of the nanoflare-scale event, which provided two viewing angles for analysis. The first moments of the spectra were also calculated in order to understand the Doppler signatures of the event.

3. Results

3.1. *Finding relevant magnetic features*

The simulation described in Section 2.1 resulted in impulsive heating events that produced local coronal plasma temperatures up to 1.47 MK. The cause of the heating was found to be large-angle reconnection between several magnetic features as shown in Fig. 1, which were identified using the aforementioned `corks` module. Corks were selected based on their proximity to current volumes that formed as a result of large-angle reconnection. Then, the same corks were used to reconstruct a time series of the relevant features.

We determined that the magnetic features did not rise coherently through the atmosphere, but rather self-order in the corona throughout the duration of the simulation (Robinson et al. 2022). Additionally, no relevant flux cancellation events are observed; the photospheric flux concentrations to which the magnetic features are rooted remain steadfast during the time leading up to the reconnection. Therefore, the reconnection event is not associated with traditional magnetoconvective processes such as flux emergence or flux cancellation.

3.2. *Buildup of the flux rope*

Because the nanoflare-scale reconnection event is powered by the gradual buildup of a magnetic flux rope, we set out to understand how the flux rope self-consistently forms.

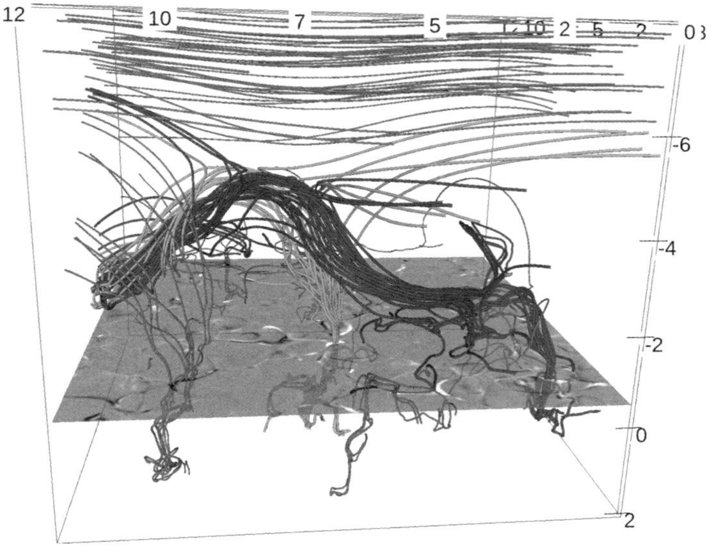

Figure 1. Visualization of relevant magnetic features at t = 11 360 s. Cyan lines represent a magnetic arcade, and red lines represent a magnetic flux rope. The overlying horizontal field (green) is nearly anti-parallel to the red and cyan features.

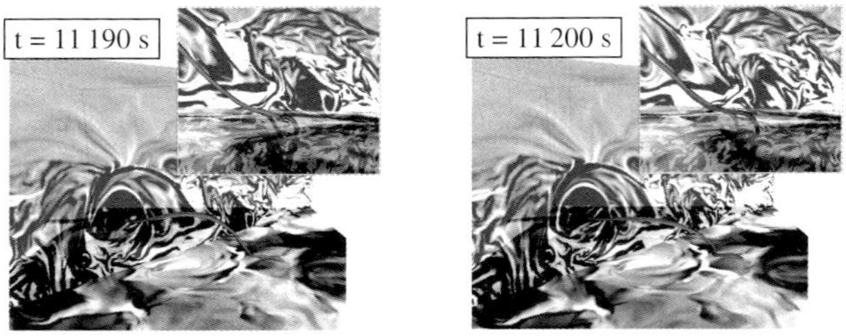

Figure 2. Field line pair at t = 11 190 s (left) and t = 11 200 s (right) that are traced by the same two corks, but switch chromospheric footpoints between the time stamps. The insets in the upper right corners offer a zoomed-in view of the footpoint switching at a different viewing angle, as evidenced by the lines switching colors.

To do this, we searched for evidence of the inverse cascade of helicity (Frisch et al. 1975; Pouquet et al. 2019), whereby the self-helicities of smaller flux bundles coalesce by reconnection, and contribute to a larger-scale helical system. To that end, we detected such a helicity cascade evidenced by not only small-scale flux bundles but by individual magnetic field line pairs undergoing component reconnection. This component reconnection was detected by isolating individual field pairs that belong to the overall flux rope, and that switch footpoints over a chromospheric slice as shown in Fig. 2. From this, we determine that the self-helicities of these magnetic field line pairs contribute to the helicities of small-scale flux bundles via component reconnection, which themselves coalesce to form the overall flux rope.

In addition to this, the flux rope tends toward a linear force-free field after it forms (that is, undergoes Taylor relaxation, see Taylor (1974)). In a constantly driven system that eventually undergoes major reconnection, the flux rope is of course unable to fully relax according to Taylor's theory; however, we see a statistical tendency toward Taylor relaxation during the time the flux rope is forming via the inverse cascade of helicity

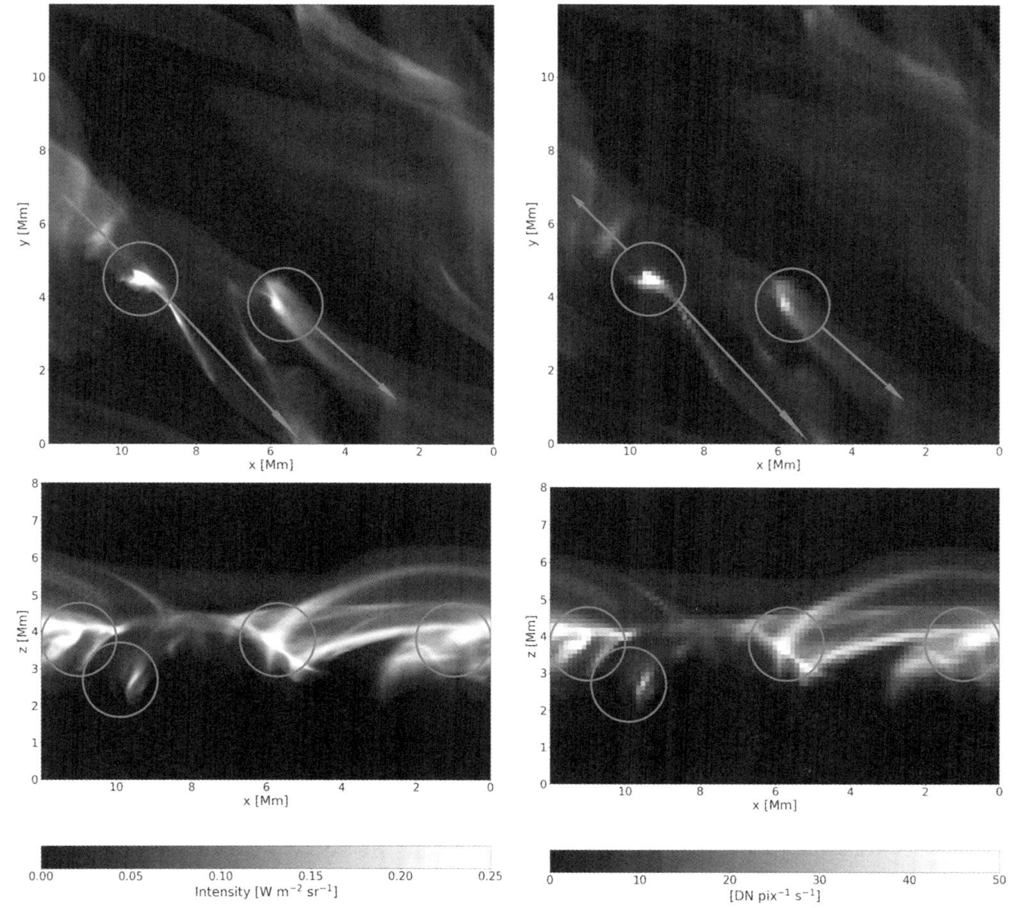

Figure 3. Synthetic intensities at the native simulation resolution for Fe IX 171 Å (left) compared to synthetic MUSE intensities (right) at t = 11 360 s. Bright points of interest are denoted with red circles, and the two jet systems in the upper panels are denoted with red arrows. Synthetic MUSE observations are degraded to the highest possible resolution of 0.167".

(Robinson et al. 2023). This is a result that had not been produced in idealized simulations because Taylor relaxation requires a significant number of reconnections and that is only possible for a more complex simulation.

3.3. *Synthetic observables*

Our final result concerns the observability of the nanoflare-scale event previously discussed. As mentioned in Section 2.2, we consider the signatures of both Fe IX and Fe XII at a reference snapshot during the nanoflare-scale event, converting their integrated intensities from physical units to telescope count rates for the Solar Dynamics Observatory Atmospheric Imaging Assembly (SDO/AIA; Pesnell, Thompson, and Chamberlin 2012; Lemen *et al.* 2012) and the Multi-silt Solar Explorer (MUSE; De Pontieu et al. 2020, 2022; Cheung et al. 2022). We determine that the count rates would be sufficient for observation in the Fe IX 171 channel for both SDO/AIA and MUSE (Robinson and Carlsson 2023). In particular, two systems of bi-directional jets emanating from the reconnection spine would be observable with both instruments, as shown for MUSE in Fig. 3. However, the integrated intensities for Fe XII at 195 Å would not be visible in the SDO/AIA 193 channel because calculated count rates are insufficient. We note that MUSE will not be designed to observe Fe XII, but the spectral resolution of MUSE is sufficient for detecting the Doppler shifts of the bi-directional jets. This serves as a support for MUSE and also contextualizes past and future observations of nanoflare-scale reconnection events that are not associated with flux emergence or flux cancellation.

4. Discussion

The described studies provide insight into the development of magnetic fields that contribute to energetic nanoflare-scale reconnection events in the quiet Sun. Not only did we simulate such an event in a self-consistent manner, but we determined that the event forms from coronal magnetic field lines rather than by more traditional magnetoconvective processes. We also demonstrate how this event would appear in observations of Fe IX 171Å, and look forward to the new capabilities of the upcoming MUSE instrument.

4.1. *Scientific discussion at the symposium*

During the symposium, the discussion focused mostly on the inverse cascade of helicity. First, it was suggested that the small-scale component reconnections may be evidence of the interchange instability. This is an interesting suggestion, but we do see evidence of topological changes rather than an interchange between the field and the surrounding plasma.

There was also a brief discussion regarding why the helicity cascade is inverse rather than forward. We of course still see a forward cascade of energy, so why not a forward cascade of helicity? In this case, we neither see nor consider any larger-scale helical systems that may disintegrate into small-scale helical systems; instead, we observe small-scale helical systems contributing to an overall large-scale system. This inverse helicity cascade is most relevant for the nanoflare-scale reconnection event because the development of the flux rope via several small-scale reconnections ultimately powers the large-scale reconnection.

Acknowledgements. Figures are adapted from Robinson et al. (2022), Robinson et al. (2023), and Robinson and Carlsson (2023). RR acknowledges funding support under ERC Syngergy grant agreement No. 810218 (Whole Sun).

References

Cheung, M.C.M., Martínez-Sykora, J., Testa, P., De Pontieu, B., Chintzoglou, G., Rempel, M., and, ...: 2022, *The Astrophysical Journal* **926**, 53. doi:10.3847/1538-4357/ac4223.
De Pontieu, B., Martínez-Sykora, J., Testa, P., Winebarger, A.R., Daw, A., Hansteen, V., and, ...: 2020, *The Astrophysical Journal* **888**, 3. doi:10.3847/1538-4357/ab5b03.
De Pontieu, B., Testa, P., Martínez-Sykora, J., Antolin, P., Karampelas, K., Hansteen, V., and, ...: 2022, *The Astrophysical Journal* **926**, 52. doi:10.3847/1538-4357/ac4222.
Del Zanna, G., Dere, K.P., Young, P.R., and Landi, E.: 2021, *The Astrophysical Journal* **909**, 38. doi:10.3847/1538-4357/abd8ce.
Frisch, U., Pouquet, A., Leorat, J., and Mazure, A.: 1975, *Journal of Fluid Mechanics* **68**, 769. doi:10.1017/S002211207500122X.
Færder, Ø.H., Nóbrega-Siverio, D., and Carlsson, M.: 2023, *Astronomy and Astrophysics* **675**, A97. doi:10.1051/0004-6361/202346447.
Gudiksen, B.V., Carlsson, M., Hansteen, V.H., Hayek, W., Leenaarts, J., and Martínez-Sykora, J.: 2011, *Astronomy and Astrophysics* **531**, A154. doi:10.1051/0004-6361/201116520.
Lemen, J.R., Title, A.M., Akin, D.J., Boerner, P.F., Chou, C., Drake, J.F., and, ...: 2012, *Solar Physics* **275**, 17. doi:10.1007/s11207-011-9776-8.
Pesnell, W.D., Thompson, B.J., and Chamberlin, P.C.: 2012, *Solar Physics* **275**, 3. doi:10.1007/s11207-011-9841-3.

Pouquet, A., Rosenberg, D., Stawarz, J.E., and Marino, R.: 2019, *Earth and Space Science* **6**, 351. doi:10.1029/2018EA000432.

Robinson, R.A., Carlsson, M., and Aulanier, G.: 2022, *Astronomy and Astrophysics* **668**, A177. doi:10.1051/0004-6361/202244750.

Robinson, R.A., Aulanier, G., and Carlsson, M.: 2023, *Astronomy and Astrophysics* **673**, A79. doi:10.1051/0004-6361/202346065.

Robinson, R.A. and Carlsson, M.: 2023, *Astronomy and Astrophysics* **677**, A36. doi:10.1051/0004-6361/202347089.

Taylor, J.B.: 1974, *Physical Review Letters* **33**, 1139. doi:10.1103/PhysRevLett.33.1139.

Dynamics of Solar and Stellar Convection Zones and Atmospheres
Proceedings IAU Symposium No. 365, 2024
A. V. Getling & L. L. Kitchatinov, eds.
doi:10.1017/S1743921323005793

Modeling line-of-sight magnetograms of emerging active regions

M. Poisson[1], M. López–Fuentes[1], C. H. Mandrini[1], F. Grings[1] and P. Démoulin[2]

[1] Instituto de Astronomía y Física del Espacio (UBA-CONICET), Buenos Aires, Argentina.
email: marianopoisson@gmail.com

[2] Observatoire de Paris (LESIA), Meudon, France.

Abstract.
Active regions (ARs) appear in the solar atmosphere as a consequence of the emergence of magnetic flux ropes (FRs). Due to the presence of twist, the photospheric line-of-sight (LOS) magnetograms of emerging ARs show an elongation of the polarities known as magnetic tongues. These tongues can affect the estimation of tilt angles during their emergence phase. In this work, we propose a Bayesian method to model LOS magnetograms of emerging ARs using a half-torus twisted FR model. We apply this model to 21 emerging ARs observed during Solar Cycle 23. We find that the Bayesian method corrects the tilt when compared to other methods, removing the spurious rotation of the polarities produced by the retraction of the tongues during the emergence. We find a variation in Joy's law with the stage of the AR emergence and the method used for its estimation.

Keywords. Sun: magnetic fields, Sun: photosphere, methods: statistical.

---

1. Introduction

Active regions (ARs) appear in the solar atmosphere as a consequence of the emergence of magnetic flux tubes (Fan (2009)). The magnetic field of these flux tubes acquires twist during their formation and rise through the convection zone (CZ), producing the so-called magnetic flux ropes (FRs). Current dynamo models indicate that these FRs originate at the base of the CZ in which the toroidal component of the magnetic field is formed and accumulated (see Howe, 2009, and references therein). Simulations show that instabilities at this depth produce the formation of coherent magnetic field structures that emerge due to the pressure unbalance between the plasma inside and outside these FRs (Nelson et al. 2013). Once they reach the solar atmosphere they are generally observed as concentrated bipolar magnetic field regions. The bipoles emerge in latitudinal bands between $-35°$ to $35°$ and are slightly inclined with respect to the east-west direction (van Driel–Gesztelyi and Green 2015). The inclination of the bipoles, known as tilt angle, and its latitudinal dependence, known as Joy's law, are fundamental to understanding the decay and dispersal of the surface magnetic flux (Wang 2017).

Simulations relate Joy's law to the effect of the Coriolis force acting over diverging flows near the FR apex during their emergence (D'Silva and Choudhuri 1993, Weber et al. 2023). The observed dispersion of the tilt, away from the average Joy's law expected behavior, results from the turbulent buffeting on emerging FRs during their transit through the CZ. The strength of the effect of the Coriolis force depends on the duration of the emergence of the FR, and therefore, the observed tilt should decrease with the increase of the FR field strength (Jha et al. 2020). This idea suggests the existence of a tilt quenching in the Babcock–Leighton dynamo that limits the source of the global axial dipole moment.

The tilt angle of ARs is generally estimated using line-of-sight (LOS) magnetograms and computing the flux-weighted center of the polarities, known as magnetic barycenters. With this method, the tilt is defined using the inclination of the line that joints the barycenters with respect to the equator. Several limitations of this method arise depending on the stage of evolution of the AR producing tilt angles and spurious rotations that may not correspond to the intrinsic inclination of the emerging FR. In particular, the twist of the FR produces an elongation of the polarities of the AR observed in LOS magnetograms during their emergence. These elongations known as magnetic tongues were first characterized by López–Fuentes et al. (2000). Poisson et al. (2020) quantified the effect of the magnetic tongues on the estimation of the tilt angle proposing a correction with a method known as the Core Field Fit Estimator (CoFFE).

In Poisson et al. (2022) we developed a method to model the observed LOS magnetograms of bipolar ARs based on a Bayesian inference scheme. The method consists of a 3D model of a half-torus FR from which synthetic LOS magnetograms can be constructed by projecting the vertical component of the field over successive horizontal planes located at different heights. This strictly magnetic field model (no plasma interactions included) with 8-free parameters can reproduce most of the global features observed during the emergence of bipolar ARs (e.g. tilt angle and magnetic tongue extension).

Due to their relevance, tilt angle estimations during the emergence of ARs have to be corrected for the tongue effect. In this work, we aim to model the twist and tilt parameters of the FRs corresponding to the emergence of 21 bipolar ARs. In Sect. 2 we summarize the data processing basics and introduce the modeling method. Sect. 3 compares the estimations of the tilt obtained with the magnetic barycenters and the Bayesian methods. In Sect. 4 we discuss the implications of our results.

2. Data processing and modeling method

We use LOS magnetograms obtained with the *Michelson Doppler Imager* (MDI; Scherrer et al. 1995) on board the *SOlar and Heliospheric Observatory* (SOHO). The 96-minute cadence magnetograms have a spatial resolution of $1.96''$ and an error per pixel of 9 G (Liu et al. 2004).

We select 21 emerging bipolar ARs from Solar Cycle 23. We limit the heliographic latitudinal and longitudinal range from $-35°$ to $35°$ with respect to the disk center to avoid projection effects near the solar limb. For each AR, we project the LOS magnetic field into the solar radial direction and rotate the magnetograms to the time the AR is closer to the central meridian using Solar SoftWare tools. We define rectangular boxes that encompass the AR evolution and we mask the background field using a Gaussian smoothing filter and a flux threshold of 20 G. The parameters of the Gaussian filter are set manually and checked visually for each case to ensure a good selection of the field of the AR. The final output of the processing algorithm provides data cubes containing the magnetograms of each AR. Fig. 1a shows as an example magnetogram 20 of the evolution of AR 8171. This magnetogram corresponds to the time of the emergence of the AR in which strongly elongated tongues are visible on the leading negative polarity.

We use the same model as the one proposed in Poisson et al. (2022). The model is defined by a half-torus magnetic field with four parameters: the small radius a, the big radius R, the axial flux Φ_A, and the number of turns of the field around the FR axis, defined as N_t. Projecting the vertical component of this field into successive horizontal planes, we construct synthetic LOS magnetograms. For that, we need at least four other parameters to define the relative position of the FR on each plane. These parameters

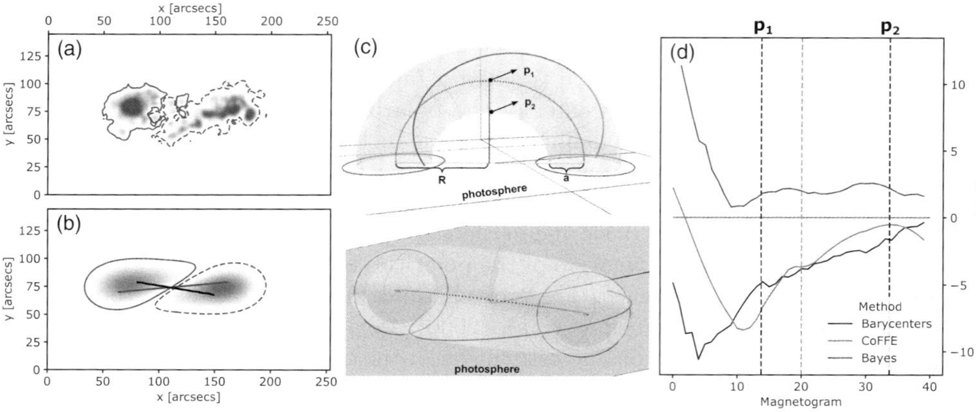

Figure 1. (a) SOHO/MDI LOS magnetogram of AR 8171. Red- and blue-shaded areas correspond to the strength of the positive and negative magnetic field. Contours indicate the field strength of −50 G (blue dashed line) and 50 G (red solid line). (b) Synthetic magnetogram of AR 8171 generated with the half-torus model. (c) Top and side views of the half-torus model. In this view, the photosphere height is set to zero. P_1 and P_2 indicate the relative positions of the photosphere used for the temporal binning described in Sect. 3. (d) The evolution of the tilt angle of AR 8171 estimated using the magnetic barycenters (black), CoFFE (blue), and the Bayesian (red) methods. The vertical green line corresponds to the magnetogram shown in panel a. Vertical black lines indicate the interpolated times in which the photospheric height is located at P_1 and P_2 in the Bayesian model.

are the horizontal Cartesian coordinates for the torus center defined as x_c and y_c, the inclination of the FR axis to the x−direction of the plane (the tilt γ), and the position of the FR apex d_0 relative to the photosphere. Fig. 1c shows different views of the half-torus model with the position of the photospheric plane at $d_0 = 1$.

We use a Bayesian inference scheme to compare the synthetic magnetograms with the observed LOS magnetograms of the selected ARs. For that we define the conditional probability distribution (or likelihood function) between the observations and the model as a normal distribution with zero as mean and a standard deviation of 50 G. The priors are all set as uniform distributions. The ranges for these distributions are defined from direct estimations done over the observed magnetograms. For instance, a, R, and Φ_A upper (lower) values are defined as a factor 1.5 (0.5) of the estimated mean size of the polarities, mean separation of the polarities, and maximum magnetic flux, respectively. A detailed description of the application of this method and the definition of priors can be found in Poisson et al. (2022). We use the No-U-Turn sampling algorithm, available in the Python PyMC5 library (Salvatier et al. 2016), to reconstruct the posterior probability distribution.

We model the full evolution of ARs defining the parameters of the half-torus (a, R, Φ_A, and N_t) as constant along time. This implies that the full data cube for each AR will have a single possible value for each of these parameters. On the other hand, the position parameters (x_c, y_c, γ, and d_0) can evolve without any temporal correlation imposed. This method generates a model in which the emergence is driven by a rigid kinetic transport of the FR as a whole across the photosphere. Fig. 1d shows the evolution of the tilt angle of AR 8171 computed using the magnetic barycenters (black), the CoFFE (blue), and the Bayesian method (red). In this case, the correction of the effect of the magnetic tongues on the tilt is achieved only with the Bayesian method, the counter-clockwise rotation obtained with the other methods is the effect of the contraction of the tongues.

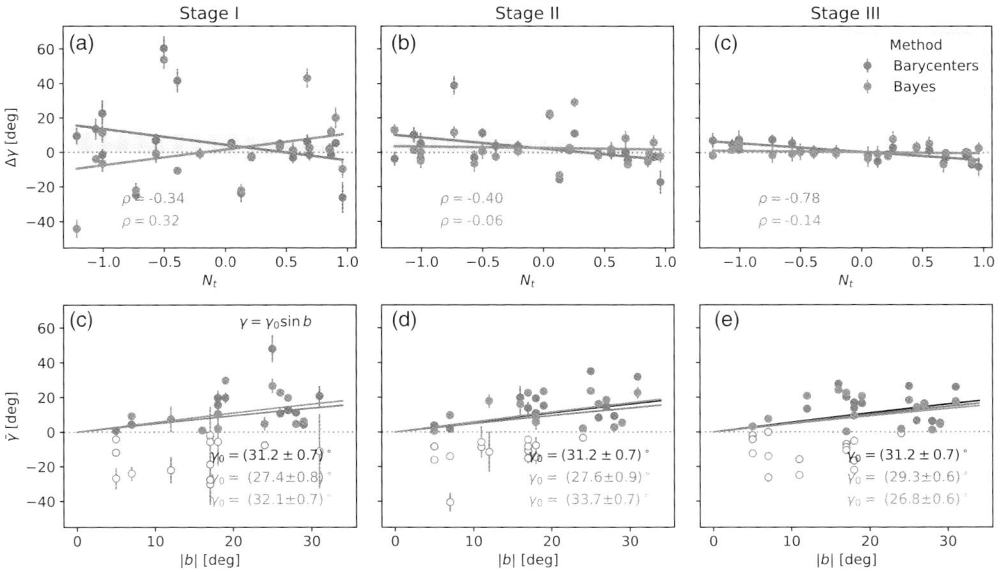

Figure 2. (a)-(c) Rotation of the bipoles $\Delta\gamma$ as a function of the number of turns N_t for the early, mid, and late stages of the evolution of 21 ARs. Blue and orange dots correspond to $\Delta\gamma$ estimated using the magnetic barycenters and the Bayesian method respectively. Linear least-squares fit and correlation coefficients ρ are displayed for each data set. (d)-(f) Mean tilt $\bar{\gamma}$ as a function of the heliographic latitude b for the early, mid, and late stages of the evolution of 21 ARs. We consider both hemispheres with positive latitudes using the absolute value of b. The black line corresponds to Joy's law obtained by Stenflo and Kosovichev (2012). Color circles correspond to ARs with negative tilt angles (inclination opposite to Joy's law). Blue and orange lines correspond to the fit of Joy's law for the tilts obtained with the magnetic barycenters and the Bayesian method respectively.

3. Results

To compare the tilt at different stages of the AR emergence we separate the evolution of the AR into three time bins. The parameter d_0 obtained using the model is a direct estimator of the stage of the FR emergence. Therefore, we divide the emergence using two relative positions of the photosphere: P_1 when its depth from the FR apex is equal to the small radius a and P_2 when this depth is equal to $2a$ (see marked points in Fig. 1c). The first bin corresponds to the span between the first observed magnetogram and P_1 (stage I). The second bin is defined in the interval between P_1 and P_2 (stage II). The third bin corresponds to the interval between P_2 and the last observed magnetogram (stage III). These bins represent the early, mid, and late stages of the emergence.

Within each of these bins, we compute the variation of the tilt $\Delta\gamma$ and the mean tilt $\bar{\gamma}$. $\Delta\gamma$ has a positive (negative) sign when the bipole rotates clockwise (counter-clockwise). The mean tilt sign is positive when the leading polarity is closer to the equator independently of the sign of the flux or the hemisphere of the AR. Figs. 2a-c show the rotation of the bipoles $\Delta\gamma$ as a function of the FR twist $N_{\rm t}$ at the three stages of the evolution for 21 ARs. The blue and orange points correspond to the tilt of each AR estimated using the magnetic barycenters and the Bayesian model, respectively. Linear least-squares fits and Pearson's correlation coefficients show the tendencies of the data at each stage of the evolution.

At stage I the correlation between the bipole rotation and the twist is low for both methods (Fig. 2a), meaning that other effects are acting on the bipoles that may produce intrinsic strong rotations apart from the magnetic tongues. Still, differences between both methods at stage I are significant, meaning that both estimations produce different results for the tilt. In the case of the tilt obtained with the magnetic barycenters, stages II and III (Fig. 2b-c) show a less dispersed tendency and a direct correlation between the twist and the rotation of the bipoles. This indicates that the contraction of the tongues produces spurious rotations. The rotations are clockwise for the negative twisted FRs and counter-clockwise for a positive twist. This tendency is not found for the tilt obtained with the Bayesian method, since the correlation coefficients of -0.06 and -0.14, suggest that the tilt angle is corrected from the effect of the magnetic tongues.

Figs. 2d-f show the mean tilt as a function of the latitude b at the three stages of evolution for 21 ARs. We consider both hemispheres together by taking the absolute value of b. The black line marks the Joy's law obtained by Stenflo and Kosovichev (2012) in which $\gamma = \gamma_0 \sin b$ with $\gamma_0 = (31.2 \pm 0.7)^\circ$. We fit the latitudinal dependence of the tilt for the analyzed dataset at the different defined stages. We fit only those ARs in agreement with Joy's law, i.e. with positive tilts (solid dots in Figs. 2d-f), since the negative tilt cannot be explained by the Babcock–Leighton model and their study requires a different analysis. From this fit, we obtain values of γ_0 shown in Figs. 2d-f. Despite the largest dispersion at stage I, the tilt angle obtained with the magnetic barycenters produces similar results of Joy's law at all three stages with a mean γ_0 of approximately 28°. In contrast, tilt angles obtained with the Bayesian method have values of γ_0 significantly lower and with lower dispersion at stage III, compared with the earlier stages. This implies that fully emerged ARs relax to a different tilt producing different Joy's law parameters. If we combine all stages as independent observations (similar to what is done when tilt angles are computed automatically from observations without considering the emergence stage of ARs), despite the ARs being repeated, we get $\gamma_0 = (30.5 \pm 0.5)^\circ$ with the Bayesian method and $\gamma_0 = (27.8 \pm 0.8)^\circ$ with the barycenters method.

4. Discussion

We model the LOS magnetograms of 21 bipolar emerging ARs observed during Solar Cycle 23. The half-torus model used here provides an estimation of the FR magnetic parameters such as the tilt angle and the twist. We find that the Bayesian method estimates the tilt by correcting the effect of the magnetic tongues. This estimation is therefore a better proxy of the intrinsic inclination of the FR than the estimation obtained using the magnetic barycenters. A proper estimation of the tilt at the early stages of the AR evolution is necessary for a correct analysis of the origin of the tilt dispersion.

The estimation of Joy's law using the corrected tilt angle obtained with the Bayesian method shows differences between the early and mid stages and the last part of the emergence. These differences indicate that the stage of AR emergence has to be considered when Joy's law is derived using statistical analyses. The lower slope and reduced dispersion of Joy's law at the last stage of the emergence suggest that the tilt relaxes to a stable value. The parameters of the fit of the latitudinal dependence of the tilt are fairly in agreement with the ones obtained by Stenflo and Kosovichev (2012), indicating that our sample is within the expected range of observed cases.

Moreover, when applying the Bayesian method we only find six estimations of the tilt that do not follow the expected Joy's orientation against nine cases when using the barycenters. The Bayesian method reduces the amount of anti-Joy's measurements compared to the estimation using the magnetic barycenters. The proper estimation of the tilt in these kinds of regions provides a better tool for a deeper analysis of the origin of these FRs.

References

D'Silva, S. and Choudhuri, A. R., 1993 *A&A*, 272, 621.

Fan, Y., 2009, *ApJ*, 697, 1529-1542.

Howe, R., 2009, *Liv. Rev. in Solar Phys.*, 6, 1.

Jha B. K., Karak B. B., Mandal S. and Banerjee D., 2020 *ApJL*, 889, L19.

Liu, Y., Zhao, X., Hoeksema, J. T., 2004 *Solar Phys.*, 219, 39.

López–Fuentes, M. C., Démoulin, P., Mandrini, C. H. and van Driel–Gesztelyi, L., 2000 *ApJ*, 544, 540.

Nelson, N. J. and Brown, B. P. and Brun, A. S., Miesch, M. S., Toomre, J., 2013, *ApJ*, 762, 2.

Poisson, M., López–Fuentes, M. C., Mandrini, C. H., Démoulin, P. and Mac Cormack, C., 2020 *A&A*, 633, A151.

Poisson, M. and Grings, F., Mandrini, C. H., López–Fuentes, M., Démoulin, P., 2022 *A&A*, 665, A101.

Salvatier, J. and Wiecki, TV and Fonnesbeck, C., 2016 *PeerJ Comp. Sci.*, 2, e55.

Scherrer, P. H., Bogart, R. S., Bush, R. I., Hoeksema, J. T., Kosovichev, A. G., Schou, J., Rosenberg, W., Springer, L., Tarbell, T. D., Title, A., Wolfson, C. J., Zayer, I. and MDI Engineering Team, 1995 *Solar Phys.*, 162, 129.

Stenflo, J. O. and Kosovichev, A. G., 2012 *ApJ*, 745, 129.

van Driel–Gesztelyi, L. and Green, L. M., 2015, *Liv. Rev. in Solar Phys.*, 12, 1.

Wang, Y. M., 2017, *Space Sci. Revs.*, 210, 1-4.

Weber, M.A., Schunker, H., Jouve, L. et al., 2023 *Space Sci. Revs.*, 219, 63

Dynamics of Solar and Stellar Convection Zones and Atmospheres
Proceedings IAU Symposium No. 365, 2024
A. V. Getling & L. L. Kitchatinov, eds.
doi:10.1017/S1743921323005136

AutoTAB: An Algorithm to Track Bipolar Magnetic Regions and Initial results from the Tracked BMRs

Anu Sreedevi[1], **Bibhuti Kumar Jha**[2], **Bidya Binay Karak**[1] and **Dipankar Banerjee**[3]

[1]Department of Physics, Indian Institute of Technology (Banaras Hindu University), Varanasi 221005, India
email: anubsreedevi.rs.phy20@itbhu.ac.in

[2]Southwest Research Institute, Boulder 80302, USA

[3]Aryabhatta Research Institute of Observational Sciences, Nainital 263002, Uttarakhand, India

Abstract. AutoTAB is a state-of-the-art, fully automatic algorithm that tracks the Bipolar Magnetic Regions (BMRs) in magnetogram observations. AutoTAB employs identified BMR regions from Line-of-Sight magnetograms from MDI and HMI (1996–2022) to track the BMRs through their evolution on the nearside of the Sun. AutoTAB enables us to create a comprehensive and unique catalog of tracked information of 9232 BMRs in the mentioned time period. This dataset is used to study the collective statistical properties of BMRs and particularly to identify the correct theory for the BMR formation. Here, we discuss the algorithm's functionality and the initial findings obtained from the AutoTAB BMRs catalog.

Keywords. Bipolar Magnetic Regions, Sunspot Groups, Solar Cycle, Joy's Law

1. Introduction

Bipolar Magnetic Regions (BMRs) are the visible proxies of magnetic solar activity. The sunspot pairs observed in white-light images of the Sun translate to opposite polarities of BMRs as seen in Line-of-Sight (LOS) magnetic field observations. The number of BMRs appearing on the solar disk shows a prominent cyclic behavior in accordance with the 11-year solar cycle. These regions are observed to emerge in a tilted fashion with respect to solar E-W direction (Hale et al. 1919). Evolution and dispersal of such regions on the surface, along with the meridional flow, causes the reversal in the poloidal field of the Sun (Babcock 1961; Leighton 1964; Karak 2020; Mordvinov et al. 2022; Golubeva et al. 2023). Consequently, the investigation of BMRs, their origin, and their evolutionary dynamics has been a compelling area of research in the past from both theoretical and observational perspectives (Karak and Miesch 2017, 2018; Biswas et al. 2023).

A regular magnetic field observation of the Sun has been performed since the early 1970s using ground-based solar observatories. In the last few decades, with the advancement of space-based observatories/telescope, such as Michelson Doppler Image (MDI; Scherrer et al. 1995) and Helioseismic and Magnetic Imager (HMI; Scherrer et al. 2012) we have been provided with the continuous full disk LOS magnetograms from 1996 onwards. Since then, magnetogram data has been used to monitor BMRs constantly to study solar magnetism and to forecast solar eruptions. Diverse methods, including automated techniques, have been developed for this purpose (Stenflo and Kosovichev 2012; Tlatov and Pevtsov 2014; Muñoz-Jaramillo et al. 2016). However, a consistent, calibrated catalog of tracked BMR information for solar magnetism study is still absent. To address this, we introduce AutoTAB, an advanced automatic tracking algorithm for BMRs (Sreedevi and Jha 2023; Sreedevi et al. 2023). In Section 2, we explain the details and workings of the algorithm.

The theory behind the formation of BMRs is still debated. According to the thin-flux tube model, BMRs are believed to be formed when magnetically buoyant flux tubes from the convection zone rise up to the solar surface (Parker 1955). The diverging flows at the top of the rising flux tubes will be subjected to Coriolis force to form the tilt in the BMRs (D'Silva and Choudhuri 1993). However, backing for this study remains limited. Here, we utilize tracked BMR data to assess the thin-flux tube model's validity in explaining BMR emergence. Many prior studies have explored this using various sunspot catalogs and before-mentioned automated methods (Wang and Sheeley 1991; Stenflo and Kosovichev 2012; Jha et al. 2020). However, some of these studies treat each BMR detection as a new BMR. This approach may introduce bias in analysis, as larger, longer-lasting BMRs contribute to a higher weightage. Our study addresses this issue by actively tracking the BMRs.

2. Data and Method

In this study, we have used the full disk LOS magnetogram data with the cadence of 96 minutes for the period of September 1996December 2022 (Cycles 23 and 24) from MDI (1996–2011; Scherrer et al. 1995) and HMI (2010–present; Scherrer et al. 2012) onboard the Solar and Heliospheric Observatory(SOHO) and Solar Dynamic Observatory (SDO), respectively. The BMRs are automatically identified using the method outlined in Stenflo and Kosovichev (2012), as adapted with slight modification by Jha et al. (2020). The identified BMRs are then saved as rectangular regions in the form of binary files. These rectangular regions also include extra pixels that may not be part of the BMR, which can potentially affect tracking efficiency, particularly in the case of closely lying BMRs. To isolate BMR boundaries more accurately, a threshold of 100 G is used, followed by the morphological operations (using *morp_close* function in IDL) with circular kernels of different sizes on these pixels. Once the boundaries of BMR are identified precisely, we extract morphological, magnetic, and location information for each identified region.

AutoTAB utilizes binary masks of identified BMRs to track them during their appearance on the near side of the Sun. The method of tracking is based on feature association similar to the one used for sunspot tracking in Jha et al. (2021). A brief description of the tracking algorithm follows: A BMR intended to be tracked is selected and isolated to a separate binary file (BF-I) and assigned with a unique AutoTAB ID. From the positional information of the BMR, we calculate the tracking period ($\mathrm{T}_{\mathrm{max}}$), i.e., the time taken for the BMR to reach the solar West limb. This indicates the duration during which AutoTAB will scan for isolated BMR. Next, we gather all binary masks within the time period $\mathrm{T}_{\mathrm{max}}$ (BF-1, BF-2, etc.). Then, BF-I is differentially rotated with respect to all the collected binary masks (BF-1, BF-2, etc.) to evaluate the overlap between each of them. If the overlap is found to be greater than 150 pixels, those regions are identified with the same AutoTAB-ID. If the overlap condition is not satisfied for the next two-day period of time, the overlap criterion is modified to 250 pixels to ensure that the BMR tracked is the same one. A detailed description of the working of the tracking algorithm and details of the pre-processing step is provided in Sreedevi et al. (2023).

Figure 1(a) is a representative example of the tracking of AutoTAB. It shows the evolution of AR08048 (AutoTAB ID 10183) observed in MDI. This BMR was first identified near the East limb (29.7° S, 44.3° E), and AutoTAB tracks it for the next seven days (in 65 observations) till it starts decaying close to the West limb. The variation of total flux and $B_{\mathrm{max}}$is shown in Figure 1(b) over the tracking period.

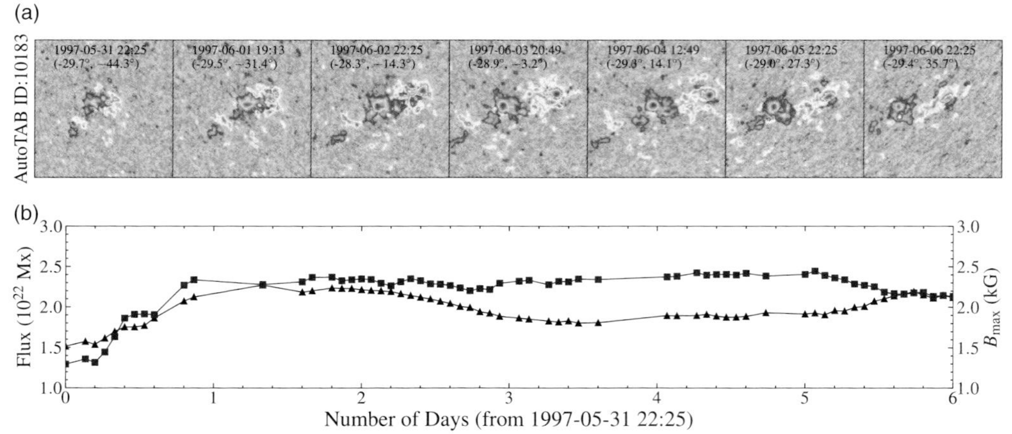

Figure 1. Panels (a) show the snapshots of the tracked BMR, AR08048 (AutoTAB-ID 10183), corresponding to each day during the tracking. Panels (b) show the evolution of total flux and $B_{\max}$ for the tracked BMR.

3. Results

AutoTAB has successfully identified and tracked 9232 BMRs in Cycles 23 and 24 for the time period between 1996–2022. On a closer look at the tracked BMRs, we find that BMRs were tracked for a broad range of lifetimes, from hours to weeks, and over large flux ranges throughout their journey on the nearside of the Sun. BMR lifetime is defined as the duration for which AutoTAB maintains tracking. In instances of poor tracking likely resulting from data corruption or absence, those BMRs were excluded from the dataset. The remaining BMRs were categorized into three distinct classes.

- Short-lived (SL): These are the BMRs that have a lifetime of less than 8 hours in the nearside of the Sun. They are generally small BMRs with small areas within the flux range in order of 10^{19} or 10^{20} Mx. Mostly, they are the ephemeral region, and a detailed study will be done for this class in the follow-up manuscript.
- Diskpassage (DP): This class accommodates such BMRs that AutoTAB has tracked in its evolutionary stages. That includes the BMRs that emerge in the nearside of the sun but continue evolving in the farside beyond 45°W, the BMRs that emerge on the far side and decay in the nearside between 45°E-W, and finally, the BMRs that emerge and decay on the far side. This class contains bigger BMRs within the flux range of 10^{22} Mx.
- Lifetime (LT): Any BMRs that emerge and decay in the nearside, between 45°E-W, constitutes this class. The lifespan of BMRs from this class is close to a week.

3.1. *Statistical Behaviour of Tracked BMRs*

To explore the statistical behavior of the tracked BMRs, we explore if the emergence of BMRs obeys the expected solar cycle behavior by looking at well-known latitudinal time distribution in the form of a butterfly diagram. Figure 2 represents the butterfly diagram, and each point represents each individually tracked BMR and the latitude of each BMR is determined at its peak flux during tracking, and the same applies to other properties ($B_{\max}$, B_{mean}, tilt). DP and LT class BMRs conform to the butterfly wing pattern across the tracked years. A definite wing-like pattern is not evident in the SL class as the points are scattered over all the latitudes at all times, but we see their latitudes limited between $\pm 60°$, suggesting that they are the byproducts of large-scale toroidal magnetic eld (Karak and Brandenburg 2016). The monthly new emergence of BMRs in each class exhibits a similar pattern. LT and DP class BMRs display consistent

Table 1. Some key parameters of different classes of tracked BMRs.

Classication	**# of BMRs**	**Flux ±ΔFlux (10^{22} Mx)**	$B_{max} \pm \Delta B_{max}$ **(G)**	$B_{mean} \pm \Delta B_{mean}$ **(G)**
Short-lived (SL)	1251	0.26 ± 0.01	541.32 ± 5.38	197.46 ± 0.66
Lifetime (LT)	3191	1.50 ± 0.02	949.20 ± 6.35	224.46 ± 0.81
Disk Passage (DP)	4710	2.05 ± 0.01	1436.83 ± 7.02	281.07 ± 0.71

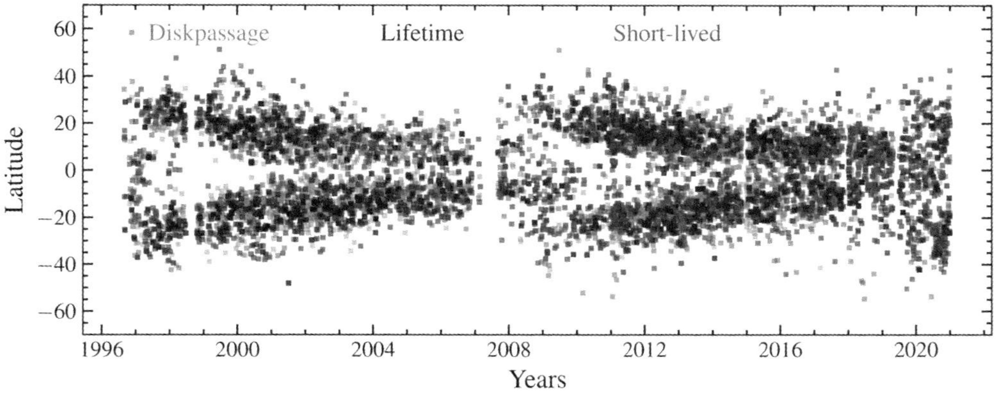

Figure 2. Latitudetime distribution (buttery diagram) of the tracked BMRs for (a) SL (red), (b) LT (blue), and (c) DP (green) classes.

solar cycle variations, aligning with fluctuations in the number of new emergences. In the case of the SL class, we could not observe such behavior, and a sufficient number of such BMRs were detected even during solar minima when the toroidal field of the sun is at a minimum (See Figure 6 in Sreedevi et al. (2023)). The number of total BMRs tracked in each class, along with their average magnetic properties, are listed in Table 1. A more detailed description of the mentioned classes and their statistics is explored in Sreedevi et al. (2023).

3.2. *Viability of thin-flux tube model for explaining BMR emergence*

In this section, we delve into an examination of whether the tracked information from AutoTAB aligns with the expected outcomes proposed by the thin-flux tube model. The thin-flux tube simulations by Fan et al. (1994) proposed the following empirical relation.

$$\gamma \propto \sin\lambda B_0^{-5/4} \Phi^{1/4}, \tag{1}$$

where B_0 is the initial magnetic field strength in the flux tube, Φ the magnetic flux inside the rising flux tubes. From this theory, it is expected that higher the initial magnetic field of the tube, the more quickly flux tubes will rise to the surface, and the Coriolis force will have less time to produce the tilt in BMRs. Furthermore, as the flux in the loop increases, tilt will also increase due to the effect of drag force. A point to note here is that, in the simulation studies, flux in the Ω-loop is independent of the initial magnetic field in the flux tube. From the tracked BMR data, we indeed find a good correlation between the measured unsigned flux and B_{mean}. Also in the observations, the magnetic field values measured for the BMRs may not be the same as the initial magnetic field of the BMR forming flux tube. Considering these, the following are some of the observational evidence we can expect from the thin-flux tube model:

- The footpoint separation to increase over the BMR's lifetime. The footpoint separation here is defined as the great circle distance between the two polarities of the BMR.

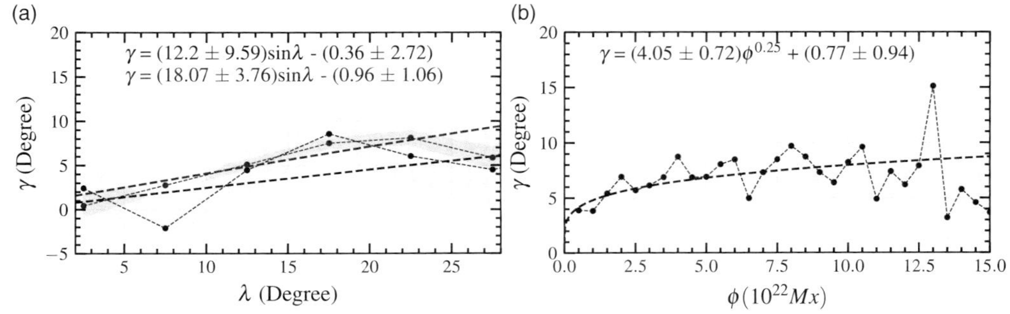

Figure 3. (a) Displays Joy's law dependence for tracked BMRs in two different flux bins (1×10^{21} Mx – 1×10^{22} Mx (blue) and 1×10^{22} Mx – 1×10^{23} Mx (red)). The dashed line represents the sinusoidal fit for both the flux bins with fitting parameters mentioned in the panel. (b) Mean tilt (and the error by shaded region) in each flux bin as a function of the absolute flux.

If the flux tubes are rising, we should expect that the distance between the polarities should increase as the BMR matures and gets saturated towards the end of life.

- If Coriolis force is the reason for tilted BMRs, then BMRs should emerge with tilt, and we can expect well-defined Joy's Law even from the initial phase of tracking.
- We can also assume that the measured unsigned flux within these regions will likely influence their tilt characteristics.

In the rest of the section, we assess each of the mentioned observations based on the findings from AutoTAB within the framework of the thin-flux tube model.

The initial inspection of the tracked data of BMRs shows the familiar tilt distribution and Joy's Law relation as shown by previous studies (Dasi–Espuig et al. 2010; Wang and Sheeley 1991) (also see Figure 4 in Karak (2023)). To evaluate the flux dependence on tilt in the data, BMRs were segregated based on their assigned flux strengths into two different bins, 1×10^{21} Mx – 1×10^{22} Mx (Bin-1) and 1×10^{22} Mx – 1×10^{23} Mx (Bin-2). Figure 3(a) shows Joy's law behavior in both the bins (Bin-1 in blue and Bin-2 in red) along with the sinusoidal fit. Notably, a greater degree of Joy's law scatter is evident for lower flux ranges, and the error bars of fitting parameters overlap, failing to show a conclusive tilt dependency on the measured flux of the BMR. In Figure 3(b), we further delve into flux dependency on tilt. Mean tilt values of BMRs in flux bins of 5×10^{21} Mx are depicted against absolute total flux, accompanied by a fitted curve of the form $\gamma = a\phi^{1/4} + b$. However, a strong tilt dependence on flux is not yet apparent. A detailed study exploring this will be presented in our follow-up manuscript.

We now proceed to evaluate the typical evolution of BMR foot points over BMR's lifetime. For this, we standardize the lifespan of LT class BMRs relative to the longest-living BMR within this class. Subsequently, we calculate the average foot point separation within each normalized lifespan interval to assess the overall progression of footprint separation, which is depicted in Figure 4(a). The figure shows a consistent rise in foot point separation during the initial phase of BMR evolution, reaching saturation at approximately half of the BMR's lifespan. The observed sustained increase is in support of the thin-flux tube model.

Finally, in order to assess whether the BMRs exhibit definite tilt at an early stage of their evolution, we compile all BMRs within the LT class. We then examine the behavior of Joy's law during the initial phase of these BMRs. To do this, we compute the average values of tilt and emerging latitude within the initial 33% of their lifespan and evaluate Joy's Law for this period, as illustrated in Figure 4(b) in blue. As for DP class BMRs, we collect the BMRs that emerge on the near side, calculate the average values of tilt

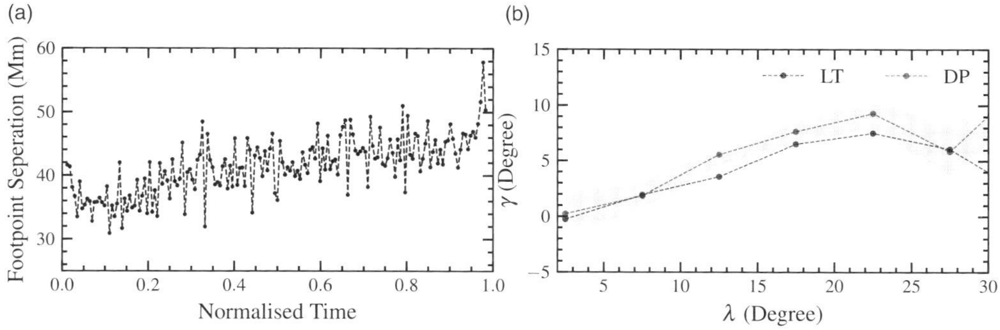

Figure 4. (a) General evolution of foot point separation of BMR over its lifetime. (b) Joys law variation at the initial phase of the BMR. Blue represents Joy's law plot of the BMRs in LT class in the first 33% of their lifetime. Red is the same as the blue curve but for BMRs in the DP class, which emerge in the nearside.

and emerging latitude over the entire tracking phase, and subsequently evaluate Joy's law (Figure 4(b) in red). The results indicate that during the initial phase of a BMR's lifespan, Joy's law is adhered to in both cases. This suggests that BMRs already exhibit a distinct tilt even in the early stages of AutoTAB tracking, which aligns with the findings of Kosovichev and Stenflo (2008) and supports the thin-flux tube model.

4. Conclusion

A statistical study of the evolving magnetic and morphological properties of BMRs is crucial in the context of understanding solar magnetism. However, a comprehensive catalog of tracked data of BMRs is limited. To address this, we introduce an automatic algorithm to track the BMRs (AutoTAB). AutoTAB utilizes binary maps of identified bipolar regions and verifies their overlap in subsequent time frames for tracking purposes. AutoTAB identifies and tracks 9232 BMRs in the period of 1996-2022. AutoTAB only tracks the BMR for their evolution in the nearside of the Sun. Consequently, the dataset encompasses BMRs tracked throughout their entire lifespan and those tracked only during their evolutionary stages.

The tracked BMRs were grouped into various classes based on their lifetime, namely, Short-lived (lives for less than 8 hr), Lifetime (emerges and disperses on the visible surface), and Disk passage (coming from and/or going to the far side of the Sun). The disk passage BMRs constitute the bigger BMRs with higher flux contents, and the short-lived group consists of mostly ephemeral regions. This dataset is further used to study the collective statistics of the BMRs. Our analysis reveals that the tracked BMRs exhibits behavior consistent with typical solar cycle patterns, with the emergence latitude of the BMRs at various times displaying butterfly wing-like patterns in latitude-time distribution plots. Furthermore, the tracked BMRs show familiar tilt distribution and tilt-latitude plots agreeing with previous studies. Further analyses were conducted to assess the validity of the thin-flux tube model for BMR formation. We observe that foot point separation of the BMRs of the LT class increases over the initial phase of their lifetime to saturate in the later phase, indicating the upward migration of magnetic field bundles from the convection zone towards the surface. Our analysis shows a well-defined Joy's law pattern in the initial phase of tracking, suggesting that BMRs emerge with a definite tilt. These findings support the thin-flux tube model. However, we do not find any strong and conclusive dependence of tilt on the flux of the BMR. These ideas will be further explored in the forthcoming article.

5. Acknowledgment

We thank IAU for the travel grant awarded to us, making it possible to attend the symposium in person in Armenia.

References

Babcock, H. W. 1961, The Topology of the Sun's Magnetic Field and the 22-YEAR Cycle. *ApJ*, 133, 572.

Biswas, A., Karak, B. B., & Kumar, P. 2023, Exploring the reliability of polar field rise rate as a precursor for an early prediction of solar cycle. *MNRAS*, 526(3), 3994–4003.

Dasi–Espuig, M., Solanki, S. K., Krivova, N. A., Cameron, R., & Peñuela, T. 2010, Sunspot group tilt angles and the strength of the solar cycle. *A&A*, 518, A7.

D'Silva, S. & Choudhuri, A. R. 1993, A theoretical model for tilts of bipolar magnetic regions. *A&A*, 272, 621.

Fan, Y., Fisher, G. H., & McClymont, A. N. 1994, Dynamics of Emerging Active Region Flux Loops. *ApJ*, 436, 907.

Golubeva, E. M., Biswas, A., Khlystova, A. I., Kumar, P., & Karak, B. B. 2023, Probing the variations in the timing of the Sun's polar magnetic field reversals through observations and surface flux transport simulations. *MNRAS*, 525(2), 1758–1768.

Hale, G. E., Ellerman, F., Nicholson, S. B., & Joy, A. H. 1919, The Magnetic Polarity of Sun-Spots. *ApJ*, 49, 153.

Jha, B. K., Karak, B. B., Mandal, S., & Banerjee, D. 2020, Magnetic Field Dependence of Bipolar Magnetic Region Tilts on the Sun: Indication of Tilt Quenching. *ApJ Letters*, 889(1), L19.

Jha, B. K., Priyadarshi, A., Mandal, S., Chatterjee, S., & Banerjee, D. 2021, Measurements of Solar Differential Rotation Using the Century Long Kodaikanal Sunspot Data. *Solar Phys.*, 296(1), 25.

Karak, B. B. 2020, Dynamo Saturation through the Latitudinal Variation of Bipolar Magnetic Regions in the Sun. *ApJ Letters*, 901(2), L35.

Karak, B. B. 2023, Models for the long-term variations of solar activity. *Living Reviews in Solar Physics*, 20(1), 3.

Karak, B. B. & Brandenburg, A. 2016, Is the Small-scale Magnetic Field Correlated with the Dynamo Cycle? *ApJ*, 816(1), 28.

Karak, B. B. & Miesch, M. 2017, Solar Cycle Variability Induced by Tilt Angle Scatter in a Babcock–Leighton Solar Dynamo Model. *ApJ*, 847, 69.

Karak, B. B. & Miesch, M. 2018, Recovery from Maunder-like Grand Minima in a Babcock–Leighton Solar Dynamo Model. *ApJ Letters*, 860, L26.

Kosovichev, A. G. & Stenflo, J. O. 2008, Tilt of Emerging Bipolar Magnetic Regions on the Sun. *ApJ Letters*, 688(2), L115.

Leighton, R. B. 1964, Transport of Magnetic Fields on the Sun. *ApJ*, 140, 1547.

Mordvinov, A. V., Karak, B. B., Banerjee, D., Golubeva, E. M., Khlystova, A. I., Zhukova, A. V., & Kumar, P. 2022, Evolution of the Sun's activity and the poleward transport of remnant magnetic flux in Cycles 21-24. *MNRAS*, 510(1), 1331–1339.

Muñoz-Jaramillo, A., Werginz, Z., Vargas–Acosta, J. P., DeLuca, M., Windmueller, J. C., Zhang, J., Longcope, D., Lamb, D., DeForest, C., Vargas-Domínguez, S., Harvey, J., & Martens, P. The best of both worlds: Using automatic detection and limited human supervision to create a homogenous magnetic catalog spanning four solar cycles. In *2016 IEEE International Conference on Big Data (Big Data* 2016,, pp. 3194–3203.

Parker, E. N. 1955, The Formation of Sunspots from the Solar Toroidal Field. *ApJ*, 121, 491.

Scherrer, P. H., Bogart, R. S., Bush, R. I., Hoeksema, J. T., Kosovichev, A. G., Schou, J., Rosenberg, W., Springer, L., Tarbell, T. D., Title, A., Wolfson, C. J., Zayer, I., & MDI Engineering Team 1995, The Solar Oscillations Investigation - Michelson Doppler Imager. *Solar Phys.*, 162, 129–188.

Scherrer, P. H., Schou, J., Bush, R. I., Kosovichev, A. G., Bogart, R. S., Hoeksema, J. T., Liu, Y., Duvall, T. L., Zhao, J., Title, A. M., Schrijver, C. J., Tarbell, T. D., & Tomczyk, S. 2012, The Helioseismic and Magnetic Imager (HMI) Investigation for the Solar Dynamics Observatory (SDO). *Solar Phys.*, 275(1-2), 207–227.

Sreedevi, A., Jha, B. K., Karak, B. B., & Banerjee, D. 2023, AutoTAB: Automatic Tracking Algorithm for Bipolar Magnetic Regions. *ApJ Supplement*, 268(2), 58.

Sreedevi, A. B. & Jha, B. K. 2023, Automatic Algorithm for Tracking Bipolar Magnetic Regions. *IAU Symposium*, 372, 97–99.

Stenflo, J. O. & Kosovichev, A. G. 2012, Bipolar Magnetic Regions on the Sun: Global Analysis of the SOHO/MDI Data Set. *ApJ*, 745(2), 129.

Tlatov, A. G. & Pevtsov, A. A. 2014, Bimodal Distribution of Magnetic Fields and Areas of Sunspots. *Solar Phys.*, 289, 1143–1152.

Wang, Y. M. & Sheeley, N. R., J. 1991, Magnetic Flux Transport and the Sun's Dipole Moment: New Twists to the Babcock–Leighton Model. *ApJ*, 375, 761.

Dynamics of Solar and Stellar Convection Zones and Atmospheres
Proceedings IAU Symposium No. 365, 2024
A. V. Getling & L. L. Kitchatinov, eds.
doi:10.1017/S1743921324000218

Flux emergence simulation and coronal response at ephemeral region scale

Zi–Fan Wang[1] **and R. H. Cameron**[2]

[1]National Astronomical Observatories, Chinese Academy of Sciences, Beijing, China
email: zfwang@nao.cas.cn

[2]Max Plank Institute for Solar System Research, Goettingen, Germany

Abstract. Flux emergence at different spatial scales and with different amounts of flux has been studied using radiative magnetohydrodynamics (rMHD) simulations. We use the radiative MHD code MURaM to simulate the emergence of an untwisted magnetic flux tube of ephemeral region scale with a density nonuniformity into a background atmosphere with a small unipolar open field. We find that the tube rises to the photosphere, forming complex loop structures seen in synthetic Atmospheric Imaging Assembly(AIA) 171 Å images. The atmosphere reaches $10^5 K$ at $3Mm$ above the surface. Our simulation provides a reference example of a less twisted ephemeral region emergence and the atmospheric response.

Keywords. Solar ephemeral region, Flux emergence

1. Introduction

The emergence of ephemeral regions is a kind of small-scale magnetic flux emergence between larger active regions and smaller intranetwork field emergences. They usually carry 10^{18} Mx to 10^{20} Mx flux and last at most 1-2 days (Harvey & Martin 1973; Hagenaar 2001). They tend to be important for flux emergence research, because they are related to both larger and smaller scale emergence in terms of physical mechanisms (Wilson et al. 1988; Illarionov et al. 2015), and play a part in the evolution of network fields (Hagenaar et al. 2003; van Driel–Gesztelyi & Green 2015). Being shorter lived, smaller in size and larger in number, ephemeral regions are good objects to study the whole process of emergence and proceeding evolution.

Previous magnetohydrodynamic (MHD) simulations of flux tube emergence show that a twist component of a magnetic flux tube is important for it to result in a successful flux emergence, because the twist can counteract deformation and fragmentation that will prevent the emergence of untwisted tubes (Schuessler 1979; Longcope et al. 1996; Moreno–Insertis & Emonet 1996; Krall et al. 1998). However, some observations suggests that flux tubes may not be twisted enough for successful emergence (Fan 2008). Hence, we try to examine the possible means of the emergence of a non-twisted tube.

In this proceeding we use numerical simulations to investigate the emergence of an ephemeral region. We consider the extreme case, introducing a non-twisted flux tube. We introduce a density nonuniformity, so that the tube is more buoyant in a part of it, and lead to its rise before complete fragmentation. We observe its emergence and the corresponding response in the corona, especially the temperature evolution.

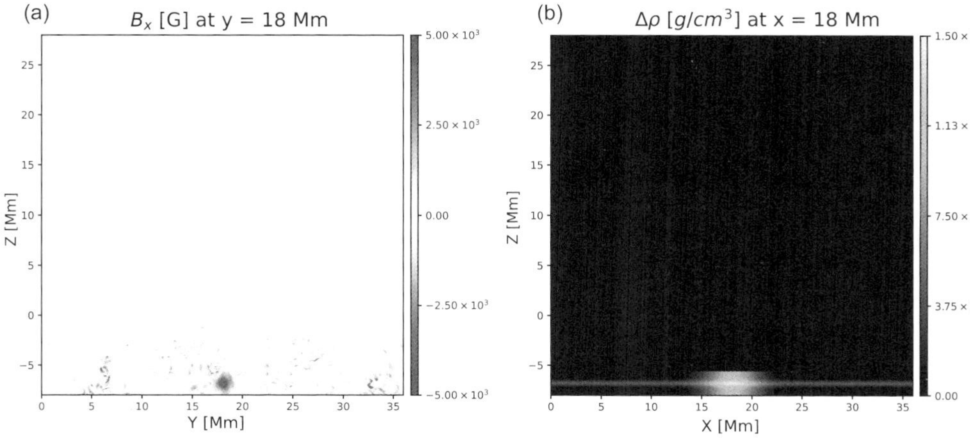

Figure 1. Initial conditions of the flux tube. Panels (a) shows B_x of an xz-slice at $y = 18$ Mm. Panel (b) shows the density decrease of introducing the tube $\Delta\rho = \rho - \rho_0$. Note that the uppermost part of the simulation domain is not presented here for convenience.

2. Methods

We use the radiative MHD code MURaM (Vögler et al. 2005; Rempel et al. 2009; Rempel 2017) to simulate the flux tube emergence and evolution. We use a 36 Mm × 36 Mm × 45 Mm Cartesian domain, with the third direction being vertical. The spatial resolution is 30 km for the vertical direction and 90 Mm for the horizontal directions, which is similar to previous simulations of active region evolution (e.g. Cheung et al. 2010; Rempel 2012; Chen et al. 2017). The boundaries of the two horizontal directions are periodic, while the upper boundary allows outflows. The photosphere where the optical depth τ is 1 is set to be 8 Mm above the bottom boundary, and is denoted as $z = 0$ Mm. A 5 G uniform magnetic field was added to the box, and is evolved until the magnetic energy reaches a statistically steady value.

We introduce a non-twisted magnetic flux tube near the bottom of the simulation domain. As shown in Figure 1 (a), the tube is parallel to the horizontal x-axis. The total flux is 1×10^{20} Mx, which is close to the flux of an ephemeral region (Harvey & Martin 1973). We then introduce a density nonuniformity to the tube, decreasing the density of part of the tube, in order to make the part rise faster. The location of the more buoyant part does not affect the results considering the periodic boundary conditions. The density nonuniformity follows a Gaussian function along the tube direction, as shown in Figure 1 (b).

We then simulate and examine the rise, emergence, and influence to the atmosphere of the flux tube. We examine the temperature, magnetic field, and the Poynting vector from the simulation. We also produce synthetic AIA images using the temperature response function from Lemen *et al.* (2012), and Boerner et al. (2012), as well as the optically-thin loss term in the corona of the MURaM model extension by Rempel (2017).

3. Results

The flux tube is simulated for 8 hours. It rises to the surface and emergence, forming loop structures and affects the properties in the atmosphere.

The time evolution of temperature in the layers of atmosphere is in agreement with the evolution of the surface field and synthetic AIA 171 Å images, and can be used to verify different periods of the simulation. As shown in Figure 2 and Figure 3, at around $t = 1.90$ hours, the temperature in the atmosphere is beginning to rise, and the emerging

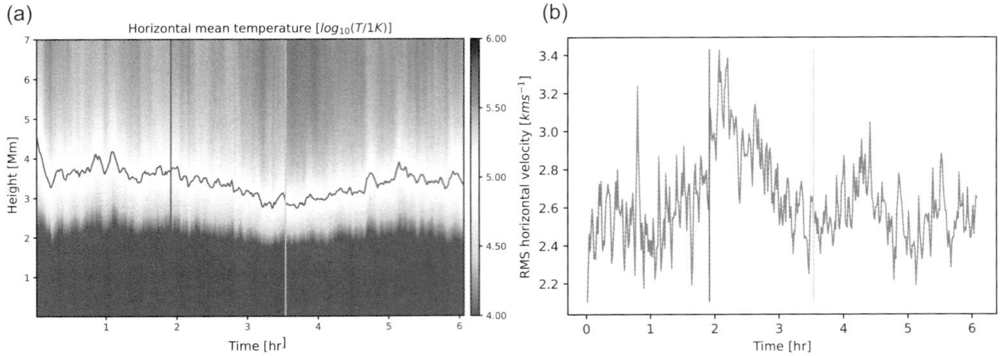

Figure 2. Time evolution of general properties during the flux emergence. Panel (a) shows the horizontally averaged temperature as a function of time and height above the photosphere. The red curve in panel (c) marks the height where the temperature is 10^5 K. Panel (b) shows the time evolution of root mean squared horizontal velocity ($v_{z,rms}$) in magnetic elements at the photosphere (at the average height of the $\tau = 1$ surface). The mean is taken over all pixels with vertical field larger in absolute value than 1×10^3 G. The red and cyan vertical lines represent 1.90 hours (red) and 3.53 hours (blue), respectively.

magnetic fields are observed at the photosphere surface. The AIA 171 Å images does not show complex structures compared to later time periods yet. After $t = 1.90$ hours, the temperature rise in the atmosphere is clearly shown, so the time period can be regarded as the main phase of the emergence. The $v_{h,rms}$ is the largest during the rapid rise of the surface flux, and falls quickly afterwards. At $t = 3.53$ hours, the photospheric vertical field at this time still possess a coherent negative polarity feature originating from the emergence, and the AIA 171 Å image shows complex structures in the atmosphere. After $t = 3.53$ hours, the transition region no longer moves downward, and begins to move back again, which marks the end of the main emergence process.

4. Discussions and conclusions

We have simulated the emergence of a flux tube at ephemeral region scale in a 3D domain with the radiative MHD model MURaM, covering it rise via magnetic buoyancy, photosphere surface emergence and atmosphere heating. During the simulation, the flux of the magnetic flux tube has emerged to the photosphere surface in the form of multiple bipoles. The emerged field generates loops reaching over 10 Mm in the atmosphere containing plasma above 10^6 K. The height of temperature reaching 10^5 K and 10^6 K during the emergence has lowered by 1 Mm in the atmosphere. After the peak, the temperature reverses to former condition as the movement of the surface field becomes less significant. The simulation describes a typical flux emergence in a partly fragmented form and with a limited effect in coronal response, which indicates the relationship between the influence to the atmosphere and the magnetic field configurations of ephemeral regions.

The flux tube we simulate is untwisted and with decreased density at its center. The emergence of twisted flux tube has been studied extensively, as an untwisted tube would distort and fragment during its rise (Schuessler 1979), disadvantageous to successful flux emergence. Twist field component keeps the tube from fragmentation and is important to flux emergence. However, our simulation shows that, for an untwisted tube, if the buoyancy at some part of it is large enough for it to rise before it fragments, a successful emergence is still possible. However, the emergence is fragmented and is limited in terms of heating in the atmosphere. How different properties of flux tubes affect atmospheric response differently is important in future study. Meanwhile, the emergence relies on the

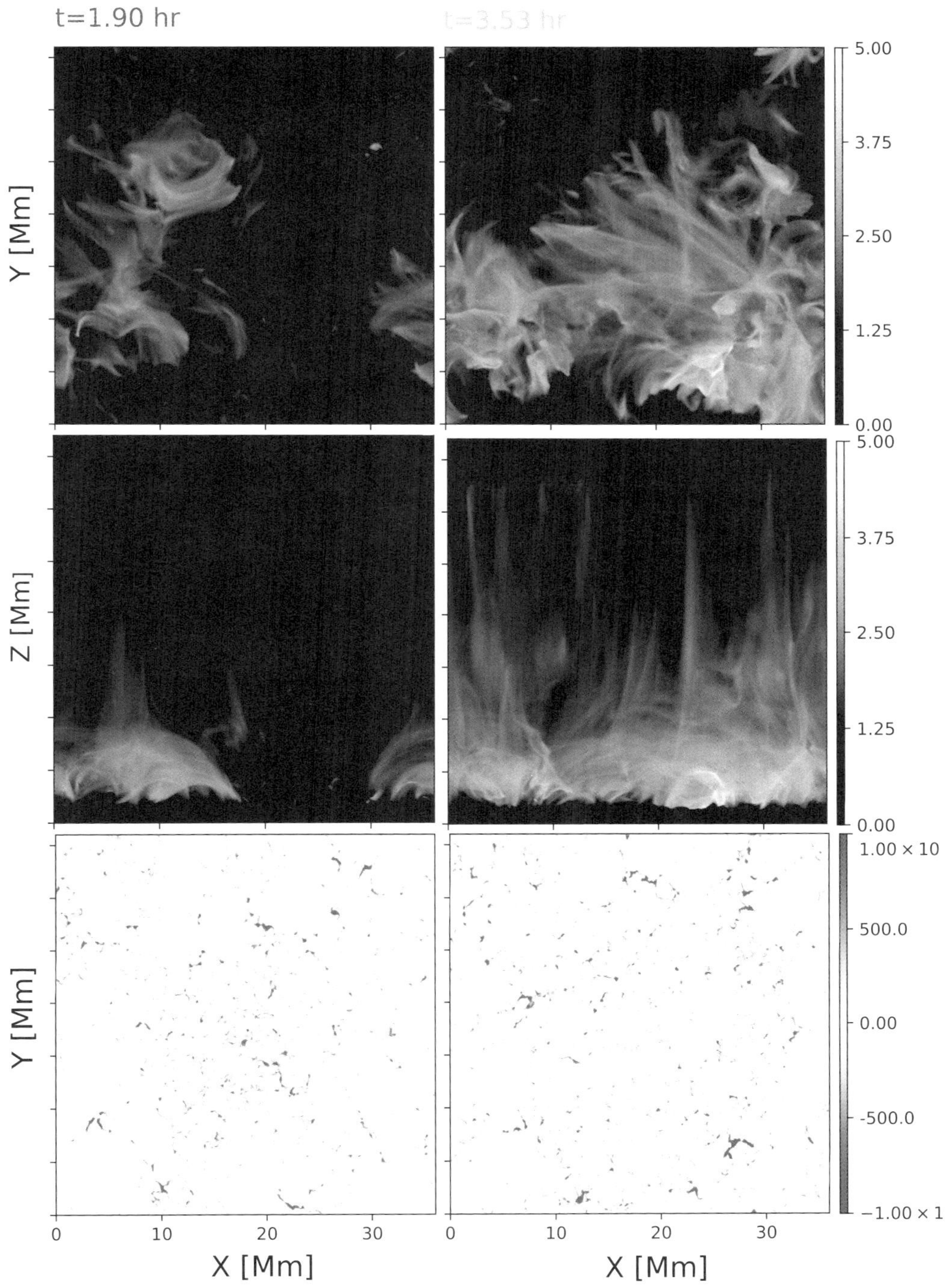

Figure 3. Synthetic 171 Å image and vertical photospheric magnetic field. The upper row shows the 171 Å image from the top, the middle row shows the 171 Å image from the yz plane, and the lower row shows the surface field. Columns 1 6 show the results at $t = 0.89$ hours, 1.90 hours, 2.52 hours, 3.15 hours, 3.53 hours, and 4.04 hours, respectively.

density nonuniformity which may be caused by turbulent motions in the real sun, and is needed to be evaluated with more detail in observational study in the future.

References

Boerner, P., Edwards, C., Lemen, J., et al. 2012, *Solar Phys.*, 275, 41. doi:10.1007/s11207-011-9804-8

Chen, F., Rempel, M., & Fan, Y. 2017, *Astrophys. J.*, 846, 149. doi:10.3847/1538-4357/aa85a0

Cheung, M. C. M., Rempel, M., Title, A. M., et al. 2010, *Astrophys. J.*, 720, 233. doi:10.1088/0004-637X/720/1/233

Fan, Y. 2008, *Astrophys. J.*, 676, 680. doi:10.1086/527317

Hagenaar, H. J. 2001, *Astrophys. J.*, 555, 448. doi:10.1086/321448

Hagenaar, H. J., Schrijver, C. J., & Title, A. M. 2003, *Astrophys. J.*, 584, 1107. doi:10.1086/345792

Hale, G. E., Ellerman, F., Nicholson, S. B., et al. 1919, *Astrophys. J.*, 49, 153

Harvey, K. L. & Martin, S. F. 1973, *Solar Phys.*, 32, 389. doi:10.1007/BF00154951

Illarionov, E., Tlatov, A., & Sokoloff, D. 2015, *Solar Phys.*, 290, 351. doi:10.1007/s11207-014-0612-9

Krall, J., Chen, J., Santoro, R., et al. 1998, *Astrophys. J.*, 500, 992. doi:10.1086/305754

Lemen, J. R., Title, A. M., Akin, D. J., et al. 2012, *Solar Phys.*, 275, 17. doi:10.1007/s11207-011-9776-8

Longcope, D. W., Fisher, G. H., & Arendt, S. 1996, *Astrophys. J.*, 464, 999. doi:10.1086/177387

Moreno–Insertis, F. & Emonet, T. 1996, *Astrophys. J. Lett.*, 472, L53. doi:10.1086/310360

Rempel, M. 2012, *Astrophys. J.*, 750, 62. doi:10.1088/0004-637X/750/1/62

Rempel, M. 2017, *Astrophys. J.*, 834, 10. doi:10.3847/1538-4357/834/1/10

Rempel, M., Schüssler, M., & Knölker, M. 2009, *Astrophys. J.*, 691, 640. doi:10.1088/0004-637X/691/1/640

Schuessler, M. 1979, *Astron. Astrophys.*, 71, 79

van Driel–Gesztelyi, L., & Green, L. M. 2015, Living Reviews in Solar Physics, 12, 1

Vögler, A., Shelyag, S., Schüssler, M., et al. 2005, *Astron. Astrophys.*, 429, 335. doi:10.1051/0004-6361:20041507

Wilson, P. R., Altrocki, R. C., Harvey, K. L., et al. 1988, *Nature*, 333, 748. doi:10.1038/333748a0

Dynamics of Solar and Stellar Convection Zones and Atmospheres
Proceedings IAU Symposium No. 365, 2024
A. V. Getling & L. L. Kitchatinov, eds.
doi:10.1017/S1743921323005148

Exploring the Quenching of Bipolar Magnetic Region Tilts using AutoTAB

Bibhuti Kumar Jha[1], Anu B. Sreedevi[2], Bidya Binay Karak[2] and Dipankar Banerjee[3]

[1] Southwest Research Institute, Boulder 80302, USA
email: maitraibibhu@gmail.com

[2]Department of Physics, Indian Institute of Technology (Banaras Hindu University), Varanasi 221005, India

[3]Aryabhatta Research Institute of Observational Sciences, Nainital 263002, Uttarakhand, India

Abstract. The tilt of the bipolar magnetic region (BMR) is crucial in the Babcock–Leighton process for the generation of the poloidal magnetic field in the Sun. We extend the work of Jha et al. (2020) and analyze the recently reported tracked BMR catalogue based on AutoTAB (Sreedevi et al. 2023) from Michelson Doppler Imager (1996–2011) and Helioseismic and Magnetic Imager (2010–2018). Using the tracked information of BMRs based on AutoTAB, we confirm that the distribution of $B_{\max}$ reported by Jha et al. (2020) is not because of the BMRs are picked multiple times at the different phases of their evolution instead it is also present if we consider each BMRs only once. Moreover, we find that the slope of Joy's law ($\langle\gamma_0\rangle$) initially increases slowly with the increase of $B_{\max}$. However, when $B_{\max}>2.5$ kG, γ_0 decreases. The decrease of observed γ_0 with $B_{\max}$ provides a hint to a nonlinear tilt quenching in the Babcock–Leighton process.

Keywords. Sun, Bipolar Magnetic Regions, Solar Dynamo

1. Introduction

The Sun and solar cycle variability have intrigued curious minds for over a century. Despite being enriched by continuous observations over the past century, solar cycle variability remains a primary puzzle in solar physics. However, it is now well established that the solar activity cycle is driven by the solar dynamo process, operating in the interior of the Sun, and it is responsible for the observed cyclic behavior of the Sun (Charbonneau 2014; Karak et al. 2014a). Solar differential rotation (Jha et al. 2021; Jha 2022) amplifies the magnetic field by twisting and stretching the existing polar field at the beginning of the solar cycle. These amplified and buoyant magnetic fields rise through the convection zone, where they experience the Coriolis force and emerge at the photosphere as tilted bipolar magnetic regions (BMRs). The residual flux from these BMRs migrates toward the poles and leads to the cancellation of existing opposite-polarity flux, initiating the onset of new flux buildup at the poles. These new opposite-polarity flux components act as the seed field for the following cycle and dictate its strength. The process of reversal and the buildup of new magnetic fields through the residual field carried to the poles by meridional flows is known as the Babcock–Leighton process (Babcock 1959, 1961).

One of the requirements for kinematic dynamo models, such as the Babcock–Leighton model, is the need for a nonlinear mechanism to suppress the exponential growth of the magnetic field (Charbonneau 2014). Flux loss due to magnetic buoyancy through the formation of BMRs (Biswas et al. 2022), latitude quenching (Karak 2020), and the tilt quenching are the three potential nonlinearities identified in the solar dynamo (Karak 2023). Lemerle et al. (2015) have first proposed that tilt quenching, which refers to the reduction of the tilt of Bipolar Magnetic Regions (BMRs) due to the presence of a strong magnetic field, can serve as the required nonlinearity in Babcock–Leighton dynamo models. Subsequently, Karak and Miesch (2017, 2018) incorporated tilt quenching in their dynamo models and achieved great success in reproducing the observed behavior of the solar cycle. More recently, Jha et al. (2020, 2023) utilized magnetogram data from the Michelson Doppler Imager (MDI) and the Helioseismic and Magnetic Imager (HMI) to provide observational evidence of tilt quenching. Additionally, Jha et al. (2020) demonstrated that the distribution of the maximum magnetic field in BMRs exhibits a double peak in their distribution.

One limitation of the work by Jha et al. (2020, 2023) is that they counted each BMR multiple times in their analysis, potentially impacting the inferences drawn from their results. To address this issue, Sreedevi et al. (2023); Sreedevi and Jha (2023) recently developed an algorithm based on feature association techniques to track BMRs in magnetograms more accurately. Here, we revisit the work done by Jha et al. (2020) and aim to determine whether the signature of tilt quenching reported by them can still be obtained from the tracked BMRs. Additionally, we investigate how the distribution of $B_{\rm max}$ changes with the tracked information of BMRs

2. Data and Method

We use the line of sight (LOS) magnetograms from the MDI (1996 – 2011) and HMI (2011 – 2022) to identify the BMRs and track them through during the period when they are in the near side of the Sun. The detection algorithm consist of identification of strong magnetic field regions followed by a moderate flux balance condition to make sure the detected regions are BMRs. This identification algorithm is similar to the one used in Stenflo and Kosovichev (2012) and Jha et al. (2020). Recently Sreedevi et al. (2023) has taken another step and developed an automatic tracking algorithm for BMRs (AutoTAB) to track the BMRs identified by Jha et al. (2020). The AutoTAB uses the features association technique, similar to one used by Jha et al. (2021) for sunspots tracking, to track the BMRs during its passage in the near side of the Sun. The AutoTAB provides a BMRs catalog with the properties of the BMRs such as total flux, location, maximum field ($B_{\rm max}$), average field. Therefore, we use these properties of the BMRs to re-analyse the results obtained in Jha et al. (2020).

3. Results

3.1. *Distribution of* $B_{\rm max}$

Jha et al. (2020) has reported that the $B_{\rm max}$ show a bimodal distribution with one peaks close to 600 G and another close to 2000 G. These peaks get well seprated when they classified the BMRs based on their signature in white-light continuum images. They reported that the peak corresponding to 600 G does not show any signature in white-light images i.e., they are not sunspots whereas other peak show a prominent signature for the same. This observed behaviour raised the question that, since in their work they have counted each BMRs multiple times it may be possible that they have picked the BMRs in their different phases and that gives rise to this observed double peak behavior. AutoTAB gives us an opportunity to track the BMRs and explore the distribution of $B_{\rm max}$ in them. In Figure 1, we show the distribution of $B_{\rm max}$, in which each BMRs are counted only once and the $B_{\rm max}$ of BMRs is determined at the point of time where $B_{\rm max}$

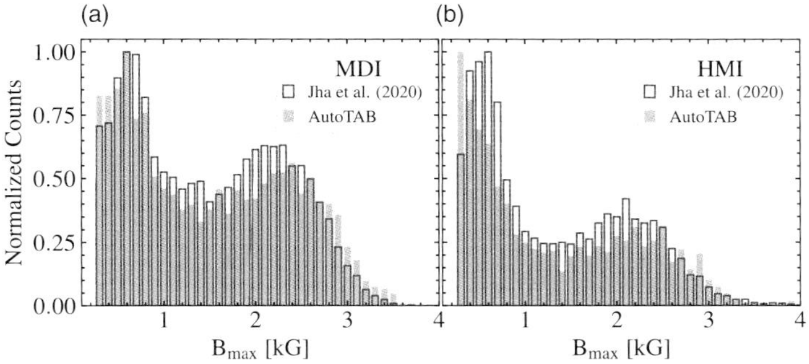

Figure 1. Normalized distribution of $B_{\max}$ for BMRs tracked using AutoTAB represented by filled bars and Jha et al. (2020), shown by unfilled bars.

of BMRs peaks during their evolution. We noted that the distribution of $B_{\max}$ in BMRs are indeed bimodal with peaks at 600 G and 2 kG. At this point we do not have the answer why the $B_{\max}$ in BMRs show such distribution, it will be worth exploring in the future. In Figure 1 we also note that at many places the normalized fraction of BMRs exceed the fraction reported by Jha et al. (2020). This is because the AutoTAB uses the data with higher cadence and for longer period.

3.2. *Tilt Quenching*

Tilt quenching is the phenomena of reduction in tilt on top of Joy's law because of the stronger magnetic field in it. The idea of tilt quenching was given by D'Silva and Choudhuri (1993), where they have explained the origin of tilt in the BMRs. Late, Lemerle et al. (2015) and Karak and Miesch (2017, 2018) has used tilt quenching in their solar cycle model. Jha et al. (2020) gave the first direct evidence of tilt quenching in observation (Figure 2(a)). In contrast with Jha et al. (2020), where they have counted each BMR many times, here we used the AutoTAB to track the BMRs and use its tilt when the flux in the BMRs is maximum during their evolution. Tracking of BMRs has significantly reduced the number of data point and hence we can not look at the $B_{\max}$ dependence of tilt in the way Jha et al. (2020) has looked at. Therefore we calculate the γ_0 of Joy's law which is $\gamma = \gamma_0 \sin\theta$ for each BMR in each $B_{\max}$ bin of size 500 G. Here, γ and θ is the tilt and latitude of BMRs, whereas γ_0 is called the amplitude of Joy's law. In Figure 2(b), we show the average of γ_0 calculate in each $B_{\max}$ bin along with the standard error as a function of $B_{\max}$. In Figure 2(b) we note that the $\langle\gamma_0\rangle$ have quite different value compared to the Figure 2(a), but still we can the similar trend in $\langle\gamma_0\rangle$ which show the downward trend after 2.5 kG instead of 2.0 kG. We emphasize here that this is the very preliminary result and detailed discussion will be reported in future publication.

4. Conclusion

In this article, we extend the work of Jha et al. (2020) by utilizing the newly developed BMR tracking algorithm AutoTAB (Sreedevi et al. 2023). AutoTAB was implemented on the LOS magnetogram data from MDI and HMI to track the BMRs during their passage on the near side of the Sun. The distribution of $B_{\max}$ in BMRs shows a bimodal distribution similar to the one reported in Jha et al. (2020), where they counted BMRs multiple times, unlike in this work. We also examine the variation of $\langle\gamma_0\rangle$ as a function of $B_{\max}$, showing a similar signature of tilt quenching as reported in Jha et al. (2020). Since we calculate the average of γ_0 for each BMR in a 500,G $B_{\max}$ bin, instead of fitting

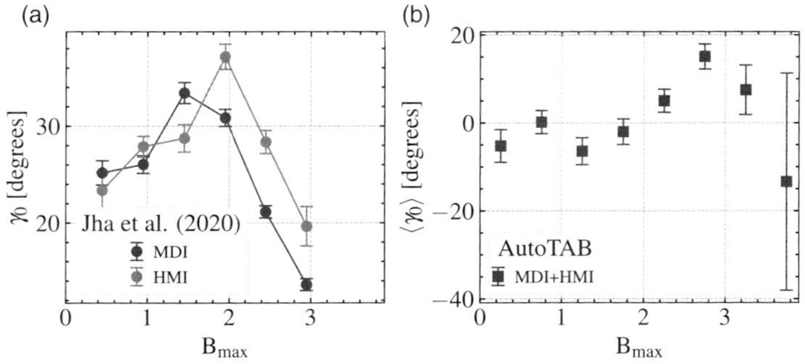

Figure 2. (a) The amplitude of Joy's law (γ_0) as a function of $B_{\max}$ for MDI and HMI, taken from Jha et al. (2020). (b) Average of amplitude of Joy's law ($\langle\gamma_0\rangle$) calculated over BMRs in 500 G $B_{\max}$ bin as a function of $B_{\max}$.

Joy's law, the values of $\langle\gamma_0\rangle$ differ from γ_0. These are preliminary results, and further details will be reported in an upcoming publication.

References

Babcock, H. D. 1959, The Sun's Polar Magnetic Field. *ApJ*, 130, 364.

Babcock, H. W. 1961, The Topology of the Sun's Magnetic Field and the 22-YEAR Cycle. *ApJ*, 133, 572.

Biswas, A., Karak, B. B., & Cameron, R. 2022, Toroidal Flux Loss due to Flux Emergence Explains why Solar Cycles Rise Differently but Decay in a Similar Way. *Phys. Rev. Lett.*, 129(24), 241102.

Charbonneau, P. 2014, Solar dynamo theory. *Ann. Rev. Astron. Astrophys.*, 52(1), 251–290.

D'Silva, S. & Choudhuri, A. R. 1993, A theoretical model for tilts of bipolar magnetic regions. *A&A*, 272, 621.

Jha, B. K. 2022,. *Long-term Study of the Sun and Its Implications to Solar Dynamo Models.* PhD thesis, Pondicherry University.

Jha, B. K., Karak, B. B., & Banerjee, D. 2023, Observational Signature of Tilt Quenching of Bipolar Magnetic Regions in the Sun. *IAU Symposium*, 372, 86–87.

Jha, B. K., Karak, B. B., Mandal, S., & Banerjee, D. 2020, Magnetic Field Dependence of Bipolar Magnetic Region Tilts on the Sun: Indication of Tilt Quenching. *ApJ Letters*, 889(1), L19.

Jha, B. K., Priyadarshi, A., Mandal, S., Chatterjee, S., & Banerjee, D. 2021, Measurements of Solar Differential Rotation Using the Century Long Kodaikanal Sunspot Data. *Solar Phys.*, 296(1), 25.

Karak, B. B., Jiang, J., Miesch, M. S., Charbonneau, P., & Choudhuri, A. R. 2014,a Flux Transport Dynamos: From Kinematics to Dynamics. *Space Sci. Rev.*, 186a, 561–602.

Karak, B. B. 2020, Dynamo Saturation through the Latitudinal Variation of Bipolar Magnetic Regions in the Sun. *ApJ Letters*, 901(2), L35.

Karak, B. B. 2023, Models for the long-term variations of solar activity. *Living Reviews in Solar Physics*, 20(1), 3.

Karak, B. B. & Miesch, M. 2017, Solar Cycle Variability Induced by Tilt Angle Scatter in a Babcock–Leighton Solar Dynamo Model. *ApJ*, 847, 69.

Karak, B. B. & Miesch, M. 2018, Recovery from Maunder-like Grand Minima in a Babcock–Leighton Solar Dynamo Model. *ApJ Letters*, 860, L26.

Lemerle, A., Charbonneau, P., & Carignan–Dugas, A. 2015, A Coupled 2 × 2D Babcock–Leighton Solar Dynamo Model. I. Surface Magnetic Flux Evolution. *ApJ*, 810, 78.

Sreedevi, A., Jha, B. K., Karak, B. B., & Banerjee, D. 2023, AutoTAB: Automatic Tracking Algorithm for Bipolar Magnetic Regions. *ApJ Supplement*, 268(2), 58.

Sreedevi, A. B. & Jha, B. K. 2023, Automatic Algorithm for Tracking Bipolar Magnetic Regions. *IAU Symposium*, 372, 97–99.

Stenflo, J. O. & Kosovichev, A. G. 2012, Bipolar Magnetic Regions on the Sun: Global Analysis of the SOHO/MDI Data Set. *ApJ*, 745, 129.

Dynamics of Solar and Stellar Convection Zones and Atmospheres
Proceedings IAU Symposium No. 365, 2024
A. V. Getling & L. L. Kitchatinov, eds.
doi:10.1017/S1743921323005240

Large-scale electric currents in processes in the solar atmosphere

Yu. A. Fursyak, V. I. Abramenko, A. S. Kutsenko and A. A. Plotnikov

Crimean astrophysical observatory of RAS, Nauchny, Crimea, Russia,
email: yuriy_fursyak@mail.ru

Abstract. In the late 80s of the 20th century, Crimean astronomers, studying the structure of transverse magnetic fields in active regions (ARs), discovered signs of the presence of large-scale vertical electric currents – global electric currents (Abramenko, Gopasyuk 1987). In 2018–2020, we finalized and adapted the method for detecting large-scale electric currents to the data of modern instruments for studying the Sun, and began studying their dynamics on time scales of 3–5 days (Fursyak *et. al* 2020). Our researches carried out during 2020–2023 showed that: 1) Large-scale electric currents with values of the order of $\sim 10^{13}$ A exist in ARs with nonzero flare activity. 2) Large-scale electric currents extend to the upper layers of the solar atmosphere in one part of the AR, and close through the chromosphere and corona in the remaining part of the AR. This assumption for the AR NOAA 12192 is confirmed by the results of numerical simulations performed in 2016 (Jiang *et al.* 2016). 3) The greater the magnitude of the large-scale electric current, the higher the probability of occurrence of M- and X- class solar flares in the AR. 4) At the final stages of AR evolution, a nonzero large-scale electric current can have a stabilizing effect on the sunspot, preventing its decay by its own magnetic field. 5) Large-scale electric currents are involved in coronal heating processes. Ohmic dissipation of a large-scale electric current is one of the mechanisms of quasi-stationary heating of coronal plasma above the AR. Our research on large-scale electric currents and the processes in which they take part continues.

Keywords. Sun, magnetic fields, electric currents, solar activity

1. Introduction

Electric currents play an extremely important role in various processes in the solar atmosphere. The first studies of electric currents on the Sun were carried out by astronomers of the Crimean Astrophysical Observatory in the 60s–70s of the 20th century (see refs in Alfven, Carlqvist 1967; Fleishman, Pevtsov 2018). Here we briefly present the results of our research conducted in 2018–2023, which, on the one hand, is a logical continuation of the early works of Crimean astronomers, and on the other hand, based on new approaches and data from modern instruments for studying the Sun.

2. Detection and Calculation of Large-Scale Electric Current

Large-scale electric currents in active regions (ARs) are distributed over a large area, so indirect techniques are required to detect them (see Fursyak *et. al* 2020).

Stage 1 involves the restoration of the potential transverse magnetic field of the AR. To solve this problem, the IDL code developed in 1982 by T. Sakurai (Sakurai 1982) is used.

At Stage 2, in each pixel of the initial magnetograms, the vector of the real (observed) transverse magnetic field of the AR is compared with the vector of the potential transverse magnetic field.

Stage 3 involves the decomposition in each pixel of the initial magnetograms of the vector of the observed transverse magnetic field into two components – potential, directed along the vector of the potential transverse magnetic field, and non-potential ($\mathbf{B}_{t\perp}$), directed across the vector of the potential field. The nonpotential component of the transverse magnetic field vector is due to electric currents in the AR.

Stage 4 involves the analysis of the distribution of the vector of the nonpotential component of the transverse magnetic field ($\mathbf{B}_{t\perp}$) in the AR. Near large developed sunspots in the AR, the vector $\mathbf{B}_{t\perp}$ forms regular vortex structures, which are due to the presence of a large-scale vertical electric current distributed over a large area. To calculate the magnitude of the large-scale electric current around the sunspot, a contour is drawn (Stage 5) so that within the contour the direction of twist of the vector $\mathbf{B}_{t\perp}$ has one predominant direction (clockwise or counterclockwise). The magnitude of the large-scale electric current for a given point in time is calculated as the resulting vertical electric current within a contour. The shape of the contour does not change over the entire time interval of AR monitoring. In order to minimize the errors caused by the evolution of the AR, the contour within which the magnitude of the large-scale electric current is calculated is tied to the center of gravity of the sunspot. The displacement of the center of gravity of the sunspot even by one pixel on the magnetogram leads to a displacement of the contour by the same number of pixels.

The analysis of the structure of the transverse magnetic fields of the studied ARs and the calculation of local vertical and large-scale electric currents were carried out using the magnetographic data of the Helioseismic and Magnetic Imager (HMI) instrument (Scherrer *et al.* 2012) aboard the Solar Dynamics Observatory (SDO) with a spatial resolution at the level of the solar photosphere of $0.5''$ (360 km) and a time resolution of 12 minutes, available from the Joint Science Operation Center (JSOC) website at http://jsoc2.stanford.edu/ajax/lookdata.html.

The obtained values of the imbalance of the vertical electric current in the studied ARs ($< 0.1\%$) suggest that a large-scale electric current goes high into the solar corona and closes within the AR. Therefore, a large-scale electric current must be associated with various processes in the atmosphere above the AR.

3. Large-scale electric currents and flare productivity of ARs

Since electric currents in ARs are carriers of the energy of nonpotential magnetic fields, the so-called "free" magnetic energy (Melrose 1991; Fleishman, Pevtsov 2018), and a large-scale electric current extends to the upper layers of the solar atmosphere, therefore, it must be closely related to the flare productivity of ARs. Indeed, as we showed in (Fursyak *et. al* 2020), large-scale electric currents with absolute values on the order of $\sim 10^{13}$ A are present in all developed ARs. However, the absolute values of the magnitude of the large-scale electric current and its temporal variations are overwhelmingly larger in the ARs with a higher level of flare activity (see Fig. 3 and Fig. 4 in Fursyak *et. al* 2020).

Initially, we suggested that large-scale electric currents are found only in developed ARs with nonzero flare productivity. However, a later study of ARs at the final stages of evolution (Fursyak, Plotnikov 2022) showed that in approximately 25–30% of cases, a nonzero large-scale electric current is also recorded in such ARs. However, the dynamics of the magnitude of the large-scale electric current in the ARs at the final stages of evolution is very low, and its absolute values ($\sim 10^{12}$ A) are several times lower than the absolute values of the large-scale electric current in developed ARs.

4. Large-scale Electric Currents in the Processes of Sunspot Stabilization

In 2022, we studied the role of electric currents in the processes of stabilization or destabilization of sunspots in ARs at the final stage of evolution (Fursyak, Plotnikov 2022). A total of 46 ARs of the 24th cycle of solar activity were analyzed. It was shown that there are no large-scale electric currents in the ARs with the magnetic flux decay rate $> 6.0 \times 10^{19}$ Mx hr^{-1} (within the calculation errors). At the same time, a nonzero large-scale electric currents were found in 37% of the ARs with magnetic flux decay rate in the sunspot $\leq 6.0 \times 10^{19}$ Mx hr^{-1} (see Fig. 5 and Fig. 6 in Fursyak, Plotnikov 2022). Thus, on the basis of statistical analysis, we concluded that a large-scale electric current by its own magnetic field can have a certain stabilizing effect on the sunspot. However, large-scale electric current is not being the only possible mechanism which can prevent the loss of magnetic flux of the sunspot.

5. Large-scale Electric Currents and Coronal Heating Processes

It is known that only a few percent of the energy of nonpotential magnetic fields of AR, which is carried by electric currents, is released in solar flares of various classes (Zaitsev *et al.* 1998). More than 90% of this energy is released in a number of other processes. It is logical to assume that a large-scale electric current extending into the upper layers of the solar atmosphere will participate in these processes, one of the most obvious of which is the heating of coronal matter above the AR.

To assess the role of large-scale electric currents in the heating of the coronal matter, data on the intensity of UV radiation in the 131, 171, 193, and 211Åchannels provided by the Atmospheric Imaging Assembly (AIA/SDO) (Lemen *et al.* 2012) were analyzed, maps of the spatial temperature distribution in the corona were obtained, the values of the average temperature in the corona above the AR were determined, and graphs of temporal changes in the average temperature during the monitoring of the AR were plotted (Fursyak 2023). In total, 9 ARs of the 24th cycle of solar activity with different levels of flare productivity were analyzed. The following results were obtained (see Fig. 3 in Fursyak 2023):

(1) Outside of solar flare events, the average temperature in the corona above the ARs during their monitoring is close to $< lgT > = 6.2$ (1.5 MK) and does not depend on the level of flare productivity of the AR. This observation suggests that the heating of coronal matter due to ohmic dissipation of large-scale electric currents occurs in a quasi-stationary regime.

(2) During solar flares of high X-ray classes, an increase in the average temperature in the corona above the AR to values $< lgT > = 6.3$–6.5 (2.0–3.2 MK) is recorded. In this case, there is no correlation between changes in the average temperature in the corona above the AR and the dynamics of the large-scale electric current. It can be assumed that the additional heating of the corona during flares is caused not only by electric currents, but also by a number of other processes at coronal heights.

(3) On the graph of time variations of the large-scale electric current and average temperature in the corona for the AR NOAA 12494, one can note a decrease in the average temperature in the corona to $< lgT > = 5.7$ (0.5–0.6 MK) with a simultaneous decrease in the value large-scale electric current to zero values. This observation indicates that the mechanism of heating of the coronal matter over the AR by ohmic dissipation of large-scale electric currents is switched off.

In ARs NOAA 12192 and 12371, which have the highest values of the large-scale electric current, the temperature distribution maps in the corona made it possible to detect in times outside the solar flare events for AR NOAA 12192 a localized region with a temperature of 10 MK, and for the AR NOAA 12371 – hot coronal loops with the same temperature (see Fig. 5 in Fursyak 2023). It is likely that such high-temperature structures mark the location of a large-scale electric current channel at coronal heights, and the high temperature in them is due to ohmic dissipation of the electric current. For the AR NOAA 12192, the existence of the electric current with large absolute value at coronal heights is confirmed by the results of numerical simulation performed in 2016 (Jiang *et al.* 2016): Fig. 2 in this article demonstrates the presence of an electric current in the AR, propagating to heights of about 30 Mm and spatially coinciding with the direction of the large-scale electric current assumed by us in (Fursyak *et. al* 2020). Attention should be paid to the characteristic scale of the electric current channel: in the AR NOAA 12192, the assumed channel of large-scale electric current has a width of the order of $\sim (3-4) \times 10^8$ cm, and in the AR NOAA 12371 it is even smaller. According to theoretical calculations performed in (Solov'ev 2022; Solov'ev *et al.* 2022), the electric current dissipation time under such conditions will be about 10^6 s, which can be considered as a quasi-stationary energy release.

6. Conclusions

Our studies carried out during 2018–2023 showed that:

(1) In well-developed ARs with nonzero flare activity, there are large-scale electric currents of the order of 10^{13} A. We have shown that the dynamics of the large-scale electric current and its absolute values are greater in ARs with a higher level of flare productivity. We also found large-scale electric currents in 25–30% of the ARs, which are at the final stage of their evolution. Temporal variations in the magnitude of the large-scale electric current in ARs that are at the final stages of evolution are insignificant, and the absolute values of the large-scale electric current are several times lower than those in developed ARs.

(2) Large-scale electric currents extend to the upper layers of the solar atmosphere in one part of the AR, and close through the chromosphere and corona in the remaining part of the AR.

(3) At the final stages of AR evolution, a nonzero large-scale electric current can have a stabilizing effect on the sunspot, preventing its decay by its own magnetic field.

(4) Large-scale electric currents are involved in coronal heating processes. Ohmic dissipation of a large-scale electric current is one of the mechanisms of quasi-stationary heating of coronal matter above the AR.

(5) In AR NOAA 12192, when plotting temperature distribution maps in the corona in times outside solar flares, a structure with a temperature of more than 10 MK was found, and in AR NOAA 12371, hot coronal loops with the same temperature were found. Probably, such high-temperature structures represent a channel of large-scale electric current at coronal heights.

References

Abramenko, V. I., Gopasyuk, S. I. 1987, *Bulletin of the Crimean Astrophysical Observatory*, 76, 163

Alfven, H., Carlqvist, P. 1967, *Solar Phys.*, 1, 220

Fleishman, G. D., Pevtsov, A. A. 2018, Electric Currents in Geospace and Beyond. Ed. by A. Keiling, O. Marghitu and M. Wheatland, Am. Geophys. Union, 43

Fursyak, Yu. A. 2023, *Astrophysics*, 66, 242

Fursyak, Yu. A., Kutsenko, A. S., Abramenko, V. I. 2020, *Solar Phys.*, 295, id. 19

Fursyak Yu. A., Plotnikov A. A. 2022, *Astrophysics*, 65, 384

Jiang, C., Wu, S. T., Yurchyshyn, V., *et al.* 2016, *ApJ*, 828, id. 62

Lemen, J. R., Title, A. M., Akin, D. J., *et al.* 2012, *Solar Phys.*, 275, 17

Melrose, D. B. 1991, *ApJ*, 381, 306

Sakurai, T. 1982, *Solar Phys.*, 76, 301

Scherrer, P. H., Schou, J., Bush, R. I., *et al.* 2012, *Solar Phys.*, 275, 207

Solov'ev, A. A. 2022, *MNRAS*, 515, 4981

Solov'ev, A. A., Riechokainen, A., Smirnova, V. V., Korol'kova, O. A. 2022, *Geomag. Aeronomy*, 62, 1021

Zaitsev, V. V., Stepanov, A. V., Urpo, S., Pohjolainen, S. 1998, *A&A*, 337, 887

Dynamics of Solar and Stellar Convection Zones and Atmospheres
Proceedings IAU Symposium No. 365, 2024
A. V. Getling & L. L. Kitchatinov, eds.
doi:10.1017/S1743921323004829

Solar activity reconstruction from the Georg Eimmart archive of 1616 – 1720

Nadezhda V. Zolotova[1] **and Mikhail V. Vokhmyanin**[2]

[1]St. Petersburg State University, Universitetskaya nab. 7/9, 199034 St. Petersburg, Russia
email: ned@geo.phys.spbu.ru

[2]Space Climate Group, Space Physics and Astronomy Research Unit, University of Oulu, PO Box 3000, 90014 Oulu, Finland

Abstract. Historical sunspot records provide piece by piece more information on solar variability on a centennial scale. In this work, we analyze sunspot observations from the archives of Georg Christoph Eimmart, which is the second-richest data set of the Maunder minimum after the archives of the Paris observatory. Comparing the dates of the blank solar disk from the database by Hoyt & Schatten (1998) with dates of observations at the Eimmart observatory, we find that spotless days reports originate from astrometric observations. A comparison of the observations by La Hire and Müller of 1719 suggests that the observations by La Hire were for astrometric purposes as well, rather than aimed at sunspot counting.

Keywords. History and philosophy of astronomy, (Sun:) activity, (Sun:) sunspots

1. Introduction

The first-order feature of the Maunder minimum known to the astrophysics community is a blank solar disk during the majority of the observations from 1645 to 1715 (Hoyt & Schatten 1998). Figure 1a shows the fractional distribution of raw daily group counts reported by all the observers during the Maunder minimum. For this period, the collection of group counts (Vaquero *et al.* 2016) currently contains 20541 reports: 89% of them (18198 observations) are spotless. The majority of the observations was carried out at the Paris observatory (Ribes & Nesme–Ribes 1993). Figure 1b shows the fraction of spotless observations reported by Jean Picard, Phillippe de La Hire, Gian Domenico Cassini, and Jacques Cassini between 1653 and 1715. Out of 9492 observations, 93% (8844 observations) are days with no sunspots. Figure 1c analyzes the second-richest archive of the Eimmart observatory, Nuremberg, Germany. Georg Christoph Eimmart, his daughter Maria Clara Eimmart, and her husband Johann Heinrich Müller from 1677 to 1709 reported 2477 spotless days which is 97% out of 2546 observations.

59 volumes of Eimmart's archiver are gathered in Collection 998 of the Manuscript Department of the National Library of Russia, St. Petersburg (MD NLR 1958). Recently, Hayakawa *et al.* (2021a,b) and Vokhmyanin & Zolotova (2023) analyzed this collection and evaluated sunspot and sunspot-group numbers, areas, and positions. Hayakawa *et al.* (2021a) argued that the majority of the observations were not sunspot observations but rather solar-altitude and solar diameter measurements instead.

Complementing these findings, we would like to hypothesize that the observations at the Paris observatory may also have been aimed primarily at astrometric purposes.

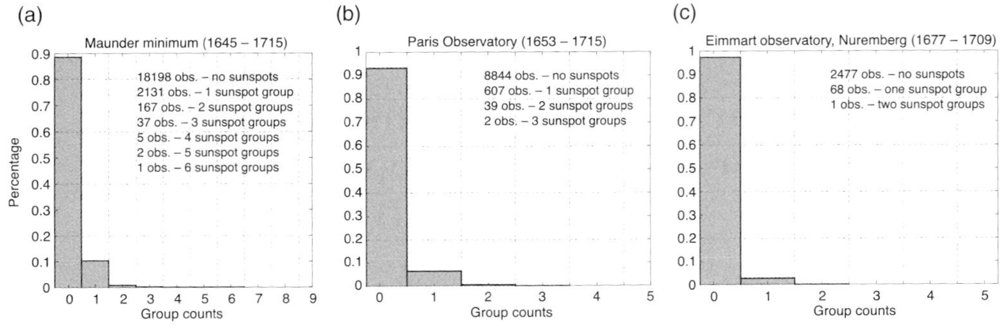

Figure 1. Fractional distribution of daily group counts: (a) all the observers during the Maunder minimum, (b) astronomers of the Paris observatory, (c) those of the Eimmart observatory.

NUMBER OF SUNSPOT GROUPS FOR THE YEAR: 1683
AS OBSERVED BY: EIMMART, G.C., NUREMBERG

Day	Jan	Feb	Mar	Apr	May	Jun	Jul	Aug	Sep	Oct	Nov	Dec
1	-99	-99	-99	-99	-99	0	-99					
2	-99	-99	-99	-99	0	-99	-99					
3	-99	-99	-99	-99	-99	0	-99					
4	-99	-99	-99	-99	-99	0	-99					
5	-99	-99	-99	-99	-99	-99	-99					
6	-99	-99	-99	-99	0	0	-99					
7	-99	-99	-99	-99	0	-99	-99					
8	-99	-99	-99	0	0	-99	-99					
9	-99	-99	-99	0	-99	-99	-99					
10	-99	-99	-99	-99	-99	0	-99					
11	-99	-99	-99	-99	-99	-99	-99					
12	-99	-99	-99	-99	-99	-99	0					
13	-99	-99	0	-99	-99	-99	-99					
14	-99	-99	-99	-99	0	0	-99					
15	-99	-99	-99	0	-99	-99	-99					
16	-99	-99	-99	-99	-99	-99	-99					
17	-99	-99	-99	-99	0	-99	0					
18	-99	0	-99	-99	-99	-99	-99					
19	-99	-99	-99	-99	-99	-99	-99					
20	-99	0	-99	-99	0	-99	0					
21	-99	0	-99	-99	-99	-99	-99					
22	-99	-99	-99	0	-99	-99	-99					
23	-99	-99	-99	0	0	-99	-99					
24	-99	0	-99	-99	-99	-99	-99					
25	-99	-99	-99	-99	-99	-99	0					
26	-99	-99	-99	-99	0	-99	-99					
27	-99	-99	-99	-99	-99	-99	-99					
28	-99	-99	-99	-99	-99	0	-99					
29	-99	-99	-99	0	-99	-99	0					
30	-99	-99	-99	-99	-99	-99	-99					
31	-99	-99	-99	-99	-99	-99	-99					

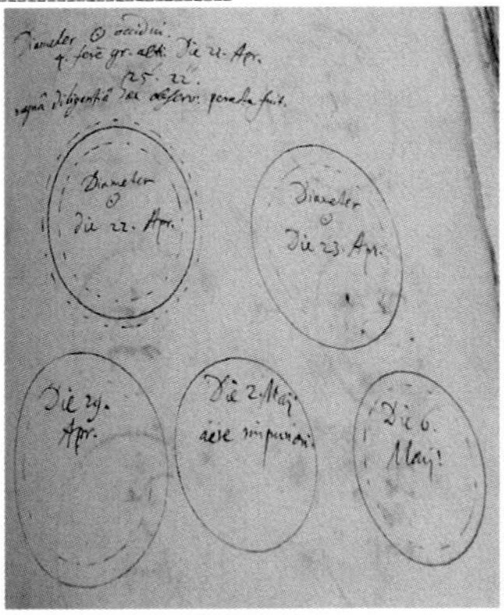

NUMBER OF SUNSPOT GROUPS FOR THE YEAR: 1704
AS OBSERVED BY: EIMMART, M-C., NUREMBERG

Day	Jan	Feb	Mar	Apr	May	Jun	Jul	Aug	Sep	Oct
1	-99	-99	-99	-99						
2	-99	-99	-99	-99						
3	-99	-99	-99	-99						
4	-99	-99	0	-99						
5	-99	-99	0	-99						
6	-99	-99	-99	-99						
7	-99	-99	-99	-99						
8	-99	-99	-99	-99						
9	-99	-99	-99	-99						
10	-99	-99	-99	-99						
11	-99	-99	-99	-99						
12	-99	-99	-99	0						
13	-99	-99	0	-99						
14	-99	-99	0	-99						
15	-99	-99	-99	-99						
16	-99	-99	-99	-99						
17	-99	-99	-99	-99						
18	-99	-99	-99	-99						
19	-99	-99	-99	-99						
20	-99	-99	-99	-99						
21	-99	-99	0	-99						
22	-99	-99	0	-99						
23	-99	-99	0	-99						
24	-99	-99	0	-99						
25	-99	-99	-99	-99						
26	-99	-99	-99	-99						
27	-99	-99	0	-99						
28	-99	-99	-99	-99						
29	-99	-99	0	-99						
30	-99	-99	-99	-99						
31	-99	-99	-99	-99						

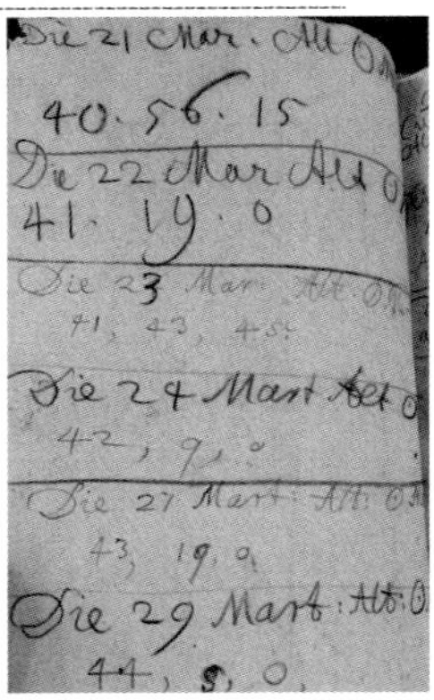

Figure 2. Examples of the annual tables of sunspot-group counts provided by Hoyt & Schatten (1998) and their apparent historical origins (Eimmart Archive Coll. 998, v. 48, f. 160 and v. 35, f. 45).

2. Archival Source of Spotless Days Reports

For the Eimmart observatory, raw daily group counts collected by Hoyt & Schatten (1998) and those from the revised collection by (Vaquero *et al.* 2016) are identical. We compare tables of group counts extracted from observation at the Eimmart observatory in 1677 – 1709 with the original observational reports. Figure 2 provides two examples: solar-diameter and solar-altitude measurements. In tables, red marks days for which we apparently found their source in the original reports. The majority of spotless days originates from solar-altitude measurements, the minor portion — from solar-diameter measurements.

The diameter measurements were carried out at local noon. Drawings are accompanied by date and weather conditions like clear or misty air. From solar-diameter measurements conducted at the Paris observatory, we know that a quadrant was fitted with telescopic sights. A similar technique might have been used at the Eimmart observatory.

Solar-altitude measurements were also made at local noon. For these purpose, gnomon and quadrant were exploited. Apparently, a telescope was not used, since it was not

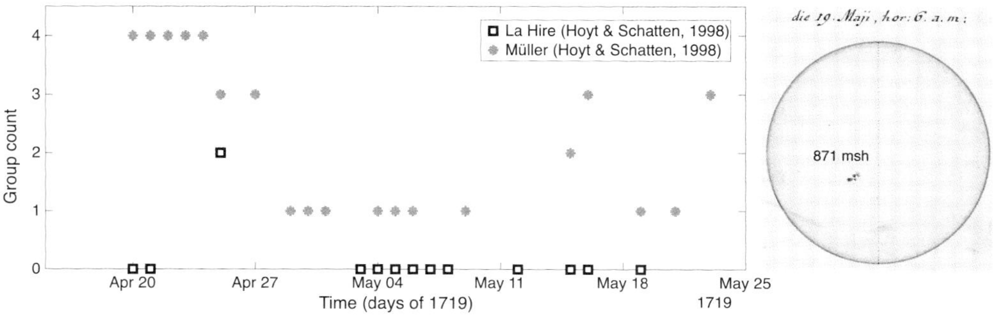

Figure 3. Sunspot-group counts defined by Hoyt & Schatten (1998) from La Hire's and Müller's observations. On the right, an example of Müller's drawing of 19 May 1719 (Eimmart Archive Coll. 998, v. 56) and the area of the sunspot group are reproduced.

mentioned. Therefore, we confirm the conclusion by Hayakawa *et al.* (2021a) that solar-altitude observations recorded without any mention of sunspots can be interpreted as spotless days with caution as rather speculative.

Figure 3 compares sunspot group counts defined from reports by Gabriel–Phillipe de La Hire and Johann Christoph Müller. In the bibliography, Hoyt & Schatten wrote that for La Hire the primary source is a re-examination of original notebooks by Elizabeth Nesme–Ribes. Analyzing the fine structure of a sunspot drawn by La Hire (apparently Phillippe de La Hire), Ribes & Nesme–Ribes (1993) suggested that any sunspot larger than one arcsecond could not be missed with the Parisian telescopes. According to these observations, during the month of 1719, the solar disk was blank, excluding one day. On the other hand, Müller drew sunspots up to 871 msh. The discrepancy in sunspot-group counts reported by La Hire and Müller can be resolved if we hypothesize that these Parisian reports were made for different purposes, not sunspot observations. Similarly to the Eimmart observatory, it might be solar-altitude or solar-diameter measurements.

3. Conclusions

We analyze archives of the Eimmart observatory stored in the National Library of Russia, St. Petersburg. Comparing the dates of the blank solar disk from the database by Hoyt & Schatten (1998) with dates of observations at the Eimmart observatory, we find that these zeros originate mainly from the solar-altitude measurements. These observations apparently were carried out without a telescope. The minor portion of zeroes originates from the solar-diameter measurements, which potentially were performed using a telescope. We confirm the conclusion by Hayakawa *et al.* (2021a) that altitude measurements can be interpreted as spotless with caution as speculative.

We also compared counts of sunspot groups from observations by La Hire and Müller. Due to reports by La Hire being spotless, while Müller drew sunspots, we hypothesize that the Parisian reports may not have been sunspot observations but rather were the solar-altitude or solar-diameter measurements. For the benefit of an open discussion, the processed drawings will be available at `https://geo.phys.spbu.ru/~ned/History.html`.

4. Acknowledgments

This work was supported Saint–Petersburg State University, project ID: 104255084.

References

Hayakawa, H., Kuroyanagi, C., Carrasco, V. M. S., *et al.* 2021a, *Astrophys. J.*, 909, 166
Hayakawa, H., Kuroyanagi, C., Carrasco, V. M. S., *et al.* 2021b, *Solar Phys.*, 296, 154
Hoyt, D.V., Schatten, H. 1998, *Solar Phys.*, 179, 189

MD NLR 1958, Inventory of the archive of the German mathematician, astronomer, artist, engraver and mechanic Georg Christoph Eimmart (16381705), Fond 998 of the Manuscript Department of the National Library of Russia

Ribes, J.C., Nesme–Ribes, E. 1993, *Astronomy Astrophys.*, 276, 549

Vaquero, J. M., Svalgaard, L., Carrasco, V. M. S., *et al.* 2016, *Solar Phys.*, 291, 3061

Vokhmyanin, M., Zolotova, N. 2023, *Solar Phys.*, 298, 113

Dynamics of Solar and Stellar Convection Zones and Atmospheres
Proceedings IAU Symposium No. 365, 2024
A. V. Getling & L. L. Kitchatinov, eds.
doi:10.1017/S1743921324001212

Topological Features in the Emerging Solar Magnetic Flux

Ilan Roth

Space Sciences Laboratory, University of California, Berkeley, CA 94720
email: ilanr@berkeley.edu

Abstract. The formation of highly structured, spatially localized complex structures during solar flux emergence facilitates adaptation of topological methods, extending the research of emerging macroscopic MHD fluxes into knots, links and braids. Combining mathematical considerations, remote images and in situ satellite observations at solar vicinity, we construct new characteristics of those braided/knotted magnetic structures, applying Braid and Knot Theory to physical configurations, deducing their topological invariants, constraining the evolution and stability while delineating the relaxation path to magnetized equilibria.

Keywords. solar corona, magnetic braids, emerging knotted magnetized structures

1. Introduction

The directly observed magnetic solar activity is a product of processes starting deep in the solar convection zone with the ensuing flux emergence into the overlying solar atmosphere. Magnetic flux bundle undergoes a lengthy evolution: (a) dynamo generation and growth, (b) buoyant ascension through a broad range of external thermodynamic parameters and (c) transport through interaction with the surrounding ambient magnetized plasma, resulting in major structural modifications Cheung (2014). Solar observations reveal filamentary nature of the emerging magnetic fields with highly structured three-dimensional magnetic loops in the overlying solar corona. Various simulations indicate that the survival of a coherent flux structure in the presence of the destructive effects of convection may require the progenitor flux tube to form twisted configuration, resulting in several observable features, like sigmoidal field lines Fan (2009). Similar features may result due to photospheric shearing flows along the polarity inversion line Leka et al. (1996). However, small-scale emerging regions close to the photosphere appear as tangled bundles of braided magnetic fields with a writhe or winding MacTaggart and Prior (2021). These indications facilitate implementation of Braid and Knot Theory allowing us, together with remote images and *in situ* satellite observations at solar vicinity to construct topological characteristics of magnetic structures and predict their dynamics as they emerge into solar atmosphere, evolve in the corona before being ejected into the solar wind.

2. Magnetic Braids and Knots

Solar braids were coined as such due to the inferred geometry of the observed solar emissions, indicating adjacent strands passing over or under, consistent with a 2-D projection of a braid Berger and Asgari–Targhi (2009). The braiding of the field lines is due to the ubiquitous small-scale motion at the photospheric leg(s) of the emerging magnetic field and interaction with the magnetized coronal plasma.

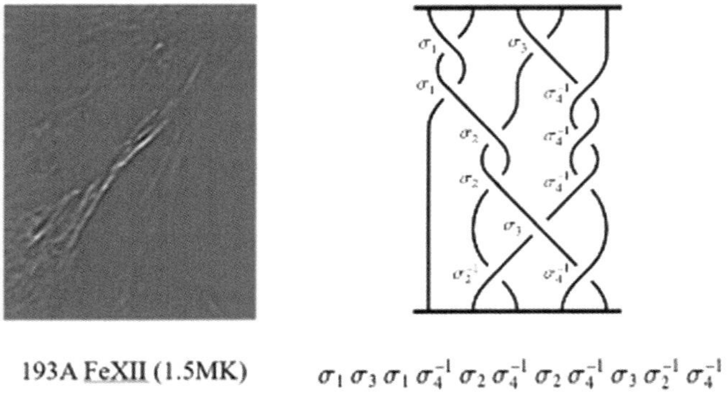

(a) (b)

Figure 1. (a) Small-scale magnetic braids observed by Hi-C rocket at 193Å Cirtain et al. (2013). (b) An example of Artin braid with conversion of braid structure into algebraic expression. Notation from top to bottom: the i-th strand passing from above over the $(i-1)$-th strand defines the crossing generator σ_i, while passing below it forms σ_i^{-1}.

Following observations of a small section of the coronal field lines Cirtain et al. (2013), magnetic braids may be depicted as a disjoint collection of crossings among a number of strands attached to foot-points at two photospheric planes ("top" and "bottom"), with one-to-one correspondence between the foot-points (Figure 1a) forming a small scale braid-like topology Thalmann et al. (2014), similarly to the Artin braids Artin (1947) - set of n disjoint strings in 3-space attached to two (horizontal) bars at the top and at the bottom (Figure 1b). The observed loops are undergoing moves preserving their topological equivalence as described by Artin braids group operations, which may be represented through algebraic expressions, while new loops light up, fade out or connect into more extended compact structures Thalmann et al. (2014). Braids and knots and their physical analogues at the solar corona form topologically related entities. Mathematical knots and field lines of magnetized plasma in the MHD approximation are described as closed loops or bundles in three-dimensional (3D) space, transformed dynamically via continuous deformation of 3D upon itself, pushed smoothly in the surrounding viscous (plasma) fluid, respectively, without self-intersection. The physical "frozen in" condition for magnetic knots is equivalent to the knot "ambient isotopy" controlled by the three Reidemeister diagrammatic moves (e.g. Adams (2004)), as shown on Figure 2: R_1 - twist (green), R_2 - poke (orange) and R_3 - slide (yellow or blue). R_j moves reduce then the complicated topological problem to a simpler diagrammatic one, relating the changes in the observed projection to the relations between the crossings in the magnetic configuration. The invariance under R_j assures that any quantity which characterizes the magnetic field knot must preserve its value while undergoing the R_j transformations, becoming a topological invariant which assigns uniqueness to each knotted magnetic configuration. Non-equivalent magnetized knots are characterized by explicit helicity/winding invariants in the form of various polynomials (e.g., Alexander, Conway, Jones, Kauffman).

3. Closure of Braids

An important relation between braids and knots/links was given through the Alexander Alexander (1928) theorem, stating that any knot may be obtained from a closure of some braid, joining orderly through simple arcs the endpoints at the upper loose ends of a braid with the lower ones. Conversely, there exist many braids which through closure become the same knot - the mapping is surjection, not injection.

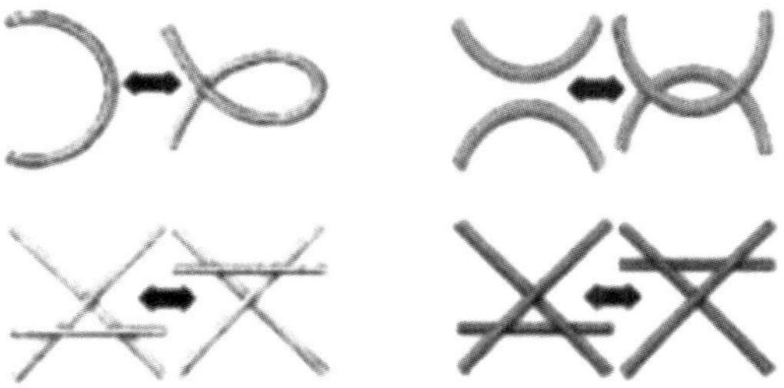

Figure 2. Reidemeister moves: R_j, $j=1$, green; $j=2$, pink; $j=3$, yellow/blue.

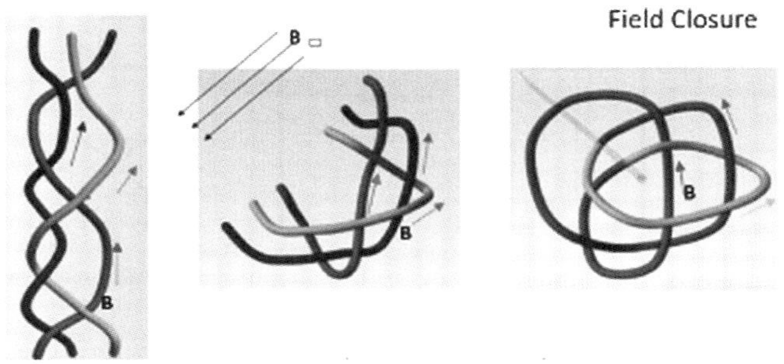

Figure 3. (a) Initial braid state. The arrows indicate the braid field polarity. (b) Skewed braid interacting with reverse polarity fields (black arrows). (c) Morphing of a braid into a knot.

Magnetic braids constitute a perturbed system in marginal equilibrium, which exhibits natural fluctuations due to interaction with the ambient field, resulting in reconfiguration of field elements with emission of excess energy in the form of fluid motion and thermal heating. In the rectilinear, non-braided configuration two oppositely directed magnetic field regions are unstable to tearing mode instability. Particle simulations with an anti-parallel, stretched magnetic field configuration like modified Harris model showed explicitly that in a nonlinear phase of tearing mode growth the fast reconnection forms a narrow x-line current layer together with 2D magnetic islands (e.g. Che et al. (2011)). The dynamic contraction of these islands may result in Fermi-like energization of electrons, which emit then the observed X-rays and gyro-synchrotron radiation in flares Drake et al. (2006). In contrast to Harris model, braided magnetic bundles in 3D may follow a different path due to their considerably sheared fields. Magnetic fields in rectilinear configuration in the presence of opposite polarity fields are reconnecting adjacent fields into closed strings of magnetic islands, while a sheared bundle of braided fields (Figure 3a), deformed into more skewed form (Figure 3b), may connect its successive strands through field closure into a new closed structure (Figure 3c), i.e. magnetic knot. Similarly to mathematical knots, these finite width strands, having topological invariants enhancing their stability, which have not been analyzed yet, become the new building blocks of heliospheric magnetic structures, which may be observed through high resolution radiation.

4. Dynamics of the Coronal Magnetic Braids

When a closure of two different braids results in the same knot, magnetic deformation leading from one braid to another describes reconfigurations preserving the knot invariant. This equivalence allows us to predict several of stable braid eigenmodes, which

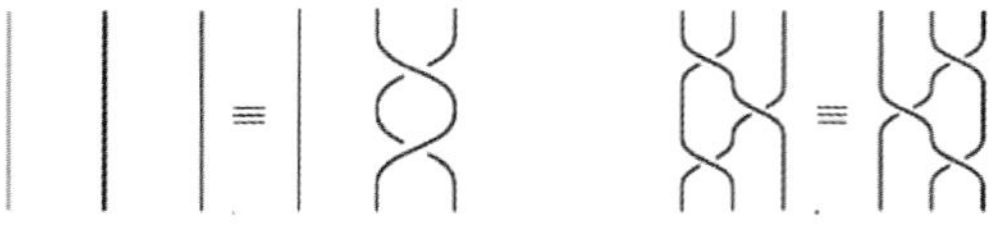

(a) $s_i s_{i-1} = 1$ (b) $s_i s_{i+1} s_i = s_{i+1} s_i s_{i+1}$

Figure 4. Operators of Artin group.

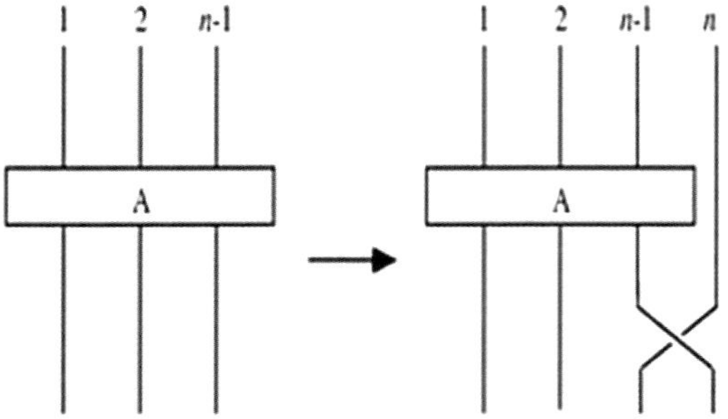

Figure 5. (Second) Markov stabilization (interchange) move.

should be seen in high resolution observations. Denoting the basic braid generators as $s_i = \sigma_i$, the natural candidates include the operations of the Arten group B_n:

(a) $s_i s_{i-1} = 1$. In this move, under local E×B Alfvenic pulse two adjacent magnetic strands interchange positions in parallel to the R_2 knot move (Figure 4a) when they are exposed to different cross forces. The braid preserves its topological characteristics while its word acquires an observable $s_i s_i^{-1}$ modification.

b) $s_i s_{i+1} s_i = s_{i+1} s_i s_{i+1}$. This move describes active magnetic strands anchored at different foot-points while one of them fluctuates with respect to its neighbors, imitating the R_3 knot move (Figure 4b). The braid changes its word to an equivalent one. Both Arten moves, preserving the topological structure of the braid, are predicted as the most prominent fluctuations in high resolution observations.

Another important convergence of multiple (infinite) braids into the same knot is based on Markov theorem Markov (1935), stating that two braids under closure become the same knot if they are related by a sequence of Markov moves. When a single active magnetic field line reconnects with n-th string of a braid A adding a new s_{n+1} crossing string, the resulting knot of $n+1$ strings formed under closure is isotopic to the closure of A - the new braid converges to the same knot under a closure (Figure 5). The addition of string generator s to A may be repeated many times, forming new braids which do not change the equivalent knot structure. Similar result applies to reconfiguration adding multiple times the generator $(s_{n+1})^{-1}$.

As a corollary of Markov move Theorem, a sequence of these operations which modify significantly the structure of magnetic braid due to coronal heating or emergence of new active field lines, form additional stable modifications of the braid due to closure into identical knot. In contrast, any large braid moves which do not satisfy these conditions are candidates for the configuration destruction.

5. Knot Sum

The fast solar wind is accelerated largely by the local coronal pressure gradients and Poynting vector due to interchange reconnection of new emerging loops at the base of the corona with open magnetic field (e.g. Fisk (1996); Fisk et al. (1999)). Since the braided reconnection events occur at clustered adjacent sites, the magnetic output into the solar

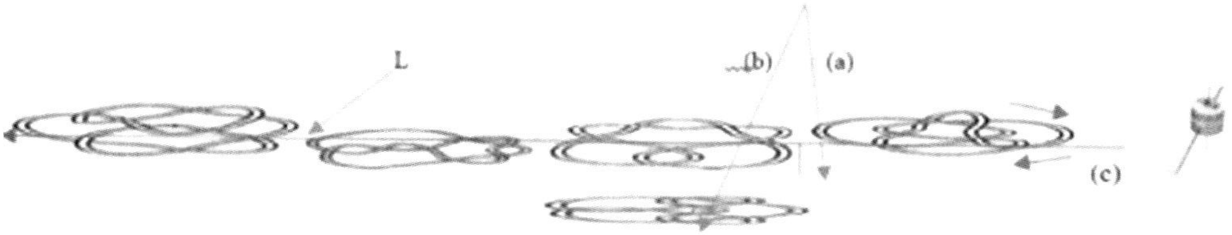

Figure 6. Example of knot sum of different prime knots embedded in the solar wind: red arrows signify the field with reversals; blue lines (a, b, c) describe various encounters.

wind consists of a sequence of propagating knots, analogously to gun discharged bullets or magnetic jetlets. The ultimate structure encountering the inner heliosphere satellites is determined by a topological procedure of mathematical knots: two magnetic knots moving along close trajectories interact when the front of the trailing knot is merging with the tail of the leading knot. This operation - Knot Sum must happen numerous times between two adjacent knots, resulting in formation of elongated composite knots. The resulting switchbacks are the observed magnetic structures ejected into the solar wind (Figure 6).

6. Summary

Application of topological concepts into the flux emerging processes allows us to characterize the structure of solar emerging configurations as braids, which through 3D interaction with an ambient magnetized plasma form robust magnetic knots whose stability is enhanced through their invariants. We then predict the observable eigenmodes of the coronal braids as Artin Group operations and stable coronal heating modifications due to Markov moves. The emerged switchbacks in the solar wind result due to Knot Sum reconnection of the propagating knots. Similar procedure may operate in various astrophysical features, like Herbig–Haro objects in the protostar outflows.

Acknowledgment

The research was carried under NASA contract NNN06AA01C.

References

Adams, C. C. 2004,. *The knot book.*
Alexander, J. W. 1928, *Transactions of the American Mathematical Society*, 30(2), 275–306.
Artin, E. 1947, Theory of braids. *Ann. of Math*, 4, 101–126.
Berger, M. A. & Asgari–Targhi, M. 2009, *The Astrophysical Journal*, 705(1), 347.
Che, H., Drake, J. F., & Swisdak, M. 2011, *Nature*, 474(7350), 184–187.
Cheung, M. C. M. & Isobe, H. 2014, *Flux emergence (theory). Living Rev. Sol. Phys.*, 11, 3.
Cirtain, J. W., Golub, L., Winebarger, A., De Pontieu, B., Kobayashi, K., Moore, R., Walsh, R. W., Korreck, K., Weber, M., McCauley, P., *et al.* 2013, *Nature*, 493(7433), 501–503.
Drake, J. F., Swisdak, M., Che, H., & Shay, M. 2006, *Nature*, 443(7111), 553–556.
Fan, Y. 2009, *Magnetic fields in the solar convection zone. Living Rev. Sol. Phys.*, 6, 4.
Fisk, L. A. 1996, *Journal of Geophysical Research: Space Physics*, 101(A7), 15547–15553.
Fisk, L. A., Schwadron, N., & Zurbuchen, T. 1999, *Journal of Geophysical Research: Space Physics*, 104(A9), 19765–19772.
Leka, K., Canfield, R., McClymont, A., & van Driel–Gesztelyi, L. 1996, *The Astrophysical Journal*, 462, 547.
MacTaggart, D. & Prior, C. In *Journal of Physics: Conference Series* 2021,, volume 1730, 012013.
Markov, A. A. 1935, *Recueil Math. Moscou.*, 1, 73–78.
Thalmann, J. K., Tiwari, S. K., & Wiegelmann, T. 2014, *The Astrophysical Journal*, 780(1), 102.

Dynamics of Solar and Stellar Convection Zones and Atmospheres
Proceedings IAU Symposium No. 365, 2024
A. V. Getling & L. L. Kitchatinov, eds.
doi:10.1017/S1743921323005252

Levels of stabilization of velocity and magnetic induction in the convective zone of the Sun

S. V. Starchenko

IZMIRAN, Kaluzhskoe Hwy 4, Troitsk, Moscow, 108840 Russia
email: sstarchenko@mail.ru

Abstract. The induction and momentum equations of solar dynamo are simplified to a dynamic system for the convective Root-Mean-Square (rms) velocity and the rms magnetic field in the solar convection zone. The study of stable stationary points of this system gives a minor excess of the critical level of the dynamo and, accordingly, moderate magnetic field typically about 1 T (10 kG). A significantly lower rms magnetic field may be possible at some parameters of the system. The stable rms velocity is about 100 m/sec, and the characteristic magnetic times are about the half-period of solar rotation or about an average lifetime of sunspots. Relative magnetic energy is of order 5 kJ/kg that is about the kinetic energy. The unstable stationary points could be near zero magnetic fields as in periods of very lower solar activity similar to the Maunder minimum.

Keywords. convective velocity, convective zone, diffusion times, levels of stabilization, magnetic field value, solar dynamo

1. Introduction

The complete system of solar magnetic dynamo equations [Gilman & Glatzmaier (1981), Lantz & Fan (1999)] is extremely complex, since it includes the induction equation for the magnetic field vector, the momentum equation for the velocity vector field of a conducting fluid, equation for entropy and equations of state. From the first decade of this century [Ghizaru *et al.* (2010)] to the present [Guerrero *et al.* (2019), Hotta *et al.* (2022)] some numerical models had been created that imitate the solar magnetic cycle and differential rotation, starting from an almost complete system. However, the key parameters (and, above all, the transport coefficients) of all such models differ by many orders of magnitude from their true values, and extremely far extrapolations to real values have to be made. Therefore, simplified models are very relevant, among which the mean field models seem to be the most reliable [Krause & Redler (1980), Moffatt & Dormy (2019), Charbonneau & Sokoloff (2023)]. But even they remain difficult enough for direct analysis and, at the same time, rely on unproven hypotheses.

The purpose of this work is to create a reliable dynamic system for the rms velocity and rms magnetic field in the interior of the solar or similar stellar convective zone with an active MHD dynamo based on first principles and on the dynamo integral equations. In this paper, I perform stability/instability analysis near the stationary points of this original system for the Sun. Many other potentially much more important manifestations of the obtained system for the Sun and other stars are supposed to be devoted to subsequent works.

In the next section, the required system of equations is derived based on the magnetic induction equation and the momentum equation. The energy and other equations of the dynamo are simply approximated by the fact that the integral power of the work of the Archimedes buoyancy force is given as, generally speaking, a function of time. The momentum equation is multiplied scalarly by the velocity vector and this product is integrated throughout the solar convection zone (CZ). This gives an evolutionary equation for the total kinetic energy, which, after dividing by the mass of CZ, sets the desired equation for the rms velocity, provided that each integral term appears to be proportional to the corresponding powers of the velocity and magnetic field. In a similar way, I multiply the induction equation by the magnetic field vector and integrate the product over the entire volume of CZ. This gives an evolutionary equation for the total magnetic energy, which, after dividing by the volume of CZ, sets the desired equation for the rms magnetic field, provided that each integral term appears to be proportional to the corresponding powers of the velocity and magnetic field.

The third, main section of this work is devoted to stable and unstable stationary points of the resulting simplified system for convective velocity and magnetic field. The stationary points themselves are defined in a standard way - as places where derivatives vanish. To assess the stability of these points, corresponding linear systems are constructed in their small neighborhood. Solutions of such systems that stabilize near these points are naturally considered stable, while those moving away (exponentially in time) from these points are considered unstable. Accordingly, we consider the location of the dynamo near stable points to be a more probable behavior. And placing dynamo near unstable points is less probable behavior. For dynamo, various estimates of physical quantities resulting from derived here equations are given. They are in agreement with known modern numerical, theoretical, and observational models of the magnetic and velocity fields. At the same time, new original relations were obtained.

A section with results is absent because this work intended to be short. One could find the major results in the abstract. For the same reason of the space shortage, I did not present proofs of stability and instability of the different stationary points here.

2. Derivation of equations

Here, I further derive the simplest dynamical system for the rms convective velocity and the rms magnetic field.

To obtain the equation for the evolution of the rms velocity $u(t)$, I integrate over the entire volume of CZ the scalar product of the velocity vector $\mathbf{U}$ and the momentum equation and, as a result of identical transformations and neglect of small terms (for details, see, for example, [Gilman & Glatzmaier (1981), Priest (2014), Starchenko (2019)]) I get

$$\frac{d}{dt}\left(\int_{r_i}^{r_o} \rho \frac{U^2}{2} dV\right) = \int_{r_i}^{r_o} \left(\rho \mathbf{A}\cdot\mathbf{U} - \frac{\mathbf{U}\times\mathbf{B}\cdot\nabla\times\mathbf{B}}{\mu_0} - \rho\nu|\nabla\times\mathbf{U}|^2\right) dV. \quad (1)$$

Here, within the integrals, r_i and r_o denote the radius of the inner and outer boundary of CZ, respectively, ρ is the average density, and ν is the coefficient of kinematic viscosity. I would especially like to note $\mathbf{A}$ - the acceleration caused by the buoyancy force of Archimedes. It generates convection, which, in turn, generates a magnetic field with vector $\mathbf{B}$.

Let us divide (1) by the mass of CZ M. We obtain on the left side of (1) directly by definition udu/dt. The first term one on the right is the specific integral power of the Archimedes buoyancy force a, which we consider being a given function of time. Next comes the specific magnetic force (or more precisely, power) of Lorentz, which, based on the degrees of velocity and magnetic field included in it, can naturally be estimated as $ub^2/(L\rho\mu_0)$. Here L is the characteristic external scale, divided, based on the corresponding vector product, by the typical sine of the angle between the velocity and magnetic field vectors. The end of (1)/M is the specific integral power of the diffusion force, which can naturally be estimated as u^2/T_u. Thus, within the framework of the developed approach, integrals are represented through their components, to which they are directly proportional. Time T_u is the diffusion time and b is the rms magnetic field. As a result, we obtain the evolutionary equation for velocity

$$udu/dt = a - ub^2/(L\rho\mu_0) - u^2/T_u. \tag{2}$$

In a similar way, I integrate over the volume the scalar product of the magnetic field vector and the induction equation and obtain (σ - conductivity):

$$\frac{d}{dt}\left(\int_{r_i}^{r_o} \frac{B^2}{2\mu_0} dV\right) = \int_{r_i}^{r_o} \left(\frac{\mathbf{U}\times\mathbf{B}\cdot\nabla\times\mathbf{B}}{\mu_0} - \frac{|\nabla\times\mathbf{B}|^2}{\sigma\mu_0^2}\right) dV. \tag{3}$$

Now dividing expression (3) by the volume of CZ and based on the considerations presented above, we finally obtain the evolutionary equation for the magnetic field (T_b is the diffusion time for the magnetic field):

$$db/dt = ub/L - b/T_b. \tag{4}$$

This equation (4) together with equation (2) constitutes the desired system.

3. Stationary points

The stationary points of the system obtained above from (2) and (4) are found by zeroing the time derivatives and then solving the corresponding algebraic systems. In this case, all the parameters considered below can, generally speaking, be time dependent, but for the order of magnitude estimates given in this section, it is natural to consider them constant.

Let's start with the stable stationary points of the system (2, 4), corresponding to a non-zero magnetic field (index - S):

$$u_S = L/T_b, \tag{5}$$

$$b_S = \pm\{\rho\mu_0[T_{ba} - L^2/(T_uT_b)]\}^{1/2}. \tag{6}$$

With a typical theoretical and practical value of the stationary velocity $u_S = 100$ m/sec [Gilman & Glatzmaier (1981), Hotta *et al.* (2022), Moffatt & Dormy (2019), Priest (2014)] and diffusion time $T_b = 1$ Msec [Moffatt & Dormy (2019), Priest (2014), Forgacs-Dajka *et al.* (2021), Fan (2021)], we have $L = 100$ Mm, which is an order of magnitude of the CZ half-depth, indicating a corresponding insignificant excess of the critical dynamo level. With the given power about $a = 10$ mW/kg (ratio of the well-known solar luminosity to mass of CZ), one obtains that for the generation, or rather, for the very existence of a stationary magnetic field, it is necessary, following from (6) and the above estimates, that the threshold condition be met

$$T_u > L^2/(T_b^2 a) = 1\text{Msec}. \tag{7}$$

The last time interval is approximately a half-month which is about an average solar spots lifetime or half-period of the solar rotation rate [Moffatt & Dormy (2019), Priest (2014), Forgacs-Dajka *et al.* (2021), Fan (2021)]. The corresponding kinetic scale or convective cell size is $u_S(1\text{Msec}) = L$. This is about the known estimates like u_ST_u for such a giant cell scale.

If condition (7) is satisfied with a margin, then the rms field b_S from (6) is about 1 T (10 kG), which corresponds to the largest magnetic field value possible in this system.

The relative magnetic energy $b^2/2\rho\mu_0$ is as well the largest about 5 kJ/kg, which is close to the relative kinetic energy value $u^2/2$. If condition (7) is satisfied almost without a margin, then the rms magnetic field could be much lower than 1 T. This could be actually the way to the following unstable stationary points.

Let's finally study unstable stationary points approaching zero magnetic fields:

$$B_0 = 0, u_0 = \pm(T_u a)^{1/2}. \tag{8}$$

Those unstable stationary points may be associated with periods of very lower solar activity similar to the Maunder minimum, and possibly (although unlikely) with an extremely rare catastrophic zeroing of the magnetic field that may produce as well catastrophic super-flare.

References

Charbonneau P., & Sokoloff D. 2023, Space Science Reviews, 219, 35

Fan Y. 2021, Living Reviews in Solar Physics, 18, 5

Forgacs-Dajka E., Dobos L., Ballai I. 2021, A&A, 653, A50

Gilman P.A., & Glatzmaier G.A. 1981, Astrophys. J. Suppl. Ser., 45, 335

Ghizaru M., Charbonneau P., Smolarkiewicz P.K. 2010, Astrophys. J. Lett., 715, L133

Guerrero G., Zaire B., Smolarkiewicz P.K., de Gouveia Dal Pino E.M., Kosovichev A.G., Mansour N.N. 2019, Astrophys. J., 880, 6

Hotta H., Kusano K., Shimada R. 2022, Astrophys. J., 933, 199

Krause F., & Redler K-H. 1980, Pergamon Press, Oxford

Lantz S.R., & Fan Y. 1999, Astrophys. J., 121, 247

Moffatt K.H., & Dormy E. 2019, Cambridge University Press, Cambridge

Priest E. 2014, Cambridge University Press, Cambridge

Starchenko S.V. 2019, Geoph. Astroph. Fluid Dyn., 113, 71

Dynamics of Solar and Stellar Convection Zones and Atmospheres
Proceedings IAU Symposium No. 365, 2024
A. V. Getling & L. L. Kitchatinov, eds.
doi:10.1017/S1743921323005392

Chromospheric activities of pre-main-sequence stars

Mai Yamashita[1], Yoichi Itoh[1], Yuhei Takagi[2] and Yumiko Oasa[3]

[1]Nishi–Harima Astronomical Observatory, Center for Astronomy, University of Hyogo, 407-2 Nishigaichi, Sayo, Sayo, Hyogo, Japan
email: yamashita@nhao.jp

[2]Subaru Telescope, National Astoronomical Observatory of Japan, 650 North A'ohoku Place, Hilo, HI 96720, U.S.A.

[3]Faculty of Education, Saitama University, 255 Shimo–Okubo, Sakura, Saitama, Saitama, Japan

Abstract. We investigated chromospheric activities of pre-main-sequence (PMS) stars. First, we studied the Ca II infrared triplet emission lines with Subaru/HDS and other spectroscopic instruments. Most PMS stars have narrow Ca II lines whose intensities are as large as the maximum of the zero-age main-sequence (ZAMS) stars. The chromosphere of PMS stars is suggested to be filled by the Ca II emitting region. Second, we found many faint chromospheric emission lines such as Mg I and Fe I for more than half of the ZAMS stars. Third, we searched the periodic light variation caused by a starspot for the 26 PMS stars. Their *TESS* light variations and Ca II emission line strengths show the positive correlation, and are located on the extensions of the superflare stars. In summary, PMS stars have very active chromosphere driven by strong dynamo process due to the fast rotation and the long convection timescale.

Keywords. stars: pre-main sequence, stars: chromospheres, stars: activity, techniques: spectroscopic, techniques: photometric

1. Introduction

Chromosphere is the region between photosphere and corona. The temperature of the chromosphere gradually increases with radial distance to the photosphere. In an active chromospheric region, atoms emit permitted lines such as Hα and Ca II. It is claimed that chromospheric activity is driven by the magnetic field, which is generated by the dynamo process.

For main-sequence stars, chromospheric activity is often discussed in relation to stellar rotation. It is claimed that chromospheric activity is driven by the magnetic field, which is generated by the dynamo process. Wilson (1978) showed that chromospheric activity analogous to that associated with the solar magnetic activity cycle is ubiquitous in stars along the lower main-sequence. As mentioned in Noyes et al. (1984), rotation is not the only parameter related to the dynamo process, but the stellar mass, spectral type, depth of convection zone, and convective turnover time (τ_c) are also related to the process. Rossby number (N_R) involves these five parameters, which is defined as

$$N_R \equiv \frac{\text{rotational period } P}{\text{convective turnover time } \tau_c}, \qquad (1)$$

where P is the stellar rotational period. According to standard dynamo theory in Parker (1955), the magnetic activity at the stellar surface is getting stronger with increasing rotation speed and strength of differential rotation.

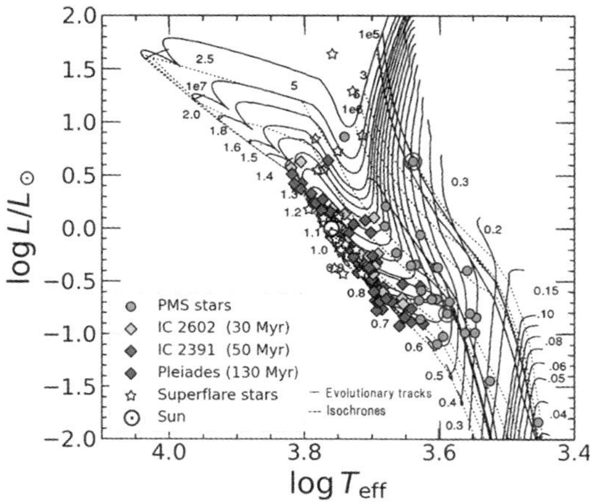

Figure 1. HR diagram of our targets. The circles, diamonds, star symbols, and odot represent the PMS stars, ZAMS stars in the three open clusters, superflare stars, and the Sun.

Marsden et al. (2009) investigated Ca II infrared triplet (IRT; $\lambda 8498, 8542, 8662$ Å) emission lines of ZAMS stars in young open clusters, IC 2391 and IC 2602. The objects having smaller the Rossby number tend to show stronger emission lines. For the stars with $\log N_{\rm R} \geq -0.8$, the ratio of the surface flux of the Ca II IRT emission line to the stellar bolometric luminosity, R' decreases with increasing $N_{\rm R}$ (the unsaturated regime). For the stars with $\log N_{\rm R} \leq -0.8$, $R'_{\rm IRT}$ is constant (the saturated regime).

Because PMS stars are fast rotators (Gallet & Bouvier 2013), they are expected to have high dynamo activities. We suppose that the low-mass PMS stars are fully-convective and have the large-scale starspots.

2. Observations and Data reduction

We investigated the Ca II IRT for 60 PMS stars (Fig. 1; $10^5 - 10^8$ yr, $0.035 - 2.5\,{\rm M}_\odot$), being associated with four molecular clouds or five moving groups. The medium- and high-resolution spectroscopy were conducted with 2 m Nayuta Telescope/MALLS at Nishi–Harima Astronomical Observatory and Subaru/HDS. Archival data obtained with the Keck/HIRES, VLT/UVES, and VLT/X-Shooter were also used. A detailed description of the data reduction methods used here is presented in Yamashita et al. (2020).

We also investigated the Mg I line of F, K, G type 52 ZAMS stars by using 3.9 m Anglo-Australian Telescope (AAT)/the University College London Echelle Spectrograph (UCLES) archive spectra. The PI is S. C. Marsden, date of the observations are 2000-03-17, 18, 19, 2001-01-06, 07, 08, and 2001-02-11, 12. The wavelength coverage was between 3522 Å and 9386 Å. The integration time for each object was between 300 s and 1200 s. A detailed description of the observations is presented in Marsden et al. (2009), and that description of the data reduction methods used here is presented in Yamashita et al. (2022a). The Mg I emission component is always buried by the photospheric absorption as shown in Fig. 2.

In this study, inactive stars with a spectral type similar to that of the target were used as template stars. For the correction of the rotational broadening, the spectra of the template stars were convolved with a Gaussian kernel to match the width of the photospheric absorption lines of each object. By using high S/N spectra of inactive field stars, we carefully subtracted the photospheric absorption component. We considered that the subtraction was reliable if nearby faint photospheric absorption lines, such as

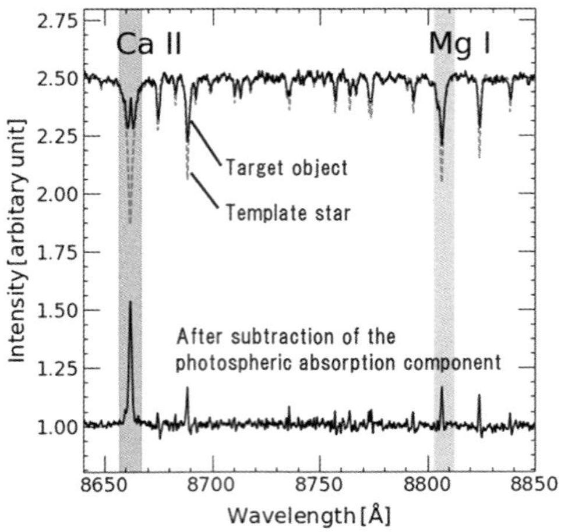

Figure 2. The procedures of the spectral subtaction of the photospheric component for the PMS star, RECX 11

Ni I, Mn I, V I, and Ti I, disappeared in the spectrum after the subtraction. Finally, we measured the equivalent widths (EQW) of the Mg I emission line. Before measuring the EQWs, the continuum component of the spectra was added to unity. To obtain the EQWs of the Mg I emission lines, the area of the emission profile was directly integrated. The typical EQW errors were estimated by multiplying the standard deviation of the continuum by the wavelength range of the Mg I emission line of the five ZAMS stars.

The PMS stars and ZAMS stars were also observed by *TESS*. We retrieved the Presearch Data Conditioning Simple Aperture Photometry (PDCSAP) fluxes for each PMS star from the Multimission Archive at the Space Telescope Science Institute. We searched periodicity of the light curve then obtained its amplitude. For ZAMS stars, we conducted principal component analysis using ELEANOR, an open-source tool for extracting light curves from *TESS* full-frame images (Feinstein et al. 2019). We calculated the standard deviation of the flux, σ_1, and removed the flux data points greater than $3\sigma_1$ above the average. We searched for periodic signals by conducting Lomb–Scargle (Scargle 1982) periodogram analysis. For each object, the period of the light curve was determined. The amplitudes of the light curves were calculated by taking the 90th percentile flux and subtracting it from the 10th percentile flux. The detail is described in Yamashita et al. (2022a).

3. Result

Seven PMS stars have broad emission lines of Ca II $\lambda8498$ Å (FWHM $> 100\,\mathrm{km \cdot s^{-1}}$), while most PMS stars exhibit narrow emission lines (FWHM $\leq 100\,\mathrm{km \cdot s^{-1}}$). The emission lines of DG Tau, DL Tau, and DR Tau are broad and strong (EQW ~ 50 Å), while those of RY Tau, SU Aur, RECX 15, and RX J1147.7-7842 are broad but not strong (EQW < 10 Å). All EQWs of the narrow emission lines are weaker than 5 Å.

Before subtracting the photospheric absorption component, the Mg I emission component is always buried by the photospheric absorption. As a result of the data reduction, the Mg I line is detected as an emission line in 45 ZAMS stars. The Mg I line of these ZAMS stars shows narrow emission, indicative of chromospheric origin. Their EQWs range from 0.02 Å to 0.63 Å. The Mg I emission lines of 33 ZAMS stars have EQWs less than 0.1 Å. 5 ZAMS stars do not show the Mg I line in emission or absorption, but like simply continuum component.

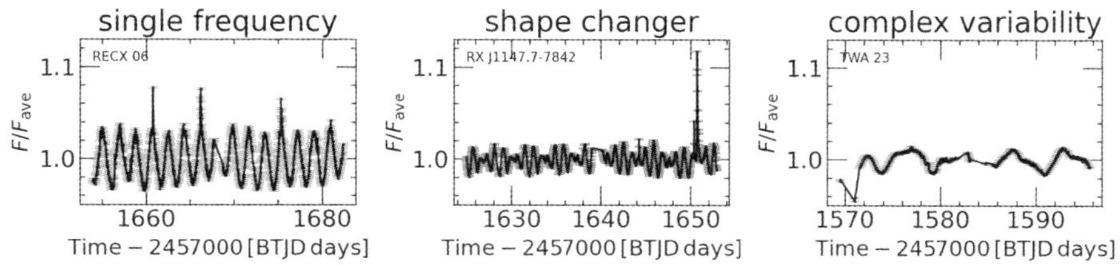

Figure 3. Three examples of the *TESS* light curves of PMS stars. The three features of the light curve, a) single frequency, b) possible shape changer, and c) complex variability, are presented.

For 26 PMS stars, the amplitudes of the light curves are $0.001 - 0.552$ mag. Fig. 3 shows three examples of light curves of the PMS stars. For the 26 PMS stars and 33 ZAMS stars in IC 2391 and IC 2602, we categorized the light curves into the following groups:

a) Single frequency: the star has a single period. A single spot or spot group is interpreted as rotating into and out of view.
b) Possible shape changer: subcategory of type a) single frequency. The star usually has a single period. Shape of the light curve may change in the observational period.
c) Complex variability; the light curve has periodicity but seems complex. It is considered that the star has multiple spot groups.

We categorized 11 PMS stars and 16 ZAMS stars as type a) single frequency, 4 PMS stars and 11 ZAMS stars as type b) possible shape changer, and 11 PMS stars and 5 ZAMS stars as type c) complex variability.

4. Discussion

Q1. How related the chromospheric activity of PMS stars with their rotation ?

We examined the relationship between R' of the Ca II IRT emission line, and N_{R} of the 60 PMS stars (Fig. 4). For calculating N_{R}, we estimated τ_{c} of the PMS stars using pre-main sequence evolutionary tracks presented in Jung & Kim (2007). In this track, τ_{c} of low-mass stars near the main sequence was not calculated. For such objects, we applied the approximation of Noyes et al. (1984). A detailed description of the calculation of N_{R} and R' is presented in Yamashita et al. (2020).

Fig. 4 shows the relationship between Rossby number and the Ca II emission line strength of PMS stars. Only three PMS stars show broad and strong emissions, indicative of large mass accretion. Most of the PMS stars have narrow and weak emissions.

Fig. 4a shows the objects with the single frequency. All of their Ca II IRT emission lines have $R' \sim 10^{-4.2}$, which is as large as the maximum R' of ZAMS stars (Marsden et al. 2009). The PMS stars have $N_{\mathrm{R}} < 10^{-0.8}$ and constant R' against N_{R}; i.e., the Ca II IRT emission lines of the PMS stars are saturated. The chromosphere of these stars is considered to be completely filled by the active region. We consider that the objects have stable spot because they show sinusoidal light curves. This indicates that their chromospheric activity is induced by the dynamo process.

In Fig. 4c, we plotted the PMS stars with complex variability. They tend to have large Rossby numbers. We consider that small spots on the photosphere cause the irregular variability.

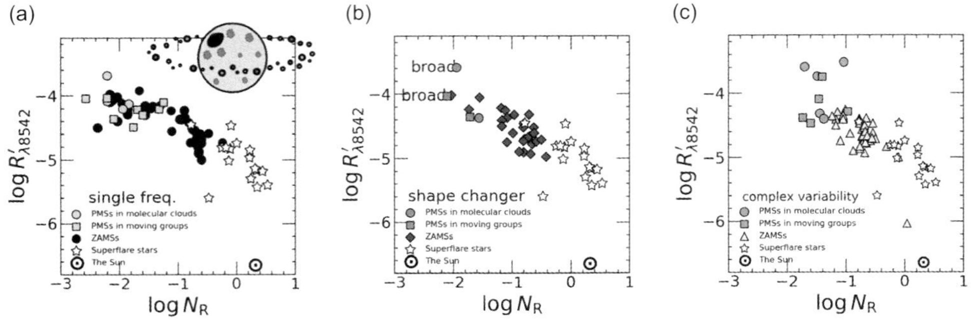

Single frequency objects Shape changer Complex variability

Figure 4. Ca II emision line strength, $R'_{\lambda8542}$, as a function of the Rossby number, $N_{\rm R}$ (Yamashita et al. 2020, 2024 in prep.). The colored symbols represent PMS stars. The filled circles in Fig. 4a, gray diamonds in Fig. 4b, open triangles in Fig. 4c show the ZAMS stars with single frequency, the possible shape changer ZAMS stars, ZAMS stars and complex variability, respectively. The star symbols and odot denote the superflare stars and the Sun (Notsu et al. 2015).

Q2. How about optically thinner emission lines ?

Because all R' of Ca II IRT emission lines were saturated against $N_{\rm R}$, it was not possible to estimate the appropriate $\tau_{\rm c}$ model for the PMS stars. PMS stars with unsaturated chromospheric emission lines should be investigated.

We examined the relationship between the ratio of the surface flux of the Mg I emission line to the stellar bolometric luminosity, $R'_{\rm MgI}$, and $N_{\rm R}$ of the ZAMS stars. In this work, we refereed $N_{\rm R}$ calculated by Marsden et al. (2009) for ZAMS stars. They used P for 21 ZAMS stars or $v \sin i$ for 31 ZAMS stars. We calculated R' for the Mg I emission line. For calculating the surface flux of the line F', a bolometric continuum flux per unit area at a stellar surface, F, was calculated at first. We used the i-band mag (the AB system) of the UCAC4 Catalogue Zacharias et al. (2013) and Marsden et al. (2009). F is given as

$$\log \frac{f}{f_0} = -\frac{2}{5} \times m_{i*}, \tag{2}$$

$$F = f \times \left(\frac{d}{R_*}\right)^2, \tag{3}$$

where f is the bolometric continuum flux of the object per unit area as observed on the Earth. m_{i*} is the apparent magnitude of the object in the i-band. The bolometric continuum flux per unit area under $m_i = 0$ mag (the AB system) condition, f_0, is $1.852 \times 10^{-12}\,{\rm W} \cdot {\rm m}^{-2} \cdot {\rm \AA}^{-1}$ Fukugita et al. (1996). d denotes the distance of an object from the Earth (*Gaia* DR2: Bailer–Jones et al. 2018). R_* is the stellar radius estimated using Stefan–Boltzmann's law with the photospheric luminosity, and $T_{\rm eff}$ of the objects in *Gaia* DR2 and Marsden et al. (2009). F was multiplied by the EQW of the Mg I emission lines.

$$F' = F \times {\rm EQW}, \tag{4}$$

For calculating R', F' are divided by $\sigma T_{\rm eff}^4$.

$$R' = \frac{F'}{\sigma T_{\rm eff}^4}, \tag{5}$$

where σ is Stefan–Boltzmann's constant. The dependence of the surface flux upon the $T_{\rm eff}$ of the objects is eliminated by this calculation.

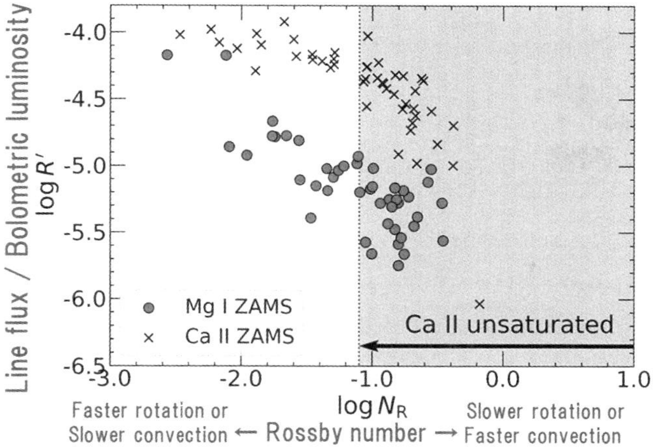

Figure 5. Relationship between ratio of surface flux of Mg I emission line (λ8807 Å) to stellar bolometric luminosity $R'_{\rm MgI}$ and Rossby number $N_{\rm R}$ of ZAMS stars (circles). Cross symbols represent $R'_{\lambda 8542}$, which is ratio of surface flux of Ca II λ8542 Å line to stellar bolometric luminosity, of ZAMS stars. (Yamashita et al. 2022a)

Fig. 5 shows the relationship between $R'_{\rm MgI}$ and $N_{\rm R}$ of the ZAMS stars. We found that the Mg I line is still not saturated in the region of $10^{-1.8} < N_{\rm R} < 10^{-0.8}$, which is the saturated region for Ca II IRT emission lines. According to the discussion by Marsden et al. (2009), the chromospheric saturation is suggested to be caused by the similar to coronal saturation. Suppose if the emitting region of Ca II are completely filled, the unsaturation of the Mg I emission line suggests that the area of the Mg I emitting region may be smaller than that of Ca II. Otherwise Mg I is optically thinner than Ca II. We conclude that the optically thin Mg I enables us to evaluate the strength of the dynamo activity for many object having strong magnetic field.

Q3. For PMS stars, how related the starspot coverage with the chromosphere ?

Notsu et al. (2015) indicated that the Sun and superflare stars show a rough positive correlation between the amplitude of the light curve and the $r_0(8542)$ index, the residual core flux normalized by the continuum level at the line core of the Ca II IRT. They claimed that the magnetic field strengths of superflare stars are higher than that of the Sun.

We investigated the amplitude of the light curve and $R'_{\lambda 8542}$ for PMS stars and ZAMS stars (Fig. 6). The ZAMS stars with a large light curve amplitude have large $R'_{\lambda 8542}$. The $R'_{\lambda 8542}$ and light curve amplitudes of ZAMS stars are approximately two orders of magnitude larger than those of the Sun. We also found that the ZAMS stars are located on the extensions of the superflare stars (Notsu et al. 2015). This result suggests that superflare stars link the properties of the Sun to those of the ZAMS stars of ages between 30 and 120 Myr.

Objects with different types of light curves seem to be unevenly distributed in Fig. 6, as represented by the colors of the symbols. Fig. 6a shows the PMS stars and ZAMS stars with single frequency. The ZAMS stars with single frequency have larger light curve amplitudes and larger $R'_{\lambda 8542}$ compared with other light curve type ZAMS stars. The PMS stars with single frequency show large light curve amplitudes and large $R'_{\lambda 8542}$. We consider that they have enormous spots of one to ten % of the surface, as well as bright chromospheric emission lines.

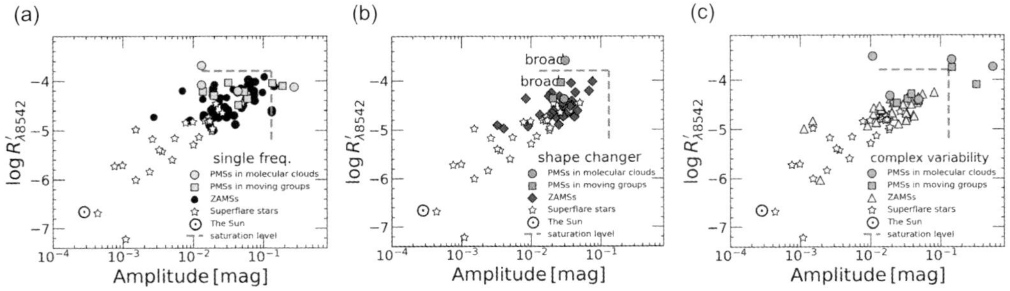

Single frequency objects Shape changer Complex variability

Figure 6. $R'_{\lambda 8542}$ as a function of the light curve amplitude of PMS stars (the colored circles and squares; this work), ZAMS stars (the black circles, gray diamonds, and gray triangles; Yamashita et al. 2022b), superflare stars (star marks; Notsu et al. 2015), and the Sun (odot; Notsu et al. 2015).

The ZAMS stars with complex variability have smaller light curve amplitudes and smaller $R'_{\lambda 8542}$ compared with other type ZAMS stars (Fig. 6c). A part of PMS stars are located near the ZAMS stars. We consider that they have small spots.

5. Acknowledgments

M. Y is supported by JSPS KAKENHI grant number 23KJ1855, and Y. I. is supported by JSPS KAKENHI grant number 22K03677. This paper includes data collected with the TESS mission, obtained from the MAST data archive at the Space Telescope Science Institute (STScI).

References

Bailer–Jones, C. A. L., Rybizki, J., Fouesneau, M., Mantelet, G., & Andrae, R. 2018, ApJ., 156, 58
Feinstein, A. D., Montet, B. T., Foreman–Mackey, D., et al. 2019, PASP, 131
Folsom et al. 2018, MNRAS, 474, 4956
Fleck, B., Deubner, F.-L., Maier, D., & Schmidt, W. 1994, IAU Symp, 154, 65
Fukugita, M., Ichikawa, T., Gunn, J. E., et al. 1996, AJ, 111, 1748
Gallet, F., & Bouvier, J. 2015, A&A, 577, 1
Hamann, F., & Persson, S. E. 1992a, Astrophys. J. Suppl., 82, 247
Jung, Y. K., & Kim, Y.-C. 2007, J. Astron. Space Sci., 24, 1
Linsky, J. L. 2017, Annu Rev Astron Astrophys, 55, 159
Marsden, S. C., Carter, B. D., & Donati, J.-F. 2009, MNRAS, 399, 888
Neuhauser, R., Strerzik, M. F., Schmitt, J. H. M. M., Wichmann, R., & Krautter, J. 1995, Astron. Astrophys., 297, 391
Notsu, Y., Honda, S., Maehara, H., et al. 2015, PASJ, 67, 1
Noyes, R. W., Hamann, F. W., Baliunas, S. L., & Vaughan, A. H. 1984, AJ, 279, 763
Parker, E. N. 1955, ApJ, 122, 293
Rebull, L. M., Stauffer, J. R., Bouvier, J., et al. 2016, AJ, 152, 114
Scargle, J. D. 1982, AJ, 263, 835
Skelly, M. B., Donati, J. F., Bouvier, J., et al. 2010, MNRAS, 403, 159
Stelzer, B., Fernandez, M., Costa, V. M., et al. 2003, A&A, 411, 517
Vernazza, J. E., Avrett, E. H., & Loeser, R. 1981, ApJS, 45, 635
Wilson, O. C. 1978, ApJ, 226, 379
Yamashita, M., Itoh, Y., & Takagi, Y. 2020, PASJ, 72, 80
Yamashita, M., & Itoh, Y. 2022a, PASJ, 74, 557
Yamashita, M., Itoh, Y. & Oasa, Y., 2022b, PASJ, 74, 1295
Zacharias, N., Finch, C. T., Girard, T. M., et al. 2013, Astrophys. J., 145, 1

Dynamics of Solar and Stellar Convection Zones and Atmospheres
Proceedings IAU Symposium No. 365, 2024
A. V. Getling & L. L. Kitchatinov, eds.
doi:10.1017/S1743921323004945

Super Flares in M Stars and Associated Characteristics of Active Regions and Magnetic Fields

M. Kumar Varnana[1], T. E. Girish[2], Thara N. Sathyan[3], V. G. Haritha[4] and G. Gopkumar[2]

[1]Department of Physics, Mar Ivanios College, Trivandrum, Kerala-695015
email: mvarnana@gmail.com

[2]Department of Physics, University College, Trivandrum, Kerala-695034
emails: tegirish5@yahoo.co.in; kggopkumar@yahoo.com

[3]Department of Physics, University College of Engineering Kariavattom, Trivandrum, Kerala-695581
email: tharasathyann@gmail.com

[4]Department of Physics, College of Engineering, Trivandrum, Kerala-695016
email: vgharithamanoj@gmail.com

Abstract. Observations of super flare occurrence (with energy 10^{33}–10^{36} erg)s in low mass stars like M dwarfs still remains as a puzzle. In this paper we have inferred the typical sizes and characteristics of magnetic fields associated with active regions in M dwarfs responsible for these super flares. This is done by extrapolation of physical conditions associated with largest solar flares. The average poloidal and toroidal magnetic fields near the surface of selected M dwarfs will be also inferred in this context.

Keywords. Super flares, M dwarfs, Solar flares, Active regions, Magnetic fields

1. Introduction

Solar flares are energetic phenomena in the solar atmosphere with profound influence on the space weather near Earth. The mechanism of solar flares is a plasma phenomena involving reconnection where stored magnetic potential energy in active regions is suddenly converted in to kinetic energy of particles and radiation (involving also enhanced EUV and X ray emission). Eighty percent of stars in our milky way are inferred to be M dwarfs which hosts many extra solar planets. Frequent occurrence of flares with energy with 2–4 orders of magnitude higher than the largest solar flares in M dwarfs (called super flares) is still a puzzle (Howard *et al.* 2020). It will be worthwhile to know the physical conditions and mechanisms associated with the super flare occurrences in M stars In this paper we have carried out order of magnitude calculations of the sizes of active regions and associated magnetic field characteristics related to super flares in M stars.

2. Magnetic fluxes associated with super flares in M stars

We can expect close relations between sizes of active regions and its magnetic field characteristics with the energy of super flares. An upper limit of magnetic flux in an active region on the Sun is found by Livadiotis and Moussas (2009) as 7.23×10^{23} Mx.

From solar observations the maximum total magnetic flux in sunspots is found to be 3.18×10^{23} Mx. This is found to be associated with maximum area of sunspots during the past 300 years and observed in 1947. The upper limit to active region flux in the Sun is expected to be connected with maximum energy of solar flares

It is understood that solar flare energy is related to energy stored in magnetic fields of active regions in the Sun (Mayfield and Lawrence 1985). In this connection we can expect a relation between solar flare energy (Ef) and total magnetic flux (Fm) in the active region/active region group associated with the solar flare.

$$Ef = f(Fm) \tag{1}$$

When the solar flare energy is found to vary in the range 10^{24} to 10^{32} ergs, the associated magnetic flux in sunspots/sunspot groups is found to vary in the range 10^{21} to 10^{23} Mx. So we can expect a non-linear relation between Ef and Fm. The energy of super flares in M dwarfs is inferred to vary between 10^{33} to 10^{36} ergs. Then a heuristic inference of associated Fm for the super flares is:

$$Fm = 10^{24} - 10^{25} \text{ Mx} \tag{2}$$

3. Inference of the sizes of active regions related to M dwarf super flares

Sakurai and Toriumi (2023) found the following linear relation between sunspot area (Sa) and associated total magnetic flux (Φ_S) based on relevant observations:

$$\text{Log}(\Phi_S) = \text{Log}(Sa) - 19.7 \tag{3}$$

Here Φ_S is expressed in units of Mx and area Sa in units of msh (millionth of the solar hemisphere)

If we assume that the relation (3) is also valid for stellar flares then we can find the area of active regions responsible for super flares in M stars

$$\text{For } \Phi_S = 10^{24} \text{ Mx we find that } Sa = 1.995 \times 10^4 \text{ MSH} \tag{4a}$$

$$\text{For } \Phi_S = 10^{25} \text{ Mx we find that } Sa = 1.995 \times 10^5 \text{ MSH} \tag{4b}$$

For $\Phi = 10^{24}$ Mx,

$$\text{we have found } S \text{ (super flares) } = 1.995 \times 10^4 \text{ MSH} = 0.02 \text{ As} \tag{5}$$

Here As is the area of solar hemisphere or disc.

For $\Phi = 10^{25}$ Mx,

$$\text{we have found } S \text{ (super flares) } = 1.995 \times 10^5 \text{ MSH} = 0.2 \text{ As} \tag{6}$$

Our calculations of S are agreement with similar values reported in published literature (Herbst *et al.* 2021). Validity of these results will also depend on the radius of M stars relative to Sun under consideration.

4. Inference of magnetic field strengths in the active regions associated with M dwarf Super Flares

The maximum magnetic field strength in active regions associated with super flares in M stars can be inferred from different methods:

4.1. *Calculations based on Flare energy*

It is understood that energy of the super flares in solar/stellar flares (E_f) is equal to the energy stored magnetic fields ($B^2/4\ \pi$) in the active regions associated with these flares

This implies that

$$E_f = B^2/4\pi \tag{7}$$

Here B is the magnetic field strength in the solar/stellar active regions.

We infer B (solar) = 4 kG for extreme solar flares. (8)

The energy of extreme flares can be assumed as

$$Ef(\text{Sun}) = 10^{32} \text{ ergs}, Ef(M \text{ dwar}) = 10^{36} \text{ ergs} \tag{9}$$

We can find that field strength B of active regions responsible for extreme super flares as

$$B(M \text{ star}) = 200 \text{ kG} \tag{10}$$

4.2. *Calculations based on internal magnetic fields of the stars*

The toroidal magnetic fields related to active regions on the stellar surface is connected to the toroidal fields in the interior or these stars. The toroidal magnetic fields in the base of solar convective zone is inferred to be of the order of 10^5 G or 0.1 MG (Fan 2021). From a recent study (Feiden and Chaboyer 2014) related to low mass star magnetic field generation, the field strength of toroidal fields in the interior of such stars (including M dwarfs) is inferred to be of the order of 10 MG or 10^7 G. This implies that toroidal fields in M stars can be 100 times more stronger than the same for Sun.

4.3. *Inference of poloidal and toroidal magnetic fields for selected M dwarfs*

In the case of the Sun poloidal fields in the sun of few Gauss in intensity is amplified by differential rotation to form toroidal fields in active regions in the photosphere of the order of few kG in intensity. The poloidal to toroidal field amplification factor for the Sun (Af) is of the order of 1000.

Remote measurements of magnetic fields in many M dwarfs (Kochukov 2021) suggest the presence of strong poloidal magnetic fields of the order of few kG. This implies that toroidal fields in these stars related to active regions are likely to be several orders stronger. As per the results in 4.1 and 4.2, poloidal to toroidal magnetic flux amplification factor for M stars can be inferred to be close to hundred. The inferred surface magnetic fields in M stars is likely to be poloidal or large scale magnetic fields (Bp) The toroidal field strength (Bt) in such stars is suggested to be given by the following relation:

$$Bt = Af \text{ Bp} = 100 \text{ Bp} \tag{11}$$

In Table 1 we have given poloidal fields of selected M stars from published values of surface fields included in the study of Kochukov (2021) inferred from a model involving stokes spectra of these stars. The average toroidal field strength related to active regions near the surface of these stars given in this table is estimated using relation (11).

5. Discussion

The magnitude of solar/stellar flare energy reflects the characteristics of active regions and associated magnetic fields responsible for these flares. Naturally superflares in M dwarfs are associated with very large active regions and super-strong magnetic fields whose details are worked out in this paper. We have inferred the magnitudes of poloidal and toroidal magnetic fields in selected M dwarfs whose surface magnetic fields details are reported in literature. The inferred values of poloidal and toridal magnetic fields for Trappist-1 by Mullan *et al.* (2018) supports our calculations given in Table 1. To account

Table 1. Average poloidal and toroidal magnetic field strengths near the surface of selected M dwarfs. Poloidal field values are derived from observations (after Kochukov 2021). Toroidal field values associated with active regions in these stars are inferred assuming a fixed poloidal to toroidal flux amplification factor ($Af = 100$).

Name of the M star	Poloidal magnetic field (kG)	Estimated mean toroidal magnetic field (kG)
GJ 2005A	2	200
GJ 65B	6.7	670
GJ 278Cb	3.2	300
GJ 412B	7.3	730
GJ 3622	1.4	140
GJ 3059	4.1	410

for the emergence of magnetic fields near the surface of M dwarfs which is two orders of magnitude higher than the solar magnetic fields, a dynamo quite different from that of our Sun may exist inside these low mass stars.

Conclusions

(1) We have made heuristic inferences of the values of active region sizes and magnetic field characteristics (field strength and total magnetic flux) associated super flares in M dwarf stars. Our results are found to be several orders of magnitude greater than the same reported for solar active regions related to large solar flares.

(2) The physical conditions and mechanisms in M stars responsible for the generation of very large active regions and super-strong magnetic fields need detailed investigations. The average poloidal and toroidal magnetic fields generated by the M dwarf dynamo is inferred to be at least hundred times stronger than the magnetic fields generated by the solar dynamo.

References

Fan, Y. 2021, *Living Rev. Sol. Phys.*, 18, 5
Feiden, G.A. and Chaboyer, B. 2014, *ApJ*, 789, 53
Herbst, K. *et. al* 2021, *ApJ*, 907, 89
Howard, W.S. *et al.* 2020, *ApJ*, 902, 115
Kochukov, O. 2021, *Astron. Astrophys Rev*, 29, 1
Livadiotis, G. and Moussas, X. 2009, *Adv. Space Res.*, 43(4), 694
Mayfield, E.B. and Lawrence, J.K. 1985, *Sol. Phys.*, 96, 293
Mullan, D.J. *et. al* 2018, *ApJ*, 869, 149
Sakurai, T. and Toriumi, S. 2023, *ApJ*, 943

Section 6. Miscellaneous

Contributed papers

Dynamics of Solar and Stellar Convection Zones and Atmospheres
Proceedings IAU Symposium No. 365, 2024
A. V. Getling & L. L. Kitchatinov, eds.
doi:10.1017/S1743921323004908

Features of the solar minimum 24/25 in the evolution of polar and non-polar coronal holes

Olga Andreeva

Crimean Astrophysical Observatory of RAS
email: olga@craocrimea.ru

Abstract. We present a study of the evolution of two types of coronal holes (CHs) in the solar minimum of 24/25, which was preceded by a prolonged minimum of 23/24 and a weak 24 solar cycle. The goal of the study is to clarify whether the behavior of CHs during this period is also unusual? The study is based on the material of observations obtained by SDO/AIA/193. The Heliophysics Events Knowledgebase was used to localize the CHs and calculate their areas. Analysis of the evolution of the areas of polar and non-polar CHs in solar minimum 24/25 revealed a number of features. The hemispheric asymmetry is evident both in solar activity indices and in the localization of maxima of polar and non-polar CH areas. The hemispheric area imbalance is minimal for polar CHs and pronounced in the regions of non-polar CHs and sunspots. This is consistent with the general concept of polar CHs as the main source of the Sun's dipole magnetic field. The areas of polar CHs significantly exceed the areas of non-polar CHs and make a significant contribution to the total area of all CHs in the solar disk. It is concluded that the dynamics of polar and nonpolar CHs suggests that the 24/25 minimum is rather close to earlier minima than to the 23/24 minimum.

Keywords. Solar Cycle, Solar minimum, Solar Corona, Coronal hole, Evolution of Coronal holes

1. Introduction

The structure of the corona is quite complex and dynamic and depends both on the spatial distribution of solar formations on its surface and on the phase of the solar cycle. During the period of activity minimum, open-field regions above the poles (polar coronal holes) reach their maximum expansion, and the number of low-latitude regions of closed strong field (active regions and sunspots) is minimal. The dipole structure of the global magnetic field in this time is very prominent and there are no significant inclusions of open flux in the equatorial zone Abramenko *et al.* (2010). Typical structure corona in the solar minimum is shown in Fig. 1.

Coronal holes (CHs) are areas of open magnetic configuration on the Sun: a structure in which magnetic lines of force freely extend into interplanetary space and are closed at infinity. These areas are characterized by much lower temperatures and lower densities compared to their surroundings, which make CHs, appear dark. The evolution of CHs is closely related to the evolution of the large-scale solar magnetic fields. CHs are a reaction of global internal processes in the Sun. Therefore, CHs are an integral part of the solar activity (SA) cycle. By analyzing the dynamics of CHs, in particular, their areas, we can judge the dynamics of the Sun and the progress of the cycle.

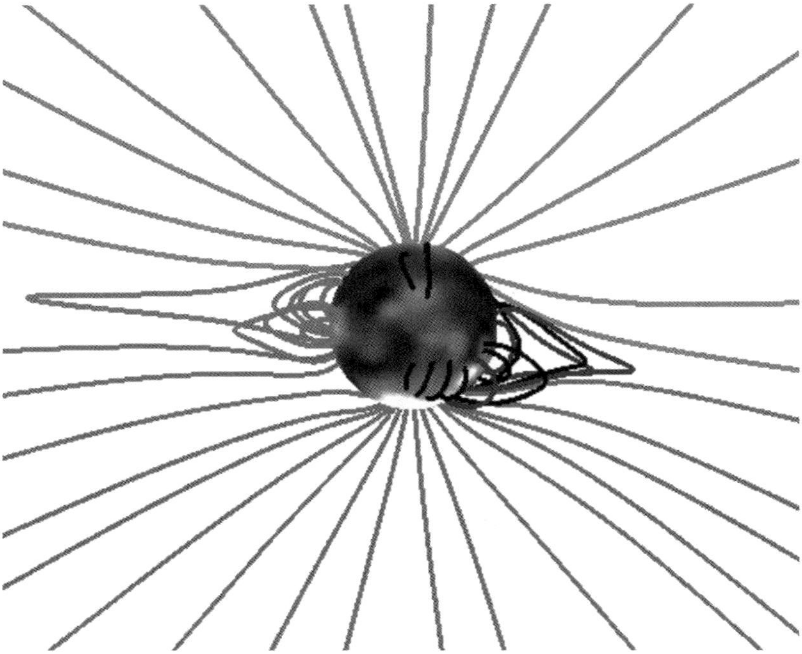

Figure 1. Typical structure corona in the solar minimum.

1.1. *The solar activity cycle 24 and minimum 24/25*

The solar activity cycle 24 (Fig. 2(a)) that recently ended was one of the weakest cycles for the last 100 years of observations (Fig. 2(b)).

It differed from the previous cycles in a number of parameters. Cycle 24 is known to have a lower maximum and less flare activity than the previous cycles. This cycle exhibited a relatively large N–S asymmetry of the polar field inversion: the sign of the field at the north pole changed more than a year earlier than at the south pole. The cycle also had a strong asymmetry of the hemispheres in many other parameters. It should be said, solar minimum 23/24 exhibits features that differ notably from features commonly seen in minimum 22/23 and earlier ones. The polar CHs were less dominant than those in previous solar minima, but the low-latitude CHs, which are not commonly seen during previous solar minima, were relatively large and persist many rotations Abramenko *et al.* (2010). Since the 24th solar cycle and the minimum preceding it had of a special character, the questions naturally arises:Is the 24/25 solar minimum (Fig. 2(a)) also unusual? This question motivates scientists to investigate this minimum in more detail. Thus, the authors of Huichao Li *et al.* (2021) considered the evolution of the sunspot number and polar field strength, compared the 24/25 minimum with the two previous minima 22/23 and 23/24 and showed:

– Minima 24/25 and 23/24 showing lower activity level than minimum 22/23. But to the end of 2020 SSN raises harply for minimum 24/25.

– The polar field strength of minimum 24/25 is stronger than that during minimum 23/24 in average, but still weaker than minimum 22/23.

– The N–S asymmetry of the polar field is more evident during minimum 24/25 than during minimum 23/24. A similar north–south asymmetry is found during minimum 22/23.

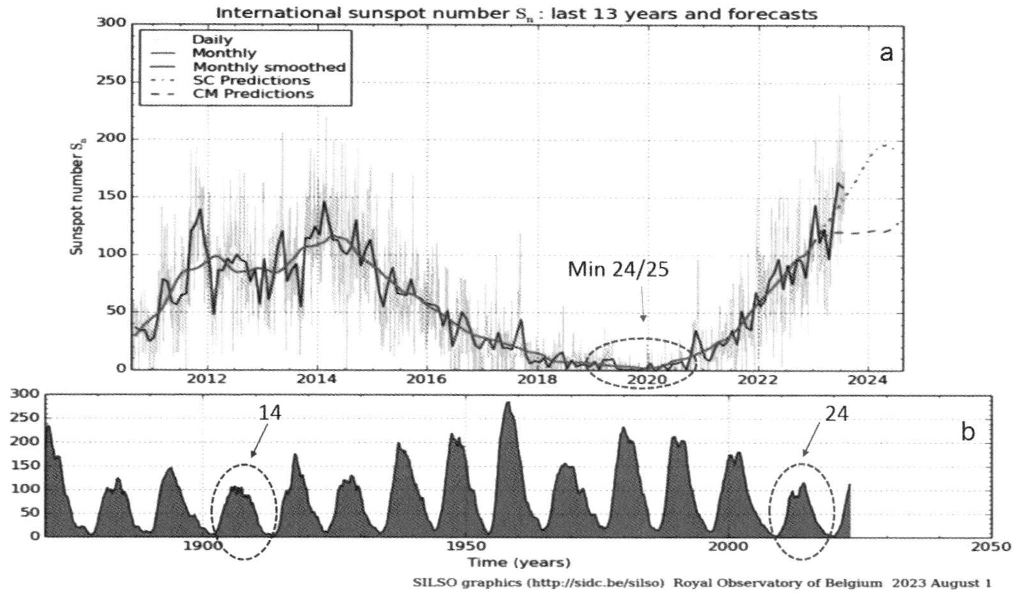

Figure 2. Daily and monthly sunspot number (last 13 years – a). Monthly 13-month smoothed sunspot number (blue) from 1870 up to the present – b.

2. Evolution two type of coronal holes during the minimum 24/25

The evolution of CHs on the scale of the solar cycle was studied earlier Andreeva (2023). Variations of the daily total area were obtained for the entire visible surface of the solar disk, separately for both hemispheres and for 2 types of coronal holes (Fig. 3(b), (c-d) and (e-f)). The asymmetry of the hemispheres in the distribution of sunspot areas and two types of coronal holes was revealed (Fig. 4(a) and (b-c)).

Here, as in our earlier works Andreeva & Malashchuk (2020), Andreeva *et al.* (2020), Andreeva *et al.* (2021), Andreeva (2023), we define as polar (CHs_pol) the CHs that begin in the polar regions, i.e., from latitude 90°, and extend to lower latitudes depending on their development and cycle phase, sometimes to middle and even low latitudes (Fig. 5 SPoCA 23027, 23064, 23632, 23717). Non-polar (CHs_nonpol) are isolated CHs, not connected with the pole. Most often, these are mid- and low-latitude CHs, but sometimes CHs_nonpol appear at relatively high latitudes (Fig. 5 SPoCA 23097, 23715, 23716).

In addition, we adopt that the area of a CH crossing the equator is divided into two parts, and each of them refers to the corresponding hemisphere, depending on its location. The contribution of such CHs is quite small because their number is less than 7% of the total number of CHs.

In this paper, we extract from the previously obtained series the interval corresponding to the 24/25 minimum (Fig. 3), analyze and evaluate the contribution of polar and nonpolar coronal holes to the evolution and hemispheric asymmetry of large-scale solar activity during this period (Fig. 6). We want, at least partially, to answer the question posed earlier: "Is the 24/25 minimum special in terms of CHs evolution?". To analyze the nature of the solar minimum, usually choose a period of time around the absolute minimum (in our case it is December 2019), when solar activity is consistently low. As can be seen in Fig. 2(a), the sunspot activity level is low in the 2-year period around the absolute minimum of solar cycle 24, so we choose the period from 01.12.2018 to

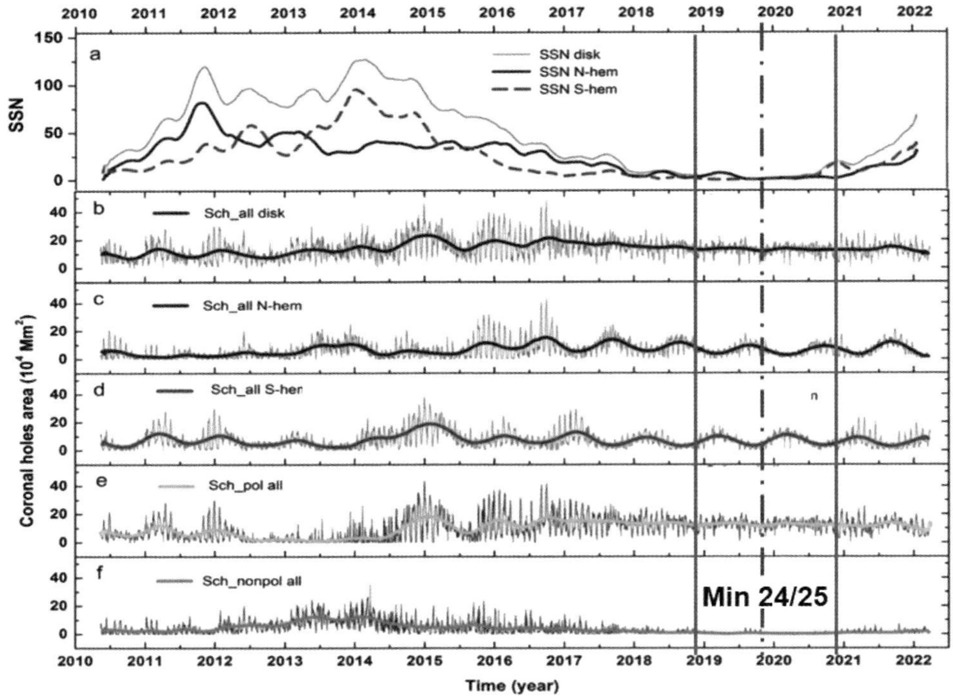

Figure 3. Variations of the SSN and daily total CH areas in 2010-2022.

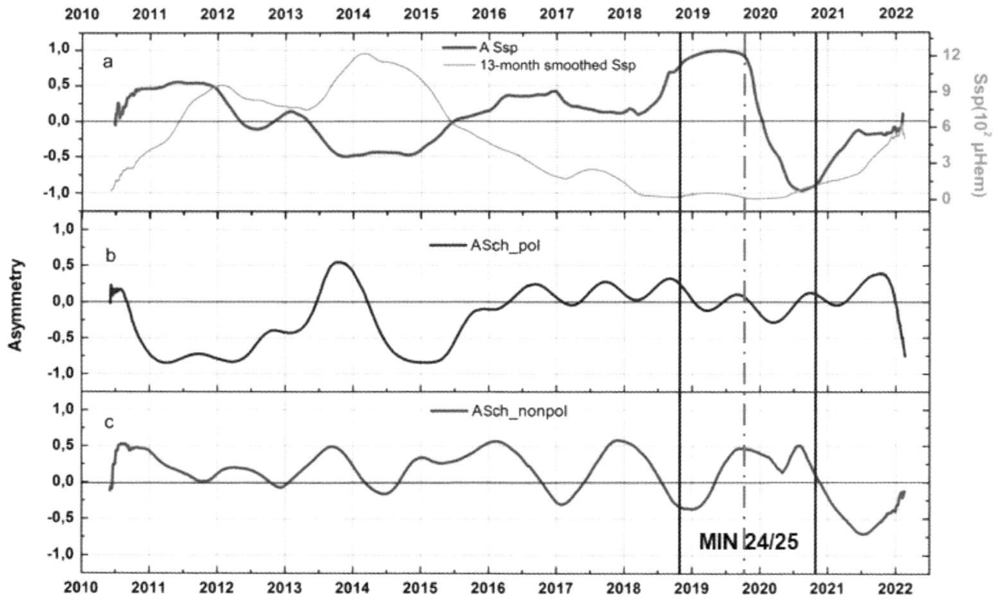

Figure 4. N-S asymmetry of the areas of sunspots and coronal holes in 2010-2022.

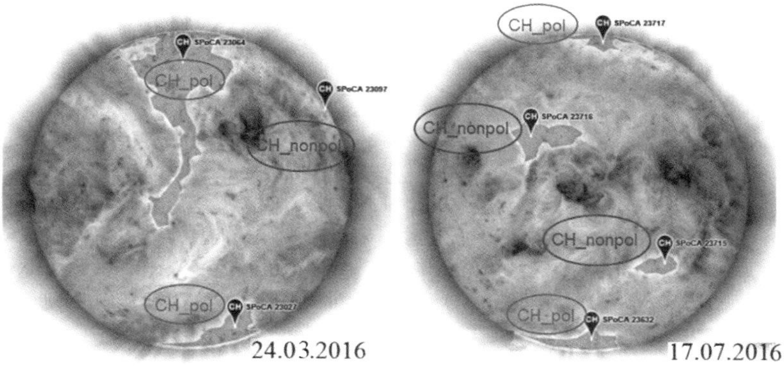

Figure 5. Inverted images of the Sun obtained by SDO/AIA in the Fe XII 19.3 nm line. The regions on the solar disk marked "CH" are coronal holes extracted from the EUV image with the SPoCA method.

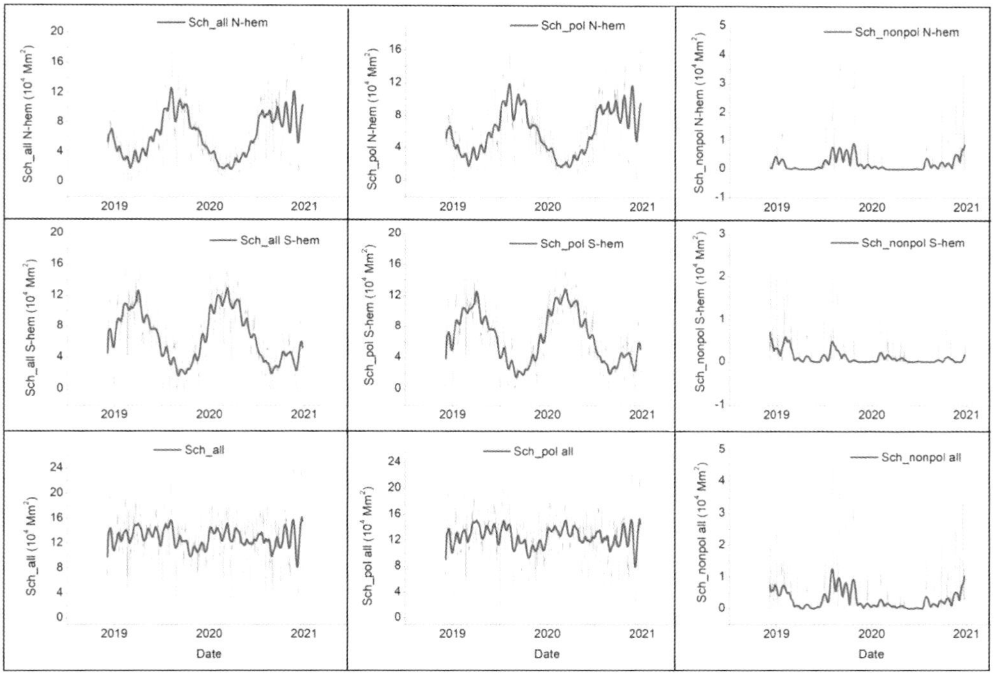

Figure 6. Evolution of polar and non-polar coronal holes during the minimum 24/25.

31.12.2020 to analyze the 24/25 minimum. The SSN data are obtained from the sunspot index and long-term solar observations (SILSO).

Figure 6 shows the dynamics of the daily areas of all, polar and nonpolar CHs (columns from left to right) observed in the N and S hemispheres and on the entire visible surface of the solar disk (rows from top to bottom) at the 24/25 minimum. There is an obvious difference in the operation of the hemispheres in generating both polar and non-polar coronal holes. At the beginning of the minimum, the southern hemisphere leads in the generation of non-polar coronal holes, while before and after the absolute minimum, the northern hemisphere leads. For polar CHs, the asymmetry in amplitude is less pronounced. We observe hemispheric asymmetry in the time of appearance of the maximum peak areas for both groups. It can be seen that polar CHs make a significant contribution to the total area of all CHs.

3. Conclusions

Analysis of the evolution of the areas of polar and non-polar coronal holes in the solar minimum 24/25 revealed a number of features:

– Hemispheric asymmetry is manifested both in the indices of solar activity and in the localization of the maximum regions of polar and non-polar coronal holes.

– The area imbalance of the hemispheres is minimal for polar coronal holes and is pronounced in the areas of non-polar coronal holes and sunspots. This is consistent with the general concept of polar coronal holes as the main source of the sun's dipole magnetic field.

– The areas of polar coronal holes are significantly larger than the areas of non-polar ones and make a significant contribution to the total area of all coronal holes on the Sun's disk.

The total area of polar coronal holes approached the average level of values of about $(12.5 \pm 0.8) \times 10^4 Mm^2$. While the area of all non-polar coronal holes is only about $0.5 \times 10^4 Mm^2$.

Of course, the concept of "unusual solar minimum" is multifaceted, but as far as the evolution of two types of coronal holes is concerned, minimum 24/25 was similar to earlier minima, in contrast to the 23/24 minimum.

References

Abramenko, V., Yurchyshyn V., Linker J., *et. al*, 2010, *Astrophysical Journal*, 712, 2, 813
Andreeva, O., Malashchuk, V., 2020, *Geomagnetism and Aeronomy*, 60, 8, 1093
Andreeva, O., Abramenko, V., Malashchuk, V., 2020, *Astrophysics*, 63, 1, 114
Andreeva, O., Abramenko, V., Malashchuk, V., 2021, *Geomagnetism and Aeronomy*, 61, 1, S1
Andreeva, O., 2023, *ASR*, 71, 4, 1915
Huichao Li, Xueshang Feng, Fengsi Wei, 2021, *ApJL*, 917, 2, L26

Dynamics of Solar and Stellar Convection Zones and Atmospheres
Proceedings IAU Symposium No. 365, 2024
A. V. Getling & L. L. Kitchatinov, eds.
doi:10.1017/S174392132300501X

Center-to-limb variation of spectral lines and their effect on full-disk observations

Alexander G. M. Pietrow[1] **and Adur Pastor Yabar**[2]

[1]Leibniz-Institut für Astrophysik Potsdam (AIP), An der Sternwarte 16, 14482 Potsdam, Germany
email: apietrow@aip.de

[2]Institute for Solar Physics, Dept. of Astronomy, Stockholm University, Albanova University Centre, SE-106 91 Stockholm, Sweden

Abstract. An accurate description of the center-to-limb variation (CLV) of stellar spectra is becoming an increasingly critical factor in both stellar and exoplanet characterization. In particular, the CLV of spectral lines is extremely challenging as its characterization requires highly detailed knowledge of the stellar physical conditions. To this end, we present the Numerical Empirical Sun-as-a-Star Integrator (NESSI) as a tool for translating high-resolution solar observations of a partial field of view into disk-integrated spectra that can be used to test common assumptions in stellar physics.

Keywords. Methods: miscellaneous, Line: formation, Sun: photosphere, Sun: chromosphere

1. Introduction

As telescopes and instruments provide ever more detailed observations of the night sky, previously negligible effects come to the forefront. This in turn requires once simple models to incorporate increasing complexity to explain the observed data, often with no clear way to test their appropriateness. One such property is the CLV of stellar spectra, which, along with its associated parameterisation, has become increasingly important for the characterisation of stars and exoplanets. For example, incorrect limb darkening curves in transit observations lead to an overestimation of the planet radius (e.g. Mandel & Agol 2002, and Chakraborty *et al.* (Submitted.), Canocchi *et al.* (Submitted)). Currently, most such studies rely on continuum limb darkening curves obtained from models such as the PHOENIX (Husser *et al.* 2013) or ATLAS (Kurucz 1970) codes, which ignore the CLV of spectral lines.

While Maxted (2023) has shown that it is a valid approximation to use continuum CLV for broadband ($\approx$ 500 nm) observations, this is not necessarily the case when studying spectra (e.g. Dineva *et al.* 2020, Reiners *et al.* 2023, and Pietrow *et al.* 2023). In addition, substructures on the disk such as sunspots and plage can also affect transits and radial velocity measurements (e.g. Meunier *et al.* 2010). For this reason, active regions must also be modelled together with their CLV, which have been shown to differ from that of the quiet Sun (e.g. Oranje 1983 and Cretignier *et al.* 2023). Therefore, theoretical models require further development so that they can account for this CLV behavior.

The Sun is an obvious test bed for such studies, as many small-scale features on its surface can be resolved due to its large angular size and proximity to Earth. However, most high spatial and temporal resolution observations are inherently limited by a small field of view, making it difficult to compare them in a Sun-as-a-star setting (Otsu *et al.*

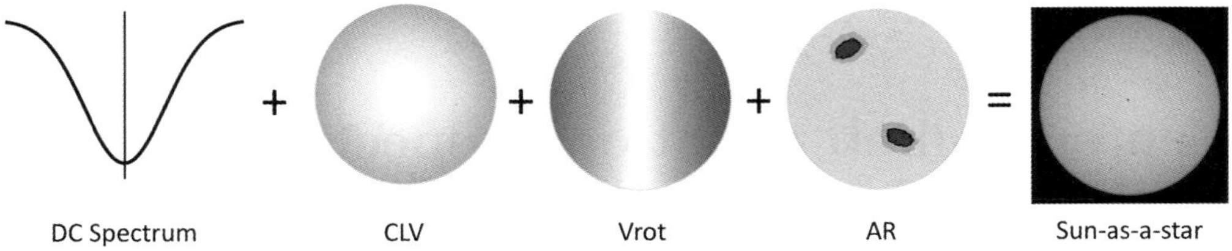

Figure 1. A cartoon showing the ingredients required for creating a disk-integrated or Sun-as-a-star profile.

2022 and Pietrow *et al.* 2023c). The NESSI code†, which is publicly available, has been developed to overcome this limitation and translate such observations into disk-integrated spectra. This allows specific aspects to be studied in isolation from other events on the solar disk.

2. Numerical Empirical Sun-as-a-Star Integrator

In its base function, NESSI works in the same way as other stellar integrator codes such as the Spot Oscillation And Planet code (SOAP, Boisse *et al.* 2012; Dumusque *et al.* 2014; Zhao & Dumusque 2023), except that it works with empirical inputs from telescopes such as the Swedish 1-m Solar Telescope (SST, Scharmer *et al.* 2003) and the Interface Region Spectrograph (IRIS, De Pontieu *et al.* 2014). Following the recipe shown in Fig. 1, the code requires a spectral profile or wavelength range of the quiet Sun (QS) at disk center (DC), along with a set of CLV curves for each wavelength point. It also requires a differential rotation velocity profile and the orientation of the rotation axis with respect to the observer, so that it can evaluate the spectra of each point on the solar disk and integrate them into a single SAAS profile. In addition, it is possible to evaluate the effect of small scale features on the integrated spectrum by replacing some grid points with these features CLV (sunspots, plage, flares, etc.).

3. Quiet Sun simulations

In order to test the method, we simulated two Sun-as-a-star (SAAS) line profiles for Hα and O I 7772 Å. Both lines need to be modeled in 3D NLTE (Leenaarts *et al.* 2012 and Bergemann *et al.* 2021), and therefore will typically not be properly recreated by stellar synthesis codes that focus on producing well-sampled spectral lines. However, since all relevant physical effects are captured in the CLV observations, this method should approximate the final integrated profiles with high accuracy.

We used the CLV from the Institut für Astrophysik Göttingen (IAG) spectral atlas (Ellwarth *et al.* 2023) and SST line CLVs (Pietrow 2023b), respectively. In both cases, we used the μ=1 line profile as the DC profile, the differential rotation described by Balthasar *et al.* (1986) and assumed a quiet Sun disc without any other feature. These lines are then compared to continuum limbdarkening curves based on Neckel & Labs (1994), and the full-disk IAG flux atlas (Reiners *et al.* 2016), which in case of the oxygen line is convolved down to the resolution of the SST data using the STIC python library (de la Cruz Rodríguez *et al.* 2019).

The results of both experiments are shown in Figs. 2 and 3, where the difference between line CLV and more traditional limb-darkening curves is shown. SAAS profiles are synthesized for both cases, which are then compared with the spectral atlas and a

† `https://github.com/apy-github/NESSI`

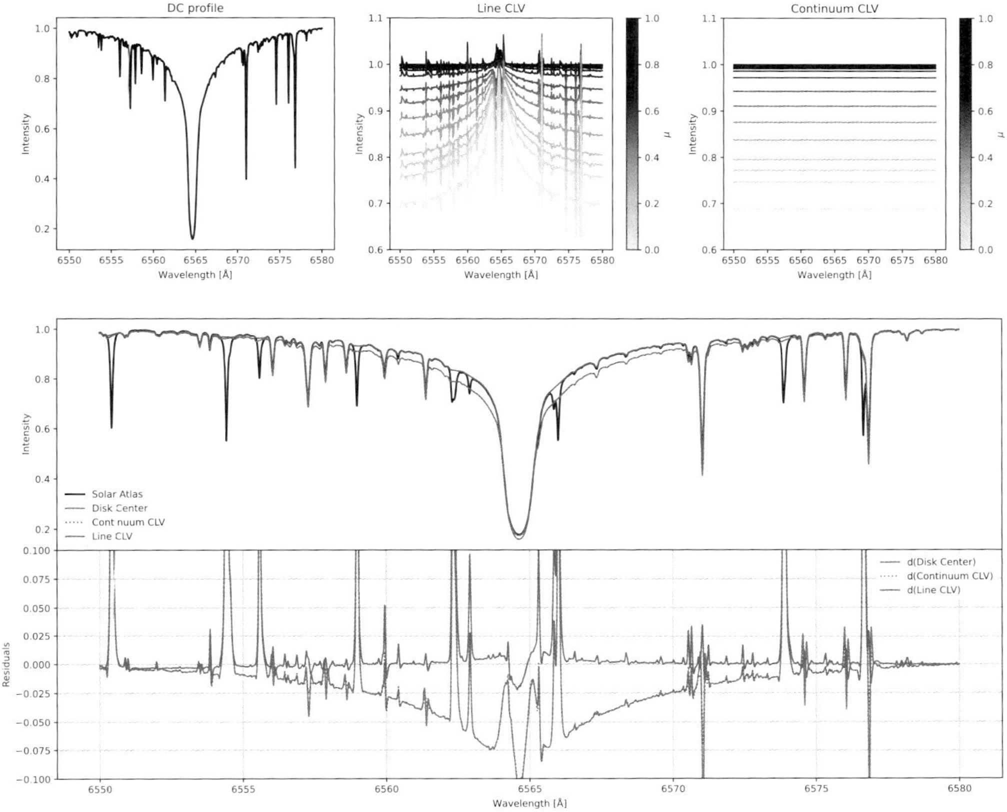

Figure 2. A Sun-as-a-star Hα profile of the QS compared with the IAG flux atlas. **Top Left:** A disk center profile. **Top Middle:** Line CLV curves. **Top Right:** Continuum CLV curves. **Middle:** A comparative plot showing the atlas, and three synthesized Sun-as-a-star profiles. **Bottom:** Difference between each profile and the atlas reference.

disk center profile, which is equivalent to assuming no limb darkening at all. Evidently, the profile constructed using the line CLV results in a much better fit in both cases, as the residuals are on the order of one percent. The larger peaks in the residual spectrum are at the locations of telluric lines that were removed in the IAG flux atlas. The gaps in these areas have been interpolated, resulting in large offsets from the atlas spectrum. Any remaining difference between this profile and the reference is likely due to the fact that the IAG flux atlas is averaged over several weeks worth of non-quiet disk observations, resulting in an unknown contribution from surface active regions. The effect of this will be investigated in the future in a comparative time-resolved study with high-resolution Sun-as-a-star telescopes such as those on HARPS-N (Dumusque *et al.* 2015) and PEPSI (Strassmeier *et al.* 2018), where the size and location of the active regions are known.

In summary, NESSI incorporates empirical line CLV into its SAAS integration, resulting in a better agreement with full-disk spectral observations compared to using continuum model CLV only. Thus, if provided with empirical/theoretical CLV for small-scale features, it facilitates a better evaluation of the impact these have over the integrated spectrum. The NESSI code is publicly available on the project github (see link on P1).

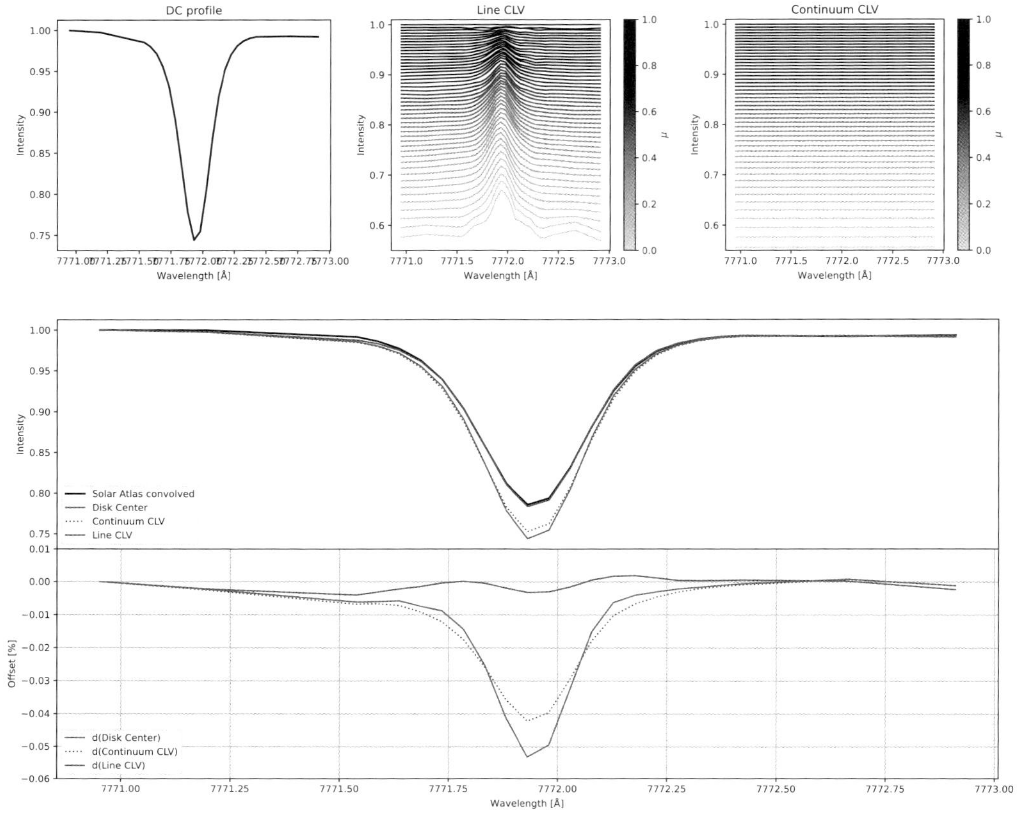

Figure 3. Same as for Fig. 2, but for the O I 7772 Å line taken with a lower spectral resolution.

Acknowledgements. We thank Dominique Petit dit de la Roche, and Malcolm Druett for their stimulating discussion on the topic. Co-funded by the European Union (ERC, MAGHEAT, 31004331). Views and opinions expressed are however those of the authors only and do not necessarily reflect those of the European Union or the European Research Council. Neither the European Union nor the granting authority can be held responsible for them. The Institute for Solar Physics is supported by a grant for research infrastructures of national importance from the Swedish Research Council (registration number 2017-00625).

References

Balthasar, H., Lustig, G., Woehl, H., & Stark, D. 1986, *A&A*, 160, 277-279
Bergemann, M., Hoppe, R., *et al.* 2021, *MNRAS*, 508, 2236-2253
Boisse, I., Bonfils, X., & Santos, N. C. 2012, *A&A*, 545, A109
Canocchi, G., Lind, K., *et al.*, Submitted to A&A
Chakraborty, H. *et al.*, submitted to A&A
Cretignier, M., Pietrow, A.G.M., Zhao, Y, & Aigrain, S., 2023, *arXiv e-prints*, 2310.15926
De Pontieu, B. *et al.* 2014, *Sol. Phys.*, 289, 2733-2779
de la Cruz Rodríguez, J., Leenaarts, J., Danilovic, S., & Uitenbroek, H. 2019, *A&A*, 623, A74.
Dineva, E. *et al.* 2020, *Solar and Stellar Magnetic Fields: Origins and Manifestations*, 354, 473
Dumusque, X., Boisse, I., & Santos, N.C. 2014, *ApJ*, 796, 132
Dumusque, X. *et al.* 2015, *ApJL*, 814, L21

Ellwarth, M., Schäfer, S., Reiners, A., & Zechmeister, M. 2023, *A&A*, 673, A19
Husser, T. *et al.* 2013, *A&A*, 553, A6
Kurucz, R. L. 1970, *SAO Special Report*, 309
Leenaarts, J., Carlsson, M., and Rouppe van der Voort, L. 2012, *ApJ*, 749, 136
Mandel, K., & Agol, E. 2002, *ApJ*, 580, L171
Maxted, P. F. L. 2023, *MNRAS*, 519, 3723
Meunier, N. and Desort, M. and Lagrange, A.-M. 2010, *A&A*, 512, A39
Neckel, H., & Labs, D. 1994, *Sol. Phys.*, 153, 91
Oranje, B. J. 1983, *A&A*, 124, 43
Otsu, T., Asai, A., Ichimoto, K., Ishii, T. T., and Namekata, K. 2022, *ApJ*, 939, 98
Pietrow, A. G. M. *et al.* 2023, *A&A*, 671, A130
Pietrow, A. G. M., Hoppe, R., Bergemann, M., & Calvo, F. 2023, *A&A*, 672, L6
Pietrow, A. G. M., Cretignier, M., *et al.* 2023, *arXiv e-prints*, 2309.03373
Reiners, A., Mrotzek, N., Lemke, U., Hinrichs, J., & Reinsch, K. 2016, *A&A*, 587, A65
Reiners, A. *et al.* 2023, *A&A*, 673, A71
Scharmer, Goran B. *et al.* 2003, *Proc. SPIE*, 4853, 341
Strassmeier, K. G., Ilyin, I., & Steffen, M. 2018, *A&A*, 612, A44
Zhao, Y. & Dumusque, X. 2023, *A&A*, 671, A11

Dynamics of Solar and Stellar Convection Zones and Atmospheres
Proceedings IAU Symposium No. 365, 2024
A. V. Getling & L. L. Kitchatinov, eds.
doi:10.1017/S1743921323004271

Characterizing Solar Spicules and their Role in Solar Wind Production using Machine Learning and the Hough Transform

R. Sadeghi and E. Tavabi

Physics Department, Payame Noor University, Tehran, Iran, 19395-3697
email: e_tavabi@pnu.ac.ir

Abstract. Solar winds originate from the Sun and can be classified as fast or slow. Fast solar winds come from coronal holes at the solar poles, while slow solar winds may originate from the equatorial region or streamers. Spicules are jet-like structures observed in the Sun's chromosphere and transition region. Some spicules exhibit rotating motion, potentially indicating vorticity and Alfvén waves. Machine learning and the Hough algorithm were used to analyze over 3000 frames of the Sun, identifying spicules and their characteristics. The study found that rotating spicules, accounting for 21% at the poles and 4% at the equator, play a role in energy transfer to the upper solar atmosphere. The observations suggest connections between spicules, mini-loops, magnetic reconnection, and the acceleration of fast solar winds. Understanding these small-scale structures is crucial for comprehending the origin and heating of the fast solar wind.

Keywords. Spicule, solar wind, machine learning, Hough Transform

1. Introduction

Spicules are spike-like structures found in the Sun's chromosphere and transition region. They exhibit spiral motion and opposite blue/red shifts. Some spicules are associated with vorticity movements and may contribute to the propagation of Alfvén rotational waves. High-resolution images reveal rotating spicules, while lower-resolution observations show evidence of twisted spicules in spectral data. Solar wind generation, especially fast solar winds, is linked to coronal holes at the solar poles. Spicules appear as single jets that split into multiple formations and exhibit spinning and rotating motions. Spectroscopic studies confirm rotation and twisting in spicules, indicating complex dynamics and the presence of torsional Alfvén waves. Twisted spicules resembling solar tornadoes have been directly detected. The twist in spicules may be transferred from erupting twisted magnetic fields to the surrounding open magnetic field through external reconnection, suggesting a connection between magnetic field dynamics and spicule behavior (Pike and Mason 1998; Moore et al. 1977; Tavabi et al. 2010, 2012b; Tavabi 2014; Tavabi et al. 2014; Ajabshirizadeh et al. 2008).

2. Observation

This study combines data from the IRIS and Hinode/SOT instruments to investigate the dynamics of solar jets and rotational spicules. The IRIS data provide high-resolution and wide-spectral-range observations, while the Hinode/SOT data offer insights into the formation and twisting of spicules. The data are calibrated using standard software for accuracy.

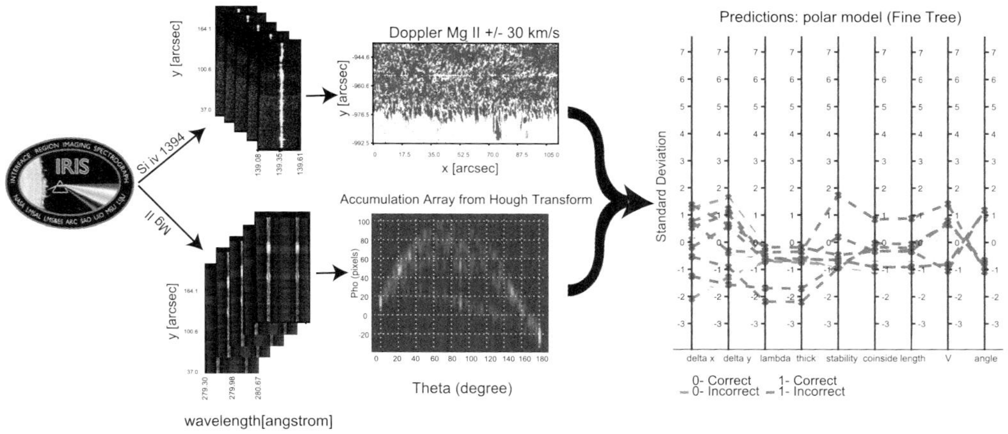

Figure 1. IRIS provided spectral data from 2013 onwards, enabling a detailed study of twisted spicules and their formation. The combination of Hinode and IRIS data, along with their spectra, deepened the understanding of these dynamic processes. Preprocessing steps were outlined to ensure the analysis-ready state of the IRIS dataset, facilitating the extraction and analysis of spicule features.

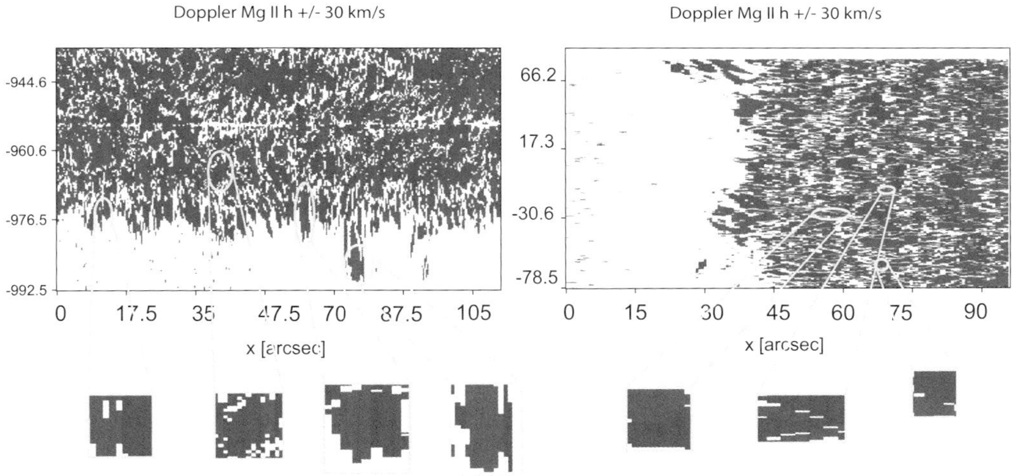

Figure 2. The labeling of twisted spicules on Doppler maps was highlighted, emphasizing their unique color pattern. Doppler maps are generated by examining spectral line shifts, revealing line-of-sight velocities. Twisted spicules exhibit distinct color patterns on the Doppler map, featuring adjacent red and blue areas without separation, indicating intertwined opposite velocity flows within the spicules.

3. Method

This study employed data from the IRIS and Hinode/SOT instruments to investigate solar jets and rotational spicules. A combination of machine learning and the Hough transform algorithm was used to separate and analyze the spicules observed in the spectra (figure 1) . The data covered various regions on the Sun, and the IRIS data had high spatial resolution and wide spectral coverage. The Hough transform algorithm identified spicule line segments in the spectra, while Doppler maps revealed the spicules' Doppler shifts (figure 2). Machine learning models were trained to classify spicules as twisted

Table 1. The summary of data received from machine learning model analysis varies depending on the study region.

	spicules	West Equator	East Equator	North Pole	South Pole
Number of	identified	1843	1152	2576	6608
	twisted	79	45	519	1434
	ratio(%)	4.30	3.91	20.15	21.71
Red rotational velocity of (km/s)	twisted	25	24	36	34
Blue rotational velocity of (km/s)	twisted	29	27	36	33
rotation angle of	identified	8.0	7.7	12.3	12.0
(degree)	twisted	7.3	7.2	12.3	12.3

Table 2. The classified data collected via machine learning model analysis.

	twisted spicules		
	average	min	max
poles			
rotational velocity (km/s)	36	24	65
rotation angle (degree)	12.3	3.5	15.0
equator			
rotational velocity (km/s)	27	24	53
rotation angle (degree)	7.3	2.1	11.5

or non-twisted based on spectral, spatial, and temporal characteristics. The accuracy of the models was evaluated using test data, and the best-performing models were used to identify twisted spicules in the analysis data (figure 1). The study provided insights into the dynamics and properties of solar jets and rotational spicules through a combination of observational data, image processing techniques, and machine learning algorithms (Sadeghi and Tavabi 2022b,a; Tavabi and Koutchmy 2012; Tavabi et al. 2012a).

4. Results

In this study, researchers analyzed a dataset of approximately 11,500 spectral images to identify and categorize spicules using trained models and the Hough transform algorithm. They found that twisted spicules accounted for around 21% of spicules in the polar regions and 4% in the equatorial region. There were slightly more twisted spicules on the west side of the Sun compared to the east side, and the number of twisted spicules was similar at the north and south poles. Tables 1, and 2 in the study provided information on the average rotational velocity, angular deviation, and Doppler velocities of the detected twisted spicules. The number of twists observed in spicules varied depending on their diameter, ranging from at least one turn for fragile structures to broader structures. The study suggested that the expansion of chromospheric mini-loops could lead to micro-eruptions similar to spicule eruptions of the spine type, based on the conservation of rotation or helicities. The structure of the transition region and low corona was found to be influenced by the complex magnetic field configuration, with higher-order multipole components rather than a simple magnetic dipole configuration. The presence of multiple magnetic mini-loop systems anchored at the photosphere resulted in a series of changes in magnetic polarity. Overall, this study provided insights into the prevalence and characteristics of twisted spicules in different regions of the Sun and highlighted the complex magnetic field configuration governing the solar atmosphere (figure 3).

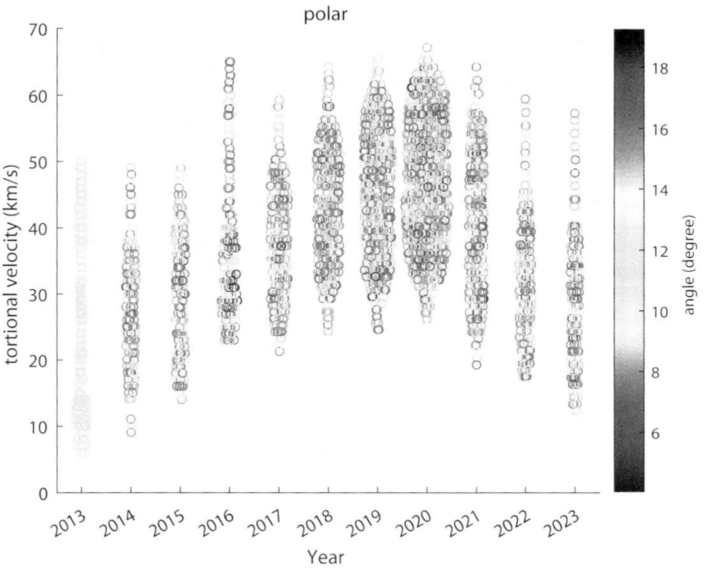

Figure 3. The significant difference in the number and velocity of rotational spicules between polar and equatorial regions has important implications for understanding their connection to material supply for fast solar winds and their acceleration mechanisms. Polar regions exhibit a higher occurrence of rotational spicules with greater velocities, suggesting they may serve as the primary source of fast solar winds. In contrast, equatorial regions have fewer rotational spicules with lower velocities, indicating distinct processes as the source of fast solar winds. Focusing on initial signals from polar activity can provide valuable insights into the activity originating from these regions.

5. Discussion

The study on spicules and their role in solar wind energy and dynamics yielded significant findings:

The transition region (TR) of the Sun exhibits strong torsional structure, with distinct spicule populations at the poles and equator, challenging previous assumptions about helix jets dominating the TR. Spicule observations indicate a higher density in polar regions compared to the equatorial regions, with a slight asymmetry favoring the east side of the Sun over the west side. Rotating spicules are more prevalent in polar regions, suggesting their significant involvement in generating solar wind energy. The abundance of rotating spicules near the solar poles and in coronal hole areas is attributed to the rupture of magnetic mini-loops, which possess magnetic helicity and drive the rotational and helical motion of spicules. The origin and formation mechanisms of spicules remain incompletely understood, but mini-loop interchange reconnection is proposed as a potential process involved in spicule formation. Detailed and systematic observations of the low-coronal region, combined with the outer corona, are crucial for comprehending the origins of fast solar wind and mini-loops. Tornado-type eruptions, including surges, consist of fine, elongated threads and are associated with helical chromospheric jets, adding complexity to the transition from the corona to the fast solar wind. Twisting spicules are associated with high-speed polar winds and play a critical role in particle acceleration and mass loading in fast solar winds. Overall, this study provides valuable insights into the intricate dynamics of spicules and their contributions to solar wind energy. It highlights the importance of further research in this field to deepen our understanding.

References

Ajabshirizadeh, A., Tavabi, E., & Koutchmy, S. L. 2008, Magneto-acoustic wave oscillations in solar spicules. *Astrophysics and Space Science*, 319, 31–35.

Moore, R., Tang, F., Bohlin, J., & Golub, L. 1977, H-alpha macrospicules-identification with euv macrospicules and with flares in x-ray bright points. *The Astrophysical Journal*, 218, 286–290.

Pike, C. & Mason, H. 1998, Rotating transition region features observed with the soho coronal diagnostic spectrometer. *Solar Physics*, 182(2), 333–348.

Sadeghi, R. & Tavabi, E. 2022,a Characteristics of chromospheric oscillation periods in magnetic bright points (mbps). *Monthly Notices of the Royal Astronomical Society*,.

Sadeghi, R. & Tavabi, E. 2022,b A new approach to kinetic energy flux at the different frequencies above the iris bright points. *The Astrophysical Journal*, 938b.

Tavabi, E. 2014, Power spectrum analysis of limb and disk spicule using hinode ca h-line broadband filter. *Astrophysics and Space Science*, 352, 43–50.

Tavabi, E. & Koutchmy, S. L. 2012, About the directional properties of solar spicules from hough transform analysis. *arXiv: Solar and Stellar Astrophysics*,.

Tavabi, E., Koutchmy, S. L., & Ajabshirizadeh, A. 2010, A statistical analysis of the sot-hinode observations of solar spicules and their wave-like behavior. *New Astronomy*, 16, 296–305.

Tavabi, E., Koutchmy, S. L., & Ajabshirizadeh, A. 2012,a Hough transform to study the magnetic confinement of solar spicules. *Journal of Modern Physics*, 3a, 1786–1791.

Tavabi, E., Koutchmy, S. L., & Ajabshirizadeh, A. 2012,b solar spicules and jets. *Eas Publications Series*, 55b, 71–78.

Tavabi, E., Koutchmy, S. L., Ajabshirizadeh, A., Maralani, A., & Zeighami, S. 2014, Alfvénic waves in polar spicules. *Astronomy and Astrophysics*, 573.

Dynamics of Solar and Stellar Convection Zones and Atmospheres
Proceedings IAU Symposium No. 365, 2024
A. V. Getling & L. L. Kitchatinov, eds.
doi:10.1017/S1743921323004283

Exploring Damping Properties of IRIS Bright Points using Deep Learning Techniques

E. Tavabi and R. Sadeghi

Physics Department, Payame Noor University, Tehran, Iran, 19395-3697
email: e_tavabi@pnu.ac.ir

Abstract. This study analyzed the Doppler shift in the solar spectrum using the Interface Region Imaging Spectrograph (*IRIS*). Two types of oscillations were investigated: long period damp and short period damp. The researchers observed periodic perturbations in the Doppler velocity oscillations of bright points (BPs) in the chromosphere and transition region (TR). Deep learning techniques were used to examine the statistical properties of damping in different solar regions. The results showed variations in damping rates, with higher damping in coronal hole areas. The study provided insights into the damping behavior of BPs and contributed to our understanding of energy dissipation processes in the solar chromosphere and TR.

Keywords. deep learning, bright points, oscillation, Doppler velocity

1. Introduction

Solar atmospheric oscillations, particularly Doppler shift oscillations, provide valuable insights into the dynamics and physical properties of the Sun. The network and internetwork regions are of particular interest due to their distinct magnetic field characteristics. However, analyzing these oscillations is challenging due to noise and instrumental effects. Recent studies have revealed oscillatory behavior in coronal BPs, and this study aims to analyze longitudinal oscillations in chromospheric BPs using Doppler maps. Deep learning algorithms are employed to overcome the challenges and uncover the underlying oscillatory signals. The study seeks to enhance our understanding of solar atmospheric oscillations and explore their connection with phenomena like spicular mass ejections and transient heating of plasma. The application of deep machine learning holds promise for accurately identifying and characterizing short-period oscillations in network and internetwork regions. By leveraging advanced data processing methodologies, researchers can gain novel insights into the physical mechanisms driving solar atmospheric oscillations(Ajabshirizadeh et al. 2011; Chandrashekhar and Sarkar 2015; Krijger et al. 2001; Tavabi et al. 2011; Rojas-Quesada 2021; Sadeghi and Tavabi 2022b; Tavabi et al. 2015; Zeighami et al. 2020; Tavabi 2018; Tavabi et al. 2022).

2. Observation

This study utilized data from the *IRIS*, including slit jaw images (SJIs) and spectral data covering a wide range of wavelengths. The data were collected from different solar regions, including equatorial coronal holes, quiet Sun, and active regions. The Mg II h and k lines were specifically studied to understand plasma at low temperatures. By integrating data from multiple sources, researchers gained a comprehensive understanding of the

Sun's activity and behavior. This interdisciplinary approach highlights the importance of combining diverse data sources in solar physics research.

3. Method

This study focused on analyzing Doppler shift oscillations observed in BPs by the *IRIS*. The researchers used spectral lines sensitive to plasma temperature and velocity to construct Doppler velocity time series. Wavelet analysis was performed to detect oscillation periods and understand the underlying physical processes. A deep learning model, based on a feedforward neural network, was trained to classify different oscillation patterns. The model utilized four components corresponding to different spectral lines observed by *IRIS*(see figure 1. Data preprocessing and segmentation were employed to handle the large volume of data. The accuracy of the model ranged from 75% to 99% in classifying oscillation patterns (see figure 2). The study provided insights into plasma behavior in the Sun's atmosphere and demonstrated the potential of deep machine learning in solar physics research (Sadeghi and Tavabi 2022a).

4. Results

In our study, we used the same methodological approach to analyze both short period damp and long period damp phenomena. The methods and techniques employed for data collection, preprocessing, and analysis were consistent across both types of damping. However, despite using the same methodology, the results of our analysis revealed distinct differences between the two types of damping. These differences were observed in various aspects such as the characteristics of the damping behavior, the statistical properties, and the quantitative measurements associated with each type (tables 1 and 2).

5. Discussion

The study investigated the presence of damping in both short-period and long-period oscillations in the Sun's atmosphere.

Short-period damping:

- Previous studies suggested longitudinal waves as a mechanism for transporting energy in the Sun's corona but found their energy insufficient.
- The study analyzed bright points in active region (AR) and coronal hole (CH) regions (see figure 3).
- Internetwork regions in AR and CH displayed shorter decay times and half-lives compared to network regions, indicating higher damping rates potentially due to small-scale magnetic fields.
- CH regions exhibited shorter decay times and half-lives compared to AR regions, likely due to lower density and weaker magnetic fields.
- Bright points in both network and internetwork regions showed damped oscillations, with varying degrees of damping observed in different regions.

Long-period damping:

- Spectral analysis revealed the presence of longitudinal oscillations with longer periods in the Sun's transition region and corona.
- These oscillations exhibited outward-propagating Doppler motions, with amplitudes ranging from 10 to 35 km/s and periods of 180–550 s.
- The oscillations in both network and internetwork bright points were expected to have sufficient energy to drive the solar wind and heat the corona.
- The study did not specifically discuss the damping behavior of long-period oscillations.

a)

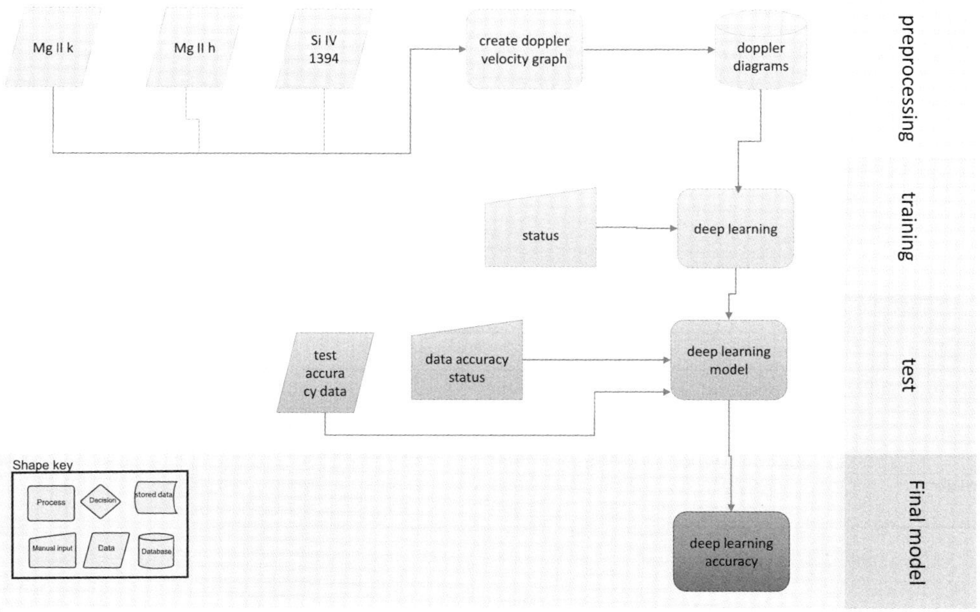

b)

	Name	type	Activ ation	Learnable	Total Learnable	States
1	sequenceinput	sequence input	6	-	0	Hidden State 500x1 Cell State 500x1
2	lstm	LSTM	500	input weights 2000x6 recurrent weights 2000x500 Bias 2000x1	1014000	-
3	fc	Fully connected	6	weights 6x500 Bias 6x1	3006	-
4	softmax	Softmax	6	-	-	-
5	classoutput	Classification Output	4	-	-	-

Figure 1. The deep machine learning examination of the finest time-series *IRIS* rasters validates the rhythmic regime of longitudinal waves discovered using Doppler maps. This research looked at the Doppler shift oscillations over time and above the *IRIS* bright spots. For each bright point, four-time series data were obtained and matched to the blue and red Doppler shifts of the Mg II h & k and Si IV spectra. The average period of Doppler velocity oscillations for network and internetwork points is 300 and 202 seconds, respectively, and bright points are now classified into six types: network bright points in an active region, internetwork bright points in an active region, network bright points in a quiet region, internetworks in a coronal hole area, and internetworks in a coronal hole region. This work employed 16 training data series to build a deep learning model with four components: Blue Si IV, Red Si IV, Blue Mg II, and Red Mg II. The model has 500 hidden layers and classifies data as moist, a single peak, rising, or none. The model's accuracy ranged from 58% to 98%, with an average of more than 80%. The model's validity was established by measuring two points in the active zone and two points in the coronal hole region.

Table 1. Short period damping parameters.

	Net-QS	Inter-QS	Net-AR	Inter-AR	Net-CH	Inter-CH
Average max shift (Km/s)	28	28	34	34	24	21
Average damp (Seconds)	216	120	131	220	150	121
Average damp/period	0.74	0.62	0.46	1.25	0.59	0.80
Average delta	0.50	0.39	0.88	0.73	0.63	0.45
damp %	62	49	32	21	70	55

Notes:

Table 2. Average of long period damp components.

region	point type	f	τ	half-life	ζ	Q	repeat frequency
coronal hole	network	0.8 m	3500	2400	0.34	1.47	1500
	internetwork	1.6 m	2500	1700	0.13	3.84	1000
active	network	2.5 m	2000	1400	0.33	1.48	4000
	internetwork	1.7 m	4800	3300	0.44	1.14	11000
quiet	network	1 m	3000	2000	0.32	1.56	1200
	internetwork	1.5 m	1900	1300	0.17	2.94	1000

Notes:

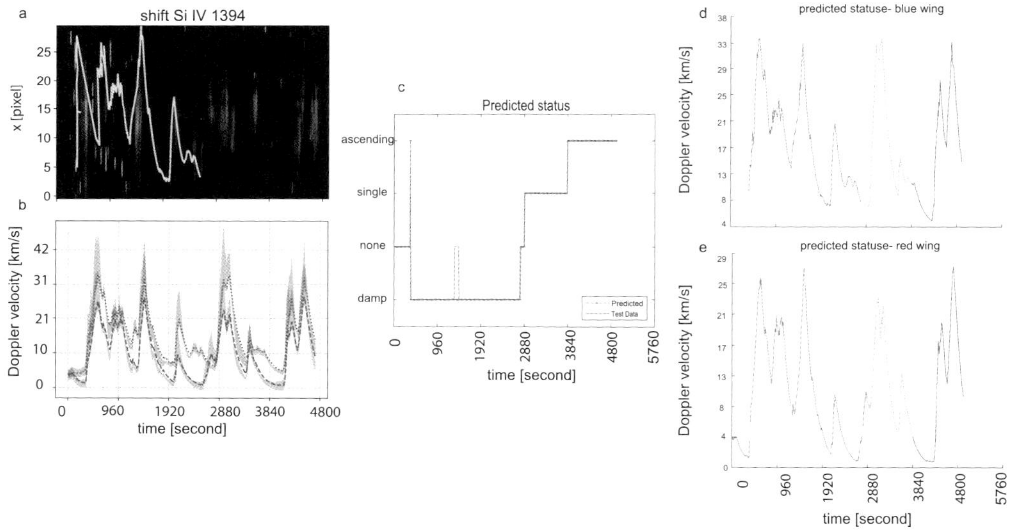

Figure 2. a) This figure illustrates the time series of the silicon 1394 Å spectrum. An indication of dampness can be observed in this image. b) The red and blue wings in this diagram depict the Doppler velocity shift. It can be noted that a blue shift corresponds to the occurrence of redshift dampening. c) In this graphic, the guide data and projected data are presented together. This information is related to the Si IV 1394 Å redshift. d, e) The predictions for two types of shifts, red and blue, are shown. Blue indicates attenuation, orange represents a single peak, and purple signifies an increasing trend.

Overall, the study provided insights into the damping behavior of both short-period and long-period oscillations in the Sun's atmosphere, contributing to our understanding of the dynamics and energy transport processes in the solar corona and solar wind.

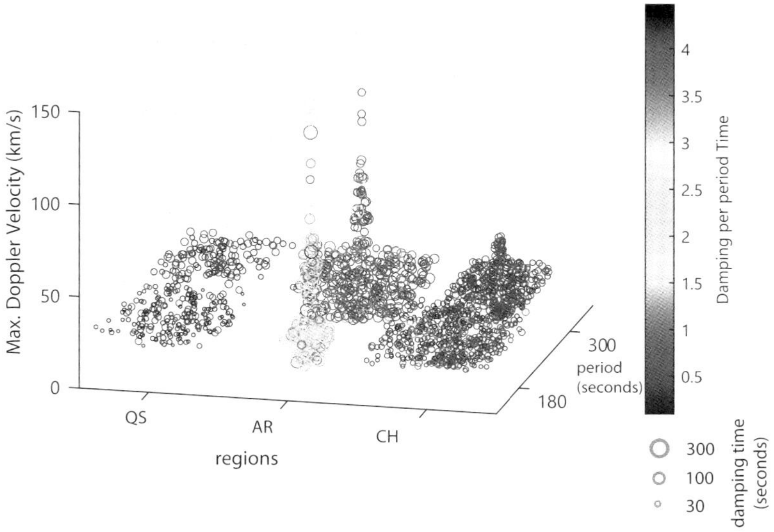

Figure 3. The swarm chart illustrates the maximum Doppler velocity for different regions (QS, AR, and CH) across two periods (300 and 180 sec.). The marker size represents the damping time, with larger markers indicating higher damping values. The chart showcases the distribution of data points and allows for visual comparison between regions and periods. The color bar indicates the corresponding damping per period time values. First, as we expected the number of net and Internetwork BPs numbers is much higher in CH, because the background is darker, also the Damp/Period ratio is the lowest at there, second, in AR we see a jump of Doppler velocity and very higher ratio of Damp/Period parameter at the net and internetwork BPs, the Damp/Period at the internetwork places is significantly growing up in some cases. Finally in the QS regions we could not see the meaningful differences between net and internetwork BPs.

References

Ajabshirizadeh, A., Tavabi, E., & Koutchmy, S. L. Oscillation of ca h jet emission observed by sot: Viscous effects 2011,.

Chandrashekhar, K. & Sarkar, A. 2015, Modeling solar coronal bright-point oscillations with multiple nanoflare heated loops. *The Astrophysical Journal*, 810.

Krijger, J. M., Rutten, R. J., Lites, B. W., Straus, T., Shine, R. A., & Tarbell, T. D. 2001, Dynamics of the solar chromosphere - iii. ultraviolet brightness oscillations from trace. *Astronomy and Astrophysics*, 379, 1052–1082.

Rojas–Quesada. Analysis of oscillations of thequiescent solar filaments 2021,.

Sadeghi, R. & Tavabi, E. 2022,a Characteristics of chromospheric oscillation periods in magnetic bright points (mbps). *Monthly Notices of the Royal Astronomical Society*,.

Sadeghi, R. & Tavabi, E. 2022,b A new approach to kinetic energy flux at the different frequencies above the iris bright points. *The Astrophysical Journal*, 938b.

Tavabi, E. 2018, Synchronized observations of bright points from the solar photosphere to the corona. *Monthly Notices of the Royal Astronomical Society*, 476, 868–874.

Tavabi, E., Koutchmy, S. L., Ajabshirizadeh, A., Zhelyazkov, I., & Mishonov, T. M. Null-point oscillations of chromospheric large cool jets 2011,.

Tavabi, E., Koutchmy, S. L., & Golub, L. 2015, Limb event brightenings and fast ejection using iris mission observations. *Solar Physics*, 290, 2871–2887.

Tavabi, E., Zeighami, S., & Heydari, M. 2022, Dynamics of explosive events observed by the interface region imaging spectrograph. *Solar Physics*, 297.

Zeighami, S., Tavabi, E., & Amirkhanlou, E. 2020, Waves propagation in network and inter-network bright points channels between the chromosphere and transition regions with iris observations. *Journal of Astrophysics and Astronomy*, 41, 1–14.

Dynamics of Solar and Stellar Convection Zones and Atmospheres
Proceedings IAU Symposium No. 365, 2024
A. V. Getling & L. L. Kitchatinov, eds.
doi:10.1017/S1743921324000127

Detecting coronal mass ejections with machine learning methods

K. Vida[1,2], **B. Seli**[1,2], **T. Szklenár**[1,2], **L. Kriskovics**[1,2], **A. Görgei**[1,2,3] **and Zs. Kővári**[1,2]

[1]Konkoly Observatory, HUN-REN Research Centre for Astronomy and Earth Sciences, Budapest, Hungary
email: vidakris@konkoly.hu

[2]CSFK, MTA Centre of Excellence, Budapest, Hungary

[3]Eötvös University, Department of Astronomy, Budapest, Hungary

Abstract. Flares on the Sun are often associated with ejected plasma: these events are known as coronal mass ejections (CMEs). These events, although are studied in detail on the Sun, have only a few dozen known examples on other stars, mainly detected using the Doppler-shifted absorption/emission features in Balmer lines and tedious manual analysis. We present a possibility to find stellar CMEs with the help of high-resolution solar spectra.

Keywords. Sun: coronal mass ejections (CMEs), Stars: activity, methods: data analysis, techniques: spectroscopic, Neural networks

---

1. Introduction

On our Sun, most of the larger flares are consistently accompanied by a coronal mass ejection (CME). CMEs are important because they release massive amounts of plasma into space, impacting space weather and potentially disrupting technological systems on Earth, such as satellites and power grids. Studying CMEs, understanding their dynamics and – in the future – developing predictive models are crucial to mitigate potential adverse effects on space-based infrastructure and communication systems. Solar CMEs are thoroughly investigated through both observational methods and modeling, as outlined in the reviews by Webb and Howard (2012) and Kilpua et al. (2017), among others. The frequency of solar CMEs ranges from 0.5 to 6 CMEs per day, depending on the phase of the solar activity cycle – these events are observed in detail on the Sun by satellites like the Solar Dynamics Observatory (SDO) or Solar and Heliospheric Observatory (SOHO).

In contrast, the observation of CMEs on other stars poses challenges, and only a handful of such events are known. While flares on these stars can be relatively easily observed through photometry, the observation of CMEs is more challenging. In-situ observations of stellar CMEs will likely remain unfeasible for the foreseeable future.

The primary approach for identifying stellar CMEs involves analyzing their Doppler signature, particularly evident in hydrogen's Balmer lines. The expelled material manifests as a flux increase at the blue wing of the spectral line or, in the case of faster events, may appear as a distinct emission feature, commonly referred to as a "bump" (or absorption if observed against the stellar disk). Odert et al. (2017) extensively discussed the potential observational signatures and constraints.

The majority of reported stellar CME detections in the literature are isolated events discovered serendipitously (e.g., Vida et al. 2016). Until recently, only a limited number

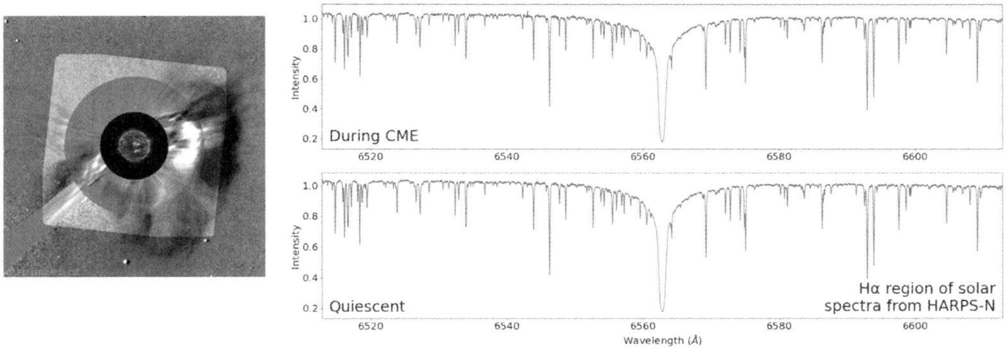

Figure 1. Left: SDO/AIA 171 and SOHO LASCO C2&LASCO C3 observations of the Sun during a massive CME event. Right: the Hα region a solar spectrum during this event compared to the quiescent Sun from HARPS-N observations.

of stellar CME detections existed, and the available data were insufficient for robust statistical analysis. Recent analysis of archive spectral observations has revealed that, although Balmer-line asymmetries are relatively common, the occurrence of successful CME events – those that actually depart from the star – appears to be less frequent than anticipated based on a straightforward scaled-up solar model (Vida et al. 2019; Leitzinger et al. 2020).

2. Using the solar paradigm to understand stellar CMEs

While a recent investigation of solar-like stars (Leitzinger et al. 2020) revealed an absence of CME events on these targets, CMEs on solar-like stars do exist (Namekata et al. 2023). Their rarity could be attributed to either the lower activity levels of these stars, or potential observational biases.

Explanations for phenomena related to stellar magnetism often draw from the solar paradigm, i.e., assuming that activities on stars are analogous to those observed on the Sun. Unlike on stars, on our Sun, we have a detailed knowledge when CMEs occur from direct satellite observations. Could high-resolution, high-cadence Sun-as-a-star observations help us to move forward? Currently, our Sun is entering a period of increased activity in Cycle 25, anticipated to reach its peak around 2025. During this phase, we expect heightened numbers of flares and increased CME activity, providing a unique opportunity to scrutinize the Sun as a star and refine our understanding of stellar CMEs.

The HARPS-N Solar Telescope is a low-cost solar telescope connected to the HARPS-N spectrograph, one of the most precise spectroscopic instruments. The instrument started observing the Sun every clear day with a 5-minute cadence from the Telecopio Nationale Galileo (TNG) facility in La Palma, Spain. The high-resolution, high-cadence Sun-as-a-star data of the first years of observations are publicly available (see Fig. 1). Using these data with the observations from the STEREO/SOHO solar satellites featuring in-situ CME measurements providing the precise timing and physical properties of CMEs we started developing a neural network to better understand these phenomena. If successful, we can use the network to analyze archival spectra possibly revealing events that would not be necessarily found by humans, on other stars.

3. The first experiments

First, we checked the optical spectra at the times of the largest solar CME events. Unfortunately, the spectra show no obvious signs of any ejected material as we observe

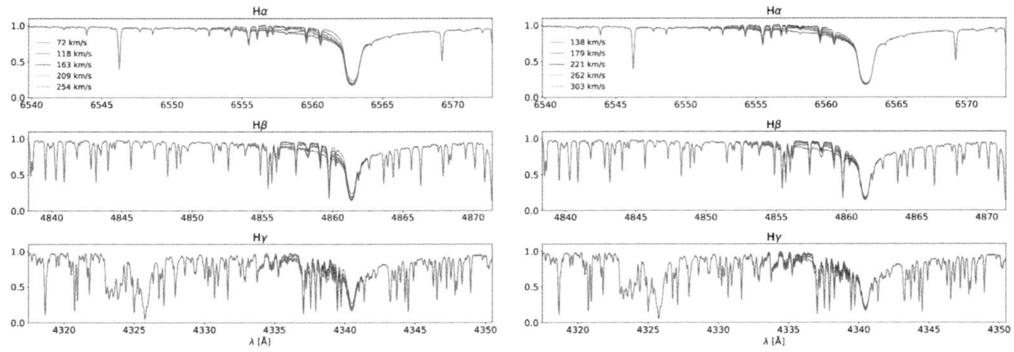

Figure 2. Two examples of injected artificial CME events for training.

these on other stars (Fig. 1) – the most significant changes in the spectra happen in the telluric lines due to airmass and humidity variations.

In our first experiments, we tried various network architectures using Keras (Chollet et al. 2015) using both the full spectral range (≈3900–6900Å) and a smaller region around the Hα line as training data, but these returned random predictions and failed to converge.

To find out what kind of neural network architecture could be optimal for this problem, we created a training set with artificial coronal mass ejections to have obvious signals. These have a typical mass of 5×10^{19}g, 5000 times larger than realistic values. The velocities of the artificial CMEs are realistic, with a maximum velocity of $\approx 400\,\mathrm{km\,s^{-1}}$ (see Fig. 2). This training set contained 50–50% spectra with/without CME signals. After testing different neural network architectures, the ones based on 1D Convolution performed the best. These types of neural networks "scan" through the data with very high accuracy and great performance. After a set of trials, one of the best-performing architecture – to our surprise – was one of the most basic runs, consisting only two `Conv1D` layers with 16 and 32 filters, respectively, followed by two `Dense` layers with 32 neurons each (see Fig. 3). During the preparation of the training, validation, and test datasets, we made sure that there was no overlap between these sets. Our test showed that the accuracy of the deeper 1D Convolution networks was similar but at the expense of speed and performance. Training our neural network was very quick, one epoch took about 2 seconds. For spectral windows of 10 000 data points (a wavelength range of ≈180Å) the whole training finished in less than 2 minutes.

While the best runs achieved over 99% accuracy in both the train and validation set, this happened only on artificial data with events 5000 times larger than realistic values where 50% of the spectra contained events and using only 50 000 data points of the spectrum (the full dataset consists of more than 200 000 points). However, with a network that works consistently on the artificial training set, we can use transfer learning in the future to use that weight file on the original scientific observations. If our experiment is successful, and we can detect coronal mass ejections on the solar spectra, we can possibly find new CME indicators, and determine what magnitude of CMEs are observable on other stars. The new algorithm can be also used to detect CMEs in archive stellar spectroscopic data.

Our results based on the artificial data are promising, but of course, there are many other tests ahead of us. Not just the size of the artificial events need to be reduced significantly, but we must utilize the performance of the 1D Convolution to be able to recognize real solar events as well.

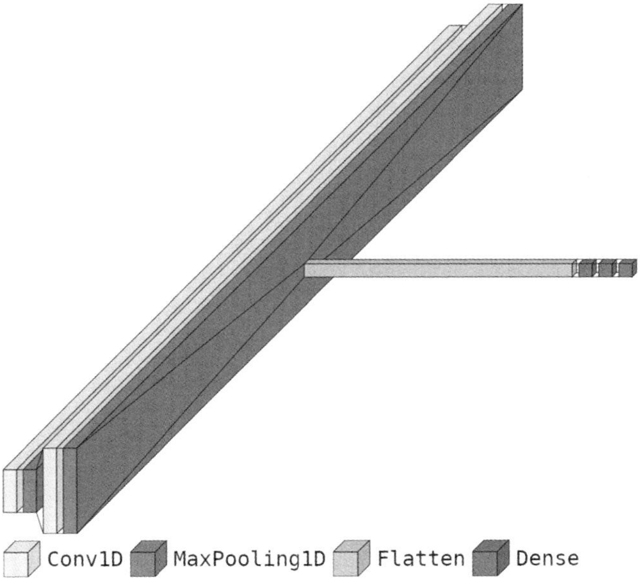

Figure 3. One of the best-performing network architectures for the artificial CME events.

Acknowledgements Authors from Konkoly Observatory acknowledge the Hungarian National Research, Development and Innovation Office (NKFIH) grants OTKA K-131508 and KKP-143986. LK acknowledges the NKFIH grants OTKA PD-134784. LK and KV are Bolyai János research fellows. KV was supported by the Bolyai+ grant ÚNKP-22-5-ELTE-1093, BS was supported by the ÚNKP-22-3 New National Excellence Program of the Ministry for Culture and Innovation from the source of the National Research, Development and Innovation Fund.

References

Chollet, F. *et al.* 2015,. Keras. `https://keras.io`.

Kilpua, E., Koskinen, H. E. J., & Pulkkinen, T. I. 2017, Coronal mass ejections and their sheath regions in interplanetary space. *Living Reviews in Solar Physics*, 14(1), 5.

Leitzinger, M., Odert, P., Greimel, R., Vida, K., Kriskovics, L., Guenther, E. W., Korhonen, H., Koller, F., Hanslmeier, A., Kővári, Z., & Lammer, H. 2020, A census of coronal mass ejections on solar-like stars. *Monthly Notices of the Royal Astronomical Society*, 493(3), 4570–4589.

Namekata, K., Airapetian, V. S., Petit, P., Maehara, H., Ikuta, K., Inoue, S., Notsu, Y., Paudel, R. R., Arzoumanian, Z., Avramova-Boncheva, A. A., Gendreau, K., Jeffers, S. V., Marsden, S., Morin, J., Neiner, C., Vidotto, A. A., & Shibata, K. 2023, Multiwavelength Campaign Observations of a Young Solar-type Star, EK Draconis. I. Discovery of Prominence Eruptions Associated with Superflares. *arXiv e-prints*,, arXiv:2311.07380.

Odert, P., Leitzinger, M., Hanslmeier, A., & Lammer, H. 2017, Stellar coronal mass ejections - I. Estimating occurrence frequencies and mass-loss rates. *Monthly Notices of the Royal Astronomical Society*, 472(1), 876–890.

Vida, K., Kriskovics, L., Oláh, K., Leitzinger, M., Odert, P., Kővári, Z., Korhonen, H., Greimel, R., Robb, R., Csák, B., & Kovács, J. 2016, Investigating magnetic activity in very stable stellar magnetic fields. Long-term photometric and spectroscopic study of the fully convective M4 dwarf V374 Pegasi. *Astronomy & Astrophysics*, 590, A11.

Vida, K., Leitzinger, M., Kriskovics, L., Seli, B., Odert, P., Kovács, O. E., Korhonen, H., & van Driel–Gesztelyi, L. 2019, The quest for stellar coronal mass ejections in late-type stars. I. Investigating Balmer-line asymmetries of single stars in Virtual Observatory data. *Astronomy & Astrophysics*, 623, A49.

Webb, D. F. & Howard, T. A. 2012, Coronal Mass Ejections: Observations. *Living Reviews in Solar Physics*, 9(1), 3.

Dynamics of Solar and Stellar Convection Zones and Atmospheres
Proceedings IAU Symposium No. 365, 2024
A. V. Getling & L. L. Kitchatinov, eds.
doi:10.1017/S1743921323005288

Sub-THz emission from stellar flares and energy release diagnostics

Yuriy T. Tsap[1], Alexander V. Stepanov[2] and Victor F. Melnikov[2]

[1] Crimen Astrophysical Observatory, Nauchny, 298409
email: yuratsap2001@gmail.com

[2] Central Astronomical Observatory at Pulkovo, St. Petersburg, 196140, Russia

Abstract. A comparative analysis of sub-THz emission of stellar flares from red dwarfs has been carried out. ALMA observations indicate that the sub-THz emission flux from stellar flares with a duration of 10 s is an order of magnitude greater than for solar flares. The sub-THz emission is linearly polarized and decreases with frequency. The degree of polarization can reach tens of percent. We show that these types of spectrum slopes and linear polarization can be caused by the synchrotron emission of ultrarelativistic electrons. The origin of the observed relationships between sub-THz, low frequency radio, and X-ray emissions of stellar flares are discussed.

Keywords. stellar flares, sub-THz emission, energy release mechanism

1. Introduction

Regular solar observations in the sub-THz range at 212 and 405 GHz (Kaufmann *et al.* 2001) began only two decades ago mainly due to the Solar Submillimeter Telescope (SST) with a 1.5-meter reflector placed in the Argentina Andes at the altitude of 2.5 km. This opened a new window for studying nonstationary solar phenomena. Despite of many researches devoted to sub-THz observations of solar flares the origin of this emission is not yet fully understood. The study of solar flare events in sub-THz range ($10^2 - 10^3$ GHz) using the unique Atacama Large Millimeter Array (ALMA) with high spatial and temporal resolution started in 2013 is still difficult due to the specifics of the ALMA tools. In this regard, sub-THz observational data on flare stars, where the energy release model is believed to be similar to the solar one (Gershberg 2005) can be very useful.

Red dwarfs are the most common stars in the Galaxy due to the smallness of their mass $< 0.3M\odot$, which, with a high degree of probability, have planets comparable in size to the Earth in the habitable zone (Wandel & Gale 2020). Although solar flares and stellar flares have much in common (Gershberg 2005), nevertheless, the energy of stellar flares is often several orders of magnitude greater than for solar flares, and the luminosity of stars during a flare can increase tens of times. Stellar flares occur much more frequently than solar flares. Besides, the correlation of light curves in different wave bands sometimes is not revealed or has a dependence different from the solar one (MacGregor *et al.* 2018, 2021; Howard *et al.* 2022). For example, a sub-THz event on May 1, 2019 at the Proxima Centauri (Prox Cen) with a short duration of about 10 s, observed with ALMA in spectral windows with a bandwidth of 2 GHz, centered on 225, 227, 239, and 241 GHz (MacGregor et. al 2021) was not accompanied by radio emissions at frequencies below a few GHz, despite the correlation with ultraviolet emission observed with the Hubble Space Telescope. In this case, the radiation flux reached 112 $\pm$

10.4 mJy, the spectral index $\alpha = -2.23 \pm 0.48$, and the lower limit of linear polarization was 0.19 ± 0.07. The millimeter event on May 6, 2019 at the Prox Cen turned out to be the weakest millimeter event so far observed by ALMA; the maximum flux value corresponding to the second radiation peak was 38 ± 5 mJy, $\alpha = -5.1 \pm 3.9$, and the minimum linear polarization was 0.18 ± 0.11. This flare, about 40 min long, was also recorded by the Chandra space observatory in the soft (0.4 – 10 keV) X-ray range. The X-ray light curve had a complex structure with a sharp increase and a slow decrease and consisted of three peaks. Millimeter short bursts observed with ALMA lasted less than 10 s. These bursts approximately corresponded in time to the soft X-ray peaks (first peak was not detected due to calibration). The flare was also observed in the optical range using the Las Cumbres Observatory Global Telescope (LCOGT) in the U band for 30 min. It is also interesting to note that millimeter bursts corresponded in time to the decaying branch of soft X-rays (Howard *et al.* 2022), which contradicts to observations of sub-THz solar events (Smirnova *et al.* 2021).

Thus, the mechanisms of energy release and radiation of stellar flares, despite their magnetic origin, may have features that differ from those predicted by the standard flare model.

Below, using the example of one flare event at the Prox Cen, we analyze the sub-THz observations of stellar flares detected by ALMA, and give probable interpretation of the revealed features.

2. Observations and interpretation of sub-THz radiation from Prox Cen

In spite that the Prox Cen is the quite typical red dwarf, it occupies a special place among the dwarf stars because it is the closest star to us. The rate of occurrence of powerful events exceeding the level of quiescent radiation by 10 times is one event per several days. In optics, the frequency of occurrence of flare events has approximately the same occurrence rate as in soft X-rays, that is, several flares per day. Moreover, according to Davenport *et al.* (2016), extrapolation suggests eight superflares per year with an energy greater than 10^{33} erg.

Let us consider a typical sub-THz flare of the Prox Cen (Fig. 1) observed by ALMA on March 24, 2017, which lasted about 1 min, while the peak flux density $F_\nu = 100$ mJy (MacGregor, 2018) corresponds to the luminosity $L_\nu = 2 \times 10^{14}$ erg/(s Hz), which is approximately an order of magnitude greater than the radio emissions flux F_ν of the most powerful solar events (see Howard et al., 2022). The negative spectrum index $\alpha = -1.8$ and linear polarization (≈ 20 %) indicate the synchrotron origin of sub-THz radiation.

Following Kaufmann *et al.* (1986), it can also be assumed that electrons first emitted softer quanta due to the synchrotron mechanism, which then became harder due to Compton scattering on relativistic electrons (inverse Compton effect). Thus, the generation of sub-THz radiation was determined by synchrotron self-Compton model, which, in particular, is used to interpret the X-ray and gamma radiation of blazars. Meanwhile, Compton losses will dominate over synchrotron losses only if the photon energy density U_{ph} exceeds the magnetic one (Kaufmann *et al.* 1986), i.e.

$$U_{ph} > \frac{B^2}{8\pi}.$$

Taking into account the magnetic origin of solar and stellar flares, the last inequality in the region of flare energy release seems unlikely because a significant part of the free energy of the magnetic field is used to the plasma heating and the large-scale MHD motions.

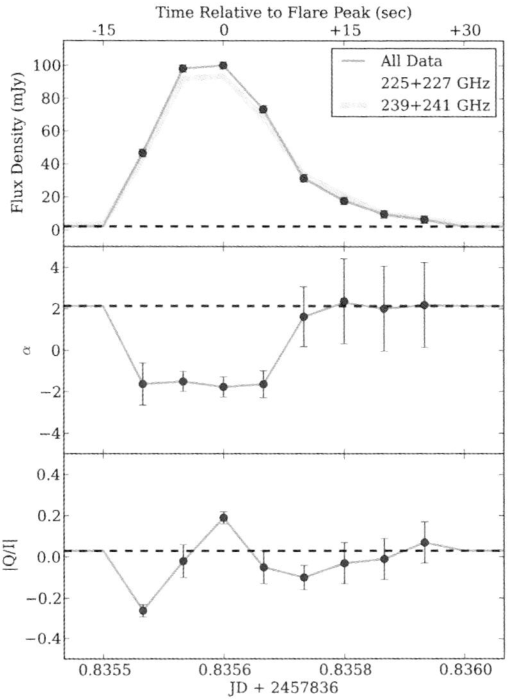

Figure 1. The flux density (top), spectral index with frequency, α, where $F_\nu \propto \nu^\alpha$ (middle), and lower limit on the fractional linear polarization ($|Q/I|$; bottom) during the observed stellar flare on 2017 March 24. The dashed line indicates the quiescent value of each parameter. In the top panel, the flux density of the $225 + 227$ GHz (green) and $239 + 241$ GHz (purple) spectral windows separately, along with the flux density for all spectral windows combined together (blue) are plotted. The figure has been taken from MacGregor *et al.* (2018).

Thus, based on the hypothesis of the synchrotron origin of the sub-THz component of stellar flares, we will do some estimates. The maximum synchrotron radiation of a relativistic electron with total energy E corresponds to the frequency (Ginzburg & Syrovatskii 1964)

$$\nu_s = 1.2 \times 10^6 B\gamma^2 \text{ [Hz]}, \tag{1}$$

where $\gamma = E/(mc^2)$ is the Lorentz factor, $mc^2 = 511$ keV is the electron rest energy, and $B(G)$ is the magnetic field. Then, from Eq. (1) it is easy to obtain the expression

$$\gamma_s \approx 10^{-3}\sqrt{\nu_s/B}. \tag{2}$$

Adopting the frequency $\nu_s = 2.3 \times 10^{11}$ Hz and $B = 500$ G, from Eq. (2) we obtain $\gamma_s \approx 20$, which corresponds to the electron energy $E_s \approx 10$ MeV. In the case of a negative spectral index of the millimeter radiation the value of E_s can be considered as an upper limit, because the frequency of the spectral maximum should be less than ν_s.

The rate of synchrotron energy loss by an electron is (Ginzburg & Syrovatskii 1964)

$$dE/dt \approx 1.6 \times 10^{-15} B^2\gamma^2 \text{ erg/s}. \tag{3}$$

For a quasi-isotropic distribution of relativistic electrons, the number of energetic particles N_s, taking Eq. (3) into account, can be estimated as

$$N_s = L_s/dE/dt = 6.25 \times 10^{14} L_s/(B^2\gamma^2), \tag{4}$$

where L_s is the luminosity in the sub-THz range. Setting $L_s \approx L_\nu \nu_s \approx 5 \times 10^{25}$ erg/s, $\gamma = 20$, and $B = 500$ G, we find from Eq. (4) that $N_s = 3 \times 10^{32}$. The resulting rough lower estimate for the number of relativistic electrons seems to be quite adequate for the following reasons.

If the electron spectrum $N(E) \propto E^{-\delta}$, the total number of accelerated electrons in the source is (Tsap *et al.* 2023)

$$N_0 \approx N_s (E_s/E_0)^{\delta - 1}, \tag{5}$$

where $\delta = 2|\alpha| + 1$ is the electron spectral index and E_0 is the lower threshold of the kinetic energy of accelerated electrons. Putting into Eq. (5) the electron energy $E_0 = 20$ keV, $E_s = 10$ MeV, the electron spectral index $\delta = 4.6$, and $N_s = 3 \times 10^{32}$, we find $N_0 \approx 10^{42}$. Note that Tsap *et al.* (2023) made an arithmetical error in the calculation of N_0. This minimum estimate of the number of energetic particles N_0 seems quite reasonable, because the total number of thermal electrons contained in the energy release region can be estimated as $N_{th} \sim n_e l^3 = 10^{41} - 10^{42}$, where we have taken the characteristic scale $l = 10^{10}$ cm, and the number density of thermal electrons $n_e = 10^{11} - 10^{12}$ cm^{-3}. Consequently, the total number of accelerated electrons N_0 and the number of thermal electrons N_{th} are the same order of magnitude.

3. Discussion and conclusion

We have shown that the synchrotron mechanism can be responsible for the sub-THz flaring emission from the Prox Cen. Because the stellar bursts exceeded solar ones by an order of magnitude, this indicates a more efficient acceleration of relativistic electrons. At the same time, we did not consider the problem of the observed weak correlation between sub-THz and low frequency radio emission. This problem is resolved if we take into account that the flare energy release occurs in denser layers of the stellar atmosphere (chromosphere/photosphere). This approach makes it possible to explain the absence of low-frequency radio emissions by a high number density of background electrons n_e. In fact, if the electron plasma frequency in the burst generation region is higher than the electron gyrofrequency, then electromagnetic radiation can escape the source only in the case when the wave frequency is greater than the plasma frequency. This suggests that for $\nu = 3 \times 10^{11}$ Hz the number density of thermal electrons $n_e > 10^{11}$ cm^{-3} and the low-frequency radio emission will be suppressed. Moreover, due to the high number density of the plasma and, accordingly, large radiation losses, the heating of the plasma by accelerated electrons to high temperatures, which is necessary for the generation of soft X-ray radiation in the energy release region, may turn out to be ineffective.

Consequently, although the sub-THz radiation flux, spectrum, and linear polarization of Prox Cen flares testify to the synchrotron origin of the radiation the analysis did not confirm the complete analogy of events with the standard model of solar flares. Moreover, there are some indications that the synchrotron mechanism can be responsible for the optical spike bursts during a giant flare of UV Ceti (Beskin *et al.* 2017).

Acknowledgments

The study was supported by the Russian Science Foundation (project no. 22-12-00308, Stepanov A.V., Tsap Yu.T.).

References

Beskin, G., Karpov, S., Plokhotnichenko, V., Stepanov, A., & Tsap, Yu. 2017, *PASA*, 34, id.e010.

Davenport,J.R.A., Kipping, D.M., Sasselov, D., Matthews, J.M., & Cameron, C. 2016, *ApJ*, 829, L31.

Gershberg, R.E. 2005, Solar-Type Activity in Main-Sequence Stars (Springer Berlin, Heidelberg)

Ginzburg, V.L. & Syrovatskii, S.I. The Origin of Cosmic Rays (Pergamon, Oxford–New York).

Howard, W.S., MacGregor, M.A., Osten, R, Forbrich, J., Cranmer, S.R. *et al.* 2022, *ApJ*, 938, 103.

Kaufmann, P., Correia, E., Costa, J.E.R., & Zodi Vaz, A.M. 1986, *A&A*, 157, 11.

Kaufmann, P., Raulin, J.-P., Correia, E., Costa, J.E.R., Guillermo, C., de Castro, Giménez, Silva, A.V.R., Levato, H., Rovira, M., Mandrini, C., Fernández-Borda, R., & Bauer, O. 2001, *IAU Symp.*, 203, 283.

MacGregor, M.A., Weinberger, A.J. Wilner, D.J., Kowalski, A.F., & Cranmer, S.R. 2018, *ApJ*, 855, L2.

MacGregor, M.A., Weinberger, A.J., Loyd, R.O., Shkolnik, E. Barclay, T. *et al.*, 2021, *ApJ*, 911, L25.

Smirnova, V.V., Tsap, Yu. T., Morgachev, A.S., Motorina, G.G, & Bárta, M. 2021, *Geomagnetism and Aeronomy*, 61, 993.

Tsap, Yu.T., Stepanov, A.V. & Kopylova Yu.G. 2023, *Geomagnetism and Aeronomy*, 63, 937.

Wandel, A. & Gale, J. 2020, *Iternational J. Astrobiology*, 19, 126.

Dynamics of Solar and Stellar Convection Zones and Atmospheres
Proceedings IAU Symposium No. 365, 2024
A. V. Getling & L. L. Kitchatinov, eds.
doi:10.1017/S1743921323005185

Thermal instability in the impulsive phase of solar flares with sub-THz component

V. V. Smirnova[1], **Yu. T. Tsap**[1], **G. G. Motorina**[2,3,4], **A. S. Morgachev**[2] and **M. Bárta**[4]

[1]Crimean Astrophysical Observatory of Russian Academy of Sciences, Nauchny, 298409, Russia
email: vvsvid.sm@gmail.com

[2]Central Astronomical Observatory at Pulkovo of Russian Academy of Sciences, St. Petersburg, 196140, Russia

[3]Ioffe Institute, Polytekhnicheskaya, 26, St. Petersburg, 194021, Russia

[4]Astronomical Institute of the Czech Academy of Sciences, Ondřejov, 25165, Czech Republic

Abstract. Coronal rain is formed in the post-impulsive phase of solar flares due to the thermal instability of coronal plasma in EUV loops. As a result, the sub-terahertz (sub-THz) emission flux in the post-impulsive phase of solar flares can be increased due to the increasing of the optical thickness of the thermal source. This suggests that sub-THz observations can be used as a diagnostic tool for coronal rain.

This work is aimed to analyse the relationship between the sub-THz radiation and variations of the temperature and the emission measure of the EUV coronal plasma during the post-impulsive phase of the SOL2022-05-04T08:45 solar flare.

Based on the two-dimensional temperature and emission measure distributions obtained from the AIA/SDO EUV intensity data, it was found that the temperature decreases whereas the emission measure reaches the maximum near the sub-THz flare peak. This circumstance and peculiarities of the radiation time profiles in different wave ranges show evidence in favor of the significant contribution of the thermal coronal loop plasma to the flare sub-THz radiation at least for some flare events. The sub-THz emission may be associated with a coronal condensation, accompanied by the formation of coronal rain.

Keywords. Sun, flares, sub-therahertz radiation

1. Introduction

Recently, it became possible to observe weak solar flares with the high spatial resolution at the sub-therahertz (sub-THz, 0.3-3 mm) radio range with the Atacama Large Millimeter/submillimeter Array (ALMA) radio interferometer. Five microflares of GOES class B have been analyzed by Skokić *et al.* (2023). Authors concluded that the sub-THz sources of the flares can be located at the tops or footpoints of hot ($T > 1$ MK) coronal loops and hence, the hot plasma can provide a significant contribution to the sub-THz radiation. This inference contradicts some works wich suggest that the sub-THz radiation is determined by the chrormospheric plasma (Trottet *et al.* 2002; Tsap *et al.* 2016, 2018). This requires additional study of chromospheric and coronal sources of sub-THz emission.

It should be stressed that the peak of sub-THz radiation from solar flares can be observed in a post-impulsive flare phase (Smirnova *et al.* 2023), when the flare-driven coronal rain is formed. Moreover, the temperature T of the coronal loops drops near the sub-THz flare maximum for some events (Figure 1). This suggests that in the post-impulsive flare phase the thermal instability can play an important role in variations of plasma parameters in the sub-THz flare source (Antolin *et al.* 2023).

The goal of this work is to analyze the relationship between sub-THz radiation and the development of the thermal instability in coronal loops observed in the SOL2022-05-04T08:45 solar flare, which can be responsible for the formation of coronal rain.

2. Methodology and data analysis

The method is based on the assumption that the evolution of the sub-THz flare emission can be determined either by the temperature T or by the emission measure $EM = n^2V$ of optically thick ($F1$) and optically thin ($F2$) thermal sources, since

$$F_1 \propto TS, F_2 \propto \frac{n^2V}{\sqrt{T}}, \tag{1}$$

where S and V - the area and the volume of the emission source, respectively. Thus, by comparing the evolution of the temperature T and the emission measure EM of the flare source with the behavior of the sub-THz radiation time profile, one can estimate the contribution of the optically thick/thin sources to the thermal sub-THz radiation.

We analyzed observations of the SOL2022-05-04T08:45 solar flare by GOES X-ray class M5.7 in various spectral ranges, including data at a frequency of 93 GHz obtained with the RT-7.5 Bauman Moscow State Technical University radio telescope (Tsap *et al.* 2016). Time profiles of the flare at different spectral bands (SXR, HXR, microwave) are presented in Figure 2. The impulsive phase of the flare was observed in HXR and microwaves whereas it was not detected in the sub-THz emission.

Previously Smirnova *et al.* (2023) calculated the expected sub-THz radiation based on the differential emission measure (DEM) of SOL2022-05-04T08:45 according to AIA/SDO data in the temperature range $0.5 \leq T \leq 32$ MK (Hannah & Kontar 2012). This allowed us to conclude that the coronal plasma makes a significant contribution to the sub-THz emission of this flare which turned out to be ~ 40 sfu.

3. Results

A detailed analysis of the evolution of the flare temperature T and the emission measure EM was carried out in the time range 08:43-09:15 UT with a step of 12 s.

Left panel of Figure 3 shows a comparison of the T and EM time profiles with the sub-THz emission profile at 93 GHz. Using AIA images (https://iris.aws.lmsal.com/data/level2/2022/05/04/) in 304 Å passband, we can conclude that the peaks of the sub-THz emission and the emission measure coincide with the beginning of coronal rain. Consequently, taking into account Equation (1), we can conclude that the sub-THz emission is determined by the optically thin source connected with the plasma condensation in coronal loops.

Right panel of Figure 3 represents the comparison of H_α and 304 Å chromospheric time profiles with the 93 GHz profile of the flare emission. The location of 93 GHz flare maximum is not coincided with the emission peaks observed in H_α and 304 Å. Thus, taking into account this circumstance and the formation temperatures of the H_α line and the emission at the 304 Å passband we can conclude, that the chromospheric heating does not occur after the sub-THz maximum. This suggests that the thermal plasma of coronal loops plays a dominant role in the generation of the sub-THz component of the SOL2022-05-04T08:45 flare.

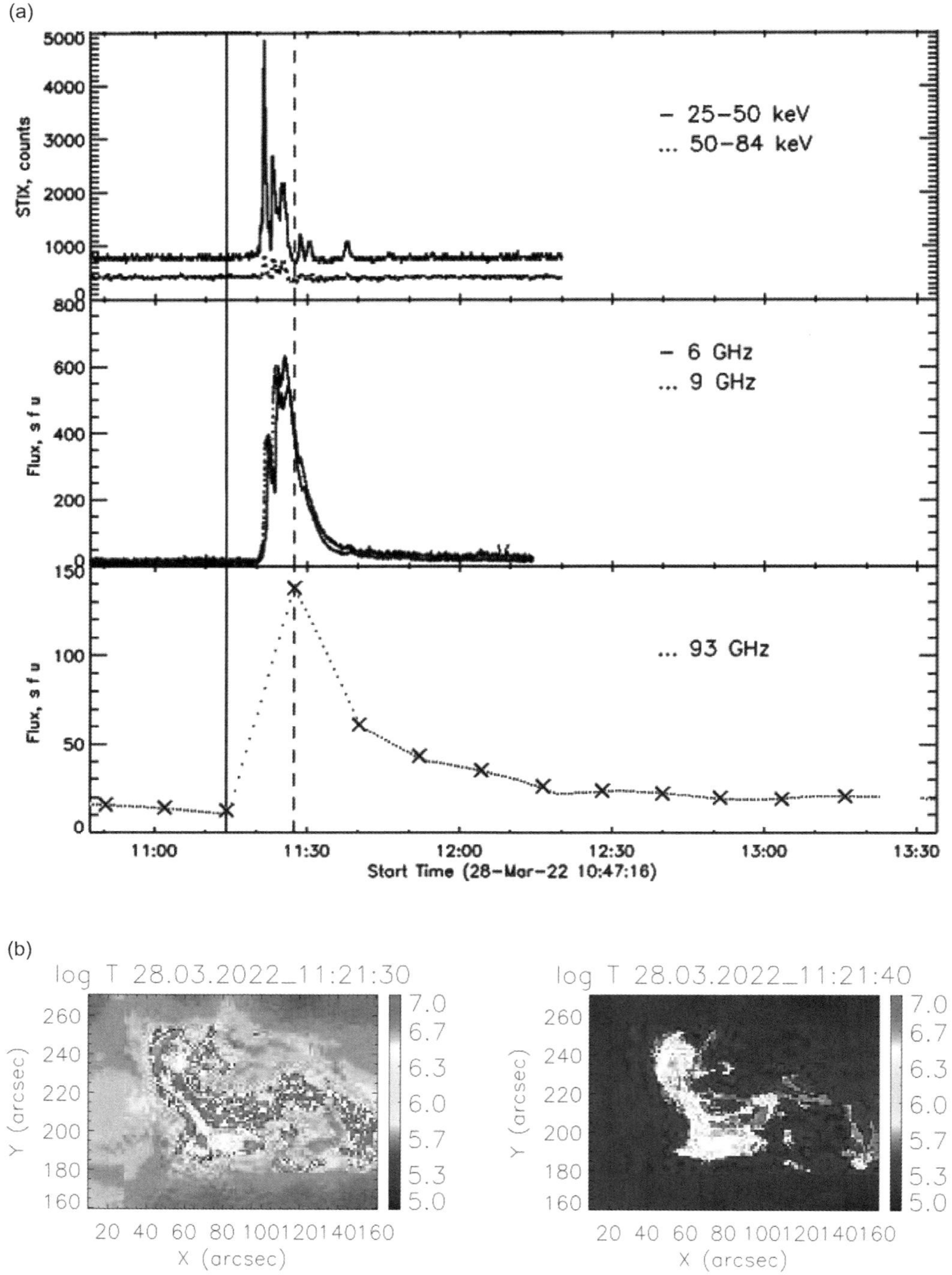

Figure 1. [a]: time profiles of the SOL2022-03-28T11:28 flare obtained with HXR, microwave, and 93 GHz observations. Solid line represents the pre-impulsive phase of the flare. Dashed line shows the maximum of the emission at 93 GHz. [b]: AIA temperature maps obtained near the maximum at 93 GHz.

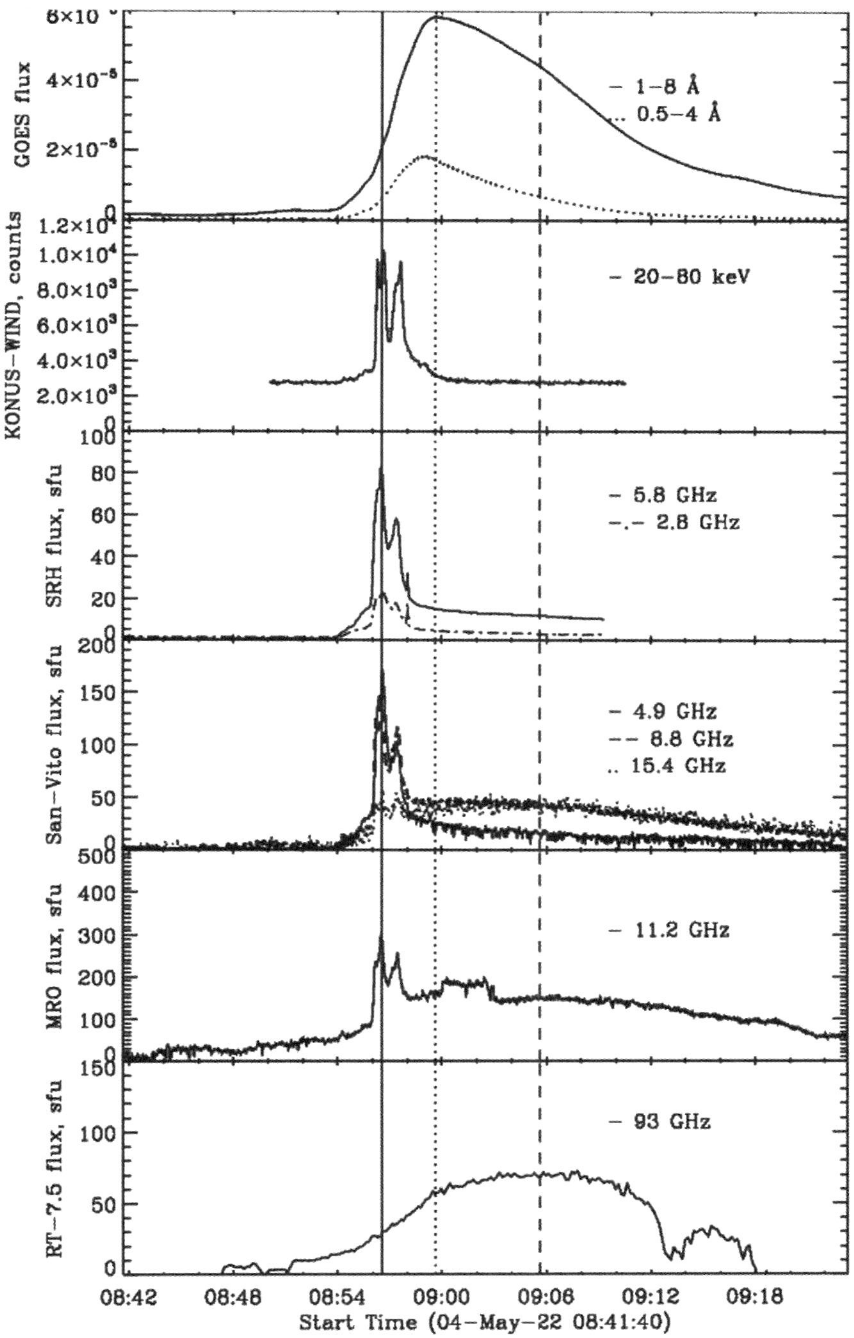

Figure 2. Time profiles of the SOL2022-05-04T08:45 flare. The vertical solid line corresponds to the peak of the impulsive phase of the flare, dashed line shows the 93 GHz flare maximum, and dotted line shows the begining of the sub-THz maximum.

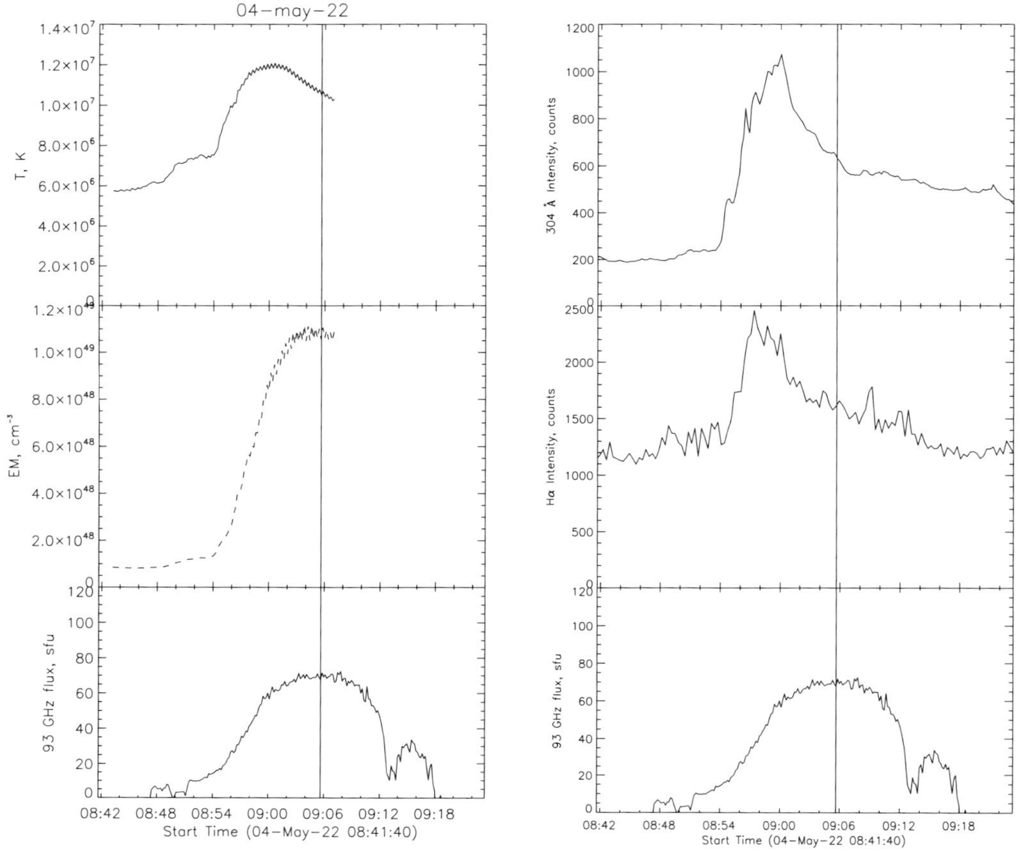

Figure 3. The comparison of time profiles of the AIA temperature T, emission measure EM (left), emission at 304 Å and H_α (right) with sub-THz observations.

4. Conclusions

(1) A multi-wave analysis of the SOL2022-05-04T08:45 flare event was carried out.

(2) The peaks of the sub-THz and emission measure time profiles of the flare are coincided, that indicates a significant contribution of optically thin coronal plasma to the sub-THz radiation.

(3) The sub-THz emission, at least for some flare events, may be associated with a coronal condensation, accompanied by the formation of coronal rain.

This work was supported by the Russian Foundation for Basic Research grant No. 20-52-26006 (Smirnova V.V.), Russian Science Foundation No. 22-12-00308 (Tsap Yu.T., Morgachev A.S.), the State Assignment No. 0040-2019-0025 (Motorina G.G.), RVO:67985815, the project LM2018106 of the Ministry of Education, Youth and Sports of the Czech Republic, Grant 21-16508J of the Grant Agency of the Czech Republic (Bárta M.).

References

Antolin, P., Dolliou, A., Auchére, F., et al. 2023, Astronomy & Astrophysics, 676, A112.
Hannah I.G. & Kontar E.P., 2012, Astronomy & Astrophysics, 539, A146.
Skokić, I., Benz, A. O., Brajša, R., et al., 2023, Astronomy & Astrophysics, 669, A156.

Smirnova V.V., Tsap Yu.T., Motorina G.G., Morgachev A.S., Bárta M., 2023, Geomagnetism & Aeronomy, 63, 5, 527.

Trottet G., Raulin J.P., Kaufmann P., 2002, Astronomy & Astrophysics, 381, 694.

Tsap Yu.T., Smirnova V.V., Morgachev A.S., Motorina G.G., Kontar E.P., Nagnibeda V.G., Strekalova P.V., 2016, Advances in Space Research, 57, 7, 1449.

Tsap Yu.T., Smirnova V.V., Motorina G.G., Morgachev A.S., Kuznetsov S.A., Nagnibeda V.G., Ryzhov V.S., 2018. Solar Physics, 293, 3, 50.

Dynamics of Solar and Stellar Convection Zones and Atmospheres
Proceedings IAU Symposium No. 365, 2024
A. V. Getling & L. L. Kitchatinov, eds.
doi:10.1017/S1743921323005276

Modeling of the solar flare chromosphere and sub-THz radiation with FLARIX and RADYN

Galina G. Motorina[1,2,3], **Yuriy T. Tsap**[4], **Jana Kašparová**[3]†, **Victoria V. Smirnova**[4], **Alexander S. Morgachev**[1] and **Miroslav Bárta**[3]

[1]Central Astronomical Observatory at Pulkovo of Russian Academy of Sciences, St. Petersburg, 196140, Russia

[2]Space Research Institute of Russian Academy of Sciences, Moscow, 117997, Russia

[3]Astronomical Institute of the Czech Academy of Sciences, 251 65 Ondřejov, Czech Republic, email: `galina.motorina@asu.cas.cz`

[4]Crimen Astrophysical Observatory, Nauchny, 298409

Abstract. The origin of the sub-terahertz (sub-THz) component of radio emission from solar flares, which is characterized by the increase flux with frequency in the 100-400 GHz range, is considered. On the basis of equations of 1D non-LTE radiation hydrodynamics we simulated the altitude distribution of the plasma density and temperature inside the flare loop caused by the interaction of non-stationary beam of accelerated electrons in the form of a triangular pulse with the chromospheric plasma. The FLARIX numerical code was used to calculate the dynamics of the flare plasma parameters at different heights which are compared with the RADYN numerical code. We found that the characteristic heights of the formation of sub-THz emission vary over a wide range with time for both codes. The main contribution to the sub-THz emission comes from the chromospheric and transition region plasma with temperatures of $10^4 - 10^5$ K.

Keywords. Sun: Flares - Sun: X-rays, EUV, Radio emission, sub-THz emission

1. Introduction

Solar flares are not yet fully understood explosive phenomena. Namely processes of particle acceleration – one of their most significant features – are still subject to many controversies. Understanding the mechanism of energy release and conversion in the flares is important not only *per se*, but also as a key to the space-weather predictions. Moreover, explosive events similar (not just phenomenologically, but by theoretical models also in their nature) to the flares happen in other astrophysical contexts, too, ranging from stellar flares to the bursts generated in the accretion discs surrounding the black-holes. Our Sun is just close enough to provide a unique laboratory for studying this kind of processes. Enhanced solar sub-terahertz (sub-THz) radiation, which is directly related to the energetic particles accelerated in the flares, represents an important and in many respects novel diagnostic tool. Understanding its generation at the Sun can shed light to the so far bit mysterious sub-THz burst at the other starts, too. Indeed, in 2017, Atacama Large Millimeter Array (ALMA) detected for the first time a number of mm-emission

† Most of the presented contribution was finished before February 2022

bursts on M-dwarfs (MacGregor et al. 2021), however, the origin of their emission still remains unclear.

A large number of studies have focused on sub-THz (0.3-3 mm) emission of solar flares (see, e.g., Krucker et al. 2013; Tsap et al. 2016; Kontar et al. 2018). Radio observations at the Solar Submillimeter Telescope (SST) at 212 & 405 GHz (Kaufmann et al. 2001) and the Köln Observatory for Submillimeter and Millimeter Astronomy (KOSMA) at 230 & 345 GHz (Lüthi et al. 2004) found that for almost half of the observed strong flares (M and X-class), the radio flux decreases with the increase of the frequency, i.e. the slope of the spectrum is negative, as expected if the gyrosynchrotron spectrum continued to higher frequencies (Lüthi et al. 2004; Trottet et al. 2002; Raulin et al. 2004). However, in the rest of the cases, the so-called sub-THz events (*STE in further*), the slope of the spectrum is positive and the fluxes can increase tens of thousands of times during the flare energy release (see, e.g., Krucker et al. 2013; Kaufmann et al. 2015). Several emission mechanisms have been proposed to explain the *STEs*: gyrosynchrotron, thermal free-free, Cherenkov or the plasma mechanism, however, all of them have their caveats (e.g., Fleishman and Kontar 2010).

A recent statistical study by Kontar et al. (2018) showed that thermal optically thick plasma of the chromosphere and transition region with temperatures $T = 2 \times (10^4 - 10^5)$ K and densities of $10^{11} - 10^{12}$ cm^{-3} can produce the observed *STEs* through the thermal free-free mechanism. Moreover, it was shown that due to rapid radiative cooling, the time profiles of sub-THz and hard X-ray emissions, up to sub-second (0.1-1 s) time scales, behave similarly. The authors revealed that there is a positive correlation between the flare ribbons area and the sub-THz fluxes, and that the spectral index $\delta \leq 2$, which is consistent with the thermal mechanism. However, validation of the model in the sub-THz frequency range is still lacking for observational confirmation. In this connection, numerical modeling can provide additional information on the contribution of the chromosphere, transition region, and corona to sub-THz radiation and thus additional constraints on the existing emission mechanisms. Previously Morgachev et al. (2020), using F-CHROMA models (Carlsson et al. 2023) calculated with RADYN code, showed that chromospheric inhomogeneities can have a significant effect on generation of sub-THz emission from solar flares.

Here we apply non-LTE models of the flare chromosphere and transition region obtained with radiation-hydrodynamic (RHD) numerical codes FLARIX (Varady et al. 2010) and RADYN (Carlsson and Stein 1992, 1997) to model the response of the chromosphere to an electron beam bombardment during solar flares. With these two codes we calculate a flare solar atmosphere and its evolution to model the thermal sub-THz emission. We also calculate the altitude distributions of the temperature, density, and contribution function, to determine the input of different atmospheric layes to the sub-THz radiation for different time moments.

2. Sub-THz radio emission from FLARIX and RADYN models

To date, it has been possible to develop numerical methods to determine the temporal evolution of flare plasma parameters as a result of solar atmosphere interaction with the accelerated electron flux in the framework of RHD (see as a recent review Kerr 2023). Recent comparison of RADYN and FLARIX (Kašparová et al. 2019) has shown that both codes are relatively well matched. However, in the context of sub-THz events, the results of the use of FLARIX has not yet been examined in detail. Thus, the aim of the work is to compare FLARIX and RADYN models in terms of sub-THz observations.

We calculated the temporal evolution of sub-THz thermal bremsstrahlung emission (Fig. 1) for 1D models of the flare chromosphere and transition region obtained with

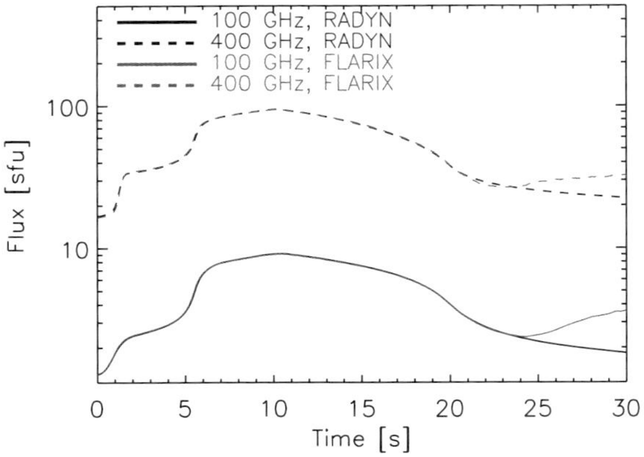

Figure 1. The time profiles of the sub-THz emission of RADYN (black) and FLARIX (red) models at frequencies 100 (solid line) and 400 GHz (dashed line).

the FLARIX and RADYN codes. The FLARIX and RADYN models were taken from (Kašparová et al. 2019). Both models describe the response of an unperturbed solar atmosphere, i.e. the model C from Vernazza et al. (1981) to a beam of non-thermal electrons. FLARIX and RADYN have differences in approaches (Fokker–Planck versus test particles), radiation losses, etc. In order to compare the results of the two simulations, the initial parameters were taken to be the same: the VAL-C undisturbed solar atmosphere and the nonthermal electron flux in the form of a symmetrical triangular time profile peaking at 10 s with duration of 20 s. The parameters of the electron beam were as follows: the spectral index of 3, the low-energy cutoff of 20 keV, the total energy of the electron flux of 10^{11} erg s^{-1} cm^{-2}.

Each model contains a set of plasma parameters varying with height: temperature, electron density, hydrogen density, degree of ionization, etc. over a time interval of 30 s with a time step of 0.1 s. The modelled height distributions of temperature, and electron density are shown in Fig. 2.

To infer the total spectral flux F_ν (its thermal (free-free) bremsstrahlung at frequencies $\nu = 100$ and 400 GHz) the model altitude dependences of plasma temperature and plasma number density at each time were used:

$$F_\nu(H) = \frac{S}{R^2} \int_0^H CF(h) dh. \tag{1}$$

Here the height begining from the photosphere $H = 10000$ km corresponds to the full source thickness, S is the source area, R is the Sun-Earth distance, and the contribution fuction of an individual chromospheric layer at the height h to the total intensity is

$$CF(h) = \eta_\nu(h) \exp\left(-\int_h^H k_\nu dh'\right), \tag{2}$$

where η_ν is the emissivity (see e.g. Morgachev et al. 2020), and k_ν is the absorption factor (Dulk 1985).

The calculated time profiles of the thermal bremsstrahlung emission at $\nu = 100$ and 400 GHz are shown in Fig. 1. The calculations of the sub-THz emission found that a spectral slope between 100 and 400 GHz remains positive at all times. The maxima of the time profiles of the electron beam and sub-THz emission coincide. The time profiles at 100 and 400 GHz for both models behave similarly: they have a rise phase during time

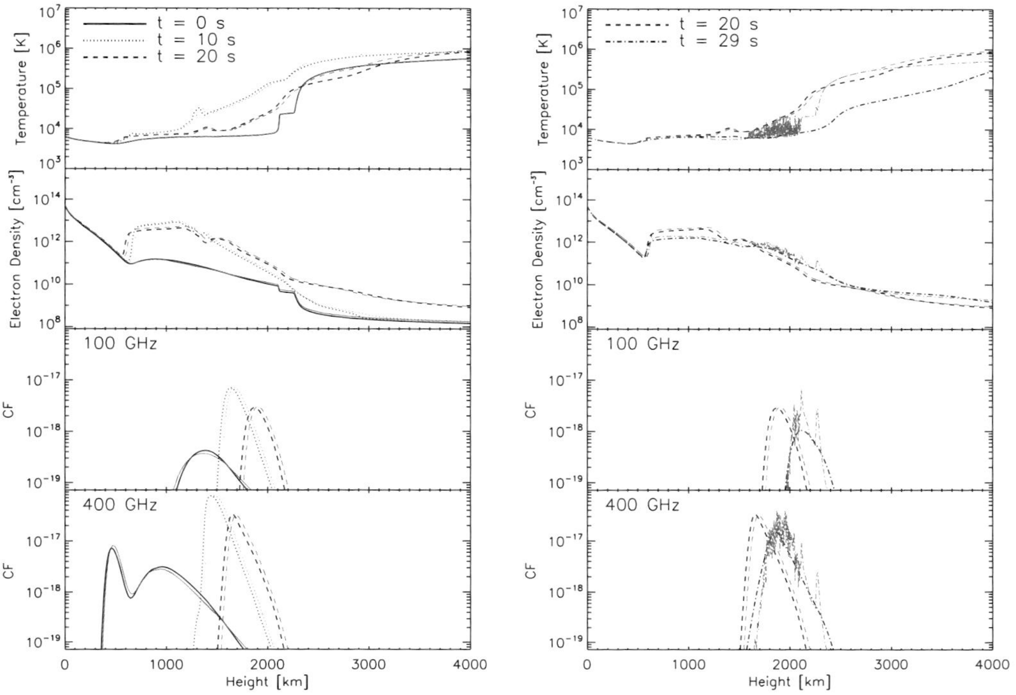

Figure 2. From top to bottom: height distribution of temperature, electron plasma density, contribution function $CF(h)$ at 100 and 400 GHz for the RADYN (black) and FLARIX (red) models at times t = 0, 10, 20 s (left panel) and t = 20, 29 s (right panel).

interval $t = 0 - 10$ s, a peak at time $t = 10$ s, and a decay phase till $t = 30$ s. However, for the FLARIX model the decay phase changes to a rise from $t \approx 24$ s till the end of the simulation. The results from both codes behave similarly till $t = 22$ s.

A good coincidence between the corresponding curves is seen in height distributions of temperature, electron density, and contribution functions till $t \approx 22$ s (Fig. 2). At around $t = 22$ s the FLARIX results start to have numerical oscillations at height ~ 2 Mm, which increase in width and grow over time. It is clear that the oscillations in temperature, electron density result in the oscillations of the contribution function $CF(h)$ (Fig. 2, right panel) and then contribute to the rise of the total spectral flux (Fig. 1). Besides the numerical oscillations a difference particularly in the temperature distribution from RADYN shows up at heights above 2 Mm.

The difference between the results of numerical calculations can be explained as follows. The used FLARIX model does not allow for temperature decrease because of the boundary condition, which is not quite correct for modeling relaxation of flaring atmosphere after $t \approx 22$ s for this simulation. This artificial condition means that the atmosphere cannot cool below its initial state. However, due to the moving shocks and condensations this is not always true. We cannot observe chromospheric condensations with FLARIX because of the temperature constraint while the RADYN temperature can significantly drop. These circumstances result in differences in simulations.

It is interesting to consider the height distribution of the contribution function $CF(h)$ from different levels of the chromosphere (Fig. 2, two bottom panels) because it determines the altitude at which sub-THz emission is generated. At frequencies 100 and 400 GHz the main contribution comes from a layer in the chromoshere and transition region of 400-500 km thickness at a height of about 2000 km with temperatures of $10^4 - 10^5$ K, which is heated by precipitating particles and moves to higher layers with time.

3. Conclusions

We calculated and compared evolution of the thermal bremsstrahlung emission in the sub-THz range for the RADYN & FLARIX models. For both models the characteristic heights of the formation of sub-THz emission is about 2000 km, which corresponds to the chromosphere and transition region, and vary over a wide range with time. The regions of formation of sub-THz emission in the RADYN & FLARIX models differ by less than 100 km. According to the evolution of the contribution function $CF(h)$, the main contribution to the sub-THz emission is made by the chromospheric and trasition region plasma with temperatures of $10^4 - 10^5$ K. The difference in the radiation flux does not exceed 15%. The maxima of the electron pulse and millimeter emission coincide. The region of low temperature and high plasma density (chromospheric condensation) with FLARIX was not observed due to the temperature constraint. Hovewer, the same results were obtained with RADYN.

New sub-THz radio observations (ALMA, RT-7.5, SST) will help to diagnose the properties of flaring plasma in the transition region/chromosphere. More numerical simulations (e.g., with FLARIX) are needed.

Acknowledgments

The study was in part supported by the RSF grant 20-72-10158 (G.G. Motorina), Russian Science Foundation grant no. 22-12-00308 (Yu.T. Tsap, A.S. Morgachev), the Russian Foundation for Basic Research grant no. 20-52-26006 (V.V. Smirnova), grant 21-16508J of the Grant Agency of the Czech Republic, the project RVO:67985815 (J. Kašparová, M. Bárta), and the project LM2018106 of the Ministry of Education, Youth and Sports of the Czech Republic (M. Bárta).

References

Carlsson, M., Fletcher, L., Allred, J., *et al.* 2023, The F-CHROMA grid of 1D RADYN flare models. *Astron. Astrophys.*, 673, A150.

Carlsson, M. & Stein, R. F. 1992, Non-LTE Radiating Acoustic Shocks and CA II K2V Bright Points. *ApJL*, 397, L59.

Carlsson, M. & Stein, R. F. 1997, Formation of Solar Calcium H and K Bright Grains. *ApJ*, 481(1), 500–514.

Dulk, G. A. 1985, Radio emission from the sun and stars. *Annual Rev. Astron. Astrophys.*, 23, 169–224.

Fleishman, G. D. & Kontar, E. P. 2010, Sub-Thz Radiation Mechanisms in Solar Flares. *ApJL*, 709(2), L127–L132.

Kaufmann, P., Raulin, J. P., Correia, E., *et al.* 2001, Rapid Submillimeter Brightenings Associated with a Large Solar Flare. *ApJ*, 548(1), L95–L98.

Kaufmann, P., White, S. M., Marcon, R., *et al.* 2015, Bright 30 THz impulsive solar bursts. *Journal of Geophysical Research (Space Physics)*, 120(6), 4155–4163.

Kašparová, J., Carlsson, M., Heinzel, P., & Varady, M. Modelling of Flare Processes: A Comparison of the Two RHD Codes FLARIX and RADYN. In Werner, K., Stehle, C., Rauch, T., & Lanz, T., editors, *Radiative Signatures from the Cosmos* 2019,, volume 519 of *Astronomical Society of the Pacific Conference Series*, 141.

Kerr, G. S. 2023, Interrogating Solar Flare Loop Models with IRIS Observations 2: Plasma Properties, Energy Transport, and Future Directions. *Frontiers in Astronomy and Space Sciences*, 9, 425.

Kontar, E. P., Motorina, G. G., Jeffrey, N. L. S., *et al.* 2018, Frequency rising sub-THz emission from solar flare ribbons. *Astronomy & Astrophysics*, 620, A95.

Krucker, S., Giménez de Castro, C. G., Hudson, H. S., *et al.* 2013, Solar flares at submillimeter wavelengths. *Astronomy and Astrophysics Review*, 21, 58.

Lüthi, T., Lüdi, A., & Magun, A. 2004, Determination of the location and effective angular size of solar flares with a 210 GHz multibeam radiometer. *Astronomy & Astrophysics*, 420, 361–370.

MacGregor, M. A., Weinberger, A. J., Loyd, R. O. P., *et al.* 2021, Discovery of an Extremely Short Duration Flare from Proxima Centauri Using Millimeter through Far-ultraviolet Observations. *ApJL*, 911(2), L25.

Morgachev, A. S., Tsap, Y. T., Smirnova, V. V., *et al.* 2020, Numerical Simulation of Sub-Terahertz Thermal Emission: RADYN Code. *Geomagnetism and Aeronomy*, 60(8), 1038–1049.

Raulin, J. P., Makhmutov, V. S., Kaufmann, P., *et al.* 2004, Analysis of the impulsive phase of a solar flare at submillimeter wavelengths. *SoPh*, 223(1-2), 181–199.

Trottet, G., Raulin, J. P., Kaufmann, P., *et al.* 2002, First detection of the impulsive and extended phases of a solar radio burst above 200 GHz. *Astronomy & Astrophysics*, 381, 694–702.

Tsap, Y. T., Smirnova, V. V., Morgachev, A. S., *et al.* 2016, On the origin of 140 GHz emission from the 4 July 2012 solar flare. *Advances in Space Research*, 57(7), 1449–1455.

Varady, M., Kasparova, J., Moravec, Z., *et al.* 2010, Modeling of Solar Flare Plasma and Its Radiation. *IEEE Transactions on Plasma Science*, 38(9), 2249–2253.

Vernazza, J. E., Avrett, E. H., & Loeser, R. 1981, Structure of the solar chromosphere. III. Models of the EUV brightness components of the quiet sun. *Astrophysical Journal, Suppl. Ser.*, 45, 635–725.

Dynamics of Solar and Stellar Convection Zones and Atmospheres
Proceedings IAU Symposium No. 365, 2024
A. V. Getling & L. L. Kitchatinov, eds.
doi:10.1017/S1743921324000267

Exploring the Relation Between Magnetic Fields, Starspots, and CMEs

Fabian Menezes[1], **Adriana Valio**[2], **Yuri Netto**[1], **Alexandre Araújo**[1], **Christina Kay**[3] **and Merav Opher**[4]

[1]Universidade Presbiteriana Mackenzie, Centro de Rádio Astronomia e Astrofísica Mackenzie (CRAAM), São Paulo, Brazil
email: `menezes.astroph@gmail.com`

[2]Universidade de São Paulo, Instituto de Astronomia, Geofísica e Ciências Atmosféricas (IAG), Departamento de Astronomia, São Paulo, Brazil

[3]NASA Goddard Space Flight Center, Heliophysics Science Division, Greenbelt, USA

[4]Boston University, Department of Astronomy, Boston, USA

Abstract. Solar-type stars, including the Sun, have magnetic fields that extend from their interiors to the surface and beyond, influencing both the stellar activity and interplanetary medium. Magnetic activity phenomena, such as coronal mass ejections (CMEs), significantly impacts space weather. These CMEs, composed of plasma clouds with magnetic fields ejected from the stellar corona, pose a potential threat to planets by affecting their magnetosphere and atmosphere. Despite advancements in detecting stellar CMEs, detection remains limited. We focus on understanding CME propagation by analyzing key parameters like position, velocities, and the configuration of stellar magnetic fields. Using spot transit mapping, we reconstruct magnetograms for Kepler-63 and Kepler-411, employing the ForeCAT model to simulate CME trajectories from these stars. Results indicate that CME deflections generally decrease with radial velocity and increase with ejection latitude. Additionally, stars with stronger magnetic fields, such as Kepler-63, tend to cause more significant CME deflections.

Keywords. Solar-type stars; Starspots; Stellar magnetic field; coronal mass ejections (CMEs)

1. Introduction

The Sun and other stars have magnetic fields that permeate their interiors and affects their evolution in many ways, according to Brun and Browning (2017). According to Parker (1955); Spruit (1997); Spruit et al. (2010); Charbonneau (2014) and Stejko et al. (2020), stellar magnetic fields originate in the depths of the convective zone of the star, specifically in the tachocline. The solar dynamo theory explains the formation of these magnetic field concentrations, initiated in the tachocline and amplified in the convective zone. Magnetic flux tubes, forming in the tachocline, float towards the photosphere due to the pressure of their intense magnetic fields. These flux tubes eventually emerge into the photosphere, giving rise to areas of concentrated magnetic field, observable as sunspots and faculae. These features serve as indicators of solar magnetic activity and topology.

Stellar magnetic activity also manifests itself as starspots and coronal mass ejections (CMEs). Starspots are cooler regions on the photosphere of stars, formed by concentrations of magnetic field. The relation between starspots and CMEs is complex and understanding it is important for predicting and mitigating the impacts of stellar activity on orbiting planets. In the case of the Sun, the most impactful events in space weather are CMEs – confined plasma clouds with a magnetic field ejected from the solar corona (Gopalswamy et al. 2009), that can cause geomagnetic storms on Earth. The study of starspots, magnetic fields, and CMEs helps unravel the dynamics of stars and their impact on space weather.

Studies by Cremades and Bothmer (2004); Kilpua et al. (2009), and Gopalswamy et al. (2009) indicate that CMEs often deflect towards local null points in the magnetic field, directed away from coronal holes. The heliospheric current sheet, marking the solar magnetic field's polarity shift, plays a crucial role. The complexity of the sheet's configuration during the solar cycle can impact CME trajectories, leading to increased deflections. Notably, slow CMEs tend to follow a deflection pattern towards the streamer belt, while fast CMEs exhibit less correlation. Kay et al. (2013, 2015) contributed to space weather predictions by developing ForeCAT, a model based on magnetohydrodynamics. ForeCAT considers magnetic forces, deflection, and rotation of CMEs, emphasizing their tendencies to deflect towards the heliospheric current sheet on a global scale, with significant deflection occurring closer to the Sun.

The extrasolar space weather can significantly differ from the space weather in the Solar System, as exoplanets may orbit a star at much closer distances, where stellar activity and magnetic field strength can be much higher. In stars with strong magnetic activity, planets in orbit may be subjected to more extreme space weather compared to Earth. The star's UV radiation, as well as the action of an intense stellar wind and the frequent impact of CMEs, can subject a planet to atmospheric erosion. This can result in a planetary surface exposed to ionizing radiation, which is harmful to biological organisms, as evidenced by Estrela and Valio (2018) and Estrela et al. (2020). In a study using the ForeCAT model and combining knowledge of the Solar System's space weather with inferred properties from other systems to study the trajectory of hypothetical CMEs launched into the interplanetary medium, Kay et al. (2016) analyzed the probability of CMEs reaching exoplanets. The authors found that this probability decreases with an increase in the inclination of the planetary orbit concerning the astrospheric current sheet.

In Menezes et al. (2023), we combined the knowledge of the space weather in the Solar System with inferred properties from other systems to reconstruct magnetograms (magnetic field maps of the photosphere) and study the trajectory of hypothetical CMEs launched into the interplanetary medium from the stars Kepler-63 and Kepler-411. Both stars exhibit strong magnetic activity (Sanchis–Ojeda et al. 2013; Sun et al. 2019) and are solar-type stars, *i.e.* FGK spectral type main-sequence stars (Izidoro et al. 2017). Also, we use solar model cases for comparison. In this present work, our primary focus is directed towards the reconstruction of magnetograms for these stars and how their topology affect the results obtained in Menezes et al. (2023).

2. Kepler Stars

Kepler-63 and Kepler-411, both exhibiting strong magnetic activity, are young stars with ages of approximately 210 $\pm$ 35 Myr and 212 $\pm$ 31 Myr, respectively (Sanchis–Ojeda et al. 2013; Estrela and Valio 2016; Sun et al. 2019; Netto and Valio 2020; Araújo and Valio 2021). Kepler-411 is a young solar-type K2V star, hosting at least 3 transiting planets (Sun et al. 2019; Araújo and Valio 2021). Kepler-63, a G-type star, has a significantly shorter magnetic activity cycle than the solar cycle, with a period of 1.27 years, and hosts a gas giant, Kepler-63b, in an almost polar orbit with respect to the star's rotation axis (Sanchis–Ojeda et al. 2013; Netto and Valio 2020).

It is worth noting that the age of a star plays a crucial role in determining various aspects, including the strength and configuration of its magnetic field, rotation, activity, and mass loss, as indicated by studies such as Skumanich (1972); Wood et al. (2002);

Charbonneau (2014); Morris (2020). Despite the relatively young age of Kepler-63 and Kepler-411, as discussed in Menezes et al. (2023), we regard them as comparable to the Sun. Consequently, we standardize the input parameters to match those of the Sun, encompassing the density and velocity of the stellar wind, the shape, size, expansion, mass, and velocity of CMEs, the source-surface height of the background magnetic field, and the relationships between intensity and magnetic field of starspots. This results in the identification of specific free parameters influencing the trajectory of CMEs: the initial position, velocities, and the strength and configuration of the magnetic field.

3. Magnetic Background Parameters

Kay et al. (2013, 2015) introduced the ForeCAT model, designed for computing the deflection and rotation of a CME influenced by Lorentz force components – pressure gradient and tension of the magnetic field. The model calculates the three-dimensional (3D) trajectories of CMEs, capturing the fundamental patterns observed in actual CMEs. We use an older, slightly less complex version of ForeCAT than the one presented in Kay et al. (2022), since we do not consider an elliptical CME cross section in Menezes et al. (2023), given how little observational constraints there are on extrasolar CMEs. The model's input parameters encompass the CME and star characteristics, in addition to magnetograms.

The simulated CME, embedded in a magnetic background, undergoes deflection and rotation, affecting its position and orientation (Kay et al. 2019). Unlike more sophisticated MHD models, ForeCAT simplifies for accurate CME reproduction, using a static solar magnetic field model to determine the magnetic tension and pressure gradient, resulting from the stellar background magnetic field, B_{SW} (Kay et al. 2015).

Stellar magnetograms are used as input to determine the B_{SW} at a source-surface height (described further) using a PFSS model (Altschuler and Newkirk 1969; Schatten et al. 1969). For the solar simulations we use a synoptic photospheric magnetograms of the Carrington rotation CR2203† generated by SDO/HMI (Pesnell et al. 2012; Scherrer et al. 2012). Nonetheless, there are no magnetograms available for Kepler-63 nor Kepler-411, not even magnetograms reconstructed by the Zeeman–Doppler Imaging technique (ZDI; Donati and Brown 1997; Vidotto et al. 2012, 2013, 2015; Yu et al. 2017). Thus, we reconstructed magnetograms for the Kepler-63 and Kepler-411, by extrapolating the magnetic field of the starspots based on their intensity and positions obtained from spot transit mapping performed by Netto and Valio (2020) and Araújo and Valio (2021), respectively.

Silva (2003) demonstrated the characterization of surface features on solar-like stars through planetary transits. During a transit, a planet may occult starspots, causing detectable variations in the light curve. Improved versions of this model allow the inference of starspot physical properties, including size, intensity, position, and temperature (Valio et al. 2017; Zaleski et al. 2019; Netto and Valio 2020; Zaleski et al. 2020; Selhorst et al. 2020; Araújo and Valio 2021; Zaleski et al. 2022). Using SOHO/MDI data, Valio et al. (2020) established linear fits for magnetic field (B) as a function of spot intensity, differentiating between warm and cold spots, however in Menezes et al. (2023), we simplified this relation to a single equation.

Then, from the position, intensity and size of the starspots of Kepler-63 and Kepler-411, we were able to reconstruct magnetograms for these stars. Since there was a great number of starspots mapped in both stars, we develop a method to choose maps with most intense magnetic field intensity and flux. These selected maps had their starspots transformed into horizontal dipoles, with the rest of the map filled with a scaled solar magnetogram.

† 2018-04-19 05:10 – 2018-05-16 10:53 UTC (solar minimum)

Table 1. Total magnetic flux of each stellar model.

Star	$\|\Phi\|_{\text{total}}$ **(Mx)**
Sun	1.13×10^{25}
Kepler-63	5.23×10^{25}
Kepler-411	3.90×10^{25}

Additionally, the multiplication factors for adjusting the solar magnetogram are discussed, taking into account the mean magnetic field strengths of the sunspots in each star's model. Menezes et al. (2023) provides more details on such methods. Table 1 lists the total magnetic flux of each stellar model.

Despite the fact that dynamo-age relationship is not linearly proportional, we considered this a reasonable hypothesis. Reconstructed magnetograms for Kepler-63 and Kepler-411 indicate total magnetic flux values ($|\Phi|_{\text{total}}$) of 5.23×10^{25} and 3.90×10^{25} Mx, respectively. A study by Coffaro et al. (2022) measured the X-ray luminosity ($\log L_X$) for Kepler-63 as approximately 29, suggesting a $|\Phi|_{\text{total}}$ of around 10^{25} Mx using the relation between average surface magnetic flux and X-ray luminosity. This reconstructed value closely aligns with the one obtained in this work.

By utilizing stellar magnetograms, one can compute the magnetic force fields generated by B_{SW} through the PFSS model. We generated the maps of B_{SW} for all simulations at $R_{\text{SS}} = 2.5\ R_\star$, and their respective deflection force fields, with the PFSS model.

Figure 1 summarizes the process of mapping stellar spots, reconstructing magnetograms, calculating B_{SW}, and extrapolating magnetic field lines, along with 3D visualizations of the respective maps. In the case of the Sun, the intensity map normalized by the flux at the center of the solar disk, $F_{\star_{\text{ctr}}}$, serves only as an illustration, as its magnetogram was directly generated from SDO/HMI data. The figure shows 3D visualizations of starspot intensity maps (top row), magnetograms (second row), B_{SW} at $2,5\ R_\star$ (third row), and extrapolated magnetic field lines (bottom row). The color scales represent intensity (normalized by $F_{\star_{\text{ctr}}}$) and magnetic field strength (in G). The left column panels correspond to the Sun; the center column panels correspond to Kepler-63; the right column panels correspond to Kepler-411. It is worth noting that B_{SW} maps, tend to approach the Astrospheric Current Sheet shape, and have higher intensities located at the poles due to coronal holes.

4. CME and Stellar Parameters

Kay et al. (2015) describes the model's CME geometry as a 3D rigid half torus, representing the flux rope's behavior as a function of radial distance. While CMEs may undergo deformation and erosion during propagation (Savani et al. 2011; Riley and Crooker 2004; Riley et al. 2004; Nieves–Chinchilla et al. 2018), observations and simulations indicate that they generally keep this shape up to 1 AU (Burlaga et al. 1981; Klein and Burlaga 1982; Cane and Richardson 2003; Vandas et al. 2002). In Menezes et al. (2023) the shape parameters, grid points and initial inclination ($\text{inc}_\circ$, measured clockwise from the equatorial plane) are kept the same for all simulations.

Other important initial CME parameters are angular width (size), mass, and velocity. Empirical models determine these parameters as functions of radial distance (Kay et al. 2013, 2015, 2016; Menezes et al. 2023). The CME trajectories are simulated from 1.1 $R_\star$ to 160 $R_\star$, since after this point the deflections are very low or null, keeping the trajectories stable. The parameters to calculate angular width and mass over radial propagation are the same for all simulations, however for velocity we used 3 configurations: low-velocity with initial velocity of$v_\circ = 50$ km/s and final velocity of$v_f = 400$ km/s, mid-velocity

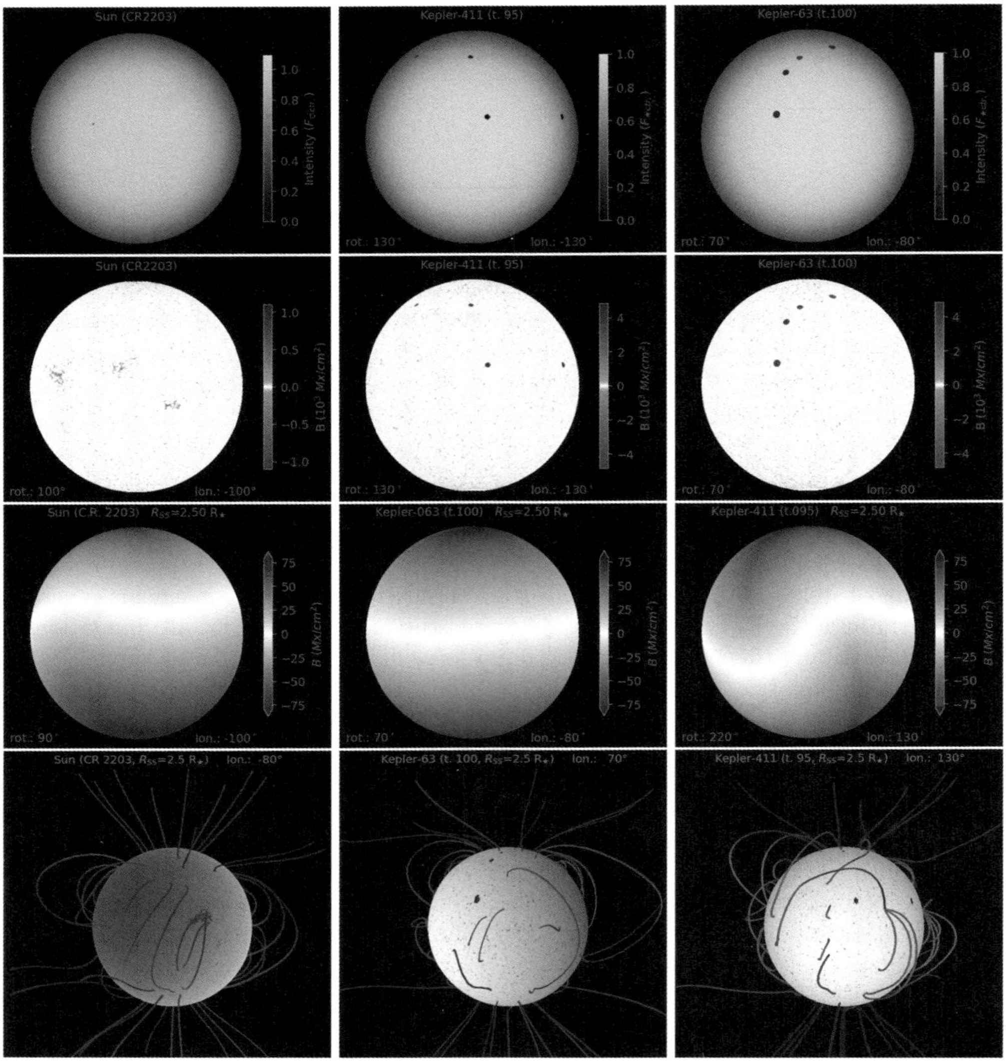

Figure 1. 3D visualizations of starspot intensity maps (top row), magnetograms (second row), B_{SW} at $2,5\ R_\star$ (third row), and extrapolated magnetic field lines (bottom row). The color scales represent intensity (normalized by $F_{\star_{\mathrm{ctr}}}$) and magnetic field strength (in G). The left column panels correspond to the Sun; the center column panels correspond to Kepler-63; the right column panels correspond to Kepler-411.

with $v_\circ = 70$ km/s and $v_f = 650$ km/s, and high-velocity with $v_\circ = 90$ km/s and $v_f = 900$ km/s.

Finally, the other free parameters are the initial latitude, $\mathbf{lat}_\circ$, and longitude, $\mathbf{lon}_\circ$, of the CMEs. The initial position ($\mathbf{lon}_\circ$, $\mathbf{lat}_\circ$, $\mathrm{inc}_\circ$) of the CME is based on regions of concentrated B field, lining up the torus with polarity inversion lines, locations where flux ropes would be likely to form. We set 3 different values of $\mathbf{lat}_\circ$ for each model, so that there are low-, mid- and high-latitude configuration. In Table 2, we list the starting position of CMEs for all simulations.

The stellar radius, $R_\star$, and rotation rate, $\Omega_\star$, also must be set. As stated in Menezes et al. (2023), for the Sun the values are respectively 7.0×10^{10} cm and 2.8×10^{-6} rad/s; for Kepler-63, 6.3×10^{10} cm and 1.3×10^{-5} rad/s (Sanchis–Ojeda et al.

Table 2. Initial position of CMEs for each star model and latitudinal region.

Region	Sun		Kepler-63		Kepler-411	
	$\mathbf{lat}_\circ$	$\mathbf{lon}_\circ$	$\mathbf{lat}_\circ$	$\mathbf{lon}_\circ$	$\mathbf{lat}_\circ$	$\mathbf{lon}_\circ$
High	45,0°	70,0°	47,6°	97,8°	49,3°	48,9°
Mid	30,0°	70,0°	36,9°	89,3°	30,0°	80,0°
Low	10,0°	70,0°	11,8°	86,5°	11,1°	110,4°

2013; Netto and Valio 2020); and for Kepler-411, 5.5×10^{10} cm and 6.9×10^{-6} rad/s (Araújo and Valio 2021). The model calculates the non-radial drag effects between a CME and the solar wind, using and adapted of the expression of Cargill et al. (1996) and Cargill (2004), with drag coefficient of $C_{\rm d} = 1$.

5. Results, Discussion and Conclusions

Our work focuses on unraveling CME trajectories influenced by different stellar magnetic backgrounds (Menezes et al. 2023). We reconstructed photospheric magnetograms for the solar-like stars Kepler-63 and Kepler-411 by extrapolating the magnetic field (B) of their respective mapped starspots (Valio et al. 2020; Netto and Valio 2020; Araújo and Valio 2021), and then computed the $B_{\rm SW}$ maps for all stellar models. We performed 27 simulations of CME trajectories for the Sun, Kepler-63, and Kepler-411, using the ForeCAT model. All simulations shared identical input parameters, except for initial positions ($\mathbf{lon}_\circ$ and $\mathbf{lat}_\circ$), velocities ($v_\circ$ and v_f), stellar parameters ($R_\star$ and $\Omega_\star$), and background magnetic fields, $B_{\rm SW}$. Furthermore, we calculated the total variations of longitude, Δlon, latitude, Δlat, and rotation, Δrot, over the complete CME trajectories ($1.1 - 160\ R_\star$) – final position values minus initial position values. In Figure 2, the total variations are listed by star, initial latitude, and velocities.

The most intense longitude deflections occur within approximately 2 $R_\star$, with rotation exhibiting more intense variations near the stellar surface. Latitude deflections are most pronounced in the solar model for low latitudes, up to about 3 $R_\star$, and in the Kepler-63 models for both low and mid latitudes, up to approximately 2 $R_\star$. In other simulations, latitude deflections are more significant up to around 6 $R_\star$, after which they persist with lower values. This behavior is attributed to the influence of concentrated magnetic field regions, such as spots or active regions, leading to CME trajectory deflection and rotation closer to the stellar photosphere (Kay et al. 2013, 2015; Menezes et al. 2023).

The latitude variations follow a gradual trend along the propagation, aligning with the approach of the Astrospheric Current Sheet by the solar wind, and CMEs tend to deviate toward regions of minimal magnetic energy. Although low-latitude solar simulations exhibit more intense latitude deflections up to around 2 $R_\star$, the absolute values of total latitude variations (Δlat) are very low, less than 1°. In contrast, Kepler-63 simulations show more significant Δlat values, ranging from approximately 13° to 17°° for low-latitude simulations and 56° to 95° for mid-latitude simulations.

All CMEs generally exhibit latitude variations toward the Astrospheric Current Sheet, except for Kepler-63 mid-latitude simulations. The magnetic forces responsible for CME deflection decay rapidly over radial distance, resulting in most deflection and rotation occurring near the stellar surface. The magnetic pressure gradient is the dominant force contributing to CME deflection toward the Astrospheric Current Sheet. However, intense magnetic fields of spots can cause greater deflection, leading to cases where CMEs propagate toward high latitudes.

The absolute total variations in deflections and rotations are small for solar simulations, intermediate for Kepler-411 simulations, and intense in Kepler-63 model. This trend agrees with the photospheric magnetic field flux of the stars Table 1. Kepler-63

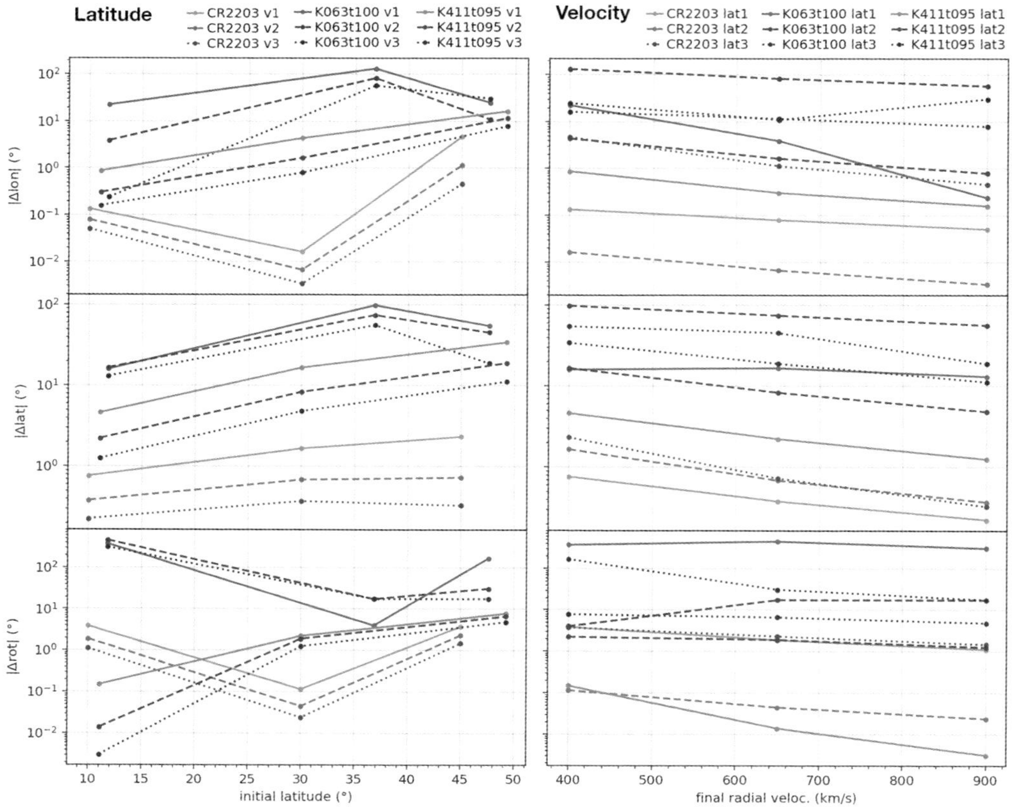

Figure 2. Total variations of the simulated CME trajectories in longitude, Δlon, latitude, Δlat, and rotation, Δrot, as functions of initial latitudes (**lat**$_\circ$, left panels) and final velocities (v_f, right panels). In both columns, the colors represent the simulations of the Sun (orange), Kepler-63 (blue), and Kepler-411 (green). In the left panels, the solid, dashed and dotted lines represent respectively low-, mid- and high-latitudes simulations. In the right panels the same as the left, but respectively for velocities.

simulations show significantly larger CME deflections and rotations compared to solar CMEs, driven by the stronger ambient magnetic field and more concentrated magnetic field in active regions. The absolute total variations of deflections and rotations increase with initial latitudes, influenced by concentrated field regions and the tendency of CMEs to deviate toward the Astrospheric Current Sheet. In most cases, Δlon, Δlat, and Δrot are inversely proportional to initial and final velocities ($v_\circ$ and v_f). However, Kepler-63 simulations exhibit deviations from these trends, attributed to the unique spatial configuration of large spots on its surface.

Kay et al. (2016) reported that planets with orbits closely aligned with the equatorial plane of their host stars are more susceptible to being impacted by CMEs. However in our work, Kepler-63's CMEs of mid-latitude and low- to mid-velocities were ejected from the northern hemisphere, deflecting their trajectory toward the south pole. Given Kepler-63b's near polar orbit (Sanchis–Ojeda et al. 2013; Netto and Valio 2020), the likelihood of it being affected by CMEs increases. Another intriguing feature of this star is the presence of a large and long-lived polar spot (Sanchis–Ojeda et al. 2013), prompting consideration of whether it serves as a CME overproducing region. Considering the elevated latitudes and the tendency for high-velocity CMEs to follow more radial trajectories, Kepler-63b emerges as a potential frequent target for CME impacts. Additionally, it is noteworthy that although active stars may theoretically generate numerous CMEs (directed towards the poles, as observed in Kepler-63, or deflected to the Astrospheric Current Sheet, as seen in Kepler-411), the robust magnetic field could prevent CME eruption through magnetic confinement (Alvarado-Gómez et al. 2018; Sun et al. 2022), thereby diminishing CME detection rates.

The impact of CMEs on space weather near planets depends of factors such as CME rotation and the direction of its B field. The direction of the B field resulting from the CME is crucial, as similar directions tend to cause compression in a planet's B field, while opposite directions can lead to significant changes through reconnection. The simulations indicate that the rotational displacement (Δrot) of CMEs is highest in the case of Kepler-63, with some CMEs rotating more than 360 degrees.

Overall, more homogeneous magnetic field, like that of the Sun, results in CMEs with lower deflections, while a magnetogram like that of Kepler-63 generates less stable trajectories. Additionally, the velocity of CMEs influences their trajectories, with faster CMEs exhibiting less deflections on their trajectories. The current understanding of stellar CMEs is still incipient, and despite using certain approximations and hypotheses, simulations for Kepler-63 and Kepler-411 were conducted. The findings indicate that free parameters, such as velocity, latitude, and specially magnetic topology, significantly affect CME trajectories. In the future we intend to explore additional parameters like initial inclination, angular width, and mass of CMEs, as well as using different magnetic backgrounds for different star models.

Acknowledgements F. M. acknowledges financial support from the Fundo Mackenzie de Pesquisa e Inovação (MackPesquisa), the Coordenação de Aperfeiçoamento de Pessoal de Nível Superior (CAPES), and the Fundação de Amparo à Pesquisa do Estado de São Paulo (FAPESP; process numbers 2022/12024-0 and 2013/10559-5). A. A. acknowledges financial support from the Conselho Nacional de Desenvolvimento Científico e Tecnológico (CNPq; #150817/2022-3). The CR2203 synoptic photospheric magnetograms is a courtesy of NASA/SDO and the AIA, EVE, and HMI science teams.

References

M. D. Altschuler and G. Newkirk. *Solar Physics*, 9(1):131–149, Sept. 1969. DOI: 10.1007/BF00145734.

J. D. Alvarado-Gómez, J. J. Drake, O. Cohen, S. P. Moschou, and C. Garraffo. *Astrophysical Journal*, 862(2):93, Aug. 2018. DOI: 10.3847/1538-4357/aacb7f.

A. Araújo and A. Valio. *Astrophysical Journal Letters*, 907(1):L5, Jan. 2021. DOI: 10.3847/2041-8213/abd3a7.

S. V. Berdyugina. *Living Reviews in Solar Physics*, 2(1):8, Dec. 2005. DOI: 10.12942/lrsp-2005-8.

A. S. Brun and M. K. Browning. *Living Reviews in Solar Physics*, 14(1):4, Sept. 2017. DOI: 10.1007/s41116-017-0007-8.

L. Burlaga, E. Sittler, F. Mariani, and R. Schwenn. *Journal of Geophysical Research*, 86(A8): 6673–6684, Aug. 1981. DOI: 10.1029/JA086iA08p06673.

H. V. Cane and I. G. Richardson. *Journal of Geophysical Research (Space Physics)*, 108(A4): 1156, Apr. 2003. DOI: 10.1029/2002JA009817.

P. J. Cargill. *Solar Physics*, 221(1):135–149, May 2004. DOI: 10.1023/B:SOLA.0000033366.10725.a2.

P. J. Cargill, J. Chen, D. S. Spicer, and S. T. Zalesak. *Journal of Geophysical Research*, 101 (A3):4855–4870, Mar. 1996. DOI: 10.1029/95JA03769.

P. Charbonneau. *Annual Review of Astron. Astrophys*, 52:251–290, Aug. 2014. DOI: 10.1146/annurev-astro-081913-040012.

M. Coffaro, B. Stelzer, and S. Orlando. *Astronomy and Astrophysics*, 661:A79, May 2022. DOI: 10.1051/0004-6361/202142298.

H. Cremades and V. Bothmer. *Astronomy and Astrophysics*, 422:307–322, July 2004. DOI: 10.1051/0004-6361:20035776.

J. F. Donati and S. F. Brown. *Astronomy and Astrophysics*, 326:1135–1142, Oct. 1997.

R. Estrela and A. Valio. *Astrophysical Journal*, 831:57, Nov. 2016. DOI: 10.3847/0004-637X/831/1/57.

R. Estrela and A. Valio. *Astrobiology*, 18(11):1414–1424, Nov 2018. DOI: 10.1089/ast.2017.1724.

R. Estrela, S. Palit, and A. Valio. *Astrobiology*, 20(12):1465–1475, Dec. 2020. DOI: 10.1089/ast.2019.2126.

N. Gopalswamy, S. Yashiro, G. Michalek, G. Stenborg, A. Vourlidas, S. Freeland, and R. Howard. *Earth Moon and Planets*, 104(1-4):295–313, Apr. 2009. DOI: 10.1007/s11038-008-9282-7.

M. Guhathakurta, E. C. Sittler, and L. Ofman. *Journal of Geophysical Research (Space Physics)*, 111(A11):A11215, Nov. 2006. DOI: 10.1029/2006JA011931.

A. Izidoro, M. Ogihara, S. N. Raymond, A. Morbidelli, A. Pierens, B. Bitsch, C. Cossou, and F. Hersant. *Monthly Notices of the Royal Astronomical Society*, 470(2):1750–1770, Sept. 2017. DOI: 10.1093/mnras/stx1232.

C. Kay, M. Opher, and R. M. Evans. *Astrophysical Journal*, 775(1):5, Sep 2013. DOI: 10.1088/0004-637X/775/1/5.

C. Kay, M. Opher, and R. M. Evans. *Astrophysical Journal*, 805(2):168, Jun 2015. DOI: 10.1088/0004-637X/805/2/168.

C. Kay, M. Opher, and M. Kornbleuth. *Astrophysical Journal*, 826(2):195, Aug 2016. DOI: 10.3847/0004-637X/826/2/195.

C. Kay, V. S. Airapetian, T. Lüftinger, and O. Kochukhov. *Astrophysical Journal Letters*, 886 (2):L37, Dec. 2019. DOI: 10.3847/2041-8213/ab551f.

C. Kay, M. L. Mays, and Y. M. Collado–Vega. *Space Weather*, 20(4):e02914, Apr. 2022. DOI: 10.1029/2021SW002914.

E. K. J. Kilpua, J. Pomoell, A. Vourlidas, R. Vainio, J. Luhmann, Y. Li, P. Schroeder, A. B. Galvin, and K. Simunac. *Annales Geophysicae*, 27(12):4491–4503, Dec. 2009. DOI: 10.5194/angeo-27-4491-2009.

L. W. Klein and L. F. Burlaga. *Journal of Geophysical Research*, 87(A2):613–624, Feb. 1982. DOI: 10.1029/JA087iA02p00613.

P. Lang, M. Jardine, J. Morin, J.-F. Donati, S. Jeffers, A. A. Vidotto, and R. Fares. *Monthly Notices of the Royal Astronomical Society*, 439:2122–2131, Apr. 2014. DOI: 10.1093/mnras/stu091.

R. M. MacQueen, A. J. Hundhausen, and C. W. Conover. *Journal of Geophysical Research*, 91 (A1):31–38, Jan. 1986. DOI: 10.1029/JA091iA01p00031.

F. Menezes, C. L. Selhorst, C. G. Giménez de Castro, and A. Valio. *Astrophysical Journal*, 910 (1):77, Mar. 2021. DOI: 10.3847/1538-4357/abe41c.

F. Menezes, A. Valio, Y. Netto, A. Araújo, C. Kay, and M. Opher. *Monthly Notices of the Royal Astronomical Society*, 522(3):4392–4403, July 2023. DOI: 10.1093/mnras/stad1078.

B. M. Morris. *Astrophysical Journal*, 893(1):67, Apr. 2020. DOI: 10.3847/1538-4357/ab79a0.

Y. Netto and A. Valio. *Astronomy and Astrophysics*, 635:A78, Mar. 2020. DOI: 10.1051/0004-6361/201936219.

T. Nieves–Chinchilla, M. G. Linton, M. A. Hidalgo, and A. Vourlidas. *Astrophysical Journal*, 861(2):139, July 2018. DOI: 10.3847/1538-4357/aac951.

E. N. Parker. *Astrophysical Journal*, 122:293, Sept. 1955. DOI: 10.1086/146087.

W. D. Pesnell, B. J. Thompson, and P. C. Chamberlin. *Solar Physics*, 275(1-2):3–15, Jan. 2012. DOI: 10.1007/s11207-011-9841-3.

P. Riley and N. U. Crooker. *Astrophysical Journal*, 600(2):1035–1042, Jan. 2004. DOI: 10.1086/379974.

P. Riley, J. A. Linker, R. Lionello, Z. Mikić, D. Odstrcil, M. A. Hidalgo, C. Cid, Q. Hu, R. P. Lepping, B. J. Lynch, and A. Rees. *Journal of Atmospheric and Solar-Terrestrial Physics*, 66(15-16):1321–1331, Oct. 2004. DOI: 10.1016/j.jastp.2004.03.019.

R. Sanchis–Ojeda, J. N. Winn, G. W. Marcy, A. W. Howard, H. Isaacson, J. A. Johnson, G. Torres, S. Albrecht, T. L. Campante, W. J. Chaplin, G. R. Davies, M. N. Lund, J. A.

Carter, R. I. Dawson, L. A. Buchhave, M. E. Everett, D. A. Fischer, J. C. Geary, R. L. Gilliland, E. P. Horch, S. B. Howell, and D. W. Latham. *Astrophysical Journal*, 775(1): 54, Sep 2013. DOI: 10.1088/0004-637X/775/1/54.

N. P. Savani, M. J. Owens, A. P. Rouillard, R. J. Forsyth, K. Kusano, D. Shiota, and R. Kataoka. *Astrophysical Journal*, 731(2):109, Apr. 2011. DOI: 10.1088/0004-637X/731/2/109.

K. H. Schatten, J. M. Wilcox, and N. F. Ness. *Solar Physics*, 6(3):442–455, Mar. 1969. DOI: 10.1007/BF00146478.

P. H. Scherrer, J. Schou, R. I. Bush, A. G. Kosovichev, R. S. Bogart, J. T. Hoeksema, Y. Liu, T. L. Duvall, J. Zhao, A. M. Title, C. J. Schrijver, T. D. Tarbell, and S. Tomczyk. *Solar Physics*, 275(1-2):207–227, Jan. 2012. DOI: 10.1007/s11207-011-9834-2.

C. L. Selhorst, C. L. Barbosa, P. J. A. Simões, A. A. Vidotto, and A. Valio. *Astrophysical Journal*, 895(1):62, May 2020. DOI: 10.3847/1538-4357/ab89a4.

A. V. R. Silva. *Astrophysical Journal Letters*, 585:L147–L150, Mar. 2003. DOI: 10.1086/374324.

A. Skumanich. *Astrophysical Journal*, 171:565, Feb. 1972. DOI: 10.1086/151310.

H. Spruit. *Memorie della Società Astronomia Italiana*, 68:397, 1997.

H. C. Spruit, G. B. Scharmer, and M. G. Löfdahl. *Astronomy and Astrophysics*, 521:A72, Oct. 2010. DOI: 10.1051/0004-6361/200912519.

A. M. Stejko, G. Guerrero, A. e. G. Kosovichev, and P. K. Smolarkiewicz. *Astrophysical Journal*, 888(1):16, Jan. 2020. DOI: 10.3847/1538-4357/ab5854.

L. Sun, P. Ioannidis, S. Gu, J. H. M. M. Schmitt, X. Wang, and M. B. N. Kouwenhoven. *Astronomy and Astrophysics*, 624:A15, Apr. 2019. DOI: 10.1051/0004-6361/201834275.

X. Sun, T. Török, and M. L. DeRosa. *Monthly Notices of the Royal Astronomical Society*, 509 (4):5075–5085, Feb. 2022. DOI: 10.1093/mnras/stab3249.

A. Valio, R. Estrela, Y. Netto, J. P. Bravo, and J. R. de Medeiros. *Astrophysical Journal*, 835: 294, Feb. 2017. DOI: 10.3847/1538-4357/835/2/294.

A. Valio, E. Spagiari, M. Marengoni, and C. L. Selhorst. *Solar Physics*, 295(9):120, Sept. 2020. DOI: 10.1007/s11207-020-01691-3.

M. Vandas, D. OdstrčIl, and S. Watari. *Journal of Geophysical Research (Space Physics)*, 107 (A9):1236, Sept. 2002. DOI: 10.1029/2001JA005068.

A. A. Vidotto, M. Jardine, M. Opher, J. F. Donati, and T. I. Gombosi. *Monthly Notices of the Royal Astronomical Society*, 412(1):351–362, Mar. 2011. DOI: 10.1111/j.1365-2966.2010.17908.x.

A. A. Vidotto, R. Fares, M. Jardine, J.-F. Donati, M. Opher, C. Moutou, C. Catala, and T. I. Gombosi. *Monthly Notices of the Royal Astronomical Society*, 423:3285–3298, July 2012. DOI: 10.1111/j.1365-2966.2012.21122.x.

A. A. Vidotto, M. Jardine, J. Morin, J.-F. Donati, P. Lang, and A. J. B. Russell. *Astronomy and Astrophysics*, 557:A67, Sept. 2013. DOI: 10.1051/0004-6361/201321504.

A. A. Vidotto, M. Jardine, J. Morin, J. F. Donati, M. Opher, and T. I. Gombosi. *Monthly Notices of the Royal Astronomical Society*, 438:1162–1175, Feb. 2014. DOI: 10.1093/mnras/stt2265.

A. A. Vidotto, R. Fares, M. Jardine, C. Moutou, and J.-F. Donati. *Monthly Notices of the Royal Astronomical Society*, 449:4117–4130, June 2015. DOI: 10.1093/mnras/stv618.

B. E. Wood, H.-R. Müller, G. P. Zank, and J. L. Linsky. *Astrophysical Journal*, 574(1):412–425, July 2002. DOI: 10.1086/340797.

L. Yu, J.-F. Donati, E. M. Hébrard, C. Moutou, L. Malo, K. Grankin, G. Hussain, A. Collier Cameron, A. A. Vidotto, C. Baruteau, S. H. P. Alencar, J. Bouvier, P. Petit, M. Takami, G. Herczeg, S. G. Gregory, M. Jardine, J. Morin, F. Ménard, and Matysse Collaboration. *Monthly Notices of the Royal Astronomical Society*, 467:1342–1359, May 2017. DOI: 10.1093/mnras/stx009.

S. M. Zaleski, A. Valio, S. C. Marsden, and B. D. Carter. *Monthly Notices of the Royal Astronomical Society*, 484(1):618–630, Mar. 2019. DOI: 10.1093/mnras/sty3474.

S. M. Zaleski, A. Valio, B. D. Carter, and S. C. Marsden. *Monthly Notices of the Royal Astronomical Society*, 492(4):5141–5151, Mar. 2020. DOI: 10.1093/mnras/staa103.

S. M. Zaleski, A. Valio, B. D. Carter, and S. C. Marsden. *Monthly Notices of the Royal Astronomical Society*, 510(4):5348–5361, Mar. 2022. DOI: 10.1093/mnras/stab3788.

Dynamics of Solar and Stellar Convection Zones and Atmospheres
Proceedings IAU Symposium No. 365, 2024
A. V. Getling & L. L. Kitchatinov, eds.
doi:10.1017/S174392132400019X

Stellar activity in open clusters

A. Görgei[1,2,3], K. Vida[1,2], B. Seli[1,2,3], L. Kriskovics[1,2]

[1]Konkoly Observatory, HUN-REN Research Centre for Astronomy and Earth Sciences, Budapest, Hungary
email: gorgei.anna@csfk.org

[2]CSFK, MTA Centre of Excellence, Budapest, Hungary

[3]Eötvös Loránd University, Budapest, Hungary

Abstract. Stellar activity depends on multiple parameters one of which is the age of the star. The members of open clusters are good targets to observe the activity at a given age of the stars since their ages are more precisely determined than that of field stars. Choosing multiple clusters, each with different age, gives us insight to the change in activity during the lifetime of stars. With the analysis of these stars we can also refine the parameters of gyrochronology (Barnes 2003), which is a method for estimating the age of low-mass, main sequence stars from their rotation periods.

Keywords. Stars: activity, rotation, methods: data analysis, techniques: photometry

1. Introduction

Studying stellar activity within open clusters provides a unique laboratory for understanding these astrophysical processes. The proximity and similar ages of cluster members allow us to investigate the impact of stellar environment on the magnetic activity of individual stars while studying clusters of different ages offers a perspective on how this behaviour evolves over time.

2. Cluster membership

In this work, the target of the analysis was the Hyades cluster – this is one of the closest (47pc) young cluster, allowing us to obtain high signal-to-noise observations even of the intrinsically fainter targets. The cluster was observed by the TESS (Ricker et al. 2015) in four sectors. These are S05, S32, S43 and S44. The short cadence TESSLightCurveFile data were also used.

The determination of the cluster membership was done with the HDBSCAN algorithm (McInnes et al. 2017) using Gaia DR3 (Gaia Collaboration et al. 2016, Gaia Collaboration et al. 2023) parallax, proper motion and coordinates. HDBSCAN is a density-based, hierarchical clustering method, the tree of groups in the data can be built as a result of hierarchical clustering. The following steps describe how the method works. First, the data is transformed according to its density and then the distance-weighted graph is built. Then the cluster hierarchy of connected components is determined and compressed according to the minimum cluster size. Finally, the method identifies stable clusters. For this a stability parameters is defined:

$$\sum_{\mathrm{p}\in\mathrm{cluster}} \lambda_{\mathrm{p}} - \lambda_{\mathrm{cluster\,birth}} \quad \text{where} \quad \lambda = \frac{1}{\mathrm{distance}} \tag{1}$$

and p refers to nodes in the built tree.

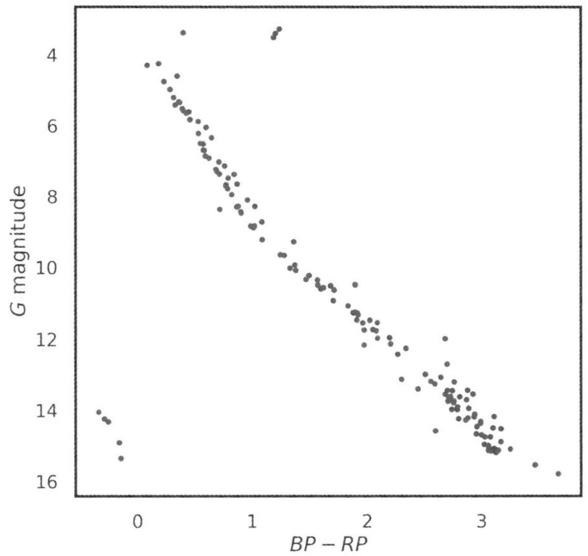

Figure 1. The color–magnitude diagram of cluster members with short candence data.

A cluster is considered stable if its stability is greater than the sum of its subclusters stability. This stability test criteria ensure that the artificial groups resulting from its operation are not included in the selected real groups. After clustering and comparing the coordinates with the TESS field of view 426 stars make up our sample (see Fig. 1).

3. Rotation

Gyrochronology is based on the fact that the rotation of a star changes with its age. This change occurs because the star loses angular momentum through magnetized winds during its lifetime. Rotational period determination was done with the Lomb–Scargle algorithm (VanderPlas 2018). The comparison with archival rotational values is plotted in Fig. 2. Between now and the time of the presentation of the poster, Núñez et al. (2023) published their result of rotational periods based on the same TESS data. The values obtained by my analysis show a good match with theirs.

4. Flare detection

Flare detection on the short cadence light curves was done using the FLATW'RM2 (Vida et al. 2021) program which is based on deep learning and is specialized to analyse TESS and Kepler light curves. In total over 600 flares were found on 135 stars (Fig. 3).

5. Flare statistics

Flare energies are determined based on the bolometric luminosity of the star and the area under the flare. The bolometric luminosities are calculated from Gaia DR3 effective temperatures and surface gravity with the Gaiadr3 Bcg package (Creevey et al. 2023). This analysis gives the cumulative flare frequency diagram as the result (see Fig. 4).

6. Future work

In the continuation of the project we will look at the cluster stars in new TESS sectors to get better flare statistics. LOFAR measurements of some of the member stars were

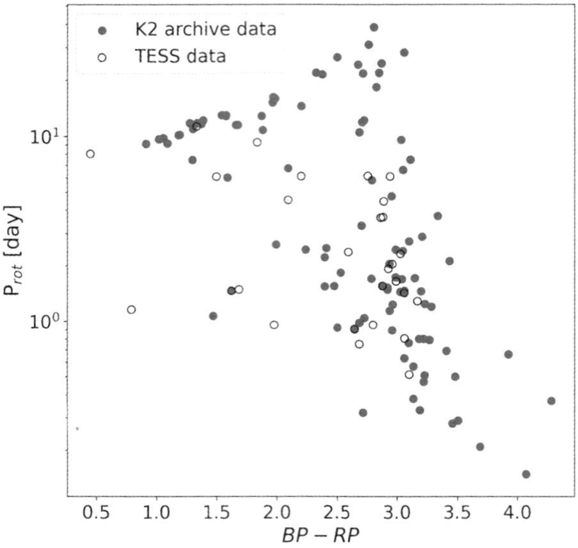

Figure 2. The color–rotation period of the cluster based on TESS short-cadence data overplotted with archived K2 (Douglas et al. 2019) data.

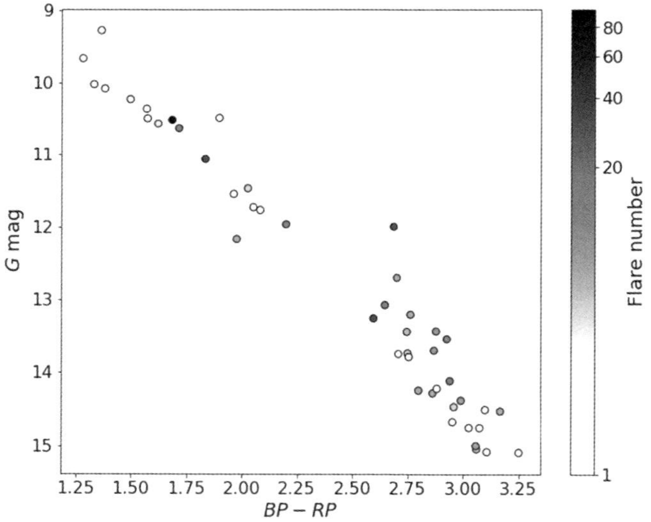

Figure 3. Color–magnitude diagram. Colors indicate the number of flares detected on each star.

gathered simultaneously with TESS sectors S70 and S71, so comparing those to the optical data and looking for flare and (possibly) CME signals is our aim too.

Acknowledgements This work was supported by the Hungarian National Research, Development and Innovation Office grants KKP-143986 (Élvonal) and OTKA K131508. Authors acknowledge the financial support of the Austrian-Hungarian Action Foundation (95öu5, 101öu13, 112öu1). LK and KV are Bolyai János research Fellows. KV is supported by the Bolyai+ grant ÚNKP-22-5-ELTE-1093, BS is supported by the ÚNKP-22-3 New National Excellence Program of the Ministry for Culture and Innovation from the source of the National Research, Development and Innovation Fund. This work has made use of

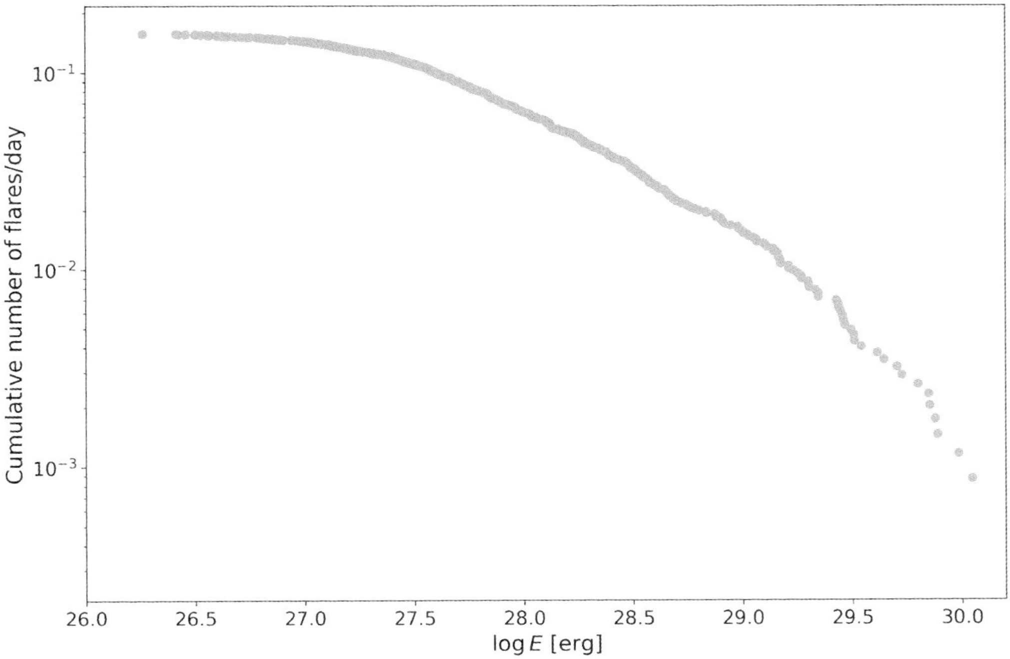

Figure 4. Cumulative flare frequency diagram for the cluster.

data from the European Space Agency (ESA) mission *Gaia* (`https://www.cosmos.esa.int/gaia`), processed by the *Gaia* Data Processing and Analysis Consortium (DPAC, `https://www.cosmos.esa.int/web/gaia/dpac/consortium`). Funding for the DPAC has been provided by national institutions, in particular the institutions participating in the *Gaia* Multilateral Agreement. This paper includes data collected by the TESS mission. Funding for the TESS mission is provided by the NASA's Science Mission Directorate.

References

Barnes, S. A. 2003, On the Rotational Evolution of Solar- and Late-Type Stars, Its Magnetic Origins, and the Possibility of Stellar Gyrochronology. *Astrophys. J.*, 586(1), 464–479.

Creevey, O. L., Sordo, R., Pailler, F., Frémat, Y., Heiter, U., Thévenin, F., Andrae, R., & Fouesneau, M. e. a. 2023, Gaia Data Release 3. Astrophysical parameters inference system (Apsis). I. Methods and content overview. *A&A*, 674, A26.

Douglas, S. T., Curtis, J. L., Agüeros, M. A., Cargile, P. A., Brewer, J. M., Meibom, S., & Jansen, T. 2019, K2 Rotation Periods for Low-mass Hyads and a Quantitative Comparison of the Distribution of Slow Rotators in the Hyades and Praesepe. *Astrophys. J.*, 879(2), 100.

Gaia Collaboration, Prusti, T., & de Bruijne, J. H. J. e. a. 2016, The Gaia mission. *A&A*, 595, A1.

Gaia Collaboration, Vallenari, A., Brown, A. G. A., Prusti, T., & de Bruijne, J. H. J. e. a. 2023, Gaia Data Release 3. Summary of the content and survey properties. *A&A*, 674, A1.

McInnes, L., Healy, J., & Astels, S. 2017, hdbscan: Hierarchical density based clustering. *The Journal of Open Source Software*, 2(11).

Núñez, A., Agüeros, M. A., Curtis, J. L., Covey, K. R., Douglas, S. T., Chu, S. R., DeLaurentiis, S., Wang, M., & Drake, J. J. 2023, The Factory and the Beehive. V. Chromospheric and Coronal Activity and Its Dependence on Rotation in Praesepe and the Hyades. *arXiv e-prints*,, arXiv:2311.18690.

Ricker, G. R., Winn, J. N., & Vanderspek, R. e. a. 2015, Transiting Exoplanet Survey Satellite (TESS). *Journal of Astronomical Telescopes, Instruments, and Systems*, 1, 014003.
VanderPlas, J. T. 2018, Understanding the Lomb-Scargle Periodogram. *Astrophysical Journal Supplement Series*, 236(1), 16.
Vida, K., Bódi, A., Szklenár, T., & Seli, B. 2021, Finding flares in Kepler and TESS data with recurrent deep neural networks. *Astron. Astrophys.*, 652, A107.

Author index